Springer Series in Translational Stroke Research

Series Editor

John Zhang
Loma Linda, CA, USA

More information about this series at http://www.springer.com/series/10064

Jun Chen • Zao C. Xu • Xiao Ming Xu
John H. Zhang
Editors

Animal Models of Acute Neurological Injury

Second Edition

Springer

Editors
Jun Chen
Department of Neurology
University of Pittsburgh School of Medicine
Pittsburgh, PA, USA

Zao C. Xu
Department of Anatomy and Cell Biology
Indiana University School of Medicine
Indianapolis, IN, USA

Xiao Ming Xu
Indiana University School of Medicine
SNRI SCBIRG
Indianapolis, IN, USA

John H. Zhang
Center for Neuroscience Research
Loma Linda University School of Medicine
Loma Linda, CA, USA

ISSN 2363-958X ISSN 2363-9598 (electronic)
Springer Series in Translational Stroke Research
ISBN 978-3-030-16080-7 ISBN 978-3-030-16082-1 (eBook)
https://doi.org/10.1007/978-3-030-16082-1

This Springer imprint is published by the registered company Springer Nature Switzerland AG
The registered company address is: Gewerbestrasse 11, 6330 Cham, Switzerland

Preface

Acute neurological injuries are devastating events that affect all individuals across the spectrum of human life. Understanding the mechanisms of cell death after these incidents will help develop pharmacological agents and therapeutic interventions to reduce or prevent the damage. Many studies have been conducted, and new exciting discoveries have been reported. However, only limited treatments are available today for the victims of acute neurological injuries. More investigations are needed to explore the underlying mechanisms of neuronal injury following the insult and to translate them into preclinical and clinical settings. Reliable animal models are the key for experimental neurobiology. To date, many animal models for acute neurological injuries have been developed. But large variations in experimental outcomes exist between laboratories even using the same animal model, making the comparison and replication impossible and jeopardizing advances in the field. This variation is largely due to the lack of a standard manual detailing the methodology in animal models.

Since publishing the first edition of this book in 2009, much progress has been made in the field. The advance in translational medicine has generated new challenges to neuroscientists. The reproducibility of the results of preclinical research is one of the recognized obstacles, which further emphasizes the importance of standardizing the animal models. In this second edition, we have included 21 new and revised chapters to better illustrate the earlier models and to introduce new animal models that reflect the expansion of preclinical research in acute neurological disorders.

Compared to other very good texts on the market that are more focused on theoretical details, this book is an illustrated workbench manual of animal models. We have selected the most commonly used models that have been proven to be reliable and reproducible. The contributors of this book are the experts in the field who have many years of hands-on experiences in these animal models; some of them are the original developers of the models. References and descriptions have been kept to a minimum for clarity. Our purpose for preparing this edition is to enable someone with little experience to produce animal models for clinical or basic science research using this manual as a guide. We hope this book will be useful for students, postdoctoral fellows, clinical investigators, and basic science researchers who would like to study acute neurological disorders.

Pittsburgh, PA, USA Jun Chen
Indianapolis, IN, USA Zao C. Xu
Indianapolis, IN, USA Xiao Ming Xu
Loma Linda, CA, USA John H. Zhang

Contents

Part I
Animal Care and Surgery

Animal Protocol

Laurel D. Schantz

Abstract This chapter outlines the important points that must be considered in developing an animal protocol using a model of acute neurological injury. It highlights the issues that should be considered in designing such a study as well as the issues that must be addressed when writing a protocol for submission to the Institutional Animal Care and Use Committee (IACUC). The key issues for regulatory compliance are addressed.

Keywords Animal protocol · IACUC · Regulatory compliance

Regulatory Oversight

In the United States, the use of animals in research has extensive regulatory oversight. The U.S. Animal Welfare Act (AWA) and the Animal Welfare Regulations (AWRs) are administered by the United States Department of Agriculture (USDA). These regulations cover warm-blooded vertebrates except birds, rats of the genus *Rattus*, mice of the genus *Mus* and farm animals used in production agriculture [1]. The Public Health Service Policy on Humane Care and Use of Laboratory Animals (PHS Policy) has its statutory mandate in the U.S. Health Research Extension Act (HREA) and is administered by the Office of Laboratory Animal Welfare (OLAW) of the National Institutes of Health (NIH). It applies to all institutions that use live vertebrates in research supported by any component of PHS. Along with the AWA and AWRs, the PHS Policy also mandates compliance with the U.S. Government Principles for the Utilization and Care of Vertebrate Animals Used in Testing, Research and Training (U.S. Government Principles), and the National Research Council's Guide to the Care and Use of Laboratory Animals (The Guide) [2, 3]. In addition to governmental oversight, many research institutions seek accreditation by the Association for Assessment and Accreditation of Laboratory Animal Care,

---

L. D. Schantz, DVM, DACLAM (✉)

J. Chen et al. (eds.), *Animal Models of Acute Neurological Injury*, Springer Series in Translational Stroke Research, https://doi.org/10.1007/978-3-030-16082-1_1

International (AAALAC International). AAALAC International is a private, non-profit organization that provides a program of voluntary accreditation to institutions that use animals in research. AAALAC International uses The Guide as a primary reference document when assessing animal care programs [4].

The AWA, AWRs, U.S. Government Principles and The Guide give the responsibility for oversight and evaluation of the animal care program and use of animals in research to the Institutional Animal Care and Use Committee (IACUC). The IACUC is tasked with assessing every protocol involving animal use in research, testing and education. The topics that it is required to evaluate include the rationale, purpose, and uniqueness of animal use, the justification of the species and number of animals, lack of non-animal alternatives, training of personnel, use of sedation, anesthesia, analgesia, appropriate methods of euthanasia, development of humane end points, conduct of operative procedures, thorough post-procedural monitoring and care of the animals, unusual husbandry or housing requirements, and safety of the working environment [5].

By its very nature, use of an animal model of acute neurological injury presents a challenge to design studies that achieve the desired experimental outcomes and minimize animal pain and distress. However, this is not only the regulatory mandate; it is the goal of everyone involved in such work. To accomplish this goal, developing a team approach incorporating the investigative staff, the veterinary staff and animal care staff is most useful. Involving all of the necessary team members in the design phase of the protocol development usually makes acceptance of the final written animal protocol by the IACUC much more expedient.

Designing the Animal Protocol

When using an animal model for the first time, there are several factors that must be considered as having the potential to introduce undesirable variables into the study. Training of personnel, special surgical methods, anesthesia monitoring, postoperative care and monitoring, analgesia, identification of humane endpoints, method of euthanasia, outcome assessment tool development, and identification of any hazards must all be considered.

Provision must be made for training of individuals participating in the animal work. Several institutions, including the NIH Office of Animal Care and Use (OACU), offer tutorial videos and documents on their websites that describe many of the common rodent procedures and discuss best practices as applied to differing research requirements [6]. Particularly, the skills of individual surgeons have a dramatic impact on model outcomes. If a surgeon has not had experience with the procedures to be carried out in the study, then it is important that accommodations be made to provide such experience prior to beginning work on experimental animals. Obviously, initial practice should be given using cadavers and it is best guided by a surgeon with experience in using such procedures. If there is no one available with the experience in the procedure to provide the "hands-on" training, often arrangements can be made to visit another facility to observe the surgery being performed by experienced surgeons as a survival procedure. Practice on cadavers must be done until the surgeon is very comfortable with the new procedures. Training on live animals as part of non-survival surgeries is the next step. Once live animals are needed even for training purposes, the work must be covered by a written animal protocol approved by the IACUC. Laboratory Animal Resource Centers may have animals available for training purposes [7, 8]. Often the work is then covered under an animal protocol which is the responsibility of the institutional veterinarian. Sometimes an amendment must be written to add the actual surgical procedure that will be practiced and/or the name of the surgeon who will participate. Sometimes the training animal protocol is broad enough to cover non-survival procedures as long as the veterinarian is present and anesthesia/euthanasia is performed by animal care personnel. If such an institutional animal protocol is not available, then the training on live animals will need to be included in the written animal protocol covering the experimental work.

The surgical procedure itself must be described in detail in the animal protocol. For survival surgeries, particular attention must be given to describing how the requirements for establishing and maintaining asepsis during surgery will be met. If the procedure will not be done in a designated operating room, the area where surgery will be done will need to be inspected by representatives of the IACUC prior to performing the work [5]. It is a good idea when possible to arrange for this inspection prior to submitting the animal protocol. During the inspection, a thorough discussion of not only the steps taken to insure asepsis, but also those concerning animal welfare during transportation and/or recovery can lead to improving the perioperative plan for animal care. If a procedure will require special consideration due to equipment that cannot be moved or equipment that cannot be sterilized by conventional means, then justification for the exceptions to required asepsis should be provided. If a procedure is new to an institution and there will need to be a break the aseptic technique due to specialized equipment or techniques, then a pilot study may be required with veterinary observation and/or outcome reporting to the IACUC to allow evaluation of the impact on animal welfare of the deviation from aseptic practices. The IACUC inspection team can be a valuable resource in the design of such a pilot study. The pilot study would then need to be described in the animal protocol.

Anesthetic choice will be dictated by the needs of the study. If the anesthesia chosen is not one commonly used in standard veterinary care for the species involved, it may require justification in the written animal protocol. Regardless of anesthetic choice, anesthetic monitoring will need to be considered. Proper monitoring of an anesthetized animal is often critical to successful outcomes of surgical procedures. The parameters that will be monitored need to be determined on the basis of animal species, anesthetic choice and expected effects of the surgical manipulation. Availability of the equipment necessary for monitoring and record keeping should be considered. For prolonged studies done on anesthetized animals, the means necessary to monitor and maintain normal physiologic functions within the study parameters must be considered. The veterinarian can be helpful in forming this plan and providing training, should it be needed. Because proper monitoring and care of the anesthetized animal ensures that the animal does not experience pain or distress during the surgical procedure, it must be described in detail in the animal protocol.

The plan for postoperative care and monitoring should also be considered in the study design. Animals recovering from acute neurological injury must often be considered "critical care" patients, especially in the immediate post-operative period. Frequency of monitoring, parameters to be monitored and any special equipment required must be considered. The veterinarian can be helpful in developing this part of the plan and in training personnel in recognition of clinical signs or critical values that require immediate action. Attention should be paid to maintaining body temperature and hydration status as well as other physiologic parameters. The use of specific analgesics, sedatives and other drugs should be considered. Those drugs that would be helpful to the animal and not interfere with the experimental outcomes should be identified. Having a well-thoughtout plan for addressing potential postoperative complications is not only beneficial to enhancing better study outcomes, but is also required as part of the written animal protocol, as it can have a major impact on animal welfare [9].

The chronic care of the animals following injury must also be considered carefully. There may be housing or husbandry accommodations that could be made to enhance the animals' well-being without compromising experimental goals. Again, the veterinarian and animal care staff can be valuable resources for identifying potential enhancements that might apply to a specific model and species. Some examples might include providing softer and/or deeper bedding, offering moistened chow in a floor feeder, or making water more easily accessible from the cage floor. If appetite suppression is anticipated, then highly palatable diets could be provided to help prevent weight loss. Increased or decreased amounts of environmental enrichment could be provided. Decreased exposure to loud sounds or bright lights might help in some cases [4, 10]. It is usually a good idea to initiate special diets or environmental changes prior to creating the injury model in an animal. Most animals benefit from the opportunity to acclimate to these changes prior to injury. Prior acclimation usually speeds up acceptance of the changes and maximizes their benefit to the animals [11]. Providing such special care and describing it in the animal protocol indicate the importance of maximizing animal welfare while conducting the study.

Animal monitoring protocols must be developed to provide chronic monitoring of experimental and health issues in the animals. These could be a combination of clinical observations, physical measurements, and behavioral measurements. The parameters measured should be specific to the species of animal used and the expected experimental outcomes. Some examples of parameters to measure might include body temperature, body weight, body condition score, respiratory rate, heart rate, normal/abnormal posturing, normal/abnormal gait, home cage activity, open field activity, coat/skin condition, skin turgor, food consumption, urine output, fecal output, approach response, touch response, sound response, etc. Specific monitoring forms may be developed for the study to include the most applicable parameters for assessing the experimental goals and animal health. Established behavioral species-specific screens may be used and/or modified appropriately for the study. Monitoring should begin prior to injury induction. Baseline measurements should be established in all parameters that will be evaluated in the post-injury period. Since many of the observations could be subjective, identifying and training the personnel involved in the animal monitoring prior to beginning the study is important. Research, animal care and veterinary staff can all be involved in this phase of the study. It is valuable to standardize observations as much as possible between observers through group discussions and training sessions prior to beginning experimental work [4, 12]. Monitoring frequency should be planned to increase, if needed, during periods of expected increase in severity of clinical symptoms from the injury model being studied. Thorough record keeping is also important to document animal monitoring. Finally, potential complications should be considered and interventions planned that would be in the best interest of both the experiment and the animal. Contact information for the vital research staff and investigator should be available to facilitate communication when critical decisions about animal welfare and experimental needs must be made. The plan for health and experimental monitoring should be described in detail in the written animal protocol. Inclusion of a sample monitoring form is often helpful.

Development of experimental endpoints is inherent in study design. These end points must also take into account the animal's welfare as much as possible. The earliest possible experimental end point should be chosen to limit animal pain and distress. Humane end points also must be defined clearly in the animal protocol. These are endpoints that are developed and used to reduce the severity and duration of pain and/or distress [13, 14]. Often, these are defined using clinical, behavioral and physiological signs that are believed to indicate a level of animal pain or distress that is unacceptable and should be relieved by euthanasia. A study that does not have humane endpoints and requires that death of the animals is the final data point must provide stringent scientific justification for the necessity [15]. To be convincing, the study should include data from a pilot study indicating that no identifiable clinical sign—or group of signs—could be found that were irrevocably linked to death [16]. If such a study has not been done, then the animal protocol would do well to include provisions to perform this pilot study to determine if humane end points can be identified using the monitoring form already developed for the study.

The AWRs and The Guide state that the method of euthanasia must be consistent with the current version of the Report of the American Veterinary Medical Association (AVMA) Panel on Euthanasia unless a deviation is justified for scientific or medical reasons [1, 5]. The method of euthanasia must also be chosen to provide the least interference with experimental data collection. The Report of the AVMA Panel on Euthanasia recommends anesthetizing rodents prior to euthanasia by physical means such as decapitation or cervical dislocation [17, 18]. Thus, if use of anesthetics would interfere with experimental end points, euthanasia of rodents by physical means alone must be rigorously justified in the written animal protocol. If scientific justification is not available but interference is suspected, a pilot study should be designed and included in the animal protocol to determine whether or not anesthetics used in euthanasia interferes with experimental data. Training of personnel who will perform euthanasia is paramount—especially if euthanasia will be performed by physical means without anesthesia [17, 18]. If the personnel involved require training, practice should be done first using cadavers and then deeply anesthetized animals until the personnel are completely comfortable with the technique. Veterinary staff should be able to provide training in means of euthanasia should it be required. Again, the experience of personnel with these techniques should be documented in the animal protocol.

In determining the experimental data points, the best paradigms for outcome assessment are chosen by the investigator as part of the study design. These could include quantitative or observational behavioral screening tests as discussed for animal monitoring. It could also include performance of behavioral tasks, blood or urine sample collection, physiologic measurements, tissue collection, etc. The data collection may require specialized equipment such as photocell arrays or implanted mini transmitters, or specialized environment such as metabolic caging or reverse light cycle housing. Data collection in some species may require a form of prolonged restraint. All these raise their own set of concerns for animal welfare that will need to be addressed in the study design period and in the written animal protocol.

In general, with any experimental manipulation that has the potential to cause more than momentary pain or distress to an animal, there is some necessary information that must be provided in the written animal protocol. Thus, it must be considered in designing the study. First, scientific justification for the need to use the manipulation must be provided. Second, a thorough and detailed description of what happens to the animal during the manipulation must be included. If an apparatus other than the animal's home cage is used, that apparatus should be described including dimensions of the primary enclosure. Unless scientific justification is provided for an exception, the surfaces of a primary enclosure should be smooth and sealed to allow sanitization and prevent injury [5]. Third, a description of monitoring the animal during the manipulation and how it will be documented should be provided. Fourth, the plan for stopping or intervening in the manipulation should the animal show evidence of unacceptable pain or distress should be described. The parameters that would lead to recognition of unacceptable pain or distress and the interventions that would be taken on the animal's behalf should be included. Finally, personnel performing the task must have sufficient experience with it or have received training. Developing this information during the design phase should allow the investigator to provide the IACUC with the information required to properly evaluate the animal welfare concerns involved in the animal protocol.

The IACUC is also tasked with evaluating hazards posed to personnel working with the animals or other animals in the institution as part of its review of animal protocols [19]. There are some species that pose inherent zoonotic risks, such as working with non-human primates. All animals can pose some physical risks to personnel. If a study involves working with a species that will be new to the investigator, the animal care staff and veterinarian can provide valuable training in proper handling and treatment procedures to the study participants to help limit risks to both personnel and animals. It is best to seek this training as part of the study design phase. Experience with the species in question will need to be documented in the animal protocol. If a study involves a chemical, infectious, and/or radiation hazard, it must be identified in the animal protocol [20]. The institution usually has other committees that must also review the use of hazards in research and are required to approve such use before experimental work can begin. Once the hazards are identified in the study design process, application for their use should be made to the appropriate committees. Personnel participating in the animal work will probably need to be enrolled in the institution's Occupational Health program if this has not already been done. This will help prevent a delay in starting experimental work.

Writing the Animal Protocol

Each IACUC will have developed an application form to guide investigators in writing a protocol for animal use in research at that institution. All the applicable information discussed in the design phase will be requested on this form. In addition to this information, a few more areas of concern must be addressed.

There will be a requirement to provide evidence that the proposed work will not duplicate any previously performed studies and that alternatives are not available to the use of animals or painful procedures [1, 5]. These justifications are most often required in the form of a recent and thorough literature search coupled with the assurance of the investigator based on his expertise in the field. (It is of note that, when performing a literature search for alternatives to animal use or use of painful procedures, having the word "alternative" as a keyword potentially makes the search more productive.)

Because IACUCs are required to have nonscientist members, a statement explaining the purpose, benefit, and brief overview of the work should be included using terminology that can be easily understood by the lay public. Brief but sufficient background information should also be provided for a scientist to understand the justification for the approach taken and the study design. It is easy for the investigator who is intimately involved with a study to assume that reviewers also have an in-depth understanding of the subject. Unfortunately, this is not always the case when a large institution is involved. It is very helpful for the investigator to take a step-by-step approach when explaining what will happen to the animals in a study and when each step will happen in the study timeline. Details such as dose, approximate volume, route of administration and frequency of treatment should be provided for any substance given to the animals. Likewise, if body fluid samples will be taken from the animals, the method of collection, volume collected and frequency of collection should be specified. These details allow the IACUC members to determine if standard veterinary methods are being followed or if exceptions need to be considered.

The choice of appropriate species, strain and genotype of animal for use in the study must be explained. Some strain choices might be dictated by ability to study must be explained. Some strain choices might be dictated by ability to study genetically manipulated animals to further refine the experimental design. The number of animals needed for the work must also be explained. The U.S. Government Principles state that investigators should use the minimum number of animals required to obtain valid results [3]. The Guide states that "Whenever possible the number of animals requested should be justified statistically" [5]. If statistical justification cannot be performed, a pilot study might be necessary to provide sufficient information to complete statistical calculations. Alternatively, documentation of published works using the same model could be provided and sample group size choice for the study based on group sizes that provided statistical significance in those studies could be considered sufficient justification for the requested animal numbers. Previously published works using the same model would at least provide a good sample size estimate for a pilot study if such is required.

All animals that might be needed for successful completion of the study must be included in the total number of animals requested for the study. An estimate of the percentage amount of the expected attrition among experimental animals should be available for an established model and must be included in the overall number of animals requested for the study. Attrition would include animals that must be euthanized or removed from the study prior to final data collection owing to humane concerns or incorrect model development within those animals. Beyond experimental animal needs, there might be the need to include training animals in the study, if another animal protocol is not available for this purpose. An investigator or veterinarian experienced in the procedures being developed should be able to provide a justifiable number of animals needed to become competent in the techniques involved. Finally, if a breeding colony must be maintained to produce the experimental animals—such as genetically manipulated rodents—those animals that are born as part of the colony but not used experimentally will also need to be counted.

Writing an animal protocol for a study of acute neurological injury can be challenging. However, involving veterinary staff, animal care staff, IACUC staff, and other appropriate institutional committees, if necessary, in the planning process, should help to make the process more efficient. Addressing all the key issues and providing necessary detail in describing everything done with the animals will also help the IACUC in its review process. It may be necessary to perform one or more pilot studies as part of initiating a new model at an institution. These could be described in a single animal protocol or written as separate protocols to facilitate work on gathering the pilot information. With everyone working together, the goal of meeting the highest scientific standards and the highest animal welfare standards should be achieved.

References

1. CFR (Code of Federal Regulations). Title 7: Agriculture. Chapter 54: Transportation, sale, and handling of certain animals. Washington, DC: Office of Federal Register; 2013. Retrieved 28 Nov 2015 from the US Government Printing Office database on the World Wide Web: http://www.gpo.gov/fdsys/pkg/USCODE-2013-title7/html/USCODE-2013-title7-chap54.htm.
2. CFR (Code of Federal Regulations). Title 9: Animals and animal products. Chapter 1: Subchapter a: animal welfare. Washington, DC: Office of Federal Register; 2013. Retrieved 28 Nov 2015 from the US Government Printing Office database on the World Wide Web: http://www.gpo.gov/fdsys/pkg/CFR-2013-title9-vol1/xml/CFR-2013-title9-vol1-chapI-subchapA.xml.
3. OLAW [Office of Laboratory Animal Welfare], USDHHS [US Department of Health and Human Services], NIH [National Institutes of Health]. Public Health Service Policy on Humane Care and Use of Laboratory Animals (2015 reprint) (pub no. 15–8013). Washington, DC: U.S. Government Printing Office; 2015. Retrieved 28 Nov 2015 from NIH Office of Extramural Resources database (Grants and Funding) on the World Wide Web: https://grants.nih.gov/grants/olaw/references/phspol.htm.
4. APHIS [Animal and Plant Health Inspection Service], USDA [US Department of Agriculture]. Animal care policy manual. Washington, DC: U.S. Government Printing Office; 2015. Retrieved 28 Nov 2015 from USDA APHIS database (Publications, Forms and Guidance Documents) on the World Wide Web: https://www.aphis.usda.gov/animal_welfare/downloads/Animal%20Care%20Policy%20Manual.pdf.
5. Health Research Extension Act of 1985, Public Law No. 99-158. "Animals in research". 1985. Retrieved 28 Nov 2015 from NIH Office of Extramural Resources database (Grants and Funding) on the World Wide Web: https://grants.nih.gov/grants/olaw/references/hrea1985.htm.
6. IRAC (Interagency Research Animal Committee). U.S. Government Principles for the utilization and care of vertebrate animals used in testing, research, and training. Washington, DC: Office of Science and Technology Policy [FR Doc. 85-12059]; 1985.
7. NRC (National Research Council). Guidelines for the care and use of mammals in neuroscience and behavioral research. Washington, DC: National Academic Press; 2003.
8. NRC (National Research Council). The guide for the care and use of laboratory animals. Washington, DC: National Academic Press; 2010.
9. NIH (National Institutes of Health). In: Morrison AR, Evans HL, Ator NA, Nakamura RK, editors. Methods and welfare considerations in behavioral research with animals: report of a National Institutes of Health workshop [NIH Publication No. 02-5083]. Washington, DC: US Government Printing Office; 2002.
10. Stokes WS. Reducing unrelieved pain and distress in laboratory animals using humane endpoints. ILAR J. 2000;41(2):59–61.
11. ARENA-OLAW (Applied Research Ethics National Association – Office of Laboratory Animal Welfare). Institutional animal care and use committee guidebook. 2nd ed. Washington, DC: US Government Printing Press; 2002.
12. Morton DB. A systematic approach for establishing humane endpoints. ILAR J. 2000;41(2):80–6.
13. AVMA (American Veterinary Medical Association). AVMA guidelines for euthanasia of animals. 2013th ed. Schaumburg, IL: AVMA; 2013; 102p.
14. NRC (National Research Council). Occupational health and safety in the care and use of research animals. Washington, DC: National Academics Press; 1997.
15. Dyson MC, Rush HG. Institutional training programs for research personnel conducted by laboratory-animal veterinarians. J Vet Med Educ. 2012;39(2):160–8.
16. Whitcomb TL, Taylor EW. Teaching laboratory rodent research techniques under the tenets of situated learning improves student confidence and promotes collaboration. J Am Assoc Lab Anim Sci. 2014;53(4):368–75.

17. Choe BI, Lee GH. Individual and collective responsibility to enhance regulatory compliance of the Three Rs. BMB Rep. 2014;47(4):179–83.
18. Medina LV, Hrapkiewicz K, Tear M, Anderson LC. Fundamental training for individuals involved in the care and use of laboratory animals: a review and update of the 1991 NRC Core Training Module. ILAR J. 2007;48(2):96–108.
19. Carstens E, Moberg GP. Recognizing pain and distress in laboratory animals. ILAR J. 2000;41(2):62–71.
20. Schallert T, Woodlee MT, Fleming SM. Experimental focal ischemic injury: behavior-brain interactions and issues of animal handling and housing. ILAR J. 2003;44(2):130–43.

Anesthesia and Analgesia for Research Animals

Lisa J. Brossia-Root, Tara M. Cotroneo, and Gerry Hish

Abstract This chapter provides a general overview of anesthesia and analgesia of animals for acute neurologic surgical models. Injectable and inhalant anesthetic agents are discussed, as well as analgesia to control post-operative pain. The impact of anesthetic and analgesic agents on three common acute neurologic surgical models is reviewed. Common agents and doses listed by species are provided.

Keywords Animal models · Anesthesia · Analgesia · Peri-operative monitoring

Introduction

The provision of proper anesthesia and analgesia to research animals is a legal and moral obligation and ensures animal welfare. The optimal anesthetic and analgesic will vary depending on the surgical procedure and the species of animal. All anesthetic medications are likely to provide some degree of neuroprotection hence may alter experimental outcomes. It is incumbent upon the investigator to evaluate the effects on their own models in order to account for this. It is worth noting that the effects of anesthetic and analgesic drugs on the nervous system may vary according to dose, route of administration, or individual animal strain. It may be beneficial to utilize pilot studies on a limited number of animals to determine whether any alterations induced by these agents are scientifically or clinically relevant to their model. With this in mind, the information provided in this chapter is intended to give investigators a general overview of potential drug effects and serve as a starting point for further investigation.

L. J. Brossia-Root (✉)
Department of Laboratory Animal Resources, University of Toledo, Toledo, OH, USA
e-mail: lisa.root@utoledo.edu

T. M. Cotroneo · G. Hish
Division of Laboratory Animal Resources, Wayne State University, Detroit, MI, USA

J. Chen et al. (eds.), *Animal Models of Acute Neurological Injury*, Springer Series in Translational Stroke Research, https://doi.org/10.1007/978-3-030-16082-1_2

Pre-anesthetic Considerations and Monitoring

Maintenance of normal body temperature is imperative, and cooling can occur very rapidly in small animals, especially those breathing compresses gasses or during long anesthetic events. Thermal support is best provided by isothermal heating pads, water blankets, or radiant heat sources throughout the surgical and recovery periods. Care should be taken to ensure that overheating does not occur and equipment should be maintained in order to avoid cutaneous burns.

Monitoring should start during the pre-operative period. Baseline physiologic parameters should be obtained prior to induction of anesthesia. Parameters may include body weight and temperature, heart rate, blood pressure, as well as respiratory rate and quality. If a pain scale will be used to detect post-operative pain, a pre-operative baseline evaluation may be useful.

Intra-operative monitoring parameters may include temperature, heart rate, blood pressure, electrocardiogram (ECG), blood gas analysis, and biphasal electroencephalogram (EEG) activity. As the animal emerges from anesthesia, temperature, heart rate, respiratory rate, and behaviors that may indicate pain should be recorded. Adverse events should be identified, addressed quickly, and recorded. Chronic post-operative records may include pain scale scores, body weight, food and water consumption, incision integrity, and other observations related to the procedure performed.

A surgical plane of anesthesia includes muscle relaxation, unconsciousness, and antinociception, and this level must be confirmed prior to the initiation of surgery [1]. Depth of anesthesia may be determined by muscular tone, eye position, or response to noxious stimulus (such as extremity pinch). These parameters may vary by species; confirm anesthetic depth by multiple methods prior to the skin incision and throughout surgery.

Injectable Anesthetics

Bartiturates

The barbiturates include pentobarbital, thiopental, and thiobarbital and produce hypnosis and unconsciousness by acting at the γ-aminobutyric acid (GABA) receptor [2, 3]. They inhibit the release of acetylcholine, norepinephrine, and glutamate, and, at high doses, barbiturates inhibit the uptake of calcium at nerve endings. Like most other GABA agonists they induce hypnosis but have no intrinsic analgesic properties and may not produce a sufficient plane of anesthesia when used as a sole agent for invasive surgeries. As a class, the barbiturates are highly lipid soluble, entering the central nervous system (CNS) quickly following IV administration but then being redistributed to other tissues. Despite the rapid onset of action, the long half-life of barbiturates may result in a prolonged anesthetic recovery period, particularly when re-dosing after an initial bolus [2]. Residual barbiturates cause CNS depression which may obscure post-operative evaluations. In most species (especially cats), barbiturates cause dose-dependent respiratory depression and reduce intestinal tract motility [3].

Thiopental is an ultra-short acting barbiturate, with onset of effect within 1 min of IV administration. It enters the CNS very rapidly, but it is quickly redistributed to the muscle and adipose tissue and quickly metabolized by the liver resulting in a relatively short (15 min) duration of anesthesia [2]. Because of this, thiopental has been used to induce anesthesia in large animals (cats, dogs, primates, and swine) followed by maintenance with an inhalant anesthetic. While all barbiturates are irritating to the tissues due to their high pH, thiopental is particularly problematic and must be given intravenously, preferably via an intravenous catheter as extravasation can result in tissue necrosis. Pentobarbital is the most commonly used barbiturate in laboratory animals, particularly in rodents, however recent problems with acquisition of the commercial product, the lack of significant analgesic properties and the ready availability of safer, more effective injectable anesthetics has reduced its use [2]. While it is still irritating to the tissues, pentobarbital can be injected intraperitoneally (IP) and compared to thiopental it is less lipophilic and has a longer duration of action. Pentobarbital can have significant cardiovascular effects, such as tachycardia, decreased myocardial contractility and stroke volume, and may cause hypotension in dogs [2, 3]. While careful titration by IV administration can produce a stable surgical plane of anesthesia, when administered to mice as a single IP bolus, the dose required to induce surgical anesthesia can result in severe cardiopulmonary depression and respiratory arrest [2]. Additionally, mouse strain has a significant impact on the sensitivity and response to pentobarbital [2].

Ketamine

Ketamine is an *N*-methyl-D-aspartate (NMDA) antagonist that weakly inhibits GABA, activates the limbic system, and suppresses the thalamoneocortical system. It increases sympathetic tone, thus increasing cardiac output, heart rate, and blood pressure; however, it induces relatively less respiratory depression than many other anesthetics in most species. Similar to the barbiturates, ketamine is very lipophilic resulting in a rapid onset and limited duration of action due to tissue redistribution. When used as a sole agent, ketamine will produce sedation without awareness but it does not provide muscle relaxation and animals may not reach a surgical plane of anesthesia. As a result of its effects on cardiac output, ketamine also increases cerebral blood flow (CBF), oxygen requirement, intracranial pressure (ICP), and EEG activity. The analgesic effects of ketamine are due primarily to its action at NMDA receptors and may be more efficacious for reducing musculoskeletal versus visceral pain [4].

As previously noted, ketamine alone is generally insufficient for surgical anesthesia but it is widely used in veterinary medicine for chemical restraint in large species, especially non-human primates. In these species, ketamine can cause increased salivation and anticholinergics like atropine or glycopyrrolate can be administered concurrently to guard against airway obstruction [2]. Ketamine is frequently used in combination with other drugs to produce more significant anesthesia. The administration of benzodiazepenes or α2 adrenergic receptor agonists with ketamine is common in laboratory animal medicine and serves to improve muscle relaxation and analgesia, permitting short to moderate duration surgical events.

Propofol

Propofol is a short-acting hypnotic agent that acts on GABA receptors to produce sedation and anesthesia. Following intravenous administration the onset of anesthesia is rapid, making it a popular choice for induction in larger species. It produces dose-dependent hypotension that is largely the result of vasodilation and it is associated with respiratory depression, especially when administered rapidly or at high doses. Because of the risk of apnea, careful monitoring is recommended and ventilatory support should be available. Propofol suppresses EEG activity and decreases CBF and ICP [5].

One significant advantage of propofol that is associated with its high rate of clearance is that recovery following a single bolus is rapid and smooth. Unlike the barbiturates, propofol does not accumulate with repeated administration and serial boluses or a continuous rate infusion can be used successfully with minimal impact on the time to recovery. Similar to barbiturates, propofol lacks significant analgesic effect and when used alone is best suited to non-invasive diagnostic procedures or minor surgical manipulations.

Propofol is available as a soybean oil based emulsion for intravenous use. The lipid nature of the commercial preparation has the potential to support microbial growth and special handling is necessary to prevent contamination of the vial. In most instances, the vial should be discarded after 4 h unless it is specifically labeled for a longer in use shelf life (e.g. Propoflo 28). Irrespective of the formulation, aseptic technique should be observed when preparing individual doses for use. In addition, the emulsion must be shaken prior to each use to ensure even mixing of the active drug and consistent dosing over time.

Sedative/Tranquilizers

There are two categories of injectable tranquilizers that are used frequently in veterinary species either alone for sedation and mild procedures or in combination with other anesthetics to produce a surgical plane of anesthesia.

The α2 adrenergic receptor agonists include xylazine and dexmedetomidine, and these compounds have both sedative and analgesic properties in veterinary patients, significantly increasing the effects of other anesthetics. Administration of the α2 agonists results in significant effects on the cardiovascular system and is characterized by peripheral and cardiac vasoconstriction, bradycardia, and respiratory depression. In rodents, the α2 agonists are most commonly combined with ketamine to produce a surgical plane of anesthesia. In larger species, xylazine and dexmedetomidine are generally used as preanesthetics or in combination with an opioid to facilitate minor procedures. Ruminants are particularly sensitive to the effects of xylazine and recommended doses are approximately 10% of that used in other species [3]. Ruminants are also prone to hypersalivation and pretreatment with atropine can help mitigate this effect and combat bradycardia [3]. In contrast, anticholinergics are not recommended with the administration of dexmedetomidine in dogs due to the potential for increased arterial pressure and an increased incidence of cardiac arrhythmia. Drugs in this category are also associated with emesis in several species, particularly when administered subcutaneously in cats. An advantage of this class of drugs is that α2 receptor antagonists are readily available as reversal agents and include yohimbine, tolazoline, and atipazmazole.

The benzodiazepines include diazepam and midazolam, and they are used in veterinary anesthesia for their anxiolytic, sedative, and muscle relaxing properties. While their sedative properties can be significant in certain species, diazepam and midazolam are generally used for their ability to potentiate the effects of other anesthetics or analgesics rather than as sole agents. When used alone in dogs and cats it is possible that excitement may be seen, rather than sedation. While the cardiovascular effects of these drugs are generally minimal, respiratory depression can occur and both diazepam and midazolam have been used clinically as anti-convulsants which may confound certain neurologic models. Flumazenil is available as a reversal agent for both diazepam and midazolam.

Local Anesthetics

Local anesthetics act at the level of the nerve by binding voltage gated sodium channels, preventing depolarization and nerve transmission. Therefore, they have the advantage of providing anesthesia at the site of tissue manipulation without concurrent generalized CNS effects. Administration can range from the simple infusion of anesthetic at the site of incision to more extensive regional anesthesia when applied to specific nerves or epidurally. The reader is referred to Hall et al. [6] for more information on specific nerve blocks. The two most commonly used local anesthetics are lidocaine and bupivacaine, and they differ clinically by their speed of onset and their duration of action. While lidocaine produces anesthesia within 3 min and lasts for approximately 60 min, bupivacaine's time to onset is at least 15 min and it provides 4–6 h of effect. The addition of epinephrine (5 μg/ml) to either lidocaine or bupivacaine prior to infusion will lead to an up to two-fold increase in duration of effect [3, 5]. Because excessive serum levels can produce life-threatening cardiac and CNS depression, local anesthetic doses should be carefully calculated based on body weight. Lidocaine can be mixed with sodium bicarbonate at a 1:10 ratio to reduce the pain associated with injection in conscious patients [2].

Gaseous Agents

Inhalant agents produce a generalized reversible depression of the CNS via unknown mechanisms. The solubility of the agent determines the rate of induction and recovery; those with low blood/gas partition coefficients have the most rapid inductions and recoveries. Inhalants produce respiratory and cardiovascular depression, both of which may be offset by sympathetic stimulation during surgery [7]. Many of the commonly used gas anesthetics have been shown to have neuroprotective properties [8, 9]. It is important to consider the impact on the research model prior to choosing an anesthetic agent.

Inhalant anesthetics are commonly administered via a nose cone or endotracheal tube. If the surgical site is the neck or skull, inhalant anesthetics may require a special apparatus to maintain a patent airway and to allow access to the incision site. A stereotaxic apparatus may have an integrated nose cone for anesthetic gas delivery built onto the incisor bar. Waste gas management must be addressed to reduce occupational exposure. Low flow delivery systems may help reduce waste anesthetic exposure.

Halothane had neuroprotective properties [10] and had been used historically but is no longer readily available in the US. It has been replaced with isoflurane and sevoflurane, two inhalant anesthetics with better safety and pharmacologic profiles.

Isoflurane

Isoflurane is one of the most commonly used inhalant anesthetics in veterinary medicine. Its high volatility and low solubility allow for rapid induction and recovery, as well as rapid responses to changes in isoflurane concentration during maintenance anesthesia. Isoflurane absorbed via the lungs, distributed to CNS, and eliminated via exhalation. The drug undergoes minimal systemic metabolism [3]. Isoflurane produces a dose-dependent depression of cardiovascular parameters, but less so than halothane. Isoflurane has significant vasodilatory effects and causes an increase in cerebral blood flow [11]. However, this can be prevented with hyperventilation prior to initiation of anesthesia [11]. Isoflurane reduces brain oxygen requirement by reducing the CNS metabolic rate. Isoflurane suppresses burst activity on electroencephalograms, thus interfering with diagnostic testing [7].

Sevoflurane

Sevoflurane has similar characteristics as isoflurane (rapid induction and recovery, less cardiovascular depression than halothane). Compared to isoflurane, sevoflurane has a lower solubility in the blood resulting in faster induction and rapid changes in anesthetic depth [11].

Other Inhalants

Nitrous oxide is an adjunct anesthetic. Veterinary applications are limited. A mixture of nitrous oxide and oxygen (30% N_2O: 70% O_2) is commonly used as a delivery gas for inhalant anesthetics in certain brain injury models [8]. Nitrous oxide has strong analgesic properties. In addition, it increases CBF and oxygen requirement [7]. Desflurane has a very low solubility in the blood, resulting in a very quick induction and recovery. It requires a special temperature-regulated vaporizer.

Hypothermia

Hypothermia has been used as anesthesia for neonatal rats up to 7 days of age [12]. It can be induced by submersion of the rat pup in ice water within a protective latex sleeve. The process of cooling and rewarming may be painful [13]. Therefore, alternative anesthetics, such as isoflurane, should be used whenever possible for animal welfare reasons.

Analgesia

It is clearly stated in applicable regulations and guidelines that animals must not experience pain unless it is scientifically-justified and approved in the institutional animal care and use committee (IACUC) protocol [14]. The rule of thumb is that if a procedure is considered painful in a human, then it is painful in an animal [15]. Justification for withholding an analgesic should include objective data from a pilot study, published adverse effects of analgesia on experimental data, plans for monitoring for excessive pain, and criteria for withdrawal from study or euthanasia when pain is excessive. Pain causes multiple physiologic and behavioral effects which could confound an experiment. The proper use of analgesia can control these effects, but analgesic agent may also impact experimental variables [16, 17]. When considering analgesic regimens, it is important to consider the anticipated cause, location, duration, and the severity of the pain [18, 19]. Opioid analgesics are considered optimal for moderate to severe pain, whereas non-steroidal anti-inflammatory drugs (NSAIDs) may be more appropriate for mild-moderate pain. In many cases, a multimodal approach (utilizing multiple types of analgesics) may be most effective [19].

Opioids

Opioids are categorized according to what receptor they target as well as what their action is at that receptor. Opiate receptors include mu (found in pain-regulating areas in the brain), kappa (deep cerebral cortex and spine), and delta (in the limbic areas). Opiates are agonists, partial agonists, or antagonists at these receptors [20].

Morphine, fentanyl, and oxymorphone are mu-receptor agonists. The primary effects of opioids in this group are analgesia, cough suppression, respiratory depression, sedation, intestinal effects, urine retention, and dependence. Secondarily, they may cause euphoria, bradycardia, and hypotension [3]. They cause increased intracranial pressure and should be used with extreme caution in patients with head injuries or respiratory disease [3]. They can be used for short-term analgesia, sedation, and as an anesthetic adjunct. They can be given epidurally. Some species may react differently to morphine. Cats and swine may develop excitement after morphine administration, whereas dogs may become depressed [21]. Oxymorphone has 10 times more potency than morphine, and fentanyl 80–100 times more than morphine [3]. Fentanyl may be administered parenterally or by transdermal patch or solution. For example, a 20 kg pig could be prescribed a 50 or 100 mcg/hour patch, placed 12–24 h before surgery on the lateral thorax [3]. The duration of effect of transdermal patches varies by species, but therapeutic levels of fentanyl may be achieved for 3–5 days.

Buprenorphine is a partial mu-receptor agonist and kappa-receptor antagonist that is 25–50 times more potent than morphine. It has a high affinity for the mu-receptors of the CNS [21], allowing for a relatively long duration of action (8–12 h in rodents) [3]. It decreases blood pressure and heart rate. It may cause less respiratory depression than other opiates. Sustained-release formulations are available and work well in rodents and other species [22–26].

Butorphanol is 4–7 times more potent than morphine. It is a kappa-receptor agonist and a weak mu-receptor antagonist. Its analgesic effect is short-lived, which limits its use for post-operative analgesia [3].

Non-steroidal Anti-inflammatory Drugs (NSAIDs)

NSAIDs are used therapeutically for analgesia, anti-inflammation, and antipyrexia. NSAIDs reduce inflammation by inhibiting the synthesis of prostaglandins by binding directly to cyclo-oxygenase (COX). There are two isoforms of COX, COX-1 and COX-2. In general, COX-1 is found in many normal tissues in the body, whereas COX-2 is considered the inducible isoform found at sites of inflammation [27]. For the most part, modern NSAID agents are safe drugs with only a small percentage of patients experiencing serious adverse health effects. They should not be prescribed to patients receiving systemic corticosteroids, or patients with renal, gastrointestinal, or hepatic disease. Anti-inflammatory agents that are more COX-2 specific may have fewer adverse side effects [27].

Aspirin blocks COX-1, making it anti-inflammatory, antipyrexic, and prevents platelet aggregation [3]. Newer NSAIDs are COX-2 selective or more COX-1 sparing than aspirin. Carprofen is one example, and it is prescribed to treat acute musculoskeletal pain and post-operative pain [27]. Meloxicam is COX-2 preferential, and prescribed for mild to moderate post-operative pain, and acute and chronic musculoskeletal disorders [3]. These agents may be given parenterally or orally, and the dosage is typically once a day.

Other Analgesics

Acetaminophen is an analgesic and antipyretic with few anti-inflammatory effects. It is a relatively weak analgesic but may be used in combination with opioids. It cannot be given to cats [3].

While not strictly related to analgesia, environmental and social enrichment is often regarded as a non-pharmacologic method of pain management. It bears mention that enrichment has various effects on neurologic measures in acute models of neurological injuries [28, 29].

Special Considerations—Anesthesia and Analgesia for Animal Models of Neurologic Diseases

Cerebral Ischemia and Hemorrhage Models

A variety of animal models exist to study the different pathologic mechanisms and types of strokes. These models are commonly classified based on the extent of disease (global vs. focal) and etiology (induced vs. spontaneous). Global models of ischemia reduces cerebral blood flow to most of the brain mimicking disease following cardiac arrest or systemic hypotension and focal models reduce blood flow to a specific brain region similar to what occurs during ischemic stroke [8, 30]. In addition, animal models are used to study sub-arachnoid hemorrhage and intracerebral hemorrhage [31]. The majority of induced animal models involve surgical manipulation. Therefore, appropriate anesthesia, analgesia, surgical monitoring protocols are critical to model outcome.

Physiologic Effects of Anesthesia and the Impact on Cerebral Ischemia and Hemorrhage Models

Stroke models are sensitive to physiologic fluctuations. Throughout the surgical period it is important to minimize changes in parameters such as temperature, blood pressure, blood glucose, and blood gas concentrations [8].

Anesthesia can introduce variations in body temperature due to decreased thermoregulation. In addition, ischemic injury can lead to increased cerebral temperature [32]. Hyperthermia has been associated with increased cerebral injury and hypothermia is neuroprotective [33]. Body temperature should be monitored closely throughout the surgical procedure. Ideally, a system which monitors rectal temperature and adjusts the external heating device accordingly should be implemented to maintain precise homeostasis. Additionally, cerebral temperature should be monitored during the ischemic period due to variation between core temperature measurements and brain temperature [32].

Mechanical ventilation is recommended if respiratory deficits are anticipated during the surgical procedure [8]. This provides more control over blood oxygen and carbon dioxide levels and reduces model variation. Blood gas monitoring should be considered during the surgical period and ventilation parameters adjusted as necessary. However, feasibility of monitoring should be based on species used and model implemented [8].

Blood glucose levels can directly impact outcomes in models of cerebral ischemia. Hyperglycemia has been shown to exacerbate brain damage in animal models [34–36]. In addition, certain anesthetic agents may directly impact blood glucose levels [37, 38]. Routine monitoring of blood glucose is recommend before, during, and after ischemia induction to help with model consistency and data analysis [8].

Many commonly used anesthetics induce hemodynamic changes [3]. This can have a direct impact on blood pressure during the surgical period. Blood pressure fluctuations can impact stroke model outcomes due to changes in cerebral blood flow [8]. Monitoring can be implemented via direct methods (telemetry devices, arterial cannulation) or indirect methods (tail cuffs, oscillometric monitors, Doppler ultrasonic monitor). The method implemented should be based on the model and frequency of blood pressure monitoring required for the study.

Constant monitoring throughout the anesthetic and recover period is very important for consistent outcomes in models of stroke. Anesthetics can affect many physiologic parameters which has a direct impact on severity of cerebral ischemia. Detailed surgical records can help ensure model uniformity and aid with data analysis.

Anesthetic Considerations

Surgical models of stroke require the use of anesthetics in order to successfully induce the cerebral injury. These models differ from the natural course of disease which patients are seldom anesthetized at the time of ischemic injury [39]. Many of the commonly used anesthetic agents provide neuroprotection during cerebral ischemic injury [8]. The degree of protection can vary based on anesthetic, animal species, and model.

Volatile inhalant gas anesthetics (isoflurane, sevoflurane) are safe and recommended for use in animal surgical procedures. However, it has been documented that gas anesthetics reduce acute cerebral injury in animal models of stroke and improve survival [9, 40]. The impact of gas anesthetics on chronic cerebral injury (>7 days) is less well established and may be less severe [9, 41]. Multiple mechanisms have been suggested to explain the neuroprotective properties of these anesthetics and their protective effects are likely multifactorial [9].

Most of the commonly utilized injectable anesthetic agents have demonstrated some degree of neuroprotection in animal models of stroke. Barbituates have been shown to provide neuroprotection in models of focal ischemia [42]. Ketamine is commonly administered in conjunction with xylazine. Ketamine has been shown to improve outcomes in focal models of ischemia [43]. In addition, xylazine induces acute episodes of hyperglycemia after administration which can negatively impact stroke models [8, 44]. Propofol has also shown neuroprotection in models of global and focal ischemia. However, the degree of effect seems to be dependent on dose, species, and surgical model [45].

In general, the choice of anesthetics should be thoroughly researched during the experimental design process. Appropriate control animals should always be used when implementing treatment studies to effectively distinguish the anesthetic effect from the treatment. Physiologic parameters should be monitored and controlled as much as possible. Limiting length of anesthetic exposure and providing consistent control of anesthetic depth will reduce variation between animals. Inhalant anesthetics, despite neuroprotective properties, may be more desirable for certain projects due to more accurate anesthetic control and reduced anesthetic duration [46].

Analgesic Considerations

Animals undergoing surgical manipulation must be provided analgesics unless scientific justification is provided and approved by the IACUC committee. The most commonly used analgesics include opioids (buprenorphine) and non-steroidal anti-inflammatories (meloxicam, carprofen). Limited studies have been performed to assess the impact of clinical doses of analgesics in surgical models of stroke. It is well established that inflammation plays a role in cerebral injury during stroke [47–49]. Therefore, analgesics with anti-inflammatory properties may be less desirable in these models. It has been demonstrated that administration of meloxicam decreased infarct size in a focal model of cerebral ischemia [17]. Alternatively, studies have shown that buprenorphine administration does not significant impact models of global and focal cerebral ischemia at relevant clinical doses [17, 50]. Therefore, buprenorphine may be a more appropriate analgesic due to limited study impact.

Traumatic Brain Injury Models

There are multiple methods to produce an experimental traumatic brain injury (TBI) in animal models. Open-head injury models include controlled cortical impact (CCI) and fluid percussion (FC). Here, the mechanical impact is applied directly to the surface of the brain via craniotomy. In closed-head injury models, a mechanical force is applied to the intact skull. Impact can be direct via weight drop or piston impact, or indirect via a blast wave or inertial loading. These methods can be applied to various species of animals, from small rodents (mice and rats) to larger species (pig, dog, non-human primate). Neonatal animals can be used when studying the impact of TBI on the developing brain. Review articles are available to help you choose the best model to achieve your research goal [16, 51].

Anesthetic Considerations

Multiple factors can affect the outcome of TBI experiments, including age, gender, body temperature, and anesthetic and analgesic agents. The choice of anesthetic and analgesic regimen is just as important as choosing the right TBI method. There is no one-size-fits-all approach to anesthesia and analgesia, and many articles have identified the effects of these agents on various parameters [52–57]. The use of proper control groups is essential. Controls may include naïve animals (no manipulations), anesthesia only, and if applicable, sham surgery [58, 59]. Studies have shown that anesthesia alone is enough to induce brain injury in neonates [60, 61]. Pilot studies should be performed to identify variables that may be controlled in large scale experiments. Inhalant agents, especially isoflurane and sevoflurane, are commonly used in TBI models. They are advantageous because of their safety and ease of use. They may be used in combination with nitrous oxide. Injectable agents may include pentobarbital, fentanyl, propofol, alone or in combination with other agents. These agents may be given as a bolus or a continuous rate infusion. Ketamine may increase intracranial pressure [3], and may affect post-TBI behavior and histologic outcomes [16].

Analgesic Considerations

Surgical procedures are painful and analgesics must be used unless justified in the IACUC protocol. A combination of an opioid and a non-steroidal anti-inflammatory are recommended for major surgical procedures, such as a craniotomy. Buprenorphine (including sustained-release formulations) and meloxicam or carprofen are recommended combinations. Few studies have been published on the effects of analgesics on TBI models. Meloxicam has been shown to be neuroprotective by preserving the blood brain barrier and reducing brain edema following TBI [62]. And it worsens outcomes in a swine inertial loading model [63]. Even if a systemic analgesic cannot be used, the surgical incision should be treated with local anesthetic. Local agents such as bupivacaine and lidocaine are injected subcutaneously around the surgical site prior to the incision. Non-surgical closed-head TBI methods can also induce pain through effects of the primary mechanical trauma. Blast overpressure may rupture the tympanic membrane, which is a painful condition [64]. Blast overpressure may also have adverse secondary effects on the lung if proper shielding is not utilized to protect thoracic organs [65]. Mechanical ventilation is recommended if respiratory deficits are anticipated following the TBI procedure.

Spinal Cord Injury Models

There is significant interest in the development of validated, translational large and small animal models of spinal cord injury (SCI). Spontaneous animal models of human SCI do not readily develop without experimental manipulation. The protection afforded by the vertebral column often necessitates surgical exposure of the spinal cord in order to induce injury, irrespective of the disease being modeled. Because of the invasive nature of these procedures, appropriate anesthetics and analgesics must be used, most of which have some neuroprotective properties. Careful consideration of the experimental aims and an understanding of how different drugs affect the central and peripheral nervous system are critical to ensure that the integrity of the data is preserved while still meeting obligations for appropriate animal care.

Physiologic Effects of Anesthesia and the Impact on SCI Models

General anesthetics alter the cardiovascular and respiratory systems, and homeostatic responses such as thermoregulation. During surgical procedures, animals should be regularly monitored for cardiac function, ventilation, and blood loss to prevent hypoxemia and hypovolemia secondary to hemorrhage. Yanagawa et al. [66] found that rats subjected to hypoxia and mild hypotension concurrent with an induced spinal cord contusion experienced significantly more histologic damage in the cord gray matter though it did not affect the degree of neurologic deficit. Studies have demonstrated a significant correlation between systemic blood pressure and spinal cord blood flow [67] and the variable cardiovascular effects of different anesthetics has a demonstrated effect on functional outcomes in a spinal contusion model using Long-Evans rats [68]. While general anesthetics variably depress the respiratory center of the brain, spinal injury to high or mid-cervical cord can also affect ventillatory compromise by interrupting the function of the phrenic nerve and therefore the diaphram [69]. Acidosis resulting from hypoventilation rapidly translates to similar changes in the pH and pCO_2 of the cerebrospinal fluid (CSF) as well [70] and ventilation changes can affect CSF pressure which has been shown to impact the degree of injury in a rabbit model [71]. The provision of mechanical or intermittent positive pressure ventilation and the delivery of 100% oxygen can help avoid hypoxia and ensure that animals are normocapnic, minimizing the potential for variation in experimental outcomes.

Hypothermia has been widely investigated as an interventional therapy for minimizing neurologic damage and several animal studies support a neuroprotective effect of hypothermia. Inoue et al. [72] observed a significant improvement in both functional and histologic measures of injury in a rabbit model comparing local spinal and systemic hypothermia during injury. While combined hypothermia provided the greatest degree of protection, even modest systemic hypothermia alone was sufficient to impact the outcomes. Similar effects have been observed in rats [73] and mice [74].

Anesthetic and Analgesic Agents Impact Spinal Injury Models

Both injectable and inhalant anesthetic agents may affect SCI models. Ketamine is one of the most widely used anesthetics in laboratory animal medicine and has demonstrated neuroprotective properties in a variety of spinal cord disease models including contusion [75] and compression [76] models with mixed results in ischemic models [77, 78]. The mechanism of this protection has not been fully elucidated, however, ketamine's antagonism at NMDA receptors reduces the release of glutamate, an excitatory neuropeptide and contributor to ischemic neuronal damage [77, 78]. Ketamine also has demonstrated effects on oxidative stress, reducing lipid peroxidation [76] and maintaining levels of superoxide dismutase [77]. Ketamine also produces an increase in cerebral blood flow and intracranial pressure but this can be mitigated by concurrent use of a sedative/tranquilizer or hypnotic anesthetic [5].

Agents frequently combined with ketamine to produce a surgical plane of anesthesia may also impact SCI models. Dexmedetomidine, a newer $\alpha 2$ agonist, has been found to confer neuroprotection in two rodent models of spinal injury. In a mouse model of ischemia induced by temporary thoracic aortic occlusion, dexmedetomidine led to the preservation of neurologic function and reduced neuronal degeneration [79], an effect that appeared to be related to cellular protection against apoptosis. A spinal compression model in rats also demonstrated a neuroprotective effect of dexmedetomidine, inhibiting lipid peroxidation similar to methylprednisolone with better preservation of neuron populations at levels that were not significantly different than sham controls [80]. Diazepam attenuated cellular injury and edema formation when administered prior to incisional injury of the caudal thoracic spinal cord [81].

The hypnotic anesthetic propofol improved recovery of neurologic function and the histologic appearance of contusion-injured spinal cord both alone and in combination with local stem cell application [82]. Regarding inhalant anesthetics, preconditioning with isoflurane or sevoflurane provided neuroprotection in rabbit models of spinal cord ischemic injury [83, 84].

Local anesthetics are commonly used clinically as adjuncts to systemic anesthetics and can have significant sparing effects. Not surprisingly, when administered intrathecally, local anesthetics have limited the severity of neurological dysfunction and the development of cord lesions in a rat contusion model [85] and a rabbit ischemia model [86]. Regional infusion of lidocaine in an occluded aortic segment of rabbits reduced the severity of ischemic spinal cord injury, [87] however, systemic administration of lidocaine in cats resulted in no apparent neuroprotection following contusive injury [88].

Analgesic use is an important component of providing complete anesthetic care and should be initiated prior to any surgical manipulations. Depending upon the invasiveness of the procedure analgesia should be provided during at least the first 48–72 h after surgery, and generally utilizes systemic NSAIDs or opioids. There are few references in the literature to the effects of either of these two analgesic categories on animal models of SCI. A study by Hakan evaluating the use of meloxicam found that it attenuated all of the biochemical, histological, and functional effects of a spinal contusion (Tables 1 and 2) [89].

Table 1 Common anesthetic drug protocols for induction and maintenance [2–4]

Species	Drug (dosage) and route
Mouse	• Isoflurane delivered with precision vaporizer, to effect[a] • Ketamine (80–100 mg/kg) + xylazine (10 mg/kg) IP[b] • Ketamine (75 mg/kg) + dexmedetomidine (0.5 mg/kg) IP[b] • Pentobarbital (40–50 mg/kg) IP or IV injection[c]
Rat	• Isoflurane delivered with precision vaporizer, to effect[a] • Ketamine (75 mg/kg) + xylazine (10 mg/kg) IP[b] • Ketamine (75 mg/kg) + dexmedetomidine (0.25 mg/kg) IP[b] • Pentobarbital (40–50 mg/kg) IP or IV injection[c]
Rabbit	• Isoflurane delivered with precision vaporizer, to effect[a,d] • Ketamine (35 mg/kg) + xylazine (5 mg/kg) IM[b] • Ketamine (15 mg/kg) + dexmedetomidine (0.125 mg/kg) IM[b] • Pentobarbital (30–45 mg/kg) IV[c]
Cat	• General anesthesia – Premedication: Acepromazine (0.05 mg/kg) + buprenorphine (0.01 mg/kg) SC or IM – Induction: Propofol (3–6 mg/kg slow IV to effect) for rapid sequence intubation – Maintenance: Isoflurane to effect • Ketamine (5 mg/kg) + dexmedetomidine (40 μg/kg) + butorphanol (0.4 mg/kg) IM • Tiletamine (47.5 mg/kg) + zolazepam (7.5 mg/kg) IM • Pentobarbital (20–30 mg/kg) given IV[c]
Dog	• General anesthesia – Premedication: Acepromazine (0.05–0.1 mg/kg) + buprenorphine (0.01–0.03 mg/kg) SC or IM – Induction: Propofol (3–6 mg/kg slow IV to effect) for rapid sequence intubation – Maintenance: Isoflurane to effect • Ketamine (5 mg/kg) + dexmedetomidine (20 μg/kg) + butorphanol (0.4 mg/kg) IM • Pentobarbital (20–30 mg/kg) given IV[c]

(continued)

Table 1 (continued)

Species	Drug (dosage) and route
Swine	• General anesthesia – Tiletamine/zolazepam (2–7 mg/kg) + xylazine (0.2–1.0 mg/kg) IM – Isoflurane to effect after intubation • Midazolam (0.3 mg/kg) + xylazine (2–4 mg/kg) + butorphanol (0.3 mg/kg) IM
All species	• Local anesthetic total maximum safe dose: lidocaine 4 mg/kg; bupivacaine 1–2 mg/kg

These protocols are intended as general guidelines for individuals experienced with anesthetic management in animals. For additional information and for a comprehensive anesthetic reference, the author recommends Laboratory Animal Anesthesia by Paul Flecknell

IM intramuscular, *IP* intraperitoneal, *SC* subcutaneous, *IV* intravenous, *ET* endotracheal, *PO* per os (orally)

[a]Maintenance dose is usually ~2–3%. Administer with an analgesic prior to induction

[b]When re-dosing, administer 1/3 of the original ketamine dose only. Do not re-dose α2 agonist

[c]Only light to medium anesthesia (poor analgesia) and is increasingly difficult to obtain in the US

[d]Rabbits may hold breath during mask induction. Premedicate with an anxiolytic to reduce apnea

Table 2 Suggested analgesic doses [2–4]

Species	Drug	Dose and route of administration
Mouse	Buprenorphine	0.05–0.1 mg/kg SQ every 8–12 h
	Sustained-release buprenorphine	1 mg/kg SQ every 48–72 h
	Carprofen	5 mg/kg PO or SC daily
	Meloxicam	1–5 mg/kg PO or SC daily
Rat	Buprenorphine	0.01–0.05 mg/kg SQ every 8–12 h
	Sustained-release buprenorphine	0.3–0.5 mg/kg SC every 48–72 h
	Carprofen	5 mg/kg PO or SC daily
	Meloxicam	1–2 mg/kg PO or SC daily
Rabbit	Buprenorphine	0.01–0.05 mg/kg SC or IV every 6–12 h
	Fentanyl transdermal patch	1–5 μg/kg/h
	Carprofen	2–4 mg/kg SC daily, or 1–2.2 mg/kg PO daily
	Meloxicam	0.3–0.6 mg/kg SC or PO daily
Cat	Buprenorphine	0.01–0.03 mg/kg SC, IM, IV every 6 h 0.01–0.03 mg/kg PO (sublingual) every 6–12 h
	Fentanyl transdermal patch	1–5 μg/kg/h
	Butorphanol	0.2–0.5 mg/kg IM or SC every 4 h
	Meloxicam	0.1 mg/kg IV, SC or PO first day, 0.05 mg/kg PO daily thereafter
Dog	Buprenorphine	0.01–0.03 mg/kg SQ, IM or IV every 6 h
	Sustained-release buprenorphine	0.2 mg/kg SC [25]
	Fentanyl transdermal patch	1–5 μg/kg/h
	Carprofen	2.2 mg/kg SC or PO every 12 h, or 4.4 mg/kg SC or PO daily
	Meloxicam	0.2 mg/kg SC or PO first day, then 0.1 mg/kg SC or PO daily thereafter

(continued)

Table 2 (continued)

Species	Drug	Dose and route of administration
Swine	Buprenorphine	0.01–0.05 mg/kg SC, IM, IV every 8–12 h
	Buprenorphine transdermal patch	30 μg/h, change patch every 72 h [22]
	Fentanyl transdermal patch	1–5 μg/kg/h
	Carprofen	2 mg/kg every 12–24 h or 3–4 mg/kg daily SC, PO, IV, IM
	Meloxicam	0.1 mg/kg PO daily or 0.4 mg/kg SC, IV daily
	Flunixin	1–2 mg/kg SC every 24 h
	Ketoprofen	1–3 mg/kg SC, IM, or IV every 12 h

IM intramuscular, *IP* intraperitoneal, *SC* subcutaneous, *IV* intravenous, *PO* per os (orally)

References

1. National Research Council (US) Committee for the Update of the Guide for the Care and Use of Laboratory Animals. Guide for the care and use of laboratory animals. 8th ed. Washington, DC: National Academies Press; 2011.
2. Flecknell P. Laboratory animal anaesthesia. Boston, MA: Elsevier; 2015.
3. Plumb DC. Plumb's veterinary drug handbook. 8th ed. Stockholm, WI: PharmaVet Inc.; 2015.
4. Fish RE. Anesthesia and analgesia in laboratory animals. 2nd ed. London, UK: Academic Press; 2008.
5. Goodman LS, Brunton LL, Blumenthal DK, Murri N, Hilal-Dandan R. Goodman & Gilman's The pharmacological basis of therapeutics. New York: McGraw-Hill Medical; 2011.
6. Hall LW, Clarke KW, Trim CM. Veterinary anaesthesia. London, UK: Saunders; 2001.
7. Harvey RC, Paddleford RR. Anesthesia for the central nervous systems adn opthalmic surgery. In: Slatter DS, editor. Textbook of small animal surgery. 2nd ed. Philadelphia: W.B. Saunders; 1993.
8. Liu S, Zhen G, Meloni BP, Campbell K, Winn HR. Rodent stroke model guidelines for preclinical stroke trials (1st edition). J Exp Stroke Transl Med. 2009;2(2):2–27.
9. Matchett GA, Allard MW, Martin RD, Zhang JH. Neuroprotective effect of volatile anesthetic agents: molecular mechanisms. Neurol Res. 2009;31(2):128–34. https://doi.org/10.1179/174313209x393546.
10. Salzman SK, Lee WA, Sabato S, Mendez AA, Agresta CA, Kelly G. Halothane anesthesia is neuroprotective in experimental spinal cord injury: early hemodynamic mechanisms of action. Res Commun Chem Pathol Pharmacol. 1993;80(1):59–81.
11. Brunson DB. Pharmacology of inhalant anesthetics. Anesthesia and analgesia in laboratory animals. 2nd ed. San Diego, CA: Elsevier Inc.; 2008.
12. Gaertner DJ, Hallman TM, Hankenson C, Batchelder MA. Anesthesia and analgesia for laboratory rodents. In: Fish RE, Brown MJ, Danneman PJ, Karas AZ, editors. Anesthesia and analgesia in laboratory animals. 2nd ed. London, UK: Elsevier Inc.; 2008.
13. Flecknell PA, Lofgren JLS, Dyson MC, Marini RR, Swindle MM, Wilson RP. Preanesthesia, anesthesia, analgesia, and euthanasia. In: Fox JG, Anderson LC, Otto G, Pritchett-Corning KR, Whary MT, editors. Laboratory animal medicine. The American College of Laboratory Animal Medicine series. 3rd ed. London: Academic Press; 2015.
14. Animal Welfare Regulations. 2005.
15. U.S. Government Principles for the Utilization and Care of Vertebrate Animals Used in Testing, Research, and Training. Interagency Research Animal Committee. Public Health Service Policy on humane care and use of laboratory animals. Washington, DC: National Institutes of Health, Office of Laboratory Animal Welfare; 2002.

16. Rowe RK, Harrison JL, Thomas TC, Pauly JR, Adelson PD, Lifshitz J. Using anesthetics and analgesics in experimental traumatic brain injury. Lab Anim. 2013;42(8):286–91. https://doi.org/10.1038/laban.257.
17. Jacobsen KR, Fauerby N, Raida Z, Kalliokoski O, Hau J, Johansen FF, Abelson KS. Effects of buprenorphine and meloxicam analgesia on induced cerebral ischemia in C57BL/6 male mice. Comp Med. 2013;63(2):105–13.
18. Roughan JV, Flecknell PA. Behaviour-based assessment of the duration of laparotomy-induced abdominal pain and the analgesic effects of carprofen and buprenorphine in rats. Behav Pharmacol. 2004;15(7):461–72.
19. Muir WW. Selecting analgesic drugs and routes of drug administration. In: Gaynor JS, Muir WW, editors. Handbook of veterinary pain management. 2nd ed. St. Louis: Mosby; 2009.
20. Heavner JE, Cooper DM. Pharmacology of analgesics. In: Fish RE, Brown MJ, Danneman PJ, Karas AZ, editors. Anesthesia and analgesia in laboratory animals, The American College of Laboratory Animal Medicine series. 2nd ed. London, UK: Academic Press; 2008.
21. Papich MG. Saunders handbook of veterinary drugs. 2nd ed. St. Louis, MO: Saunders Elsevier; 2007.
22. Thiede AJ, Garcia KD, Stolarik DF, Ma J, Jenkins GJ, Nunamaker EA. Pharmacokinetics of sustained-release and transdermal buprenorphine in Gottingen minipigs (Sus scrofa domestica). J Am Assoc Lab Anim Sci. 2014;53(6):692–9.
23. Kendall LV, Hansen RJ, Dorsey K, Kang S, Lunghofer PJ, Gustafson DL. Pharmacokinetics of sustained-release analgesics in mice. J Am Assoc Lab Anim Sci. 2014;53(5):478–84.
24. Chum HH, Jampachairsri K, McKeon GP, Yeomans DC, Pacharinsak C, Felt SA. Antinociceptive effects of sustained-release buprenorphine in a model of incisional pain in rats (Rattus norvegicus). J Am Assoc Lab Anim Sci. 2014;53(2):193–7.
25. Nunamaker EA, Stolarik DF, Ma J, Wilsey AS, Jenkins GJ, Medina CL. Clinical efficacy of sustained-release buprenorphine with meloxicam for postoperative analgesia in beagle dogs undergoing ovariohysterectomy. J Am Assoc Lab Anim Sci. 2014;53(5):494–501.
26. Catbagan DL, Quimby JM, Mama KR, Rychel JK, Mich PM. Comparison of the efficacy and adverse effects of sustained-release buprenorphine hydrochloride following subcutaneous administration and buprenorphine hydrochloride following oral transmucosal administration in cats undergoing ovariohysterectomy. Am J Vet Res. 2011;72(4):461–6. https://doi.org/10.2460/ajvr.72.4.461.
27. Budsberg S. Nonsteroidal antiinflammatory drugs. In: Gaynor JS, Muir WW, editors. Handbook of veterinary pain management. 2nd ed. St. Louis: Mosby; 2009.
28. Berrocal Y, Pearse DD, Singh A, Andrade CM, McBroom JS, Puentes R, Eaton MJ. Social and environmental enrichment improves sensory and motor recovery after severe contusive spinal cord injury in the rat. J Neurotrauma. 2007;24(11):1761–72. https://doi.org/10.1089/neu.2007.0327.
29. Jacqmain J, Nudi ET, Fluharty S, Smith JS. Pre and post-injury environmental enrichment effects functional recovery following medial frontal cortical contusion injury in rats. Behav Brain Res. 2014;275:201–11. https://doi.org/10.1016/j.bbr.2014.08.056.
30. Traystman RJ. Animal models of focal and global cerebral ischemia. ILAR J. 2003;44(2):85–95.
31. Krafft PR, Bailey EL, Lekic T, Rolland WB, Altay O, Tang J, Wardlaw JM, Zhang JH, Sudlow CL. Etiology of stroke and choice of models. Int J Stroke. 2012;7(5):398–406. https://doi.org/10.1111/j.1747-4949.2012.00838.x.
32. DeBow S, Colbourne F. Brain temperature measurement and regulation in awake and freely moving rodents. Methods. 2003;30(2):167–71.
33. Campos F, Blanco M, Barral D, Agulla J, Ramos-Cabrer P, Castillo J. Influence of temperature on ischemic brain: basic and clinical principles. Neurochem Int. 2012;60(5):495–505. https://doi.org/10.1016/j.neuint.2012.02.003.
34. Sung JH, Shah FA, Gim SA, Koh PO. Identification of proteins in hyperglycemia and stroke animal models. J Surg Res. 2015; https://doi.org/10.1016/j.jss.2015.07.020.

35. Yamazaki Y, Harada S, Tokuyama S. Post-ischemic hyperglycemia exacerbates the development of cerebral ischemic neuronal damage through the cerebral sodium-glucose transporter. Brain Res. 2012;1489:113–20. https://doi.org/10.1016/j.brainres.2012.10.020.
36. Duverger D, MacKenzie ET. The quantification of cerebral infarction following focal ischemia in the rat: influence of strain, arterial pressure, blood glucose concentration, and age. J Cereb Blood Flow Metab. 1988;8(4):449–61. https://doi.org/10.1038/jcbfm.1988.86.
37. Saha JK, Xia JQ, Grondin JM, Engle SK, Jakubowski JA. Acute hyperglycemia induced by ketamine/xylazine anesthesia in rats: mechanisms and implications for preclinical models. Exp Biol Med. 2005;230(10):777–84.
38. Zuurbier CJ, Keijzers PJ, Koeman A, Van Wezel HB, Hollmann MW. Anesthesia's effects on plasma glucose and insulin and cardiac hexokinase at similar hemodynamics and without major surgical stress in fed rats. Anesth Analg. 2008;106(1):135–42. https://doi.org/10.1213/01.ane.0000297299.91527.74, table of contents
39. Turner RJ, Jickling GC, Sharp FR. Are underlying assumptions of current animal models of human stroke correct: from STAIRs to high hurdles? Transl Stroke Res. 2011;2(2):138–43. https://doi.org/10.1007/s12975-011-0067-3.
40. Bleilevens C, Roehl AB, Goetzenich A, Zoremba N, Kipp M, Dang J, Tolba R, Rossaint R, Hein M. Effect of anesthesia and cerebral blood flow on neuronal injury in a rat middle cerebral artery occlusion (MCAO) model. Exp Brain Res. 2013;224(2):155–64. https://doi.org/10.1007/s00221-012-3296-0.
41. Kawaguchi M, Drummond JC, Cole DJ, Kelly PJ, Spurlock MP, Patel PM. Effect of isoflurane on neuronal apoptosis in rats subjected to focal cerebral ischemia. Anesth Analg. 2004;98(3):798–805.
42. Kawaguchi M, Furuya H, Patel PM. Neuroprotective effects of anesthetic agents. J Anesth. 2005;19(2):150–6. https://doi.org/10.1007/s00540-005-0305-5.
43. Hoffman WE, Pelligrino D, Werner C, Kochs E, Albrecht RF, Schulte am Esch J. Ketamine decreases plasma catecholamines and improves outcome from incomplete cerebral ischemia in rats. Anesthesiology. 1992;76(5):755–62.
44. Saha JK, Xia J, Grondin JM, Engle SK, Jakubowski JA. Acute hyperglycemia induced by ketamine/xylazine anesthesia in rats: mechanisms and implications for preclinical models. Exp Biol Med (Maywood). 2005;230(10):777–84.
45. Adembri C, Venturi L, Pellegrini-Giampietro DE. Neuroprotective effects of propofol in acute cerebral injury. CNS Drug Rev. 2007;13(3):333–51. https://doi.org/10.1111/j.1527-3458.2007.00015.x.
46. Howells DW, Porritt MJ, Rewell SS, O'Collins V, Sena ES, van der Worp HB, Traystman RJ, Macleod MR. Different strokes for different folks: the rich diversity of animal models of focal cerebral ischemia. J Cereb Blood Flow Metab. 2010;30(8):1412–31. https://doi.org/10.1038/jcbfm.2010.66.
47. Candelario-Jalil E, Gonzalez-Falcon A, Garcia-Cabrera M, Alvarez D, Al-Dalain S, Martinez G, Leon OS, Springer JE. Assessment of the relative contribution of COX-1 and COX-2 isoforms to ischemia-induced oxidative damage and neurodegeneration following transient global cerebral ischemia. J Neurochem. 2003;86(3):545–55. https://doi.org/10.1046/j.1471-4159.2003.01812.x.
48. del Zoppo GJ, Becker KJ, Hallenbeck JM. Inflammation after stroke – is it harmful? Arch Neurol. 2001;58(4):669–72. https://doi.org/10.1001/archneur.58.4.669.
49. Iadecola C, Alexander M. Cerebral ischemia and inflammation. Curr Opin Neurol. 2001;14(1):89–94. https://doi.org/10.1097/00019052-200102000-00014.
50. Kalliokoski O, Abelson KS, Koch J, Boschian A, Thormose SF, Fauerby N, Rasmussen RS, Johansen FF, Hau J. The effect of voluntarily ingested buprenorphine on rats subjected to surgically induced global cerebral ischaemia. In Vivo. 2010;24(5):641–6.
51. Namjoshi DR, Good C, Cheng WH, Panenka W, Richards D, Cripton PA, Wellington CL. Towards clinical management of traumatic brain injury: a review of models and mechanisms from a biomechanical perspective. Dis Model Mech. 2013;6(6):1325–38. https://doi.org/10.1242/dmm.011320.

52. Thal SC, Timaru-Kast R, Wilde F, Merk P, Johnson F, Frauenknecht K, Sebastiani A, Sommer C, Staib-Lasarzik I, Werner C, Engelhard K. Propofol impairs neurogenesis and neurologic recovery and increases mortality rate in adult rats after traumatic brain injury. Crit Care Med. 2014;42(1):129–41. https://doi.org/10.1097/CCM.0b013e3182a639fd.
53. Yurdakoc A, Gunday I, Memis D. Effects of halothane, isoflurane, and sevoflurane on lipid peroxidation following experimental closed head trauma in rats. Acta Anaesthesiol Scand. 2008;52(5):658–63. https://doi.org/10.1111/j.1399-6576.2008.01635.x.
54. Statler KD, Alexander H, Vagni V, Holubkov R, Dixon CE, Clark RS, Jenkins L, Kochanek PM. Isoflurane exerts neuroprotective actions at or near the time of severe traumatic brain injury. Brain Res. 2006;1076(1):216–24. https://doi.org/10.1016/j.brainres.2005.12.106.
55. O'Connor CA, Cernak I, Vink R. Interaction between anesthesia, gender, and functional outcome task following diffuse traumatic brain injury in rats. J Neurotrauma. 2003;20(6):533–41. https://doi.org/10.1089/089771503767168465.
56. Goren S, Kahveci N, Alkan T, Goren B, Korfali E. The effects of sevoflurane and isoflurane on intracranial pressure and cerebral perfusion pressure after diffuse brain injury in rats. J Neurosurg Anesthesiol. 2001;13(2):113–9.
57. Statler KD, Alexander H, Vagni V, Dixon CE, Clark RS, Jenkins L, Kochanek PM. Comparison of seven anesthetic agents on outcome after experimental traumatic brain injury in adult, male rats. J Neurotrauma. 2006;23(1):97–108. https://doi.org/10.1089/neu.2006.23.97.
58. Kamnaksh A, Kovesdi E, Kwon SK, Wingo D, Ahmed F, Grunberg NE, Long J, Agoston DV. Factors affecting blast traumatic brain injury. J Neurotrauma. 2011;28(10):2145–53. https://doi.org/10.1089/neu.2011.1983.
59. Cole JT, Yarnell A, Kean WS, Gold E, Lewis B, Ren M, McMullen DC, Jacobowitz DM, Pollard HB, O'Neill JT, Grunberg NE, Dalgard CL, Frank JA, Watson WD. Craniotomy: true sham for traumatic brain injury, or a sham of a sham? J Neurotrauma. 2011;28(3):359–69. https://doi.org/10.1089/neu.2010.1427.
60. Turner CP, Gutierrez S, Liu C, Miller L, Chou J, Finucane B, Carnes A, Kim J, Shing E, Haddad T, Phillips A. Strategies to defeat ketamine-induced neonatal brain injury. Neuroscience. 2012;210:384–92. https://doi.org/10.1016/j.neuroscience.2012.02.015.
61. Zheng H, Dong Y, Xu Z, Crosby G, Culley DJ, Zhang Y, Xie Z. Sevoflurane anesthesia in pregnant mice induces neurotoxicity in fetal and offspring mice. Anesthesiology. 2013;118(3):516–26. https://doi.org/10.1097/ALN.0b013e3182834d5d.
62. Hakan T, Toklu HZ, Biber N, Ozevren H, Solakoglu S, Demirturk P, Aker FV. Effect of COX-2 inhibitor meloxicam against traumatic brain injury-induced biochemical, histopathological changes and blood-brain barrier permeability. Neurol Res. 2010;32(6):629–35. https://doi.org/10.1179/016164109x12464612122731.
63. Friess SH, Naim MY, Kilbaugh TJ, Ralston J, Margulies SS. Premedication with meloxicam exacerbates intracranial haemorrhage in an immature swine model of non-impact inertial head injury. Lab Anim. 2012;46(2):164–6. https://doi.org/10.1258/la.2011.011084.
64. Surgery AAoOHaN. Perforated eardrum. 2015. http://www.entnet.org/content/perforated-eardrum.
65. Shridharani JK, Wood GW, Panzer MB, Capehart BP, Nyein MK, Radovitzky RA, Bass CR. Porcine head response to blast. Front Neurol. 2012;3:70. https://doi.org/10.3389/fneur.2012.00070.
66. Yanagawa Y, Marcillo A, Garcia-Rojas R, Loor KE, Dietrich WD. Influence of post-traumatic hypoxia on behavioral recovery and histopathological outcome following moderate spinal cord injury in rats. J Neurotrauma. 2001;18(6):635–44. https://doi.org/10.1089/089771501750291873.
67. Guha A, Tator CH, Rochon J. Spinal cord blood flow and systemic blood pressure after experimental spinal cord injury in rats. Stroke. 1989;20(3):372–7.
68. Nout YS, Beattie MS, Bresnahan JC. Severity of locomotor and cardiovascular derangements after experimental high-thoracic spinal cord injury is anesthesia dependent in rats. J Neurotrauma. 2012;29(5):990–9. https://doi.org/10.1089/neu.2011.1845.

69. Lee KZ, Huang YJ, Tsai IL. Respiratory motor outputs following unilateral midcervical spinal cord injury in the adult rat. J Appl Physiol (1985). 2014;116(4):395–405. https://doi.org/10.1152/japplphysiol.01001.2013.
70. Andrews RJ, Bringas JR, Alonzo G. Cerebrospinal fluid pH and PCO2 rapidly follow arterial blood pH and PCO2 with changes in ventilation. Neurosurgery. 1994;34(3):466–70; discussion 470.
71. Horn EM, Theodore N, Assina R, Spetzler RF, Sonntag VK, Preul MC. The effects of intrathecal hypotension on tissue perfusion and pathophysiological outcome after acute spinal cord injury. Neurosurg Focus. 2008;25(5):E12. https://doi.org/10.3171/FOC.2008.25.11.E12.
72. Inoue S, Mori A, Shimizu H, Yoshitake A, Tashiro R, Kabei N, Yozu R. Combined use of an epidural cooling catheter and systemic moderate hypothermia enhances spinal cord protection against ischemic injury in rabbits. J Thorac Cardiovasc Surg. 2013;146(3):696–701. https://doi.org/10.1016/j.jtcvs.2012.11.040.
73. Yu CG, Jimenez O, Marcillo AE, Weider B, Bangerter K, Dietrich WD, Castro S, Yezierski RP. Beneficial effects of modest systemic hypothermia on locomotor function and histopathological damage following contusion-induced spinal cord injury in rats. J Neurosurg. 2000;93(1 Suppl):85–93.
74. Kang J, Albadawi H, Casey PJ, Abbruzzese TA, Patel VI, Yoo HJ, Cambria RP, Watkins MT. The effects of systemic hypothermia on a murine model of thoracic aortic ischemia reperfusion. J Vasc Surg. 2010;52(2):435–43. https://doi.org/10.1016/j.jvs.2010.03.021.
75. Tang SH, Yu JG, Li JJ, Sun JY. Neuroprotective effect of ketamine on acute spinal cord injury in rats. Genet Mol Res. 2015;14(2):3551–6. https://doi.org/10.4238/2015.April.17.4.
76. Kose EA, Bakar B, Ayva SK, Kilinc K, Apan A. Neuroprotective effects of racemic ketamine and (S)-ketamine on spinal cord injury in rat. Injury. 2012;43(7):1124–30. https://doi.org/10.1016/j.injury.2012.02.022.
77. Yu QJ, Zhou QS, Huang HB, Wang YL, Tian SF, Duan DM. Protective effect of ketamine on ischemic spinal cord injury in rabbits. Ann Vasc Surg. 2008;22(3):432–9. https://doi.org/10.1016/j.avsg.2008.03.003.
78. Lips J, de Haan P, Bodewits P, Vanicky I, Dzoljic M, Jacobs MJ, Kalkman CJ. Neuroprotective effects of riluzole and ketamine during transient spinal cord ischemia in the rabbit. Anesthesiology. 2000;93(5):1303–11.
79. Bell MT, Puskas F, Bennett DT, Herson PS, Quillinan N, Fullerton DA, Reece TB. Dexmedetomidine, an alpha-2a adrenergic agonist, promotes ischemic tolerance in a murine model of spinal cord ischemia-reperfusion. J Thorac Cardiovasc Surg. 2014;147(1):500–6. https://doi.org/10.1016/j.jtcvs.2013.07.043.
80. Gul S, Hanci V, Bahadir B, Acikgoz S, Bektas S, Ankarali H, Kalayci M, Acikgoz B. The effectiveness of dexmedetomidine in experimental spinal cord injury compared to methylprednisolone in rats. J Clin Neurosci. 2010;17(4):490–4. https://doi.org/10.1016/j.jocn.2009.05.041.
81. Winkler T, Sharma HS, Stalberg E, Westman J. Benzodiazepine receptors influence spinal cord evoked potentials and edema following trauma to the rat spinal cord. Acta Neurochir Suppl. 1997;70:216–9.
82. Zhou YJ, Liu JM, Wei SM, Zhang YH, Qu ZH, Chen SB. Propofol promotes spinal cord injury repair by bone marrow mesenchymal stem cell transplantation. Neural Regen Res. 2015;10(8):1305–11. https://doi.org/10.4103/1673-5374.162765.
83. Ding Q, Wang Q, Deng J, Gu Q, Hu S, Li Y, Su B, Zeng Y, Xiong L. Sevoflurane preconditioning induces rapid ischemic tolerance against spinal cord ischemia/reperfusion through activation of extracellular signal-regulated kinase in rabbits. Anesth Analg. 2009;109(4):1263–72. https://doi.org/10.1213/ane.0b013e3181b2214c.
84. Park HP, Jeon YT, Hwang JW, Kang H, Lim SW, Kim CS, Oh YS. Isoflurane preconditioning protects motor neurons from spinal cord ischemia: its dose-response effects and activation of mitochondrial adenosine triphosphate-dependent potassium channel. Neurosci Lett. 2005;387(2):90–4. https://doi.org/10.1016/j.neulet.2005.06.072.

85. Lopez S, Dadure C, Vergnes C, Capdevila X. Intrathecal bupivacaine protects against extension of lesions in an acute contusive spinal cord injury model. Eur J Anaesthesiol. 2006;23(9):793–800. https://doi.org/10.1017/s0265021506000615.
86. Breckwoldt WL, Genco CM, Connolly RJ, Cleveland RJ, Diehl JT. Spinal cord protection during aortic occlusion: efficacy of intrathecal tetracaine. Ann Thorac Surg. 1991;51(6):959–63.
87. Apaydin AZ, Buket S. Regional lidocaine infusion reduces postischemic spinal cord injury in rabbits. Tex Heart Inst J. 2001;28(3):172–6.
88. Haghighi SS, Chehrazi BB, Higgins RS, Remington WJ, Wagner FC. Effect of lidocaine treatment on acute spinal cord injury. Neurosurgery. 1987;20(4):536–41.
89. Hakan T, Toklu HZ, Biber N, Celik H, Erzik C, Ogunc AV, Cetinel S, Sener G. Meloxicam exerts neuroprotection on spinal cord trauma in rats. Int J Neurosci. 2011;121(3):142–8. https://doi.org/10.3109/00207454.2010.537415.

Rodent Surgical Procedures and Tissue Collection

Cesar Reis, Onat Akyol, Jay Malaguit, Lei Huang, Budbazar Enkhjargal, Richard Applegate II, Robert Martin, Gary Stier, and John Zhang

Abstract Collection of tissue for neurological investigation requires training and development of skills to assure precision in the sampling of not only the brain tissues but also other organs associated with each experiment. Preparation for sample analysis follows successful collection. A diverse amount of animal models have been successfully used for investigation of new pathways or specific diseases that may someday lead to improvement of future medicine. For that reason, tissue collection is extremely important to be mastered to achieve appropriate data results.

C. Reis · L. Huang
Department of Anesthesiology, Loma Linda University Medical Center, Loma Linda, CA, USA

Department of Physiology and Pharmacology, Loma Linda University School of Medicine, Loma Linda, CA, USA

O. Akyol · J. Malaguit · B. Enkhjargal
Department of Physiology and Pharmacology, Loma Linda University School of Medicine, Loma Linda, CA, USA

R. ApplegateII
Department of Anesthesiology, UC Davis School of Medicine, Sacramento, CA, USA

Department of Pain Medicine, UC Davis School of Medicine, Sacramento, CA, USA

R. Martin · G. Stier
Department of Anesthesiology, Loma Linda University Medical Center, Loma Linda, CA, USA

J. Zhang (✉)
Department of Anesthesiology, Loma Linda University Medical Center, Loma Linda, CA, USA

Department of Physiology and Pharmacology, Loma Linda University School of Medicine, Loma Linda, CA, USA

Department of Anesthesiology, Loma Linda University School of Medicine, Loma Linda, CA, USA

Department of Neurosurgery, Loma Linda University School of Medicine, Loma Linda, CA, USA

J. Chen et al. (eds.), *Animal Models of Acute Neurological Injury*, Springer Series in Translational Stroke Research, https://doi.org/10.1007/978-3-030-16082-1_3

In this chapter, we will cover steps to intubate, perfuse, remove, and separate the brain into anatomical sections as well as specific staining of the brain with triphenyltetrazolium chloride (TCC).

Keywords Rodent model · Ventilation · Tracheostomy · Rodent model · Intubation · Brain · Perfusion · Fixation · Brain dissection · Triphenyltetrazolium chloride (TCC) staining

Introduction

Obtaining tissues for study is the final goal of most animal experiments, and these are important for studying brain and neuronal disorders. Tissue collection allows measurement of neurotransmitter concentration in several brain areas, and receptors associated with neuronal diseases. The first step is to intubate the rodent and proceed with surgery. After surgery, neurobehavior tests may be done in order to evaluate effects of the experimental protocol. After neurobehavior analysis, the rat may need to be sacrificed and perfused before samples are collected. In this chapter we will describe a step-by-step method on how to intubate, perfuse and collect brain samples for analysis. We will end the chapter demonstrating how to collect tissue for TTC staining.

Airway Management in the Rodent Model

Intubation

The rat is a common species used in laboratory practice. In our lab we are very pleased to have researchers with clinical backgrounds. They initially think that since the principles of intubation are similar to those in human subjects, the skills to intubate a rat would not be difficult. They are often surprised to find out that the technical skills required for rat intubation are not similar to humans. Morphology and physiology of the airway are different than humans, requiring researchers to practice the use of several instruments such as laryngoscopes, fiber optic light sources and even the use of microscopes to achieve intubation. There are several methods to secure the airway and those include intubation under direct visualization, tracheostomy and translumination techniques. Several methods of airway management are particular to each investigator's experience and institution, resulting in different complications depending on the chosen method, understanding that each technique carry its own risks. In this section we will focus on intubation, as it is the method widely used in our lab to secure the airway. We will also demonstrate a step-by-step guide that allows a practical visualization of the procedure.

In our institution we use Sprague Dawley rats (250–500 g) that are intubated for the subarachnoid hemorrhage model, surgical brain injury, and cardiac arrest model of global ischemia. The simplest method to intubate the rodent is the Seldinger technique that uses an otoscope to perform the procedure under direct vision. It is a fast and easy method that has a high success rate [1].

Following is the list of several advantages to the use of direct visual intubation.

(a) Easy to perform using relatively inexpensive equipment.
(b) It is safe with a low number of complications in the process of securing the airway.
(c) The rodent is quickly and easily extubated.
(d) Provides a fast procedure that is acceptable by the rodent [1].

In the following section we will lay out a list of materials and the step by step procedure needed to achieve airway placement. The pitfalls that may happen after important steps are highlighted throughout the section.

Materials

1. Anesthetic equipment (Anesthesia gas blender with dual flow meter tubes and isoflurane vaporizer—Whitemore Enterprises, Rancho Cucamonga, CA).
2. Operating surface (nonporous, reusable, and cleanable), well-ventilated and well-lit working area.
3. Sterile drapes, gown, mask, hair cover, and gloves are required before starting the procedure and preparing the operating field.
4. Otoscope and a 4-mm viewing cone (Welch Allyn, Inc. 4341 State Street Road, Skaneateles Falls, New York 13153, USA).
5. Castaloy three-prong extension clamp.
6. Bougee: a metal guide wire (gauge 0.0889 cm).
7. An aluminum 15 G blunt needle over which the plastic catheter from a 14 G (Insyte) intravenous catheter has been cannulated.
8. Small-animal volume controlled ventilator, TOPO Dual Mode Ventilator, Kent Scientific, Torrington, CT.
9. Syringe (20 ml) for suction.
10. Operating microscope (Kent Scientific).
11. Standard operating surgical instruments and micro instruments if available. This should include a scalpel (disposable if), fine forceps, vascular clamps, needle holder, retractors, and microscissors.
12. Sterile sutures: 4/0 nylon monofilament suture, and 3/0 silk suture.
13. PE 250 tubing with an obliquely fashioned tip.

Procedures

Intubation is a simple procedure if a step by step guide is followed will lead to easy intubation with minimal complications. In order to accomplish a smooth procedure, lay out all materials side by side and start the process of mounting the endotracheal tube. Follow the steps outlined below to prepare the cannula and proceed to intubation.

1. Grab the stylet and the needle holder. Use the needle holder to grasp the end of the stylet and break it off.

   Note$_1$: Even though the stylet is thin, remember to cut its sharp end to prevent damage of the oropharyngeal tissues upon intubation.

   Note$_2$: Next, insert an aluminum 15 G blunt needle into it, securing the stylet onto the needle and forming an intubation tube. Next, place tape around the endotracheal tube (Fig. 1).

   Note$_3$: Place the 15 G needle gently inside the tube as the blunt tip of the needle may damage the stylet.
2. Place the rat in a chamber that will receive a mixture of 4% isoflurane with 30% oxygen and 70% medical air for 4 min (Fig. 1).

   Note: If Anesthetic I.P injection in chosen, place the rat inside a restraining cone (Braintree Scientific Inc., Braintree, MA) and then give I.P. anesthesia. Next observe the rat as it will probably fall asleep in a couple of minutes.

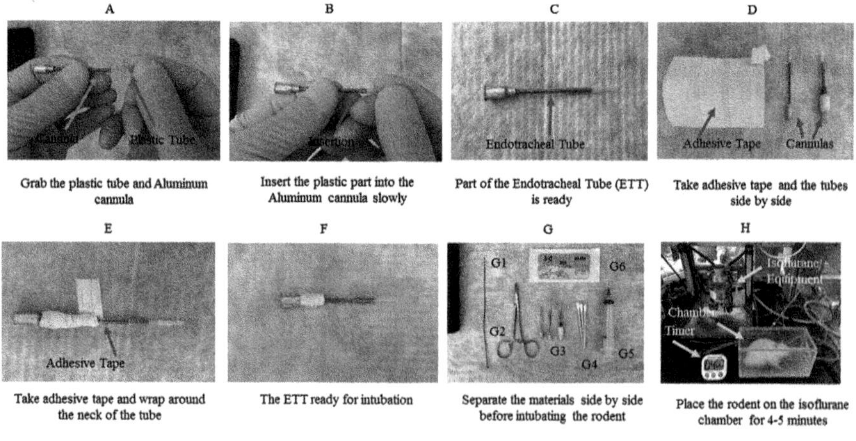

Fig. 1 A process of mounting the tube is shown on the pictures. (**a**) Get a cannula and plastic tube ready. (**b**) Insert plastic tube to cannula. (**c**) Part of Endotracheal Tube (ETT) ready. (**d**) Prepare adhesive tape. (**e**) Wrap adhesive tape around neck of the tube. (**f**) ETT is ready. (**g**) Materials needed for intubation. [G1: Guide wire; G2: Needle holder; G3: Cannulas in different stages in the production of the endotracheal tube; G4: Q-tips; G5: Syringe with suction tip; G6: 3-0 Nylon to suture the ETT to the skin (Optional)]. (**h**) Anesthesia set up for intubation which includes isoflurane chamber, rodent anesthetic induction chamber and timer. Place rat in the isoflurane induction chamber after the equipment is ready for intubation

Some animals may resist the original dose of anesthetic so giving an additional dose may be necessary.

3. After the animal is fully anesthetized, grab a Q-tip and swab out the oral cavity to remove any remains before the procedure.
4. Hold the animal with your left hand and with a cotton swab, roll the tongue out of the mouth. Grasp the tongue between the thumb and index finger and retract it upwards and laterally to either side. With another swab clean the oral cavity and look for presence a foreign body. Compress the tongue use the wooden part of the swab and try to visualize the epiglottis.
5. Using a mounted otoscope on a Castaloy three-prong extension clamp and holding the tongue to one side, insert the viewing otoscope cone into the oral cavity and attempt to directly visualize the vocal cords. The tip of the cone will push the epiglottis forward and out of the way visualizing the upper trachea (Fig. 2).

   Note$_1$: A 3 ml plastic syringe cut obliquely to create a wedge can be used as an alternative to the otoscope [2]. No light source is available in this method,

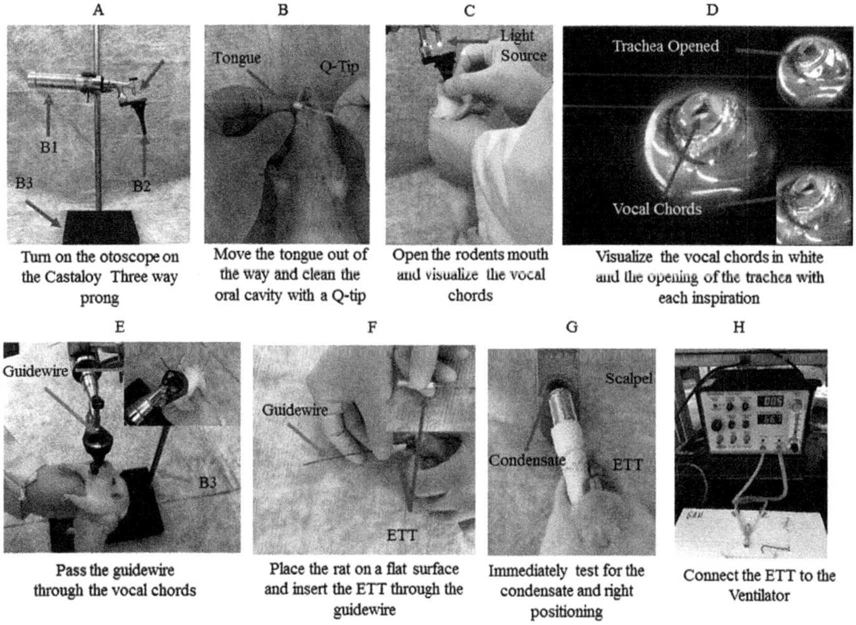

Fig. 2 (**a**) Rat intubation equipment shows a Castaloy three way prong. B1: Otoscope; B2: Viewing Cone; B3: Castaloy Three prong extension clamp. (**b**) After rat is asleep move the tongue out of the way and clean oral cavity of any debris. (**c**) Assemble the otoscope and turn the light on. Make sure the battery is charged. Open the oral cavity to visualize the vocal cords. (**d**) Visualize the opening of the trachea. (**e**) Pass guidewire through the vocal cords. (**f**) Insert Endotracheal Tube (ETT) through the guidewire and remove guidewire. (**g**) Test for condensation to confirm ETT position is correct. (**h**) Connect ETT to a ventilator

thus making intubation more difficult to perform. We do not perform this method at our institution. Furthermore, since the rodent is under isoflurane we do not use presoaked lidocaine 2% to anesthetize the posterior pharynx and glottis.

Note₂: Difficult intubation and associated multiple attempts may lead to edema of the oropharyngeal tissues and trauma to the trachea. This can lead to subsequent difficult and traumatic extubation. In cases like this, an operating microscope may aid in achieving the desired results. Beginners may benefit from the use of a microscope to practice intubation as well.

6. Keep the visualized vocal cords and the rat stable and with the left hand and insert the guide wire through the cone (gauge 0.0889 cm) past the vocal cords into the upper segment of the trachea.
7. Remove the rodent from the cone and place it on a flat surface. Then, place the endotracheal tube over the guide wire and remove the guide wire slowly.

   Note: The guide wire will direct the tube into the trachea avoiding an esophageal intubation and its associated complications. Tracheal rings are noticed in a step-like fashion as one slides the endotracheal tube over the guide wire. This will be noticed after some experience in intubating has been attained.
8. After intubation, place the metal scalpel base over the endotracheal opening and observe breathing through condensation on the blade to confirm air being exhaled from the tube. Carbon dioxide measurement through the use of a capnograph is another way to assure the right endotracheal tube location.

   Note$_1$: If no condensation is observed, palpate the stomach for possible esophageal intubation. The stomach will insufflate if air is given through a syringe. Take out the tube immediately if that happens. Withdraw the catheter and reattempt the whole process.

   Note$_2$: Do not place the rat on a ventilator before checking for appropriate intubation. After hooking the tube to the monitor, extra piece of mind can be given by seeing water vapor condense inside the three-way stopcock as well as by symmetric chest excursions.

   Note$_3$: During the entire procedure excess mucus production may lead to desaturation of the rodent during the desired experiment. A suction device will allow checks and ruling out conditions that may be causing hypoxemia. A syringe attached to the cannula will allow suction of endotracheal contents during ventilation if for any reason there is suspicion of a mucus plug or bleeding.
9. Fix the tube to the skin using 3.0 silk suture and tie it around the catheter hub.

   Note: Fixing the tube to the skin may be optional. Some models (Subarachnoid hemorrhage) for example may not require securing the tube while other models (Cardiac Model of Global Ischemia) may require this step.
10. The equipment should be set up by now (Fig. 3). The end of the three-way stopcock should be secured to the end of a catheter hub which will connect to the Endotracheal tube.

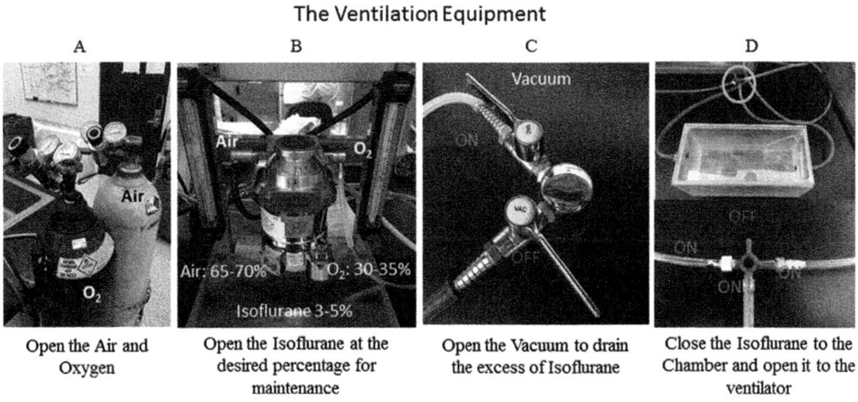

Fig. 3 Equipment necessary during surgery. (**a**) Check the tank of oxygen and the tank of Air before initiating the experiment; (**b**) Place the ventilator at the desired settings before initiating the procedure and check for isoflurane availability; (**c**) Check the vacuum that should suck out the excess of isoflurane. Make sure the vacuum is properly working to avoid isoflurane breathing by laboratory personnel; (**d**) Check the tubes that connect the chamber box to the isoflurane apparatus. After all the setup is done and intubation achieved, turn on the ventilator, open all supplies, set isoflurane at maintenance levels, and place the rat on the ventilator

11. Initiate mechanical ventilation (rate up to 85 bpm, tidal volume 2.5 ml, isoflurane 2–3%).
12. Recheck the anesthetic setting: 2–3% isoflurane and a mixture of 30% oxygen and 70% medical air to maintain the rodent during surgery.

    Note₁: Make sure appropriate ventilator parameters are set up (Fig. 3). Pneumothorax is a possible disastrous consequence of inappropriate ventilation. The appropriate parameters for the ventilation of any animal will depend largely on the animal and its size. We use Sprague-Dawley rat between 270–350 g most of the time, using tidal volumes of approximately 2.5–3 ml stroke^{-1} to keep respiratory rate between 45–55 breaths per minute with a mean airway pressure between 9–12 cm H_2O.

    Note₂: Following intubation, place the rodent on top of a heating pad and maintain its temperature at 36–37 °C or expose it to a source of light (heating lamp) during the entire procedure. If the rat has been given I.P injection maintain the mixture of 30% oxygen and 70% medical air without isoflurane. Furthermore, observe depth and frequency for breathing as needed for the entire procedure.
13. Place a rectal probe to monitor temperature.

    Note: Pay attention to the temperature while the experiment is ongoing. Giving cold room temperature fluids to the anesthetized rat may cause a drop in their temperature unless pre warmed fluids are given [3].

Transcardiac Perfusion and Fixation

Rapid and uniform tissue fixation is essential for anatomical, histological, and molecular assessments. Techniques for perfusion fixation vary depending on the tissue to be fixed and how it will be processed following fixation. Transcardiac perfusion with aldehyde-based fixatives is usually considered to be an effective paradigm to preserve the tissue. This effectiveness is due to advantages such as the utilization of the circulatory system to distribute the fixative as well as the penetration and crosslinking between aldehyde-based fixatives and tissues, resulting in minimal disruption of the structural integrity of proteins [4]. The main advantage of this technique is to effectively use the circulatory system for perfusion.

The perfusion equipment should be completely set up before the animal is anesthetized, and the line(s) cleared of air bubbles and filled with the first solution to be perfused (Fig. 4). It is critical to adjust the flow rate of perfusion not to exceed the normal rate of blood flow. A 60 ml syringe may also be used to infuse PBS or PFA/formalin and achieve the same results. After thoracotomy, insert the cannula through the apex and direct the cannula into the aorta. Following insertion, clamp the cannula into place using a hemostat. Then, cut open the right atrium with scissors. Perfusion is first performed with 4 °C phosphate buffers (PBS) or saline to remove the blood from the vascular system. This is an important step as aldehyde-based fixatives can clot blood and prevent good perfusion to regions of the brain if such clot forms. The total PBS/saline perfusion volume is about 6–8 time the blood volumes of the rodent (blood volume (ml) = 0.06 × body weight (g) + 0.77) [5]. Right after PBS rinse, PFA/formalin (4 °C) is perfused at total volumes of 100 ml PFA/formalin per 100 g body weight. The detailed procedures are described in the Step by step procedure section.

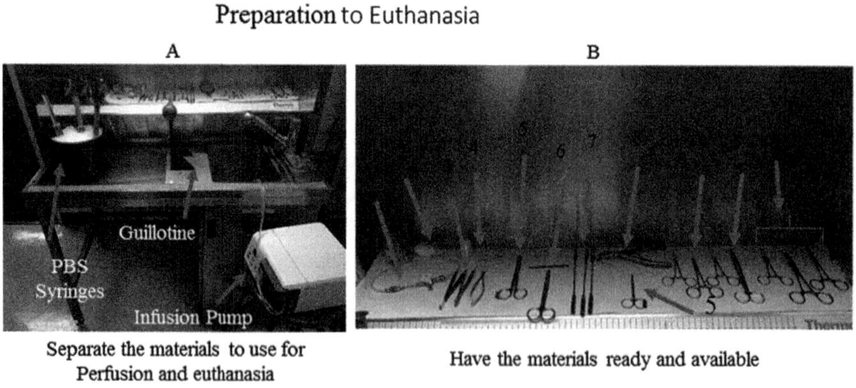

Fig. 4 Preparation for euthanasia. (**a**) Prepare the station with the guillotine, ice bucket with syringes filled with PBS, and/or the electronic perfusion pump. Notice in (**b**) the equipment laid side by side (1: Catheter line; 2: Isoflurane with gauze in a 50 ml tube; 3: Forceps; 4: Dish; 5: Scissors; 6: Blade; 7: Micro Spatulas; 8: Friedman rongeur; 9: Straight Forceps; 10: Angled Forceps)

Equipment and Materials

1. Chemical fume hood.
2. Perfusion pump or container with perfusion solution placed about 100 cm above the operating table or 60 ml syringe (for manual perfusion) attached to an IV line which can fit to the cannula.
3. Cannula (size is selected based on the size of the animal; try to use largest gauge cannula that can be inserted into the aorta).
   - 16 G × 1.5 in. aluminum hub blunt needle for rat.
   - Butterfly needles, 23 G × 1.2 in. for mouse 24 G × 25.4 mm animal feeding needle.
4. Surgical tools.
   - Blunt-ended scissor.
   - Sharp scissor.
   - Blunt-ended forceps.
   - Hemostats.
5. Solutions (Pre-chilled to 4 °C before perfusion):
   - 1× PBS or saline (to rinse the blood off).
   - 4% paraformaldehyde or 10% formalin (for tissue fixation).

Step by Step Procedures

In order to achieve good perfusion and post-operative results follow the step by step procedural list below.

1. First, Weigh the rat or mouse before initiating the procedure. Calculate the total perfusion volumes of PBS/saline and 4% PFA/10% formalin, respectively.
   Note: Incorrect weight will lead to incorrect volume of solution(s) leading to either excess or insufficient amounts being infused, possibly altering results.
2. Pre-chill perfusion solutions to 4 °C and maintain this temperature during perfusion by holding containers in an ice filled basin.
   Note: Cold solutions minimize protein degradations during perfusion.
3. Set up the perfusion equipment completely and ensure that all the line(s) have been cleared of air bubbles and filled with the first solution (usually is 1× PBS or saline) to be perfused.
   Note: Having bubbles in the lines will block the small vessels and affect the perfusion quality.
4. Use 5% isoflurane or other appropriate anesthetics to sedate the animal. Leave the rat for 4–5 min in the anesthetic chamber before proceeding to thoracotomy. Confirm a deep plane of anesthesia as that the toe pinch response is completely absent.

Note: Prepare the sacrificing station and the equipment before initiating sedation. Isoflurane will be metabolized quickly.

5. After sedation, lay the rat flat in the sacrifice station. By now all the equipment should be set up and ready to go (Fig. 5).

   Note: Not having all the equipment ready will lead to wakening of the rat or mouse and need to redo anesthesia.
6. Lift the skin just beneath the xiphoid process with a pair of blunted forceps or hemostat and make an incision through the skin with a pair of blunt-ended scissors.
7. Continue to cut through the abdominal skin and muscles along the inferior margin of the ribs bilaterally.

   Note: Prevent cutting intra-abdominal viscera especially the liver which is sitting right below the diaphragm and underneath the Xyphoid process where the cut will be made. Lifting the skin will prevent such things from happening.
8. Cut through the diaphragm along the costal margins of the rib (Fig. 6).

   Note: Cutting the diaphragm will lead to lung collapse. The following steps up to the point of perfusion of the 1× PBS/saline rinse (step 10) need to proceed as quickly as possible to minimize tissue hypoxia damage.

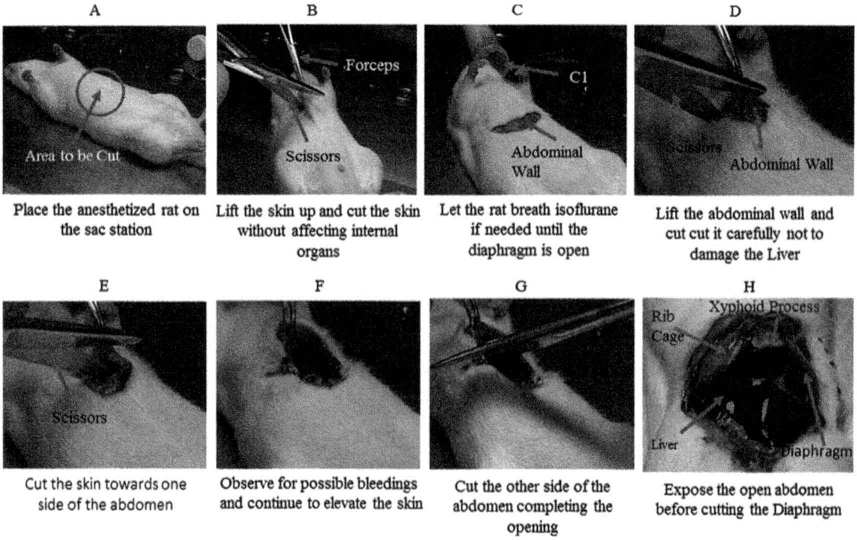

Fig. 5 Opening the abdominal cavity. Euthanasia is explained in a step-by-step manner. Lay the rodent flat on its back after sedation and start following the step-by-step procedure in the text. (**a**) Place the rat in supine position and identify the xiphisternal area. (**b**) Lift the skin up using forceps and cut the skin using scissors avoiding injury to internal organs. (**c**) Isoflurane administered during the procedure using C1: 50 ml tube with anesthetic. (**d**–**g**) Continue to cut the abdominal wall using scissors one side at a time while lifting up the skin. (**h**) Abdominal cavity and diaphragm is visualized

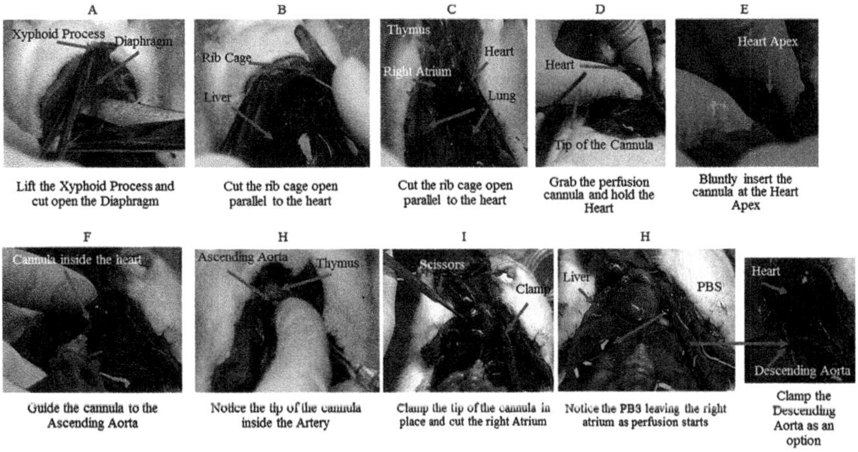

Fig. 6 Opening the Chest and perfusion. Figures show how to open the chest for perfusion. (**a**) Lift the xiphoid process and cut open the diaphragm. (**b**) Cut open the rib cage. (**c**) Reflect the rib cage and visualize the underlying organs. Right atrium is identified. (**d**–**h**) Stabilize the heart between fingers and insert the tip of a cannula into the apex of left ventricle and guide it towards ascending aorta. Make a cut in the right atrium and infuse PBS through the cannula

9. Make two additional cuts through the lateral part of the rib cage on both sides to finalize the thoracotomy. Use the hemostat attached to the Xyphoid process to retract the anterior part of the rib cage superiorly and hold it out off the field. The heart is now clearly exposed and still beating.

   Note$_1$: One can *choose* to expose the root of the aorta by carefully moving the surrounding tissues out of the way.

   Note$_2$: Clamp *descending* aorta underneath or above the diaphragm can reduce the perfusion volume to a half the total amount calculated by body weight (100 ml/100 g). In such case, use a vascular clamp or a curved Kelly hemostat to clamp the descending aorta running along the spinal cord behind the left lung, which prevents the infusion from reaching the lower half of the animal.
10. Insert the cannula bluntly into the heart or cut the heart apex to create an access point to the left ventricle with a pair of dissecting scissors or an 11 scalpel blade. An option is to stabilize the heart with forceps or use your dominant hand.

    Note$_1$: Bluntly insert the tip of the cannula into the Apex of the heart. It will lead the cannula into the left ventricle. In our experience both methods lead to successful cannulation of the ascending Aorta.

    Note$_2$: Clamping the Ascending Aorta will free the heart from being clamped. It will allow sample collection of the heart without muscle damage from clamping allowing further muscle enzyme study.

11. Insert the perfusion cannula through the apical opening into the ascending aorta. Confirm the placement of the cannula tip within the aorta about 1–3 mm above the top of heart (make sure not to insert the perfusion line/cannula through the cardiac septum to the pulmonary vein). Fix the cannula in place by clamping a straight or curved Kelly hemostat across the superior part of the ventricles or the ascending aorta.

    Note: Clamping with a hemostat will prevent backflow of the cannula from the heart during perfusion. Clamping the Ascending Aorta will free the heart from being clamped. It will allow sample collection of the heart without muscle damage from clamping and further muscle enzyme study.
12. Cut open the right atrium with sharp scissors and start the perfusion using a perfusion pump or a 60 ml syringe with manual pressure at the first sign of blood flow. The gravity perfusion method is another option to perfuse the rodent.

    Note: Maintain a slow and steady perfusion flow of around 20 ml/min.
13. First perfuse 1× PBS or saline (about total volume of 120–180 ml for a 250–350 g rat) at the selected rate.

    Note: This *rinse* solution flushes out blood in the vasculature system and flows out from incision in the right atrium. The fluid turns clear at the end of rinse.
14. Clamp the input lines from the container of PBS or saline and open the input lines in the container of the fixative solution (4% PFA or 10% formalin).
15. Continue the perfusion with fixative solution at the selected rate.

    Note: After the fixative solution enters the body, a sign of good perfusion is the occurrence of tremors ("formalin dancing") due to the aldehyde-crosslinking of nerves and muscle. If fluid is running out from animal's nostrils or mouth, reposition the cannula.
16. Upon the completion the perfusion, remove the cannula and hemostats from the animal.

    Note: Evidence of successful perfusion are the "whitening" of the rat's eyes, lightened color of liver (only applicable for the method without of Descending Aorta clamping), and stiffness of the body. After this step the animal is ready for excision of the tissue of interest (see brain extraction on Fig. 7, 1 and 2). For better results, immerse the extracted tissues in fixation solution overnight at 4 °C before proceeding to dehydration and embedding.

Methods of Brain Removal

Removal of the brain from the skull should be fast and secure enough not to damage the integrity of brain anatomy particularly while using surgical tools very close to skull bones. There are several methods of brain removal used in different laboratories. In addition to sampling, fixation allows the brain to be safely stored for

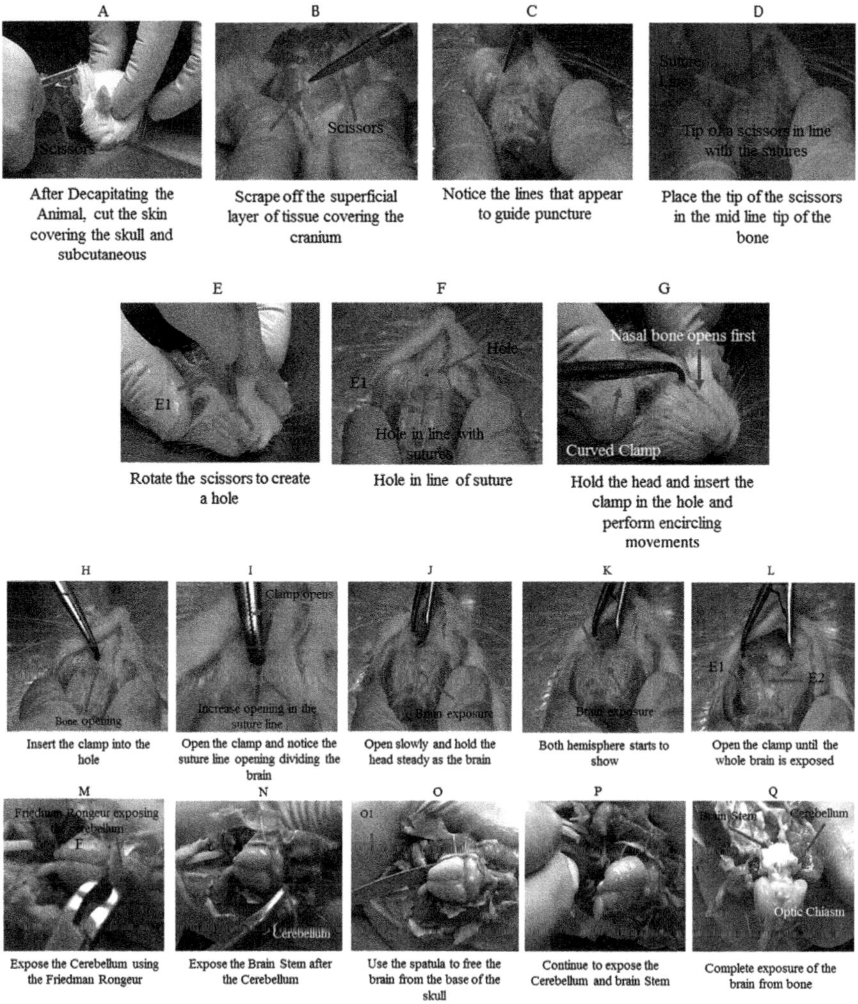

Fig. 7 Decapitation and Brain Removal. The Pictures show how to separate parts of the brain. E1: Straight line guides the place where the hole will be; (**f**) Notice the hole ready to be used as a point of opening of the entire skull. From (**a**) to (**q**) a detailed explanation on how to remove the brain from the skull

future study. Formalin may also be infused after perfusion with PBS allowing the brain to be subsequently stored in 30% sucrose solutions prior to placement in liquid nitrogen and storage in −80 °C to prevent freezing artifacts during sectioning. This section will describe and give illustrations for a particular established method of brain removal.

Materials

1. Guillotine
2. Scalpel handle and scalpel blades
3. Forceps
4. Curved Kelly hemostat
5. Spring scissors
6. Friedman rongeur
7. Spatula
8. Sucrose solution (30%)

Materials from 1–7 were bought from Fine Science Tools (USA), Inc.; Foster City, CA.

Procedures

In the section below, follow the steps to remove the brain. As in any procedure, caution should be taken to avoid possible mistakes that lead to alteration of results.

Brain Removal

1. Deeply anesthetize the animal with 4% isoflurane for 5 min before sacrifice.

   Note: Check the rodents reflex before guillotining otherwise low level of anesthesia will be noticeable immediately upon guillotining the head off. If the rat is not completely under it will start jerking on the sacrificing station, indicating more anesthesia should have been given.
2. Place the rat onto the sacrificing station and insert the head appropriately into the guillotine for decapitation of the upper cervical region just above the shoulder.

   Note: One of the pitfalls during step 2 is not placing the head fully into the guillotine window. This leads to decapitation at a segment higher than the desired one. This will lead to cutting more brain stem off than necessary, leaving more tissue behind that could be used for research and creating different results in the measurement of brain water content for example.
3. Following decapitation, expose the skull. Using a scissor, cut from the posterior decapitation border to the end of the nasal bone.
4. Next, remove the periosteum from the parietal, frontal, and nasal bones by scraping the surface of the skull with a scalpel blade.

   Note: Residual periosteum left in place before separating the skull bones apart will require more strength in using the angled clamp and possibly lead to uneven separation of the bones. More time will be needed to cut the extra bone that was left in place.

5. Make a hole on the intersection of the Anterior Frontal Suture and the Jugum limitans olfactoriae, enabling a future insertion of a right angle clamp (Fig. 7, 1 and 2).
   Note: Choose the right location to puncture the bone. Wrongly marking the intersection of the anterior frontal suture and the Jugum limitans olfactoriae before insertion of right angle clamp will lead to possible brain puncture and difficulties in opening the skull evenly.
6. Take the right angle clamp, insert in the intersection hole. With your left hand holding the skull together, use your right hand to open the angled clamp. This will split the skull along the midline separating the two halves of the parietal bones all the way from the frontal nasal bones to the lambdoid suture proximal to the intraparietal bone.
   $Note_1$: One of the problems faced by researchers in this step is seen at the beginning of training. When starting this process, one may find out how hard it is to split two separate windows exactly the same with no bony piece left covering the parietal area. With time, the right and left hand together generate appositional forces necessary to stabilize the skull in such a way that leads to cracking the midline and opening the suture upon splitting the angle clamp open.
   $Note_2$: In case the correct opening doesn't happen, use the Friedman rongeur slowly to cut the pieces left covering the brain. Avoid damaging the area underneath the skull while chipping the bone left.
7. Once the olfactory bulbs become exposed, cut them without damaging the frontal lobes.
   Note: If the final goal is to perform TTC, leave the olfactory bulb in place. It will guide you when placing the brain on the slicing matrix.
8. The frontal and parietal bones can be dissected from the skull with a Friedman rongeur to expose the superior portions of the hemispheres.
   $Note_1$: After opening the parietal bones, the occipital bones are left in place. Using a right angle clamp or Friedman rongeur continue to cut the bone in order to expose the cerebellum.
   $Note_2$: The spinal cord should be cut by now and we should be able to freely visualize all the brain structures. If not, proceed to step 9.
9. Once the underside of the brain is free, cut the spinal cord from the brain stem. The brain will easily separate from the skull base.
10. After removing the calvarium and freely visualizing the hemispheres and cerebellum, separate the meninges that are beneath the skull and lift the brain from the skull base using a spatula. In addition, cut away cranial nerves as well as vertebral arteries from the brain.
    Note: Be aware not to Damage the brain during scrape of cranial nerves and vertebral arteries.
11. Separate the cerebellum and brain stem from the bony part using a spatula.
    Note: After step 11, the brain is completely separated from the bony structures and ready to be separated into hemispheres for further studies.

Separation of the Brain into Anatomical Sections

Following brain removal, the researcher should dissect the whole brain into anatomical structures according to either the anatomical target of primary neuronal injury or the properties of animal model. Additionally, different brain anatomical regions have variable responses to injury and treatment. Use the materials below to separate the regions of interest. The brain should have been perfused with PBS for certain studies such as hemoglobin content measurement, Western blot analysis, and immunohistochemistry. If collection and separation is being performed for brain water content measurement, no perfusion is indicated. Follow the steps bellow to separate the brain in anatomical sections accordingly [6].

Materials

- Spatula or Mini Hippocampal Tool
- Curved Scissors
- Petri Dish
- Fine forceps

Procedure

1. Following perfusion with PBS and removal of the brain from the skull, separate the brain into two hemispheres, cerebellum and brain stem. Remove cerebellum from brain stem by gentle dissection from the caudal end. Make the separating cut at the cerebellar peduncles and split the hemispheres at the interhemispheric fissure.

   Note: Cutting at the wrong place will generate the wrong results.
2. After separating the hemispheres, use one of them. Place the hemi-hemisphere with its medial side facing up and cut with a curved tip scissors around the delineation seen in Fig. 8. Separate the basal ganglia from the surrounding cortex.

   Note: Another technique used is blunt dissection with a micro spatula or a hippocampus tool. It can also be used to separate basal ganglia from hemispheres and residual white matter tracts can be cut with microscissors.
3. The hippocampus is located at the caudal end of the cortex and its free edge can be gently lifted from the interior of the removed cortex.

   Note: Use the spatula *gently* and lift the hippocampus slightly and follow its anatomical location surrounding the semi-hemisphere.
4. Repeat this procedure for the opposite hemisphere, the whole brain should have six regions separated: right and left cortex, right and left basal ganglia, cerebellum, brain stem, and the two hippocampus.
5. Put each brain region inside small plastic tubes and store at −80 °C freezer for future studies.

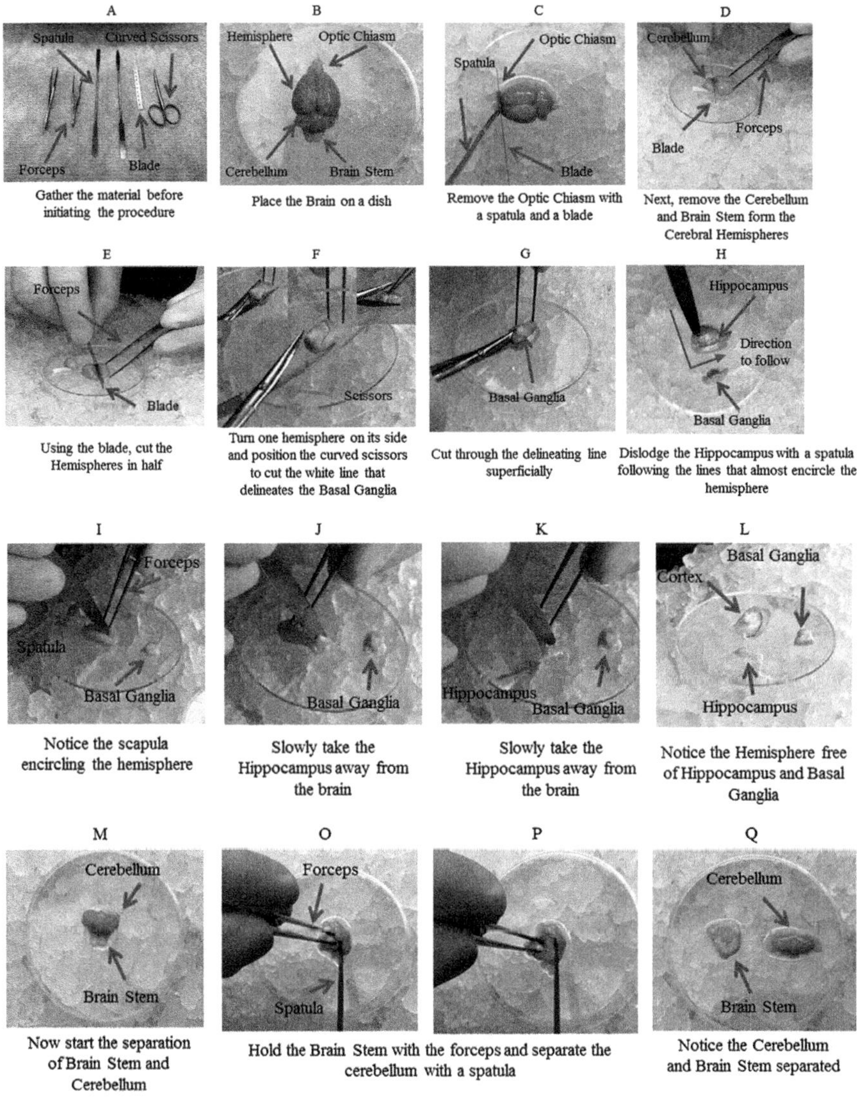

Fig. 8 Methods of brain removal. This series of pictures show the materials and the step-by-step process that will lead to separation of the brain regions. (**a**) Materials needed for brain removal. Forceps, spatula, blade, curved scissors. (**b–d**) Place brain on a petri dish over ice and identify the parts. Using a blade and spatula separate the optic chiasma, cerebellum and brain stem. (**e**) Using a blade separate the two hemispheres. (**f**) Identify white line that delineats the basal ganglia. (**g–l**) Cut along the delineating line to separate hippocampus and basal ganglia from the cortex using a spatula. (**m–q**) Using a forceps hold the brainstem and separate cerebellum away from brainstem using a spatula

Preparation of Brain Slices for Triphenyltetrazolium Chloride Staining

In this protocol, infarct volume analysis is assumed to be the endpoint for the brain sections. Therefore, transfer of the brain to the slicing matrix, sectioning and incubation of brain sections in TTC (Triphenyltetrazolium chloride) solution, subsequent washes in PBS and fixation in formalin are all conducted at room temperature. If the sections are to be used for assays post-TTC [7] (e.g. qPCR, Immunoblotting, etc.) the animal must first be perfused with cold PBS prior to brain extraction with all of the following steps conducted on ice and foregoing step 8. Variations exist pertaining to the percentage of TTC in solution (weight/volume), incubation temperature and duration [7–9]. With preliminary testing, this TTC protocol should be modified to determine the optimal conditions for your specific goals and desired outcomes.

Note: Prior to inducing the animal, pre-chill the sectioning blades on ice. This facilitates the ease of cutting.

Materials

1. TTC solution (2% weight/volume, Sigma Aldrich)
2. PBS (Sigma Aldrich)
3. Spatula
4. Razor Blades
5. Brain Matrix (Alto matrix for coronal sections, Roboz Surgical Instrument Co., Inc., Gaithersburg, MD)
6. Petri Disk
7. Formalin (10%, Protocol/Fisher Scientific)

Procedure

1. Transfer the brain (including the olfactory bulb, cerebellum and brainstem) from the skull to the brain-slicing matrix using a spatula (Fig. 9).

   Note: If the brain-slicing mold contains a small groove located at the tip of the cortex-shaped portion of the mold, place the olfactory bulb there.
2. Use the spatula to align the midsagittal line of the brain to the center cutting line of the brain-slicing matrix.
3. Stabilize the rostral portion of the brain by placing one blade between the olfactory bulb and the tip of the frontal lobes.

   Note: If no groove exists to accommodate the olfactory bulb, a blade placed in front of the frontal lobe will still help to stabilize the brain from moving

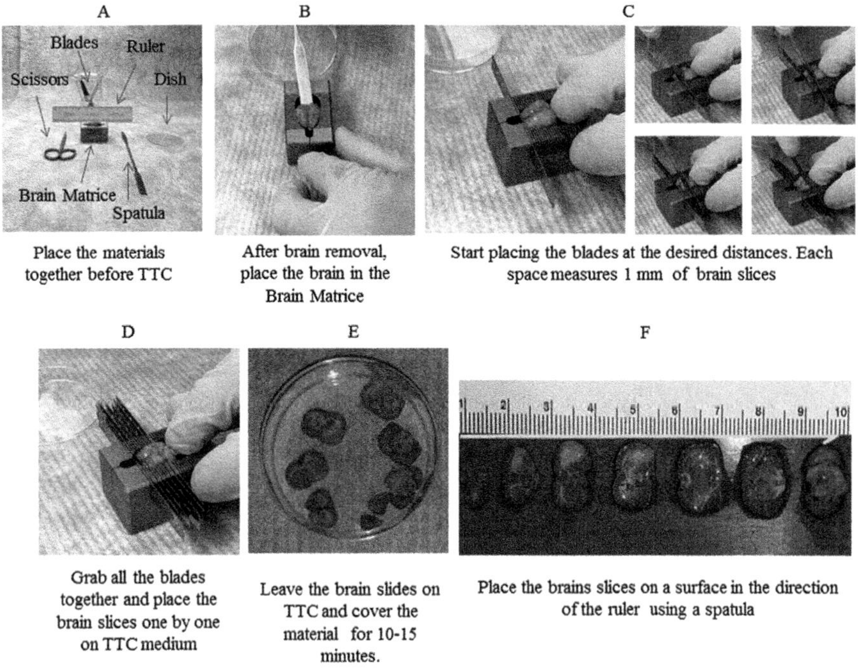

Fig. 9 TTC Preparation. In this set of pictures we demonstrate the step-by-step on how to cut the brain after its removal from the skull and placing on TTC. (**a**) Materials needed for TTC staining. Brain matrice, spatula, scissors, blades, ruler and petri dish. (**b**) Place the brain in a rodent brain matrice. (**c–d**) Cut the brain into slices by placing blades every 1 mm distance. (**e**) Place the brain slices in a petri dish with TTC stain keeping it covered for 10–15 mins. (**f**) Place the TTC stained brain slices on a surface using a spatula. Place a ruler on the side and take pictures

laterally while making coronal sections. Alternatively, you may place the frontal lobe up against the groove of the mold, skip step 3 and move to step 4.

4. Stabilize the caudal portion of the brain by placing one blade between the cerebellum and the occipital lobe.
5. Make coronal sections starting from the frontal lobe and then move caudally towards the cerebellum (Fig. 9).

   Note: Be aware of the desired thickness of your slices. The brain-slicing mold shown has 1 mm cutting intervals. The brain is being sectioned into 2 mm thick coronal slices (i.e. sectioning blades are placed in every other 1 mm sectioning groove).
6. To extract the brain slices from the mold at one time, grasp both ends of the blades between the thumb and index or middle finger and lift them straight up. The ends of the blades can be rested on the edges of the petri dish containing the TTC solution. The blades are removed one at a time, exposing a brain slice stuck to the blade behind it. Use scissors to cut between the slices as needed to free them up at their bottom, then use the spatula to gently push the slice into

the TTC solution. Alternatively, blades can be removed one at a time from the mold. To further facilitate the detachment of the slice from the blade, wet the spatula with TTC solution and ease it in between the slice and the blade.

7. While in TTC solution, cover the petri dish with an opaque material (e.g. foil, box, etc.) or place it in a darkened area to facilitate the reaction and incubate for at least 10–15 min. Most protocols incubate the brain sections for 30 min at 37 °C.

   Note: Exposure to light will slow the TTC reaction down.
8. Carefully extract the TTC solution using a 10 ml syringe and replace with approximately 10 ml PBS. Repeat the PBS washes twice.
9. Replace the last PBS wash with formalin. Brain slices can be either used for infarct volume analysis immediately after immersion in formalin or incubated for 24 h at 4 °C. Brain slices can be stored at 4 °C and reused as needed.
10. Take pictures of brain sections organized rostral-caudally and quantify using your preferred photo analysis program (Fig. 9).

Acknowledgments Thank you Dr. Weiwei Xu for providing the final TTC picture on Fig. 9.

References

1. Chen J, Xu X-M, Zhang JH. Animal models of acute neurological injuries. In: Zhang JH, editor. Animal models of acute neurological injuries. 1st ed. Totowa, NJ: Springer Science and Business Media; Humana Press; 2009.
2. Jou IM, Tsai YT, Tsai CL, Wu MH, Chang HY, Wang NS. Simplified rat intubation using a new oropharyngeal intubation wedge. J Appl Physiol (1985). 2000;89(5):1766–70.
3. Karwacki Z, Kowianski P, Morys J. General anaesthesia in rats undergoing experiments on the central nervous system. Folia Morphol (Warsz). 2001;60(4):235–42.
4. Srinivasan M, Sedmak D, Jewell S. Effect of fixatives and tissue processing on the content and integrity of nucleic acids. Am J Pathol. 2002;161(6):1961–71.
5. Lee HB, Blaufox MD. Blood volume in the rat. J Nucl Med. 1985;26(1):72–6.
6. Ayer R, Cahill J, Sugawara T, Jadhav V, Zhang JH. Rodent surgical procedures and tissue collection. In: Chen J, Xu ZC, Xu X-M, Zhang JH, editors. Animal models of acute neurological injuries. Totowa, NJ: Humana Press; 2009. p. 19–38.
7. Kramer M, Dang J, Baertling F, et al. TTC staining of damaged brain areas after MCA occlusion in the rat does not constrict quantitative gene and protein analyses. J Neurosci Methods. 2010;187(1):84–9.
8. Benedek A, Moricz K, Juranyi Z, et al. Use of TTC staining for the evaluation of tissue injury in the early phases of reperfusion after focal cerebral ischemia in rats. Brain Res. 2006;1116(1):159–65.
9. Joshi CN, Jain SK, Murthy PS. An optimized triphenyltetrazolium chloride method for identification of cerebral infarcts. Brain Res Brain Res Protoc. 2004;13(1):11–7.

Brain Monitoring

Cesar Reis, Onat Akyol, Wing Man Ho, Richard Applegate II, Gary Stier, Robert Martin, and John Zhang

Abstract In experimental neurologic injury models, evaluating the effect of any manipulation or treatment on cerebral blood flow (CBF) gives significant information about metabolic changes in rodent's brains. There are different methods of investigating brain microcirculation based on changes of H_2 clearance, radionuclide labeled microspheres, pial artery diameter, and iodoantipyrine. Another method, CBF measurement using laser Doppler, is based on measurements of absolute changes in regional blood flow of the cortex and detects real time changes in microvascular perfusion in a noninvasive manner. In this chapter we will approach the

---

C. Reis
Department of Anesthesiology, Loma Linda University Medical Center, Loma Linda, CA, USA

Department of Physiology and Pharmacology, Loma Linda University School of Medicine, Loma Linda, CA, USA

O. Akyol · W. M. Ho
Department of Physiology and Pharmacology, Loma Linda University School of Medicine, Loma Linda, CA, USA

R. ApplegateII
Department of Anesthesiology, UC Davis School of Medicine, Sacramento, CA, USA

Department of Pain Medicine, UC Davis School of Medicine, Sacramento, CA, USA

G. Stier · R. Martin
Department of Anesthesiology, Loma Linda University Medical Center, Loma Linda, CA, USA

J. Zhang (✉)
Department of Anesthesiology, Loma Linda University Medical Center, Loma Linda, CA, USA

Department of Physiology and Pharmacology, Loma Linda University School of Medicine, Loma Linda, CA, USA

Department of Anesthesiology, Loma Linda University School of Medicine, Loma Linda, CA, USA

Department of Neurosurgery, Loma Linda University School of Medicine, Loma Linda, CA, USA

J. Chen et al. (eds.), *Animal Models of Acute Neurological Injury*, Springer Series in Translational Stroke Research, https://doi.org/10.1007/978-3-030-16082-1_4

brain monitoring mechanism with laser Doppler equipment in a step-by-step fashion. We will demonstrate the potential pitfalls and how to avoid them. A rat model will be used to demonstrate brain monitoring and the associated procedures that accompany the process.

Keywords Brain monitoring · Syncope · Animal models · Brain imaging · Cerebral blood flow

Introduction

Laser Doppler blood flow measurement is a technique generated from the Doppler shift induced by reflection of laser light from moving red blood cells. The measurement of CBF provides useful information in several experimental rodent models of cerebral ischemia, subarachnoid hemorrhage, traumatic brain injury, post cardiac arrest, and syncope, and also in studies of drug effects on cerebral circulation. In addition, it is beneficial in managing cerebrovascular diseases, head injury and intracranial neoplasms [1]. CBF measurement cannot always directly indicate the severity of the injury. It allows for plotting groups against each other, thereby aiding in distinguishing treatment and non-treatment groups. It is important to point out that there is no normal absolute value for CBF with the laser Doppler technique. Measurement site, animal condition, injury type, and anatomical variations can affect CBF measurement. Outcome data should be evaluated in terms of percentage of normal for each animal to obtain the same baseline CBF values. The blood flow values are not described as ml/g/min but in units relative to the initial baseline values for each animal. Measurements of relative cerebral perfusion with laser Doppler are also important for assessment of cerebral autoregulation because the laser Doppler technique allows continuous measurement, and autoregulation change curves can be obtained as part of the study. A major limitation of Laser Doppler CBF measurement includes failure to evaluate absolute cerebral blood flow [2–4].

Model Selection

CBF measurement is used in a number of basic science research models including the effect of drugs on CBF, head injuries, cerebral ischemia, seizures, and syncope, to name a few. Stroke models in particular benefit from CBF measurement not only because this measurement shows reduced blood flow at a period of time but also confirms the existence of injury. This technique does not measure the extension of the brain injury when comparing several groups in an experiment. On the other hand, it allows plotting several groups individually and comparing each group, thereby allowing analysis of treatment and non-treatment groups. Each animal in the study will have its own CBF value due to variables that give each rodent a unique CBF, including the condition of the animal, site of injury, and anatomical variations in the rat. Thus, this model describes the data in terms of percentage of normal for each rat instead of absolute values [5].

There are several advantages to using CBF as a measurement:

1. Easily reproducible and reliable.
2. Does not require use of force; the probe is placed easily on top of the Dura.
3. It has the ability to measure non-lethally and continuously for up to several hours [6].
4. Useful in several models of experimental research.
5. Femoral catheter can be used to assess the heart rate, administer drugs, fluids, or blood, as well as measuring catecholamines or other parameters.
6. Requires minimal training.
7. Easily monitored for detecting mistakes or misplacement of the probe [5].
8. Easily monitored by one individual.
9. Data plotted into the software is readily available.
10. Assessment of cerebral autoregulation [6].

The pitfalls or problems one may face will be pointed in the *Notes* portion following the steps listed for each procedure.

Materials

Rats can be anesthetized using a variety of techniques. In our lab, among other techniques, we use intubation to maintain ventilation for long periods of time and an isoflurane apparatus for shorter experiments. Anesthetic equipment that has a nosepiece built into the frame will guarantee the delivery of oxygen and isoflurane to the rodent. Tracheostomy is used less often, as it is more invasive and has increased risk for bleeding complications and respiratory failure. It is not recommended unless extremely beneficial because of the duration of the experiment. We recommend using a built in nose piece and giving isoflurane in order to properly maintain the rodent anesthetized and ventilated. Below is a list of the equipment needed to proceed with the experiment.

1. Anesthetic equipment (Anesthesia gas blender with dual flow meter tubes and isoflurane vaporizer—Whitemore Enterprises, Rancho Cucamonga, CA).
2. Benchmark stereotactic frame with 45° ear bars (Leica Biosystems Inc., Buffalo Grove, IL).
3. Sterile drapes, gown, mask, hair cover, and gloves are required before starting the procedure and preparing the operating field.
4. MV-9 electro surgical unit (MV-9 MACAN, Chicago, IL).
5. Drill (Ideal Microdrill, CellPoint Scientific, Gaithersburg, MD).
6. Normal saline (0.9%).
7. Small, straight Doppler probe (Standard Pencil Probe, MNP100XP, AD Instruments, Colorado Springs, CO).

8. Tissue adhesive (optional), 3/0 silk ties, 4/0 vicryl and 3/0 nylon sutures.
9. Scalpel (may be disposable), fine forceps, needle holder, retractors (optional), Q-tips (preferred), suction and micro scissors on top of the operating field.
10. PE50 and PE240 polyethylene tubing (Intramedic™ Polyethylene Tubing-Non-Sterile) and a three-way stopcock.
11. Oxygen, Compressed (99.7% O_2, Gilmore Liquid Air Company, South El Monte, CA).
12. Air, Compressed, UN 1002, Medical gas (Praxair, Danbury, CT).
13. Isoflurane (Western Medical Supply, Arcadia, CA).
14. Rectal probe (TR-200 Fine Science Tools, Foster City, CA).
15. Blood Pressure Monitoring System (Digi-Med Blood Pressure Analyzer, Louisville, KY).
16. Clamp Hemostats.
17. 1.0, 5, and 10 ml syringes, and a three way stopcock.
18. Eye ointment (Rugby Lab Inc., Duluth, GE).
19. Betadine solution (Fisher Scientific, Pittsburgh, PA).
20. Alcohol prep pads (Fisher Scientific, Pittsburgh, PA).
21. Buprenorphine (Reckitt Benckiser Pharmaceuticals, Richmond, VA).
22. Surgical Microscope and a fiber optic illuminator (Kent Scientific Corporation, Torrington, CT).
23. Doppler probe (Power Lab 4/35, AD Instruments, Colorado Springs, CO).
24. Laser Doppler monitoring unit (Power Lab, AD Instruments, Colorado Springs, CO).

Intravenous Femoral Catheter

A femoral catheter allows access to the circulatory system in lab studies. It permits administration of substances, collection of blood samples, and monitoring of blood pressure and heart rate. In addition, it causes minimal stress to the animal as determined from numerous reports [7]. In this section we will describe a step-by-step method on how to place an indwelling arterial catheter. Prepare the necessary materials beforehand. After organizing all the necessary materials, follow the step-by-step list that will guide you in performing the surgery.

Procedure

1. Prepare the surgical area where the procedures will take place.
2. Place the animal in the anesthetic chamber to receive a mixture of isoflurane 4%, oxygen 30%, and air 70% for 4 min for appropriate induction (Fig. 1).

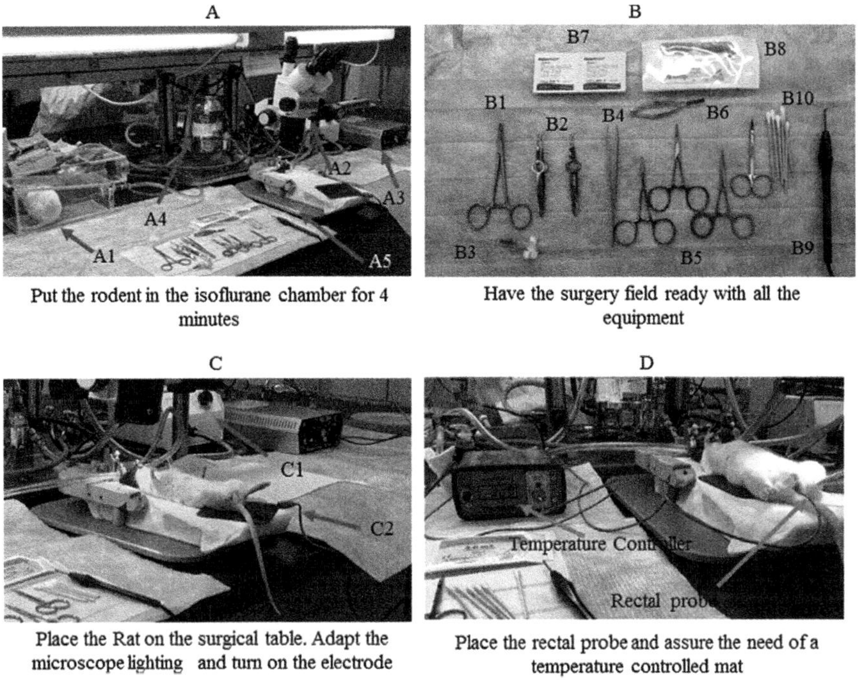

Fig. 1 (**a**) Put the rodent in the Isoflurane chamber for 4 min. Have the surgery field ready with all the equipment (a1: Rodent in the Chamber, a2: Microscope, a3: Electrosurgical unit; a4: isoflurane vaporizer; a5: rodent surgical frame); (**b**) Equipment for femoral artery catheterization (b1: Needle Holder, b2: Fine forceps, b3: Three-way stopcock with the catheter, b4: Fine Forceps, b5: Straight forceps, b6: Microscissors, b7: Alcohol pads, b8: 3/0 silk ties, b9: Cauterization Pen, b10: Q-tips). Notice the temperature pad not present in any of the pictures. Temperature may need to be controlled with a Lamp or the Pad. During anesthesia with isoflurane the rodent loses it's ability to adapt to environmental temperature. (**c**) Rat is placed on the surgical table (c1: Microscope light source, c2: Electrocautery pad). (**d**) Monitor temperature throughout the procedure using a rectal probe and temperature controller

Note: Maintain isoflurane at 2–3% after induction according to each rodent's need. In addition, give the same amount of isoflurane to all the rodents during the experiment and make note of the rodents that require a different amount.

3. Test the rodent reflexes to assure proper induction. Maintain isoflurane at 2–3% after induction according to each rodent's need. Pinch the rodent's paw to certify induction. If no movement is observed, proceed to step 4.

   Note: If rat is not appropriately anesthetized it may demonstrate erratic movement, causing difficulties in femoral access and involuntary dislodgment of the tip of the probe.

4. Before placing the rat into the frame, place a temperature pad that connects to a rectal probe. This allows the temperature to be controlled during the entire procedure.

Note: A heating lamp may be used in place of the heating pad. Heat pad dislodgment may lead to death of the rodent by overheating. Secure the probe to the surface of the table if necessary.

5. Place the rat in supine position (I.V. Femoral Line) in the stereotactic benchmark frame on top of the temperature pad with his nose placed in the frame nosepiece to continue sedation during the entire procedure.

   Note$_1$: Be careful when applying the nosepiece to the nose. Leave a small gap to guarantee no excessive pressure is being applied to the nose. Adjusting the nosepiece too tight can compress the animal's nose and lead to death from airway obstruction.

   Note$_2$: During the procedure, observe for signs of poor oxygenation. As more experience is gained, one may notice color changes right away in the color of the rodent's feet in the case of poor oxygenation. The color change is a good determinant of the oxygenation status of each animal. Bright pink feet imply good oxygenation. On the other hand, if the feet start to get dusky or blue, assess the animal and resolve any problems contributing to obstructing air exchange or assess for excess anesthesia contributing to respiratory depression.

6. Shave the right femoral area and apply betadine solution. Next, clean the surface area with sterile alcohol prep pads (Fig. 2).

   Note: Be careful not to injure the skin. Injured superficial skin may lead to the formation of skin abscesses and post-operative wound dehiscence. In our protocol we don't use lidocaine or bupivacaine in the surgical area since the rat is under isoflurane. Buprenorphine 0.01–0.03 mg/kg will be administered post operatively to control pain.

7. Apply eye ointment (Rugby Lab Inc., Duluth, GE) to the animal's eyes to avoid drying.
8. Allow exposure to the shaved, dry, and clean right femoral area and cover the rest of the area with sterile drapes.

   Note: In our lab we start catheterization in the right femoral region. If complications occur, the left side is used. Once betadine or alcohol dries out, proceed to the next step. The pictures below show the catheterization of the left femoral artery.

9. Make an incision (2–5 cm) over the region of the femoral artery using scissors or the surgical scalpel (n° 10) (Fig. 3).

   Note: The size of the incision will depend on the researcher's experience. Start with a larger incision until more familiar with the anatomy of the region and the procedure itself. Furthermore, stop minor subcutaneous bleeding before proceeding to step 10. Bleeding may complicate visualization of the inguinal anatomy.

10. After incising the skin, notice the inguinal fat pad coming into view, reflect it laterally and identify the neurovascular bundle.
11. Bluntly reflect the abdominal wall muscles to view the inguinal ligaments and expose the common femoral vessels in their entire length.

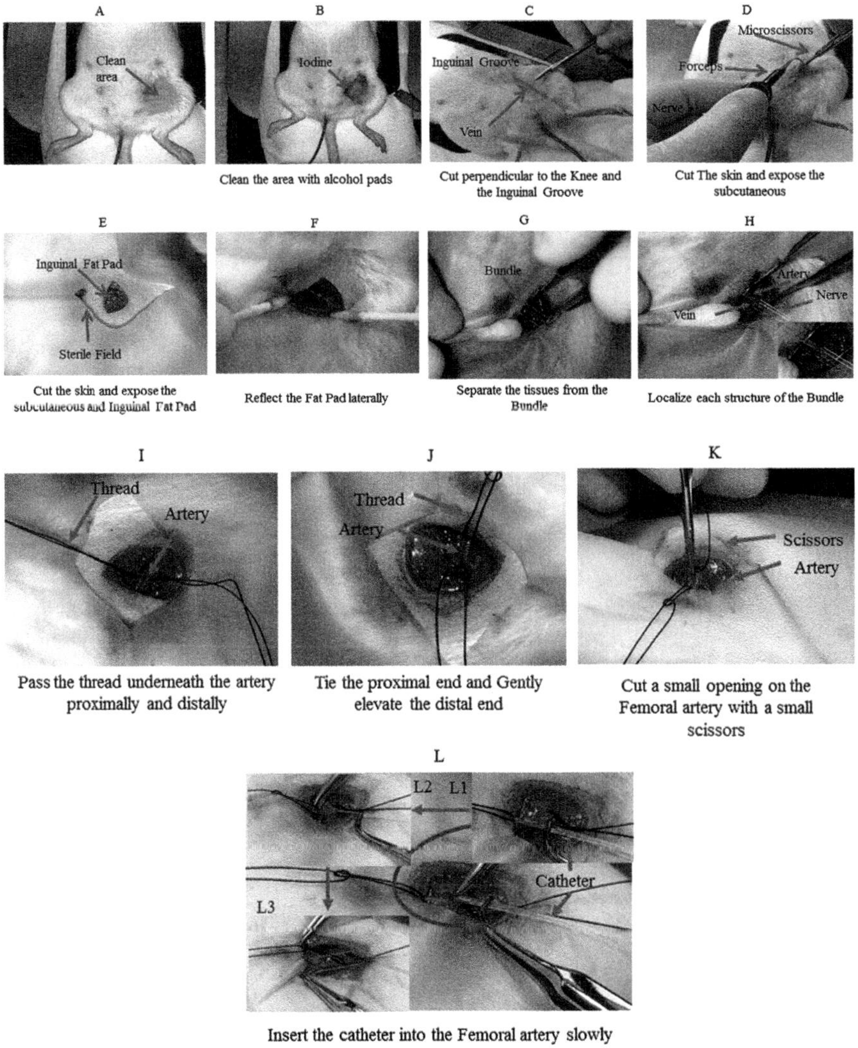

Fig. 2 Illustration of the step-by-step on How to perform a femoral artery catheterization. (**a**) After shaving apply Betadine; (**b**) Clean the area with alcohol pads; (**c**) Measure a perpendicular area from the knee to the inguinal groove; (**d**) Cut the skin and expose the subcutaneous tissue; (**e**) Inguinal Fat Pad; (**f**) Reflecting the lateral fat pad; (**g**) Separate the bundle; (**h**) Separate the artery from the bundle; (**i**) Pass the suture underneath the artery proximally and distally; (**j**) Lift up the vessel to insert the catheter; (**k**) Cut the artery; (**l**) Inset the cannula inside the artery (l1: Inserting the cannula; l2: Advancing the cannula; l3: Bigger portion of the cannula inside)

Note: The neurovascular bundle is composed of the femoral vein, femoral artery and femoral nerve. The vein has a larger diameter and a dark blue color and is located medial to the other structures. Retrolaterally to the vein, notice the presence of a structure with a red color, pulsation, and of smaller diameter

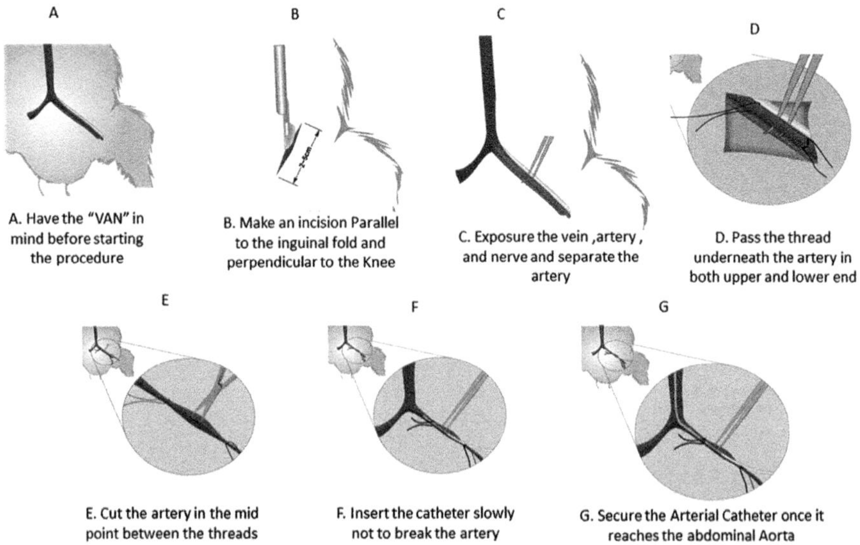

Fig. 3 These figures illustrates some of the steps necessary to gain access to the Femoral artery and insert the femoral catheter. (**a**) VAN: Vein, Artery, and Nerve. (**b**) Make a skin incision parallel to the inguinal fold and perpendicular to the knee. (**c**) Expose the Femoral vein, artery and nerve. (**d**) Pass two ties underneath the artery as shown. (**e**) Ligate the suture in the distal end and cut the artery between the threads. (**f**) Carefully insert catherter into the proximal end of Femoral artery. (**g**) Secure the tubing by tying the proximal suture

(femoral artery). Next, observe the femoral nerve, a thin and clear-colored structure located lateral to the artery. Approach it gently to avoid lower extremity movement paralysis as a result of severing the nerve [7].

Note: Memorize the mnemonic "VAN" for vein, artery, and nerve in order to appropriately identify the neurovascular bundle and select the structures from medial to lateral.

12. The perivascular sheath will be exposed. Make an incision parallel to the femoral vessels.

    Note: Keep in mind that after making an incision of the perivascular sheath, the vasa vasorum may be cut and small amounts of bleeding may occur. Though usually self-limited, if persistent hemorrhaging occurs cautiously use the electrocautery to achieve hemostasis.

13. Locate the femoral artery below the inner part of the inguinal ligament. Use electrocautery or a Q-tip to maintain hemostasis if needed.

    Note$_1$: Be careful when moving the fat pad laterally. The distal ends of the epigastric vessels enter this region. Preserve the epigastric vessels. The femoral artery measures around 1.5 cm in the rat. Anatomically it commences at the mid-point of the inguinal ligament and ends when it gives off the superficial femoral artery and deep branches.

    Note$_2$: To facilitate visualization of the artery pour 1 ml of 1% Lidocaine drops around the vessel to help distinguish the femoral artery from the femoral vein. Lidocaine vasodilates the artery and facilitates visualization of pulsations. Warm saline is another option to help dilate the artery and facilitate insertion of the catheter. In case of wall perforation, withdraw the catheter and reintroduce it at a more proximal location. Insert the catheter gently to avoid tearing the artery or breaking through the arterial wall in the process [8].

Note₃: When obtaining the artery view, notice that anatomically one branch of the epigastric artery comes off just before the division of the femoral artery.

14. Separate and isolate the artery from the vascular sheath ("VAN") and place a tie at the distal and proximal ends of the artery. Ligate the distal end of the vessel with a 3/0 silk tie and leave the proximal end loose with the tie attached to an artery (Fig. 2).
15. Make a hole in the artery using microscissors. Bleeding may start. Gently perform a proximal tie, encircling the proximal end to slightly elevate the vessel and stop possible bleeding and clean the region. Elevation of the artery can be done before making the arterial incision to help prevent bleeding.
16. Prior to beginning the catheterization procedure evaluate the catheters by flushing them with sterile saline to ensure they are patent [7].
17. Cannulate the proximal artery with PE50 polyethylene tubing.
18. Next, secure the tubing by tying the proximal end around the vessel where it is contained.

    Note$_1$: After finishing this procedure certify that there is no ongoing bleeding and the right amount of tubing has been inserted into the vessel (approximately 5 cm will reach the abdominal aorta) [7]. The tubing now can be connected to the three-way stopcock which consequently will be connected to the monitor and allow not only blood withdrawal for a diverse range of tests during and post experiment but also monitoring of the blood pressure and mean arterial pressure.

    Note$_2$: Monitor for possible bleeding during the entire procedure, as the knot can become loose. Do not suture the catheter too tight; maintain vessel patency.
19. Soak gauze with normal saline and place the gauze over the open area to establish a moist environment in the inguinal region to prevent drying of the tissues during CBF monitoring.

    Note: Preventing tissue from drying will also prevent dehydration. It is important to test the rat for dehydration during long procedures that include exposure of body surfaces. First, pinch the skin on the back and let go. It will take approximately 1 s for the skin to go flat. If several seconds are needed for the skin to go flat, it indicates the need to give fluid replacement. In our lab, we give 1 ml of normal saline subcutaneously after surgery to ensure hydration status. If longer procedures are planned, an infusion of 0.2 ml of normal saline is used to avoid dehydration during the procedures and this is not harmful. Another option is the use of a pressure infusion cuff with normal saline connected to the femoral line. This will provide small volumes of normal saline if needed, and keep the line clean at all times for collection of blood during the experiment.

At this point the rodent is ready to either undergo brain monitoring or proceed to the burr hole procedure. Please refer to the section entitled Burr Hole and Brain Monitoring before continuing to step 20.

20. After the experiment is done, undo the knot and remove the line quickly and tie off the proximal knot.

    Note: After ligation of the proximal femoral artery, collateral flow will prevent lower limb ischemia.
21. Suture the skin with 3.0 silk and reapply betadine on top of the suture line.
22. Give buprenorphine subcutaneously.
23. Place the rat in the post-op chamber in order to maintain body temperature using the heating lamp until rat is freely moving.

Burr Hole and Brain Monitoring

A microscope and a fiber optica illuminator are used to achieve greater precision in making a window. When a procedure such as subarachnoid hemorrhage or surgical brain injury is done prior to brain monitoring, choose the contralateral side to drill the window and monitor the necessary parameters. In our lab we use two different frames of the same brand: One to drill a burr hole and another to monitor the rodent during the CBF data collection.

1. In the already anesthetized animal, shave the head and apply skin disinfectants such as betadine, then clean with alcohol (Fig. 4).

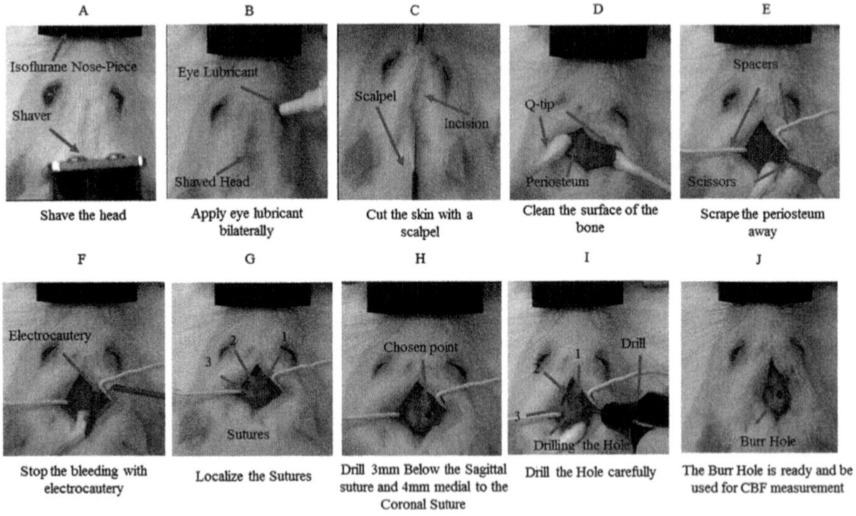

Fig. 4 Summary of the Burr hole procedure from (**a**) to (**j**); **i** (1—Coronal Suture, 2—Bregma, 3—Sagittal Suture)

2. Apply eye lubricant to the rodent's eye to prevent drying.
3. Use a scalpel to do a midline incision from the frontal to occipital bone, allowing enough exposure of the bone to create the right bony window for monitoring. Expect some skin bleeding to occur after the incision is made.
4. An electrosurgical unit with a high frequency electrical current can be used to stop the bleeding. In low bleeding cases, a Q-tip can be used to stop the bleeding.

   Note: After the bony surface area is cleaned and hemostasis has been achieved, continue to step 5.
5. Use clips to retract the skin and subcutaneous tissues laterally to visualize the underlying periosteum. Remove this very thin and transparent membrane that covers the skull.

   Note: After step 5, small amounts of bleeding may still appear from the bony surface. Use electrocautery or Q-tips to achieve hemostasis. The cleaning and drying process allows full bone exposure and visualization of the midline, coronal suture, and the right parietal region. Continue to step 6 once visualization is achieved.
6. Using the drill, make a bony window in the parietal bone of the skull. In our lab, we make a window 3 mm below the coronal suture and 4–5 mm lateral to the superior sagittal suture.

   $Note_1$: Ensure the Dura remains intact and prevent damage to the brain. Drilling with inexperienced hands may puncture the Dura and brain tissue. A square window or a circular burr (both 0.5–1 cm^2) hole are options for the type of bony window to choose for brain monitoring. The type and size of the hole may vary depending on each researcher's experience in placing the Doppler probe tip on top of the window and their experience with either type of window.

   $Note_2$: The size of the bony window may vary according to the size of the probe holder each laboratory has available. Use the holder you have available, place the holder on the bony surface and visualize the size of the window you may need to drill. In our experience, using the holder or placing the Doppler on top of the dura generates the necessary readings.

   After the bony window has been made we are ready to perform either the femoral line procedure or proceed to brain monitoring. Assuming the femoral line is in position, proceed to obtain CBF measurements.
7. Before starting the procedure make sure the Doppler probe is connected to the monitor unit and the tip is clean. Turn on the computer and software.
8. Place the Doppler lightly on top of the Dura (Fig. 5).

   $Note_1$: The Doppler tip is hollow and may easily damage the brain when placing the probe on the brain surface. Impinging the Dura with the drill may lead to direct contact of the probe with the brain tissue leading to possible bleeding, incorrect measurement, and brain damage, as the tip of the probe is hollow and uses a laser beam.

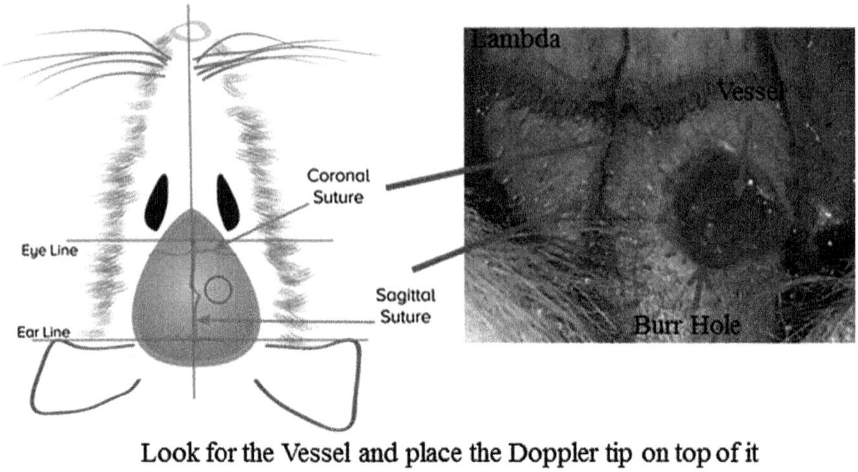

Fig. 5 After the Burr Hole is opened notice the vessel crossing the middle. This is where the Doppler tip will be placed

Note$_2$: Clean the tip before and after the experiment. Place it in water for a couple of minutes after finishing the procedure to dissolve and dislodge any possible solids that attached to it during the experiment.

Note$_3$: The laser probe generates a laser beam at a wavelength of 830 nm to a depth of approximately 0.5–1 mm. Thus, the probe can't be placed outside of the Dura more than 1 mm. The Doppler is connected to the laser Doppler monitoring unit which in turn is connected to a computer where readings can be analyzed using LabChart software, Windows version.

Note$_4$: Misplacing the probe on top of the bone instead of the dura will lead to incorrect readings because the wavelength produced goes to different depths in different tissues.

9. Observe the CBF readings starting to show on the computer screen.

   Note: There is an option to glue the probe with a probe holder over the craniotomy site to hold the probe firmly in place. In our experience, both the probe holder and the stable tip of the Doppler on top of the window will generate the necessary results as long as the rat is appropriately anesthetized and is not moving. In our lab, the frame used to monitor CBF allows the passage of the Doppler cable through a ring that holds the cable and tip erect and firm in the craniotomy site over the vessel. This allows appropriate CBF recording on the software.
10. After placing the Doppler over the craniotomy site, check the computer and the signal from the probe. Adjust the probe on top of the Dura to make sure it sits on top of the vessel. Glue (optional) or secure the holder into place after finding the correct location of the vessel. If using glue, allow enough time for the holder to stick to the bony surface. In cases where a holder is not being used, visualize the correct location of the vessel and assure the location is associated with the best possible signal (Fig. 6).

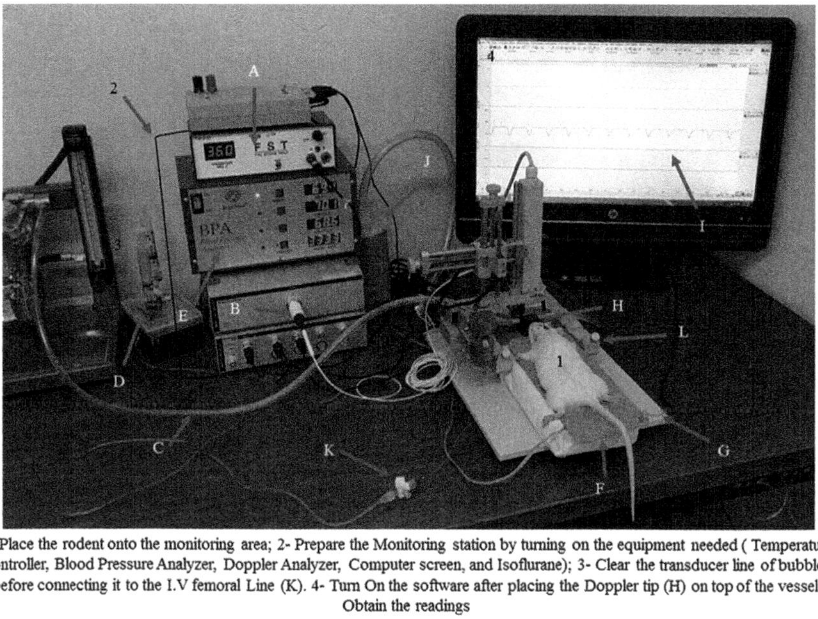

Fig. 6 Demosntration of the area and equipment in which we will be working to measure CBF. (**a**) Temperature Monitor; (**b**) Blood Pressure gaging system, (**c**) Isoflurane Tube to nosepiece; (**d**) Doppler connection; (**e**) Blood Pressure Analyzer; (**f**) Temperature Pad connection; (**g**) Rectal Probe; (**h**) Doppler Probe; (**i**) Monitor with CBF Graph; (**j**) Canister to drain Isoflurane; (**k**) Three-way-stopcock for blood pressure analyses and blood collection if needed; (**l**) Frame

Note: If an experiment (e.g. Syncope) is planned during the monitoring period, collect baseline CBF data for 3–5 min before starting the experimental data collection.

At this point, follow one of two options.

(a) Close the subcutaneous tissue and skin if you plan to monitor the rodent for a long period of time with the use of available probe holders.
(b) Leave the skin open allowing for visualization of the probe during the entire procedure if the experiment requires monitoring for short intervals. In our lab we measure CBF for 30 min after syncope induction (data not published) without consequences to the rodent after the end of the experiment.

11. Collect the data for the time needed.

Postoperative Care

The post-operative period is a crucial period in which the animal is susceptible to respiratory depression and other complications. Placing the cage close to observers allows immediate action including chest compressions and ventilation if necessary.

Make sure the following actions are undertaken:

1. Place the rodent in the postoperative cage/chamber under a lamp that provides heat to avoid hypothermia, or on a warm pad as used in the operative period. The length of anesthesia will determine the animal's ability to recover. Thus, anesthesia duration should always be kept to minimum as needed to complete experimental procedures.
2. Check reflexes and hydration status (pinch the skin on the back and let go) if there is a prolonged recovery period. The use of isoflurane allows faster recovery from anesthesia, as discussed in the **Materials** section. In addition, normal saline injection (I.P. or S.C.) can be used.

   Note: In our experience, giving the rodent 1 ml of normal saline subcutaneous after the procedure will improve the hydration status.
3. Watch the color of the feet in the post-operative period. It is a good determinant of the oxygenation status of the rodent. Dusky or blue feet will indicate the need to stimulate the respiratory system.

   Note: Respiratory depression may occur after the experiment is over.
4. Use a thin soft catheter on a 5 ml syringe to remove any mucus and stimulate breathing. Insert the thin catheter into the trachea and pull back on the syringe and withdraw the catheter slowly at the same time. A vacuum will be generated, removing mucus and stimulating breathing. Repeat tracheal suction if necessary. Lastly, start manual chest compressions and ventilate the animal if necessary [9].
5. Place the animal back into it original cage once it is fully awake.

Conclusion

In spite of new methods in brain monitoring, Doppler is still very useful and has low complication rates. It allows study of each rodent's CBF with direct plotting of the data into the software for further analysis. The step-by-step method described in this chapter is easily learned with training. The ease of use and minimal complications allows for monitoring CBF in many experimental conditions.

Acknowledgments A special thank you to Devin McBride for helping set up the equipment for the CBF pictures.

References

1. Mette D, Strunk R, Zuccarello M. Cerebral blood flow measurement in neurosurgery. Transl Stroke Res. 2011;2(2):152–8.
2. Dirnagl U, Kaplan B, Jacewicz M, Pulsinelli W. Continuous measurement of cerebral cortical blood flow by laser-Doppler flowmetry in a rat stroke model. J Cereb Blood Flow Metab. 1989;9(5):589–96.

3. Fabricius M, Lauritzen M. Laser-Doppler evaluation of rat brain microcirculation: comparison with the [14C]-iodoantipyrine method suggests discordance during cerebral blood flow increases. J Cereb Blood Flow Metab. 1996;16(1):156–61.
4. Haberl RL, Heizer ML, Ellis EF. Laser-Doppler assessment of brain microcirculation: effect of local alterations. Am J Phys. 1989;256(4 Pt 2):H1255–60.
5. Cahill J, Zhang JH. Brain Monitoring. In: Chen J, Xu ZC, Xu X-M, Zhang JH, editors. Animal models of acute neurological injuries. Totowa, NJ: Humana Press; 2009. p. 39–45.
6. Tonnesen J, Pryds A, Larsen EH, Paulson OB, Hauerberg J, Knudsen GM. Laser Doppler flowmetry is valid for measurement of cerebral blood flow autoregulation lower limit in rats. Exp Physiol. 2005;90(3):349–55.
7. Jespersen B, Knupp L, Northcott CA. Femoral arterial and venous catheterization for blood sampling, drug administration and conscious blood pressure and heart rate measurements. J Vis Exp. 2012;(59) https://doi.org/10.3791/3496.
8. Alfredo R, DL V. Catheterization of the femoral artery and vein. In: Experimental surgical models in the laboratory rat. Boca Raton, FL: CRC Press; 2009. p. 73–6.
9. Institute for Laboratory Animal Research (U.S.). Committee on Guidelines for the Use of Animals in Neuroscience and Behavioral Research, National Academies Press (U.S.). Guidelines for the care and use of mammals in neuroscience and behavioral research. Washington, DC: The National Academies Press; 2003.

Veterinary Care Methods for Rats and Mice in Experimental Spinal Cord Injury Studies: An Update

Christine D. Yarberry, Aaron H. Puckett, and Stephen M. Onifer

Abstract The incidences and causes of traumatic injury to the adult human spinal cord vary throughout the world (Lee et al., Spinal Cord 52:110–116, 2014). Regardless, spinal cord injury (SCI) can result in devastating dysfunctions that lead to tremendous and long-term physical, emotional, and financial hardships not only for the survivor but also for family and friends. For these reasons, both clinical trials and numerous experimental studies of individual and combined protective, reconstructive, and rehabilitative SCI treatments have taken place (https://clinicaltrials.gov/ct2/results?term=spinal+cord+injury&Search=Search; Alilain et al., Nature 475:196–200, 2011; Harrop et al., J Neurosurg Spine 17:230–246, 2012; Steward et al., Exp Neurol 233:597–605, 2012; Lee et al., J Neurosci 33:10591–10606, 2013; Sharp et al., Exp Neurol 248:321–337, 2013; Angeli et al., Brain 137:1394–1409, 2014; Collinger et al., Clin Transl Sci 7:52–59, 2014; Hammond et al., Arch Phys Med Rehabil 95:2342–2349, 2014; Kressler et al., Arch Phys Med Rehabil 95:1878–1887, 2014; Moreno-Duarte et al., Neuroimage 85:1003–1013, 2014; Plemel et al., Prog Neurobiol 117:54–72, 2014; Popovich et al., Exp Neurol 261:97–108, 2014, Ramer et al., Lancet Neurol 13:1241–1256, 2014; Sharp et al., Exp Neurol 254:1–11, 2014; Sharp et al., Exp Neurol 257:186–204, 2014; Ward et al., J Neurotrauma 31:819–833, 2014; Aslan et al., Arch Phys Med Rehabil 97:964–973, 2016; Bonner and Steward, Brain Res 1619:115–123, 2015; Dietrich, Top Spinal Cord Inj Rehabil 21:174–187, 2015; Dugan and Sagen, J Neurotrauma 32:622–632, 2015; Dvorak et al., J Neurotrauma 32:645–654, 2015; Evaniew et al., J Neurotrauma

C. D. Yarberry
Department of Neurosurgery, Kentucky Spinal Cord Injury Research Center, School of Medicine, University of Louisville, Louisville, KY, USA

A. H. Puckett
Division of Cardiovascular Medicine, Department of Medicine, School of Medicine, University of Louisville, Louisville, KY, USA

S. M. Onifer (✉)
Palmer Center for Chiropractic Research, Palmer College of Chiropractic, Davenport, IA, USA
e-mail: stephen.onifer@palmer.edu

J. Chen et al. (eds.), *Animal Models of Acute Neurological Injury*, Springer Series in Translational Stroke Research, https://doi.org/10.1007/978-3-030-16082-1_5

33:468–481, 2016; Hosier et al., J Neurotrauma 32:1633–1644, 2015; Lang et al., Nature 518:404–408, 2015; Nagoshi and Fehlings, Expert Opin Investig Drugs 24:645–658, 2015; Ruschel et al., Science 348:347–352, 2015; Sandrow-Feinberg and Houlé, Brain Res 1619:12–21, 2015; Siddiqui et al., Prog Brain Res 218:15–54, 2015; Squair et al., J Neurotrauma 32:609–621, 2015; Tazoe and Perez, Arch Phys Med Rehabil 96:S145–S155, 2015; Wang et al., Exp Neurol 271:368–378, 2015; Wang and Pearse, Int J Mol Sci 16:16848–16879, 2015; West et al., Neurorehabil Neural Repair 30:317–327, 2016).

Advances in acute and chronic medical care have markedly improved the quality and length of the lives of persons with traumatic SCI (Ropper et al., Pract Neurol 15:266–272, 2015). We feel that veterinary care of spinal cord injured animals should have no less of a goal. Moreover, spinal cord injured animals in optimal acute and chronic health are essential for successful experimental studies of SCI treatments. In this chapter, we will report veterinary care methods to optimize the health of experimental animals, particularly rats and mice, with tetraplegia or paraplegia after traumatic SCI.

Keywords Care · Handling · Housing · Morbidity · Mortality · Pain · Paraplegia · Tetraplegia · Spinal cord injury · Veterinary

Model Selection

There are four types of traumatic injury to the adult human spinal cord: (1) contusion evolving to cavity formation, (2) massive compression, (3) laceration, and (4) solid cord [37–39]. These can be anatomically and functionally complete or incomplete. Monkeys, pigs, dogs, cats, opossums, and guinea pigs are being utilized for mammalian models of the different types of traumatic adult human spinal cord injury (SCI) in experimental studies [40–51]. Spinal cord injured rats and mice are also frequently used because damage processes and dysfunctions are similar to those that occur in humans after SCI [52–101].

Veterinary care methods for rats, mice, and monkeys after SCI have been described [102–104]. Additionally, care methods are included in articles about experimental SCI studies. We will describe veterinary care methods for spinal cord injured rats and mice developed and utilized at the Department of Physiology (SUNY Health Science Center), The Miami Project to Cure Paralysis (University of Miami), the Kentucky Spinal Cord Injury Research Center (University of Louisville), and the Spinal Cord and Brain Injury Research Center (University of Kentucky) as well as by the faculty, veterinarians, and their technical staffs at Ohio State University. These methods were approved by the Institutional Animal Care and Use Committees. They are in accordance with the Public Health Service Policy on Humane Care and Use of Laboratory Animals and with the Policies on the Use of Animals and Humans in Neuroscience Research.

Materials and Methods

Before Surgery and SCI

1. Housing. Keeping the animal's distress to a minimum will lead to optimal performance throughout an experimental study and both less morbidity and mortality. This can start upon the arrival of the animals at the research institution. The American Association for Laboratory Animal Science suggests at least 48 h for animals to recover from their trip and to acclimate to new housing conditions [105]. Housing is typically done in and under the direction of the research institution's veterinary care facility. Rats and mice are housed in separate rooms. After weaning, males and females are housed separately. The home cage typically consists of a three tier system: a cage, a lid to hold food and a water bottle for *ad libitum* eating and drinking, and a micro-filter top that fits over the lid (Allentown Inc., Allentown, NJ; Alternative Design Manufacturing & Supply, Siloam Springs, AR; ENVIGO, Indianapolis, IN; Lab Products, Inc., Seaford, DE; SE Lab Group, Napa, CA; Tecniplast USA, West Chester, PA). Various types of cage bedding are available including moisture-absorbent ALPHA-dri® and Shepherd's™ Cob (Shepherd Specialty Papers, Watertown, TN), TEK-FRESH (Harlan TEKLAD, Madison, WI), and wood chips (P. J. Murphy Forest Products, Montville, NJ). The dimensions of the cage depend upon the storage rack and the space available in the room for the maximum number of cages. The maximum number of animals housed in a cage is dictated by their sizes and guidelines for the minimal amount of cage floor space required by that size of animal [106]. The housing density to that maximum number will be dictated by the experimental study. The cages are stored on racks with or without water systems and air systems (Alternative Design Manufacturing & Supply, Siloam Springs, AR; Lab Products, Inc., Seaford, DE; Tecniplast USA, West Chester, PA). Typically, the cages are changed at least twice a week. The rooms are usually on a 12 h or 10 and 14 h dark/light cycle, at 75 °F, and under positive or negative air pressure. All experimental and veterinary care methods typically are performed during the light cycle.
2. Physical Examination. Pathogen-free animals are purchased from commercial vendors (Charles River Laboratories, Raleigh, NC; ENVIGO, Indianapolis, IN; The Jackson Laboratory, Bar Harbor, ME; TACONIC, Cambridge City, IN). It is also recommended that the animals undergo a physical examination during the acclimation period. The results should be recorded (Table 1). Veterinarians and the vendors should be consulted with before using an animal that exhibits any complications.
3. Acclimation to Handling and Apparatuses. Experimental SCI studies typically require more frequent handling of the animals for daily veterinary care and functional assessments than would normally occur at the vendors and animal care facilities. Handling and being out of their home cage are distressing to the animals. As a result, scratching and biting can occur. This can be reduced by

Table 1 Physical examination checklist

Project Name:				
ID#:	Strain:	Sex:	Date of Birth:	Arrival Date:
Examination Date:		Examiner:	Body Weight (g):	
Check for the following health issues. Circle signs that are present. If an animal exhibits any of these signs, seek veterinarians and vendors assistance to decide whether to remove the animal from the study				
• Discharges from the eyes, ears, nose, and mouth				
• Dehydration (lift the skin over the back and determine whether it quickly returns to its original position)				
• Constipation or diarrhea				
• Fractures or amputations				
• Bites, scratches, wounds, abscesses, or calluses				
• Hair loss or overgrooming				
• Labored breathing				

acclimation. Lab coats and gloves should be worn as well as masks if necessary. First aid for bites and scratches, and the possible allergic reactions to them, should be readily available. Regarding handling, it is strongly recommended that this be done frequently till the animals do not vocalize. The animals should not be lifted or transferred by their tails, especially after SCI. This should be done firmly but not impair breathing. Frequently transferring the animals to the apparatuses used for behavior or electrophysiological training and assessments (Columbus Instruments, Columbus, OH; IITC Life Sciences, Woodland Hills, CA; Stoelting Co., Wood Dale, IL) so that they can explore them are also highly recommended.

(a) Rats can be handled the following ways.

- Gently approach the rat while it rests on bedding. Place thumb and index finger of a gloved hand over right and left sides of the body. Gently pick up the rat. It will brace itself with the forepaws on handler's thumb. Be cautious of the rat's ability to maneuver itself and bite.
- Gently approach the rat while it rests on bedding. Cover it from head to base of tail with a cloth or paper towel leaving excess cloth over its head like a hood. Place the gloved thumb and index finger over the towel and the right and left sides of the rat's body. The thumb will be under its forepaws. The excess towel and thumb position will decrease the likelihood of the rat biting the handler. Rats that are accustomed to being handled do not require the towel.
- Gently approach the rat while it rests on bedding. When stationary, place the rat's head between the gloved index and middle fingers. Lift the rat and apply gentle pressure to be certain it does not wiggle excessively. Use the opposite hand to support the hindquarters and tail. This "V" method is useful when giving intramuscular injections in the hindquarters.

(b) Mice can be handled the following way.

- Wait until the mouse is in a corner and at rest. Gently approach it while cupping both gloved hands. If the mouse is not in a corner, slowly move it towards one with cupped hands. Wait for the mouse to move onto the hands then close them. If it does not move onto the hands, gently close them around it. Transport the mouse out of the cage and place it on a wire cage top. Gently grip the tip of its tail with one hand. The mouse will grasp the cage top with both forelimbs. At this time, use the thumb and index finger of the opposite hand to grasp the skin between the mouse's shoulder blades. To avoid excessive movement, gently place the tail between the ring and last fingers.

4. Food Restriction. Optimal training for behavior assessments that involve obtaining food or a food reward require that the animal be fasted overnight or every other overnight or continuously food restricted throughout the training period [107, 108]. For continuous food restriction, it is recommended that the animals be housed individually and that their body weights be measured every day and maintained between 80–90% of littermates housed individually and eating *ad libitum*. It is also recommended that *ad libitum* eating occur for at least one week before surgery and SCI. Food restriction after SCI may be detrimental especially to animals with severe cervical SCI. Moreover, we found that *ad libitum* eating did not alter the food retrieval performances of sham-operated rats compared to their baseline performances [107].

Surgery and SCI

Surgery can be performed once the animals are acclimated to housing and handling and once the baseline assessments are performed. Specific survival surgery methods are described in the other chapters. In general, survival surgery requires that aseptic methods be utilized to prevent infections. These include using sterile instruments, sterile materials, and sterile drapes for the surgery sites as well as sterile gloves, masks, caps, clean lab coats, and disinfected surgery areas. Instruments, materials, and gloves should be changed when moving from skin to muscle, bone, and nervous system tissues. Anesthetics/tranquilizers, analgesics, antibiotics, ointments/topicals, and fluids should be used (Table 2). Normothermia should be maintained throughout surgery.

Sterilization

1. According to manufacturer's instructions (Fine Science Tools, Inc., Foster City, CA; Roboz Surgical Instrument Co., Inc., Gaithersburg, MD), instruments should be cleaned with a neutral pH detergent in distilled or deionized water and

Table 2 Product names and manufacturer information

	Product name	Manufacturer information
Anesthetics/ tranquilizers	Isoflurane	Baxter Healthcare Corporation, Deerfield, IL
	Nembutal® Sodium Solution	Abbott Laboratories, North, Chicago, IL
	Ketaset®	Fort Dodge Animal Health, Fort Dodge, IA
	Acepromazine Maleate	Boehringer Ingelheim Vetmedica, Inc., St. Joseph, MO
Analgesics	Xylazine	IVX Animal Health, Inc., St. Joseph, MO
	Buprenorphine	Reckitt Benckiser Healthcare, Ltd., Hull, England
Antibiotics	Gentozen® (Gentamicin Sulfate Solution)	Phoenix Scientific, Inc., St. Joseph, MO
	Penicillin G Procaine	Bimeda, Inc., Irwindale, CA
	Baytril	Bayer Healthcare LLC, Shawnee Mission, KS
	Clavamox®	Pfizer, New York, NY
Ointments/ topicals	Dermachlor®	Henry Schein Animal Health, Columbus, OH
	Lacri-Lube®	Allergan, Irvine, CA
	Bacitracin Zinc Ointment USP	Fougera, Melville, NY
Fluids	0.9% Sodium Chloride Solution	Hospira, Inc., Lake Forest, IL
	Lactated Ringers Solution	B. Braun Medical, Inc., Irvine, CA
	5% Dextrose in Lactated Ringers Solution	B. Braun Medical, Inc., Irvine, CA

a soft nylon toothbrush or ultrasonic cleaner. They should be thoroughly rinsed with distilled or deionized water and dried.

2. Surgery instruments and materials should be heat sterilized in pouches with an autoclave, gas sterilized, or cold sterilized with a disinfectant not caustic to animal tissues. After cleaning and rinsing with sterile distilled or deionized water, instruments should be heat sterilized with a glass bead sterilizer between animals on the same day.
3. Many of the materials indicated above and below as well as in Table 2 can be obtained from Henry Schein® Animal Health (https://www.henryscheinvet.com).

Preparation

1. It is recommended that these methods be performed in a location that is different than the surgery area. The animal should be gently weighed and this, as well as other surgical information, should be recorded (Table 3).
2. Anesthetics are required for preparation and surgery. A Drug Enforcement Agency License may be necessary for their purchase (http://www.deadiversion.usdoj.gov/drugreg/index.html). Inhalational gas anesthetic (Isoflurane) mixed with oxygen and nitrous oxide can be delivered to a covered induction chamber containing the animal. For this and for surgery, safe gas delivery and evacuation are required (Smiths Medical, Dublin, OH). Sterile chemical anesthetics

Table 3 Surgery sheet

Project Name: ____________________

ID#: ________ **Date:** ________ **Surgeon:** ________ **Investigator:** ________

Strain: ________ **Sex:** ________ **Starting Weight:** ________g

Anesthesia (SC, IP, IM, IV, INH)
Nembutal® Sodium Solution (____mg/kg IP): ________mL **Additional:** ________mL ________mL
Ketaset® / Xylazine (____mg/kg / ____mg/kg IP): ________mL **Additional:** ________mL ________mL
Isoflurane (0.5–4.0 INH): ________ **O2 (INH):** ________L/min
Anesthesia Level (blink reflex, paw withdrawal, pads/mucous membrane colors): ________

Vital Signs (temperature, blood pressure, heart rate, blood flow, oxygen saturation, respiratory rate, expired carbon dioxide): ________

Lacri-lube – Eyes ________

Gentozen® (____mg/mL SC): ________mL **Fluids (L. Ringers, Dextrose-L. Ringers, Saline):** ________mL SC

Topical: Bacitracin – Incision site ________

Procedure: ________

Illustration of procedure done or damage that may have occurred:

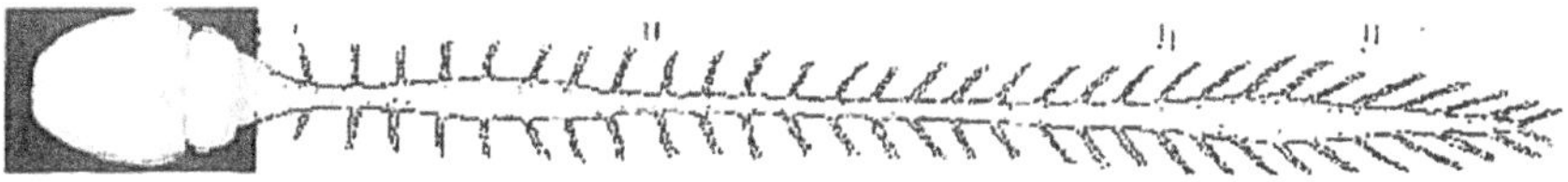

Comments: ________

Complications: ________

Additional Procedures / Surgeries:

Date: ________ **Weight:** ________g **Anesthesia** ________ **Procedure:** ________
Date: ________ **Weight:** ________g **Anesthesia** ________ **Procedure:** ________
Date: ________ **Weight:** ________g **Anesthesia** ________ **Procedure:** ________
Date: ________ **Weight:** ________g **Anesthesia** ________ **Procedure:** ________

(Nembutal® Sodium Solution, Ketaset®) can be injected intraperitoneally (IP) or intramuscularly (IM). Sterile chemical analgesics (Xylazine) and tranquilizers (Acepromazine Maleate) can be mixed with the chemical anesthetics. Antibiotics (Gentozen®) should also be administered at this time. The animal should be placed in a covered and empty cage to prevent aspiration of bedding or food. To prevent hypothermia, the induction chamber and cage should be placed on a cloth towel over a water circulating pad set at 37.5 °C or an electric heating pad at the lowest setting. The absence of the blink reflex to touching the cornea or limb withdrawal to toes or digit pinch should be used to ensure that the animal is deeply anesthetized before proceeding and during surgery. Respiratory rate and mucous membrane or paw pads color should also be monitored for the depth of anesthesia. Gray or blue coloration of the mucous membranes around the eyes and mouth or paw pads indicates inadequate tissue oxygenation. If necessary, "mouth-to-mouth" resuscitation can be given to rats using a 10 or 20 mL syringe with the plunger removed. If the animal sounds and/or feels raspy or is choking, pull the tongue to one side of the mouth. Use a gavage feeding needle (Fine Science Tools, Inc.) to carefully suction fluids or objects that are obstructing the airway. Do not attempt this on an unanesthetized animal.

3. Hair at and around the surgery site should be removed by careful shaving and vacuuming. Plucking the hair or a depilatory cream also can be used. Shaving the abdomen and hindquarters of animals can be done to prevent the skin from being damaged by urine-soaked hair.
4. Beginning at the surgery site then moving outward in increasingly larger circles, the skin should be scrubbed with gauze or a cotton-tipped applicator containing Chlorhexidine Gluconate 4% (Dermachlor®) then a 10% diluted Dermachlor® solution.
5. Ophthalmic ointment (Lacri-Lube®) should be applied to the eyes to prevent the corneas from drying. It is recommended that the ointment be put on to a clean cotton-tipped applicator then the eye to prevent contamination.
6. One-half the final volume of sterile, core temperature fluids (0.9% sodium chloride or saline solution, lactated Ringers solution) can be injected subcutaneously (SC) to prevent dehydration and to account for blood loss during surgery. It is recommended that the fluids be injected away from the surgery site at two locations adjacent to the body's midline.
7. An identification system should be administered at this time. Permanent markers can be used to put numbers or codes on the tail. This needs to be repeated throughout the experimental study. Non-surgical implantation of transponders (BioMedic Data Systems, Inc., Seaford, DE) may also be used.
8. The animal's core temperature should be maintained at 37.5 °C during surgery. This can be accomplished with a homeothermic blanket system (Harvard Appartus, Inc., Holliston, MA). A water circulating heating pad (Gaymar Industries Inc., Orchard Park, NY) and thermometer with a flexible rectal probe (Physitemp Instruments, Inc., Clifton, NJ) are also recommended. Electric heating pads, warmers (Grabber Performance Group, Grand Rapids, MI), and heat lamps can be used with extreme caution. A towel or drape should be between the animal and the blanket, pads, or warmers.
9. A sterile drape with a hole in it should be placed over the animal. This will expose the surgery site and prevent hair from entering it.

Surgery and SCI

1. It is essential that the animal's depth of anesthesia and core temperature be monitored as described above. Additionally, monitoring of blood pressure, heart rate, blood flow, oxygen saturation, respiratory rate, and expired carbon dioxide is suggested (PhysioSuite®: Kent Scientific Corp., Torrington, CT; MouseOx Plus: Starr Life Sciences Corp., Oakmont, PA).
2. The surgery site should be flushed with sterile, core temperature saline to prevent dehydration of the exposed tissues.

3. Immediately following contusive cervical SCI and during incision closure, we have found that inhalation of 100% oxygen can be used to compensate for potential respiratory dysfunction [109].
4. Muscle and skin incisions should be sutured in layers. Absorbable materials, such as chromic gut, and round needles can be used to suture muscles. Skin can be sutured with stainless steel wound clips or staples. Non-absorbable materials, such as nylon, and cutting needles can also be used. Cyanoacrylate adhesives are also available but should be used with caution and for small incisions.
5. Ointments containing antibiotics (Bacitracin Zinc Ointment USP) should be applied to sutured skin incisions.
6. The remaining one-half of the sterile, core temperature fluids (saline solution, lactated Ringers solution with or without 5% dextrose) can be injected subcutaneously as described above.

Acutely and the First Week After Surgery and SCI

Housing

1. After manually expressing their bladders (see below), rats and mice should be placed on a clean paper towel or drape in a clean, empty, and covered home cage in a recovery room or their housing rooms. The animals should not be handled by the tail. They should lie on their side and the head should face toward the cage front. The cage should be placed on a cloth towel over a water-circulating pad set at 37.5 °C or an electric heating pad at the lowest setting. Importantly, systemic hypothermia [34] or hyperthermia [110] after SCI can be either therapeutic or harmful, respectively. Should a recovery room be used, cages of mice and rats should be separated from each other and the environmental conditions should be the same as those in the housing rooms.
2. The animals should be watched closely and turned until in sternal recumbency. This may not occur in animals with cervical SCI. Instead, the blink reflex should be monitored. When sternal recumbency or a blink reflex is observed, the animal should be transferred to a home cage prepared as described above. Information about the surgery should be listed on the cage card for identification and for suture or wound clip removal.
3. Importantly, the number of animals in a cage, the cage size, and whether to add objects (tunnels, running wheels, etc.…) to the cage must be carefully considered given the experimental study's goals, the equivocal results for the functional effects of environmental enrichment, and home cage activity's effects on spontaneous recovery [111–121]. Animals that are housed together should be separated when mutilation and extreme domination behaviors occur.
4. We recommend that the soft and moisture absorbent ALPHA-dri® bedding be used. One half of the cage should be on a water-circulating or an electric heating pad so that the animal can move away from it.

Veterinary Care

1. It is our contention that spinal cord injured animals should be treated no differently than spinal cord injured humans. Thus, the animals should undergo physical examination and veterinary care once daily or more times as necessary. The frequency directly depends on the site and severity of SCI. This should also be done for animals that undergo sham methods. An Animal Care Sheet (Table 4) is of utmost benefit after surgery and SCI. It can be used to document the day-to-day health status (activity/lethargy, body weight, hydration, bowel, bladder, incisions, scratches, abscesses, callouses, autophagia, etc.…) and veterinary care that is being provided (food supplements, fluids, bladder expression, antibiotics, analgesics, etc.…) as well as aid the veterinarians when consultations are required.
2. Hydration should be assessed daily in the morning. This can be done as described above and by presenting the spout of a water bottle to the animal. A water bottle with a curved spout should be used after SCI. An additional water bottle can be placed on the bedding. Sterile, core temperature saline solution or lactated Ringers solution (3–5 mL SC for rats and 0.5–1 mL SC for mice) can be injected subcutaneously as described above. The bladder status (see below) should also guide fluid administration.
3. Body weight should be measured daily in the morning. Weight loss after surgery and SCI naturally occurs. Food and cereal can be broken into bite-size pieces and scattered on the bedding. When weight loss persists, poor hydration and/or nutrition may be the causes. In addition to the fluids indicated above, commercially available, core temperature liquid diets and precision pellets can be used (BIO-SERV, Frenchtown, NJ). A mash of food soaked in water can be placed on a paper towel or in glass dishes. Apples and potatoes can also be used for both nutrition and hydration. The mash, fruits, and vegetables should be examined daily and changed when contaminated. Since these will increase the moisture in the covered cage, more frequent cage changes are recommended. Even though they are initially hesistant, we have also found that animals that are severe tetraplegics or lethargic will drink core temperature ENSURE® from a 1 mL syringe or glass dish [109]. Resistance by the animal indicates when this should stop.
4. Bladder function and urine should be evaluated at least once daily in the morning to avoid urinary tract infections, renal complications, and urine scald. Not only do these negatively affect the health of the animal, they also could alter behavior. It is very helpful if a bladder and its location are seen in a euthanized animal before proceeding with this method.

    (a) The bladder should be located and gently palpated with the thumb and fingers while the animal is on the bedding or being handled. Its size and firmness should be recorded as well as whether urine is on the hair and skin. If the bladder contains urine, the animal should be handled over an empty cage. Typically, the fingers and/or thumb of the free hand are used to gently press the bladder ventral-to-dorsal or to gently squeeze it lateral-to-lateral.

Table 4 Animal care sheet

Month/Year:
• Date, AM/PM:
Rat/Mouse ID #:
Surgery/Surgeon's Initials/Surgery Date:
• Activity: None, Hypo, OK
• Forelimb/ Hindlimb Flaccid Paralysis:
• Forelimb/Hindlimb Support:
• Locomotion:
• Nares: Blood, OK
• Ears/Paws: Pale Color, OK
• Eyes: Ptosis, Crusty, OK
• L,R Forepaw Nails or Digits 1 5 Chewing:
• L,R Hindpaw Nails or Digits 1–5 Chewing:
• L,R Hindpaw Callouses:
• L,R Forelimb Hair Loss:
• L,R Hindlimb Hair Loss:
• Body Weight (g):
• Feces: None, Solid, Moist, Dry, Runny
• Bladder: Voiding or S/M/L
• Urine: Clear, Concentrated, Bloody, Cloudy, Crystals, on Abdomen
• Thirsty/Dehydrated:
• Fluids: Dextrose-L. Ringers, L. Ringers, Saline
• Incision: Infected, OK
• Suture/Clips: Off at 7–10 days Post-Op
• Implant: Infected, OK
• Scratching: Location
• Bacitracin: SID, PRN
• Gentozen®: 0.1 mL SC daily × 7–10 days
• Baytril®: 0.1 mL SC daily × 3 days for bloody urine
• Buprenorphine: 0.1 mL SC daily for 48 h
• Comments:

Return the animal to its home cage when the bladder is empty. Tremendous care must be taken, especially in males where there is more resistance to urine flow, to not fatally rupture the bladder. When expression is difficult, the bladder can be massaged, the urethra should be checked for plugs or blood clots, or the animal returned to its home cage before trying again.

(b) Voided or expressed urine should be examined for color, clarity, and odor. It is normally dilute and free of cloudiness, crystals, puss, blood, and a strong odor. Any change of the urine should be aggressively treated. Fluid administration is recommended especially for concentrated urine and the dehydration this suggests. Urine that is cloudy and has crystals or a strong odor can be treated with Gentozin® (50 mg/mL for rats, 0.2 mg/mL for mice, 0.1 mL SC once daily for 7–10 days). Core temperature Clavamox® can also be used for rats (62.5 mg/mL, 0.1 mL orally from a 1 mL syringe for 5 days). Blood may be in the urine initially after SCI. Should this stop then reappear, more gentle and frequent expression should be tried to reduce bladder damage. Baytril® can also be used (22.7 mg/mL for rats, 5 mg/mL for mice, 0.1 mL SC daily for 3 days). Vitamin C tablets can also be administered to acidify the urine. Consult with the veterinarian should these conditions persist beyond a couple of days.

(c) When crystals appear on the hair, treatment should be as described above. Urine scald occurs when urine persistently keeps the hair and skin moist and hindlimb weight support is limited. Moreover, this can lead to lead to a vagina or penis infection. We have found that covering the shaved abdomen with diaper rash ointment or Desitin Ointment (Pfizer Consumer Healthcare, Morris Plains, NJ) is helpful. The affected area should be washed with warm soapy water and thoroughly dried. More frequent bladder expression and bedding changes are also recommended.

5. Bowel function and feces should be examined at least once daily in the morning. Feces should be solid and moist or dry. While infrequent, no new feces suggest constipation. Abdomen palpation with the fingers and hydration should ensue. The anus should be checked for plugs. We have found that a core temperature mixture of sterile water and Epsom Salts (2 cups/gallon of water, Cumberland Swan, Smyrna, TN) administered orally from a 1 mL syringe to be useful. Very moist feces indicate diarrhea and that fluid intake and bladder status should be managed. Discolored feces typically are due to nutritional supplements. Blood may infrequently be seen on the feces. Consult with the veterinarians should these conditions persist beyond a couple of days.
6. Antibiotics should be delivered, but cautiously, during this time. The dosage and administration days depend on which antibiotics are being used. For example, Gentozin® (50 mg/mL for rats, 0.2 mg/mL for mice, 0.1 mL SC) can be delivered every other day for 5 days. Diarrhea can occur due to changes of intestinal flora. Antibiotic ointment can also be applied daily to skin incisions.
7. Analgesics can be delivered at this time if necessary and appropriate [122]. We administer the mixed opioid agonist-antagonist buprenorphine (0.05 mg/mL for rats, 0.015 mg/mL for mice, 0.1 mL SC) 12–24 h after surgery, so as to not fatally depress the respiratory rate, and as needed.
8. If skin incisions are healing, wound clips and sutures should be removed at 7–10 days. Afterwards, they can be a source of irritation, scratching, and infection. This can be done without anesthesia in gentled animals or with gas anesthesia.
9. Issues that also may arise during this time include excessive grooming, scratches, wounds, and autophagia. Bitter ChewGuard (Summit Hill Laboratories, Navesink, NJ) can be sprayed when excessive grooming occurs before the skin is opened. Gumabones (BIO-SERV) can also be an effective deterrent. Scratches and superficial wounds can be cleansed with 3% hydrogen peroxide and thoroughly rinsed with sterile water. New-Skin® liquid bandage (Medtech, Jackson, WY) can be applied to the treated scratch or wound. This approach can also be

used for callouses that are infrequently found on the hindpaw heels. We have found that topical application of New-Skin® and metronidazol is effective for preventing and abolishing autophagia [123]. Clean bedding and applying antibiotic ointment then wrapping with gauze, Vetrap™ (3M, St. Paul, MN), and tape are approaches that we have found to be effective for deep wounds caused by autophagia. The bandages and bedding should be changed frequently. Digits and paws swelling is also an infrequent occurrence. They should be soaked twice a day for 10–15 min in a mixture of warm water and Epsom Salts until the swelling subsides. Ulcers at injection sites or from pressure at bony prominences and abscesses at incision sites are infrequently seen. Animals may contort their body due to them. Callouses can be infrequently found on the hindpaws. In consultation with the veterinarians, antibiotics, such as Penicillin G Procaine (300,000 units/mL, 0.1 mL SC daily for 3–5 days) and analgesics, such as buprenorphine, may need to be administered when these and the other issues arise. When severe, euthanasia should be performed, for example with Fatal-Plus® (0.88 mL/kg, IP [124]; Vortech Pharmaceuticals Ltd., Dearborn, MI).

Chronically After Surgery and SCI

1. The cages should be removed from the water-circulating or electric heating pads when the animals spend little time over them. This typically happens within the first week post-surgery or -SCI. Additionally, the core temperature can be checked. The pads can be removed when the animals are normothermic. Should an animal be found to be cold to the touch at anytime or shivering, the pads should be used.
2. The physical examination and all of the veterinary care methods described above should continue to be used daily. Particular attention should be paid to hydration and nutrition, bladder and bowel function, and skin condition. Even though reflexive voiding may occur, it is recommended that the decision to stop expressing the bladder be very conservative.
3. Physical therapy is used for persons with SCI [125]. Careful consideration should be given to using physical therapy approaches for experimental SCI. For example, chronic daily hindlimb stretching after experimental SCI impaired locomotion recovery and knee joint range of motion [126].

Acknowledgements C.D.Y. was supported by NIH/NIGMS/P30 GM103507 (S.R. Whittemore, Kentucky Spinal Cord Injury Research Center, University of Louisville). Sincere appreciation is extended to Dr. Russell G. Durkovic, Deyanira Santiago, Dr. Frances M. Doepel, Dr. William W. King, Dr. Jeffrey Smiley, the MASCIS Group of Investigators (*Unpublished Document*), as well as the faculty, veterinarians, and their technical staffs at Ohio State University (Basso DM, Beattie MS, Bergdall V, Bresnahan JC, Buford JA, Jakeman LB, Kostyk SK, McTigue DM, and Popovich, PG, *Unpublished Document*, Spinal Cord Injury Training Manual, NINDS, Facilities of Research Excellence in Spinal Cord Injury, Copyright Ohio State University, 2004–2008) for their guidance in providing the highest quality of veterinary care to spinal cord injured animals.

References

1. Lee BB, Cripps RA, Fitzharris M, Wing PC. The global map for traumatic spinal cord epidemiology: update 2011, global incidence rate. Spinal Cord. 2014;52:110–6.
2. https://clinicaltrials.gov/ct2/results?term=spinal+cord+injury&Search=Search.
3. Alilain WJ, Horn KP, Hu H, Dick TE, Silver J. Functional regeneration of respiratory pathways after spinal cord injury. Nature. 2011;475:196–200.
4. Harrop JS, Hashimoto R, Norvell D, Raich A, Aarabi B, Grossman RG, Guest JD, Tator CH, Chapman J, Fehlings MG. Evaluation of clinical experience using cell-based therapies in patients with spinal cord injury: a systematic review. J Neurosurg Spine. 2012;17:230–46.
5. Steward O, Popovich PG, Dietrich WD, Kleitman N. Replication and reproducibility in spinal cord injury research. Exp Neurol. 2012;233:597–605.
6. Lee YS, Lin CY, Jiang HH, Depaul M, Lin VW, Silver J. Nerve regeneration restores supraspinal control of bladder function after complete spinal cord injury. J Neurosci. 2013;33:10591–606.
7. Sharp KG, Yee KM, Stiles TL, Aguilar RM, Steward O. A re-assessment of the effects of treatment with a non-steroidal anti-inflammatory (ibuprofen) on promoting axon regeneration via RhoA inhibition after spinal cord injury. Exp Neurol. 2013;248:321–37.
8. Angeli CA, Edgerton VR, Gerasimenko YP, Harkema SJ. Altering spinal cord excitability enables voluntary movements after chronic complete paralysis in humans. Brain. 2014;137:1394–409.
9. Collinger JL, Kryger MA, Barbara R, Betler T, Bowsher K, Brown EH, Clanton ST, Degenhart AD, Foldes ST, Gaunt RA, Gyulai FE, Harchick EA, Harrington D, Helder JB, Hemmes T, Johannes MS, Katyal KD, Ling GS, McMorland AJ, Palko K, Para MP, Scheuermann J, Schwartz AB, Skidmore ER, Solzbacher F, Srikameswaran AV, Swanson DP, Swetz S, Tyler-Kabara EC, Velliste M, Wang W, Weber DJ, Wodlinger B, Boninger ML. Collaborative approach in the development of high-performance brain-computer interfaces for a neuroprosthetic arm: translation from animal models to human control. Clin Transl Sci. 2014;7:52–9.
10. Hammond ER, Metcalf HM, McDonald JW, Sadowsky CL. Bone mass in individuals with chronic spinal cord injury: associations with activity-based therapy, neurologic and functional status, a retrospective study. Arch Phys Med Rehabil. 2014;95:2342–9.
11. Kressler J, Thomas CK, Field-Fote EC, Sanchez J, Widerström-Noga E, Cilien DC, Gant K, Ginnety K, Gonzalez H, Martinez A, Anderson KD, Nash MS. Understanding therapeutic benefits of overground bionic ambulation: exploratory case series in persons with chronic, complete spinal cord injury. Arch Phys Med Rehabil. 2014;95:1878–87.
12. Moreno-Duarte I, Morse LR, Alam M, Bikson M, Zafonte R, Fregni F. Targeted therapies using electrical and magnetic neural stimulation for the treatment of chronic pain in spinal cord injury. Neuroimage. 2014;85:1003–13.
13. Plemel JR, Keough MB, Duncan GJ, Sparling JS, Yong VW, Stys PK, Tetzlaff W. Remyelination after spinal cord injury: is it a target for repair? Prog Neurobiol. 2014;117:54–72.
14. Popovich PG, Tovar CA, Lemeshow S, Yin Q, Jakeman LB. Independent evaluation of the anatomical and behavioral effects of Taxol in rat models of spinal cord injury. Exp Neurol. 2014;261:97–108.
15. Ramer LM, Ramer MS, Bradbury EJ. Restoring function after spinal cord injury: towards clinical translation of experimental strategies. Lancet Neurol. 2014;13:1241–56.
16. Sharp KG, Yee KM, Steward O. A re-assessment of treatment with a tyrosine kinase inhibitor (imatinib) on tissue sparing and functional recovery after spinal cord injury. Exp Neurol. 2014;254:1–11.
17. Sharp KG, Yee KM, Steward O. A re-assessment of long distance growth and connectivity of neural stem cells after severe spinal cord injury. Exp Neurol. 2014;257:186–204.
18. Ward PJ, Herrity AN, Smith RR, Willhite A, Harrison BJ, Petruska JC, Harkema SJ, Hubscher CH. Novel multi-system functional gains via task specific training in spinal cord injured male rats. J Neurotrauma. 2014;31:819–33.

19. Aslan SC, Randall DC, Krassioukov AV, Phillips A, Ovechkin AV. Resistive respiratory training improves blood pressure regulation in individuals with chronic spinal cord injury. Arch Phys Med Rehabil. 2016;97(6):964–73.
20. Bonner JF, Steward O. Repair of spinal cord injury with neuronal relays: from fetal grafts to neural stem cells. Brain Res. 2015;1619:115–23.
21. Dietrich WD. Protection and repair after spinal cord injury: accomplishments and future directions. Top Spinal Cord Inj Rehabil. 2015;21:174–87.
22. Dugan EA, Sagen J. An intensive locomotor training paradigm improves neuropathic pain following spinal cord compression injury in rats. J Neurotrauma. 2015;32:622–32.
23. Dvorak MF, Noonan VK, Fallah N, Fisher CG, Finkelstein J, Kwon BK, Rivers CS, Ahn H, Paquet J, Tsai EC, Townson A, Attabib N, Bailey CS, Christie SD, Drew B, Fourney DR, Fox R, Hurlbert RJ, Johnson MG, Linassi AG, Parent S, Fehlings MG, RHSCIR Network. The influence of time from injury to surgery on motor recovery and length of hospital stay in acute traumatic spinal cord injury: an observational Canadian cohort study. J Neurotrauma. 2015;32:645–54.
24. Evaniew N, Belley-Côté EP, Fallah N, Noonan VK, Rivers CS, Dvorak MF. Methylprednisolone for the treatment of patients with acute spinal cord injuries: a systematic review and meta-analysis. J Neurotrauma. 2016;33(5):468–81.
25. Hosier H, Peterson D, Tsymbalyuk O, Keledjian K, Smith BR, Ivanova S, Gerzanich V, Popovich PG, Simard JM. A direct comparison of three clinically relevant treatments in a rat model of cervical spinal cord injury. J Neurotrauma. 2015;32:1633–44.
26. Lang BT, Cregg JM, DePaul MA, Tran AP, Xu K, Dyck SM, Madalena KM, Brown BP, Weng YL, Li S, Karimi-Abdolrezaee S, Busch SA, Shen Y, Silver J. Modulation of the proteoglycan receptor PTPσ promotes recovery after spinal cord injury. Nature. 2015;518:404–8.
27. Nagoshi N, Fehlings MG. Investigational drugs for the treatment of spinal cord injury: review of preclinical studies and evaluation of clinical trials from Phase I to II. Expert Opin Investig Drugs. 2015;24:645–58.
28. Ruschel J, Hellal F, Flynn KC, Dupraz S, Elliott DA, Tedeschi A, Bates M, Sliwinski C, Brook G, Dobrindt K, Peitz M, Brüstle O, Norenberg MD, Blesch A, Weidner N, Bunge MB, Bixby JL, Bradke F. Axonal regeneration. Systemic administration of epothilone B promotes axon regeneration after spinal cord injury. Science. 2015;348:347–52.
29. Sandrow-Feinberg HR, Houlé JD. Exercise after spinal cord injury as an agent for neuroprotection, regeneration and rehabilitation. Brain Res. 2015;1619:12–21.
30. Siddiqui AM, Khazaei M, Fehlings MG. Translating mechanisms of neuroprotection, regeneration, and repair to treatment of spinal cord injury. Prog Brain Res. 2015;218:15–54.
31. Squair JW, West CR, Krassioukov AV. Neuroprotection, plasticity manipulation, and regenerative strategies to improve cardiovascular function following spinal cord injury. J Neurotrauma. 2015;32:609–21.
32. Tazoe T, Perez MA. Effects of repetitive transcranial magnetic stimulation on recovery of function after spinal cord injury. Arch Phys Med Rehabil. 2015;96:S145–55.
33. Wang H, Liu NK, Zhang YP, Deng L, Lu QB, Shields CB, Walker MJ, Li J, Xu XM. Treadmill training induced lumbar motoneuron dendritic plasticity and behavior recovery in adult rats after a thoracic contusive spinal cord injury. Exp Neurol. 2015;271:368–78.
34. Wang J, Pearse DD. Therapeutic hypothermia in spinal cord injury: the status of its use and open questions. Int J Mol Sci. 2015;16:16848–79.
35. West CR, Crawford MA, Laher I, Ramer MS, Krassioukov AV. Passive hind-limb cycling reduces the severity of autonomic dysreflexia after experimental spinal cord injury. Neurorehabil Neural Repair. 2016;30(4):317–27. pii:1545968315593807.
36. Ropper AE, Neal MT, Theodore N. Acute management of traumatic cervical spinal cord injury. Pract Neurol. 2015;15:266–72.
37. Bunge RP, Puckett WR, Becerra JL, Marcillo A, Quencer RM. Observations on the pathology of human spinal cord injury. A review and classification of 22 new cases with details from a case of chronic cord compression with extensive focal demyelination. Adv Neurol. 1993;59:75–89.

38. Bunge RP, Puckett WR, Hiester ED. Observations on the pathology of several types of human spinal cord injury, with emphasis on the astrocyte response to penetrating injuries. Adv Neurol. 1997;72:305–15.
39. Norenberg MD, Smith J, Marcillo A. The pathology of human spinal cord injury: defining the problems. J Neurotrauma. 2004;21:429–40.
40. Côté MP, Hanna A, Lemay MA, Ollivier-Lanvin K, Santi L, Miller K, Monaghan R, Houlé JD. Peripheral nerve grafts after cervical spinal cord injury in adult cats. Exp Neurol. 2010;225:173–82.
41. Nout YS, Rosenzweig ES, Brock JH, Strand SC, Moseanko R, Hawbecker S, Zdunowski S, Nielson JL, Roy RR, Courtine G, Ferguson AR, Edgerton VR, Beattie MS, Bresnahan JC, Tuszynski MH. Animal models of neurologic disorders: a nonhuman primate model of spinal cord injury. Neurotherapeutics. 2012;9:380–92.
42. Darian-Smith C, Lilak A, Garner J, Irvine KA. Corticospinal sprouting occurs selectively following dorsal rhizotomy in the macaque monkey. J Comp Neurol. 2013;521:2359–72.
43. Lee JH, Jones CF, Okon EB, Anderson L, Tigchelaar S, Kooner P, Godbey T, Chua B, Gray G, Hildebrandt R, Cripton P, Tetzlaff W, Kwon BK. A novel porcine model of traumatic thoracic spinal cord injury. J Neurotrauma. 2013;30:142–59.
44. Wu W, Wu W, Zou J, Shi F, Yang S, Liu Y, Lu P, Ma Z, Zhu H, Xu XM. Axonal and glial responses to a mid-thoracic spinal cord hemisection in the Macaca fascicularis monkey. J Neurotrauma. 2013;30:826–39.
45. Qi HX, Kaas JH, Reed JL. The reactivation of somatosensory cortex and behavioral recovery after sensory loss in mature primates. Front Syst Neurosci. 2014;8:84.
46. Saunders NR, Noor NM, Dziegielewska KM, Wheaton BJ, Liddelow SA, Steer DL, Ek CJ, Habgood MD, Wakefield MJ, Lindsay H, Truettner J, Miller RD, Smith AI, Dietrich WD. Age-dependent transcriptome and proteome following transection of neonatal spinal cord of Monodelphis domestica (South American grey short-tailed opossum). PLoS One. 2014;9:e99080.
47. Yates JR, Gay EA, Heyes MP, Blight AR. Effects of methylprednisolone and 4-chloro-3-hydroxyanthranilic acid in experimental spinal cord injury in the guinea pig appear to be mediated by different and potentially complementary mechanisms. Spinal Cord. 2014;52:662–6.
48. Kwon BK, Streijger F, Hill CE, Anderson AJ, Bacon M, Beattie MS, Blesch A, Bradbury EJ, Brown A, Bresnahan JC, Case CC, Colburn RW, David S, Fawcett JW, Ferguson AR, Fischer I, Floyd CL, Gensel JC, Houle JD, Jakeman LB, Jeffery ND, Jones LA, Kleitman N, Kocsis J, Lu P, Magnuson DS, Marsala M, Moore SW, Mothe AJ, Oudega M, Plant GW, Rabchevsky AS, Schwab JM, Silver J, Steward O, Xu XM, Guest JD, Tetzlaff W. Large animal and primate models of spinal cord injury for the testing of novel therapies. Exp Neurol. 2015;269:154–68.
49. McMahill BG, Borjesson DL, Sieber-Blum M, Nolta JA, Sturges BK. Stem cells in canine spinal cord injury—promise for regenerative therapy in a large animal model of human disease. Stem Cell Rev. 2015;11:180–93.
50. Rossignol S, Martinez M, Escalona M, Kundu A, Delivet-Mongrain H, Alluin O, Gossard JP. The "beneficial" effects of locomotor training after various types of spinal lesions in cats and rats. Prog Brain Res. 2015;218:173–98.
51. Mondello SE, Jefferson SC, Tester NJ, Howland DR. Impact of treatment duration and lesion size on effectiveness of chondroitinase treatment post-SCI. Exp Neurol. 2015;267:64–77.
52. Metz GA, Curt A, van de Meent H, Klusman I, Schwab ME, Dietz V. Validation of the weight-drop contusion model in rats: a comparative study of human spinal cord injury. J Neurotrauma. 2000;17:1–17.
53. Onifer SM, Rabchevsky AG, Scheff SW. Rat models of traumatic spinal cord injury to assess motor recovery. ILAR J. 2007;48:385–95.
54. Kastner A, Gauthier P. Are rodents an appropriate pre-clinical model for treating spinal cord injury? Examples from the respiratory system. Exp Neurol. 2008;213:249–56.
55. Lee JK, Zheng B. Axon regeneration after spinal cord injury: insight from genetically modified mouse models. Restor Neurol Neurosci. 2008;26:175–82.

56. Mihai G, Nout YS, Tovar CA, Miller BA, Schmalbrock P, Bresnahan JC, Beattie MS. Longitudinal comparison of two severities of unilateral cervical spinal cord injury using magnetic resonance imaging in rats. J Neurotrauma. 2008;25:1–18.
57. Plemel JR, Duncan G, Chen KW, Shannon C, Park S, Sparling JS, Tetzlaff W. A graded forceps crush spinal cord injury model in mice. J Neurotrauma. 2008;25:350–70.
58. Zhang YP, Burke DA, Shields LB, Chekmenev SY, Dincman T, Zhang Y, Zheng Y, Smith RR, Benton RL, DeVries WH, Hu X, Magnuson DS, Whittemore SR, Shields CB. Spinal cord contusion based on precise vertebral stabilization and tissue displacement measured by combined assessment to discriminate small functional differences. J Neurotrauma. 2008;25:1227–40.
59. Anderson KD, Sharp KG, Steward O. Bilateral cervical contusion spinal cord injury in rats. Exp Neurol. 2009;220:9–22.
60. Beare JE, Morehouse JR, DeVries WH, Enzmann GU, Burke DA, Magnuson DS, Whittemore SR. Gait analysis in normal and spinal contused mice using the TreadScan system. J Neurotrauma. 2009;26:2045–56.
61. Hill RL, Zhang YP, Burke DA, Devries WH, Zhang Y, Magnuson DS, Whittemore SR, Shields CB. Anatomical and functional outcomes following a precise, graded, dorsal laceration spinal cord injury in C57BL/6 mice. J Neurotrauma. 2009;26:1–15.
62. Kim JH, Tu TW, Bayly PV, Song SK. Impact speed does not determine severity of spinal cord injury in mice with fixed impact displacement. J Neurotrauma. 2009;26:1395–404.
63. Kouyoumdjian P, Lonjon N, Prieto M, Haton H, Privat A, Asencio G, Perrin FE, Gaviria M. A remotely controlled model of spinal cord compression injury in mice: toward real-time analysis. J Neurosurg Spine. 2009;11:461–70.
64. Marques SA, Garcez VF, Del Bel EA, Martinez AM. A simple, inexpensive and easily reproducible model of spinal cord injury in mice: morphological and functional assessment. J Neurosci Methods. 2009;177:183–93.
65. Steuer I, Guertin PA. Spinal cord injury research in mice: 2008 review. ScientificWorldJournal. 2009;9:490–8.
66. Aguilar RM, Steward O. A bilateral cervical contusion injury model in mice: assessment of gripping strength as a measure of forelimb motor function. Exp Neurol. 2010;221:38–53.
67. Byrnes KR, Fricke ST, Faden AI. Neuropathological differences between rats and mice after spinal cord injury. J Magn Reson Imaging. 2010;32:836–46.
68. Dunham KA, Siriphorn A, Chompoopong S, Floyd CL. Characterization of a graded cervical hemicontusion spinal cord injury model in adult male rats. J Neurotrauma. 2010;27:2091–106.
69. Pajoohesh-Ganji A, Byrnes KR, Fatemi G, Faden AI. A combined scoring method to assess behavioral recovery after mouse spinal cord injury. Neurosci Res. 2010;67:117–25.
70. Lee JH, Streijger F, Tigchelaar S, Maloon M, Liu J, Tetzlaff W, Kwon BK. A contusive model of unilateral cervical spinal cord injury using the infinite horizon impactor. J Vis Exp. 2012;(65)
71. Marcol W, Slusarczyk W, Gzik M, Larysz-Brysz M, Bobrowski M, Grynkiewicz-Bylina B, Rosicka P, Kalita K, Węglarz W, Barski JJ, Kotulska K, Labuzek K, Lewin-Kowalik J. Air gun impactor—a novel model of graded white matter spinal cord injury in rodents. J Reconstr Microsurg. 2012;28:561–8.
72. Sharp K, Boroujerdi A, Steward O, Luo ZD. A rat chronic pain model of spinal cord contusion injury. Methods Mol Biol. 2012;851:195–203.
73. Hilton BJ, Assinck P, Duncan GJ, Lu D, Lo S, Tetzlaff W. Dorsolateral funiculus lesioning of the mouse cervical spinal cord at C4 but not at C6 results in sustained forelimb motor deficits. J Neurotrauma. 2013;30:1070–83.
74. Inoue T, Lin A, Ma X, McKenna SL, Creasey GH, Manley GT, Ferguson AR, Bresnahan JC, Beattie MS. Combined SCI and TBI: recovery of forelimb function after unilateral cervical spinal cord injury (SCI) is retarded by contralateral traumatic brain injury (TBI), and ipsilateral TBI balances the effects of SCI on paw placement. Exp Neurol. 2013;248:136–47.
75. Krishna V, Andrews H, Jin X, Yu J, Varma A, Wen X, Kindy M. A contusion model of severe spinal cord injury in rats. J Vis Exp. 2013;(78)

76. Kuypers NJ, James KT, Enzmann GU, Magnuson DS, Whittemore SR. Functional consequences of ethidium bromide demyelination of the mouse ventral spinal cord. Exp Neurol. 2013;247:615–22.
77. Nichols NL, Punzo AM, Duncan ID, Mitchell GS, Johnson RA. Cervical spinal demyelination with ethidium bromide impairs respiratory (phrenic) activity and forelimb motor behavior in rats. Neuroscience. 2013;229:77–87.
78. Rouleau P, Guertin PA. A valuable animal model of spinal cord injury to study motor dysfunctions, comorbid conditions, and aging associated diseases. Curr Pharm Des. 2013;19:4437–47.
79. Streijger F, Beernink TM, Lee JH, Bhatnagar T, Park S, Kwon BK, Tetzlaff W. Characterization of a cervical spinal cord hemicontusion injury in mice using the infinite horizon impactor. J Neurotrauma. 2013;30:869–83.
80. Vierck CJ, Cannon RL, Acosta-Rua AJ. Evaluation of lateral spinal hemisection as a preclinical model of spinal cord injury pain. Exp Brain Res. 2013;228:305–12.
81. Zhang YP, Walker MJ, Shields LB, Wang X, Walker CL, Xu XM, Shields CB. Controlled cervical laceration injury in mice. J Vis Exp. 2013;(75):e50030.
82. Cheriyan T, Ryan DJ, Weinreb JH, Cheriyan J, Paul JC, Lafage V, Kirsch T, Errico TJ. Spinal cord injury models: a review. Spinal Cord. 2014;52:588–95.
83. Forgione N, Karadimas SK, Foltz WD, Satkunendrarajah K, Lip A, Fehlings MG. Bilateral contusion-compression model of incomplete traumatic cervical spinal cord injury. J Neurotrauma. 2014;31:1776–88.
84. Jin Y, Bouyer J, Haas C, Fischer I. Behavioral and anatomical consequences of repetitive mild thoracic spinal cord contusion injury in the rat. Exp Neurol. 2014;257:57–69.
85. Keomani E, Deramaudt TB, Petitjean M, Bonay M, Lofaso F, Vinit S. A murine model of cervical spinal cord injury to study post-lesional respiratory neuroplasticity. J Vis Exp. 2014;(87)
86. Luedtke K, Bouchard SM, Woller SA, Funk MK, Aceves M, Hook MA. Assessment of depression in a rodent model of spinal cord injury. J Neurotrauma. 2014;31:1107–21.
87. Marques SA, de Almeida FM, Mostacada K, Martinez AM. A highly reproducible mouse model of compression spinal cord injury. Methods Mol Biol. 2014;1162:149–56.
88. Oakden W, Kwiecien JM, O'Reilly MA, Lake EM, Akens MK, Aubert I, Whyne C, Finkelstein J, Hynynen K, Stanisz GJ. A non-surgical model of cervical spinal cord injury induced with focused ultrasound and microbubbles. J Neurosci Methods. 2014;235:92–100.
89. Okada SL, Stivers NS, Stys PK, Stirling DP. An ex vivo laser-induced spinal cord injury model to assess mechanisms of axonal degeneration in real-time. J Vis Exp. 2014;(93):e52173.
90. Sahinkaya FR, Milich LM, McTigue DM. Changes in NG2 cells and oligodendrocytes in a new model of intraspinal hemorrhage. Exp Neurol. 2014;255:113–26.
91. Surey S, Berry M, Logan A, Bicknell R, Ahmed Z. Differential cavitation, angiogenesis and wound-healing responses in injured mouse and rat spinal cords. Neuroscience. 2014;275:62–80.
92. Watson JL, Hala TJ, Putatunda R, Sannie D, Lepore AC. Persistent at-level thermal hyperalgesia and tactile allodynia accompany chronic neuronal and astrocyte activation in superficial dorsal horn following mouse cervical contusion spinal cord injury. PLoS One. 2014;9:e109099.
93. Carpenter RS, Kigerl KA, Marbourg JM, Gaudet AD, Huey D, Niewiesk S, Popovich PG. Traumatic spinal cord injury in mice with human immune systems. Exp Neurol. 2015;271:432–44.
94. del Mar N, von Buttlar X, Yu AS, Guley NH, Reiner A, Honig MG. A novel closed-body model of spinal cord injury caused by high-pressure air blasts produces extensive axonal injury and motor impairments. Exp Neurol. 2015;271:53–71.
95. McDonough A, Monterrubio A, Ariza J, Martínez-Cerdeño V. Calibrated forceps model of spinal cord compression injury. J Vis Exp. 2015;(98)
96. Mondello SE, Sunshine MD, Fischedick AE, Moritz CT, Horner PJ. A cervical hemicontusion spinal cord injury model for the investigation of novel therapeutics targeting proximal and distal forelimb functional recovery. J Neurotrauma. 2015;32:1994–2007.

97. Moonen G, Satkunendrarajah K, Wilcox JT, Badner A, Mothe A, Foltz W, Fehlings MG, Tator CH. A new acute impact-compression lumbar spinal cord injury model in the rodent. J Neurotrauma. 2016;33(3):278–89.
98. Ropper AE, Zeng X, Anderson JE, Yu D, Han I, Haragopal H, Teng YD. An efficient device to experimentally model compression injury of mammalian spinal cord. Exp Neurol. 2015;271:515–23.
99. Vierck CJ, Baastrup C, Maersk-Moller C, Roth M, Cannon RL, Finnerup NB, Yezierski RP. A preclinical model of hyperalgesia following spinal stenosis/compression. Eur J Pain. 2015;19:1158–67.
100. Wen J, Sun D, Tan J, Young W. A consistent, quantifiable, and graded rat lumbosacral spinal cord injury model. J Neurotrauma. 2015;32:875–92.
101. Bhatnagar T, Liu J, Yung A, Cripton PA, Kozlowski P, Tetzlaff W, Oxland TR. Relating histopathology and mechanical strain in experimental contusion spinal cord injury in a rat model. J Neurotrauma 2016; 33(18):1685-1695.
102. Santos-Benito FF, Munoz-Quiles C, Ramon-Cueto A. Long-term care of paraplegic laboratory mammals. J Neurotrauma. 2006;23:521–36.
103. Puckett AH, Nunn CD, Onifer SM. Veterinary care methods for rats and mice in experimental spinal cord injury studies. In: Chen J, Xu Z, Xu X-M, Zhang J, editors. Animal models of acute neurological injuries, Contemporary neuroscience. Totowa, NJ: The Humana Press Inc.; 2008. p. 47–60.
104. Ramsey JB, Ramer LM, Inskip JA, Alan N, Ramer MS, Krassioukov AV. Care of rats with complete high-thoracic spinal cord injury. J Neurotrauma. 2010;27:1709–22.
105. Laboratory animal technician training manual. Animal health maintenance. Memphis, TN: Sheridan Books. 2005. p. 88–9.
106. National Research Council. Guide for the care and use of laboratory animals. 8th ed. Washington, DC: National Academy Press, 2011. p. 55–8
107. Onifer SM, Rodríquez JF, Santiago DI, Benitez JC, Kim DT, Brunschwig J-P, Pacheco JT, Perrone JV, Llorente O, Martinez-Arizala A. Cervical spinal cord injury in the adult rat: Assessment of forelimb dysfunction. Restor Neurol Neurosci. 1997;11:211–23.
108. Titsworth WL, Onifer SM, Liu N, Xu X-M. Focal phospholipases A2 group III injections induce cervical white matter injury and functional deficits with delayed recovery concomitant with Schwann cell remyelination. Exp Neurol. 2007;207:150–62.
109. Onifer SM, Nunn CD, Decker JA, Payne BN, Wagoner MR, Puckett AH, Massey JM, Armstrong J, Kaddumi EG, Fentress KG, Wells MJ, Calloway CC, Schnell JT, Whitaker CM, Burke DA, Hubscher CH. Loss and spontaneous recovery of forelimb evoked potentials in both the adult rat cuneate nucleus and somatosensory cortex following contusive cervical spinal cord injury. Exp Neurol. 2007;207:238–47.
110. Yu CG, Jagid J, Ruenes G, Dietrich WD, Marcillo AE, Yezierski RP. Detrimental effects of systemic hyperthermia on locomotor function and histopathological outcome after traumatic spinal cord injury in the rat. Neurosurgery. 2001;49:152–8.
111. Lankhorst AJ, ter Laak MP, van Laar TJ, van Meeteren NL, de Groot JC, Schrama LH, Hamers FP, Gispen WH. Effects of enriched housing on functional recovery after spinal cord contusive injury in the adult rat. J Neurotrauma. 2001;18:203–15.
112. Woerly S, Doan VD, Evans-Martin F, Paramore CG, Peduzzi JD. Spinal cord reconstruction using NeuroGel implants and functional recovery after chronic injury. J Neurosci Res. 2001;66:1187–97.
113. Van Meeteren NL, Eggers R, Lankhorst AJ, Gispen WH, Hamers FP. Locomotor recovery after spinal cord contusion injury in rats is improved by spontaneous exercise. J Neurotrauma. 2003;20:1029–37.
114. Koopmans GC, Deumens R, Honig WM, Hamers FP, Steinbusch HW, Joosten EA. The assessment of locomotor function in spinal cord injured rats: the importance of objective analysis of coordination. J Neurotrauma. 2005;22:214–25.
115. Erschbamer MK, Pham TM, Zwart MC, Baumans V, Olson L. Neither environmental enrichment nor voluntary wheel running enhances recovery from incomplete spinal cord injury in rats. Exp Neurol. 2006;201:154–64.

116. Moon LD, Leasure JL, Gage FH, Bunge MB. Motor enrichment sustains hindlimb movement recovered after spinal cord injury and glial transplantation. Restor Neurol Neurosci. 2006;24:147–61.
117. Berrocal Y, Pearse DD, Singh A, Andrade CM, McBroom JS, Puentes R, Eaton MJ. Social and environmental enrichment improves sensory and motor recovery after severe contusive spinal cord injury in the rat. J Neurotrauma. 2007;24:1761–72.
118. Burke DA, Magnuson DS, Nunn CD, Fentress KG, Wilson ML, Shum-Siu AH, Moore MC, Turner LE, King WW, Onifer SM. Use of environmentally enriched housing for rats with spinal cord injury: the need for standardization. J Am Assoc Lab Anim Sci. 2007;46:34–41.
119. Fischer FR, Peduzzi JD. Functional recovery in rats with chronic spinal cord injuries after exposure to an enriched environment. J Spinal Cord Med. 2007;30:147–55.
120. Koopmans GC, Deumens R, Honig WM, Hamers FP, Mey J, van Kleef M, Joosten EA. Functional recovery, serotonergic sprouting, and endogenous progenitor fates in response to delayed environmental enrichment after spinal cord injury. J Neurotrauma. 2012;29:514–27.
121. Fouad K, Hurd C, Magnuson DS. Functional testing in animal models of spinal cord injury: not as straight forward as one would think. Front Integr Neurosci. 2013;7:85.
122. Santiago JM, Rosas O, Torrado AI, González MM, Kalyan-Masih PO, Miranda JD. Molecular, anatomical, physiological, and behavioral studies of rats with buprenorphine after spinal cord injury. J Neurotrauma. 2009;26:1783–93.
123. Zhang YP, Onifer SM, Burke DA, Shields CB. A topical mixture for preventing, abolishing, and treating autophagia and self-mutilation in laboratory rats. Contemp Top Lab Anim Sci. 2001;40:35–6.
124. Onifer SM, Quintero JE, Gerhardt GA. Cutaneous and electrically evoked glutamate signaling in the adult rat somatosensory system. J Neurosci Methods. 2012;208:146–54.
125. Gómara-Toldrà N, Sliwinski M, Dijkers MP. Physical therapy after spinal cord injury: a systematic review of treatments focused on participation. J Spinal Cord Med. 2014;37:371–9.
126. Caudle KL, Atkinson DA, Brown EH, Donaldson K, Seibt E, Chea T, Smith E, Chung K, Shum-Siu A, Cron CC, Magnuson DS. Hindlimb stretching alters locomotor function after spinal cord injury in the adult rat. Neurorehabil Neural Repair. 2015;29:268–77.

Part II
Cerebral Ischemic Animal Models

Four-Vessel Occlusion Model in Rats

Ping Deng and Zao C. Xu

Abstract Four-vessel occlusion model in rat induces transient forebrain ischemia that resembles cardiac arrest in clinical situations. In this model, two vertebral arteries are coagulated and two common carotid arteries are reversibly occluded. This model produces reliable outcome with selective, delayed cell death. For example, CA1 pyramidal neurons in the hippocampus die 2–3 days after ischemia whereas CA3 neurons survive the same insult. The complications of this model include mortality and seizure. The methods of quantitative analysis of ischemic outcome are introduced. The advantages and limitations of this model are discussed.

Keywords Wistar rat · Transient global ischemia · Hippocampus · Striatum · Ischemic depolarization

Model Selection

Four-vessel occlusion (4-VO) rat model has been widely employed to investigate the mechanisms underlying brain damage after transient global ischemia, which mimics the pathological conditions of cardiac arrest. 4-VO is produced by permanent coagulation of two vertebral arteries and temporary ligation of two common carotid arteries. This model induces transient forebrain ischemia because the blood supply to the hindbrain remains relatively intact [1]. Such insult causes severe, reversible forebrain ischemia with consistent neuronal damage in a species that is large enough to allow physiological monitoring and maintenance with relative low costs. The original model involves a 2-day procedure inducing an ischemic insult in awaked animals. Several modified 4-VO models have been developed in the following decades. For example, to perform the 2-day procedure in anesthetized animals [2]; to perform

P. Deng · Z. C. Xu (✉)
Department of Anatomy & Cell Biology, Indiana University School of Medicine, Indianapolis, IN, USA
e-mail: pideng@iupui.edu; zcxu@iupui.edu

J. Chen et al. (eds.), *Animal Models of Acute Neurological Injury*, Springer Series in Translational Stroke Research, https://doi.org/10.1007/978-3-030-16082-1_6

all procedures on the same day in anesthetized rats with maintaining brain temperature and recording of ischemic depolarization (ID) [3]; to coagulate vertebral arteries from ventral approach [4] etc.

Materials

Animal

The choice of animal for 4-VO model is male adult Wistar rats. The strain is an important factor affecting the success rate. 4-VO induces severe ischemia in ~80% of Wistar rats [1] whereas only 50–60% of Sprague-Dawley rats exhibit successful ischemia [5]. It is worthwhile to point out that even Wistar rats from different suppliers may exhibit different outcomes. It is therefore necessary to purchase rats with high success rate from the same supplier. The age of the animal is another factor influences the outcome. Neuronal sensitivity to ischemia significantly decreases in neonatal rats due to the developmental nature. The success rate of complete ischemia is significantly reduced in aged rats due to the increase of blood vessel collaterals to the brain. The rats with 200–300 mg in body weight have higher success rate and more consistent outcomes for 4-VO model.

Anesthesia

Anesthetics used for 4-VO models include ether, barbiturates, and isoflurane, etc. Each anesthetic has its side effects, such as lowering the body temperature and metabolism, blocking certain glutamate receptors or ion channels, which might complicate the ischemic pathophysiology and therefore the data interpretation. Isoflurane is recommended because of its relative fewer side effects and rapid recovery.

Equipment

Anesthesia

Isoflurane mixer (Matrx, VIP 3000)[1]
Nasal masks (Fig. 1)

[1] Equipment in parenthesis is the one used in our lab.

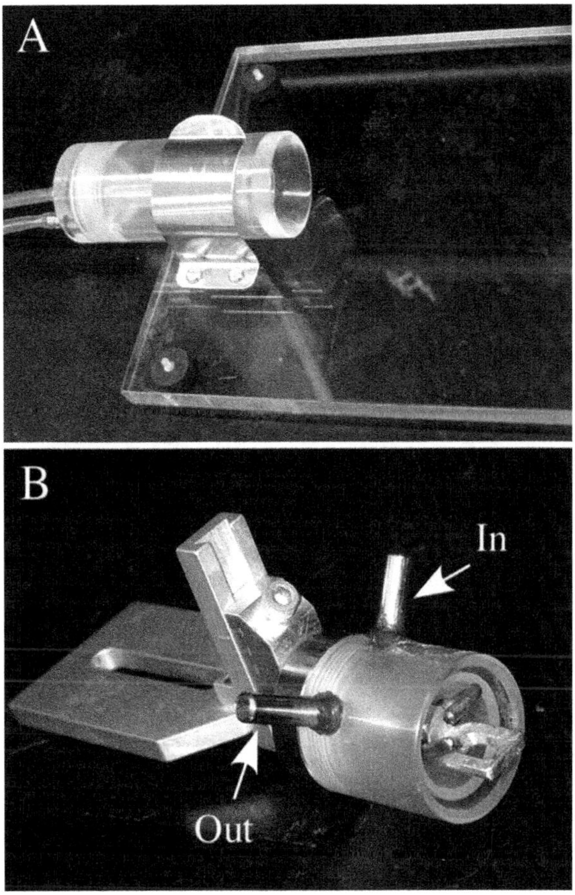

Fig. 1 (**a**) Nasal mask for surgery on ventral neck. (**b**) Nasal mask for stereotaxic apparatus. Both masks have an inlet for anesthesia and an outlet for retrieval of anesthesia by connecting to a vacuum

Surgery

Surgical instruments (scissors, forceps, knives etc.)
Stereotaxic device (KOPF 1430)
Dissecting microscope (Nikon, SMZ-1B)
Driller (Dremel, Multipro, model 395, type 5)
Vessel cauterizer (Cordless soldering iron, ISOTIP, model 7700 with micro-tip)
Buttons (made of Teflon, Fig. 2)
Plastic tube (made of 1 cc syringe, Fig. 2)
Laboratory tubing (SILASTIC, CAT NO. 508-003, OD 1.19 mm 0.025 in., ID 0.64 mm, 0.047 in.)
Hook for occlusion (Ligation aid, model 18062, Fine Science Tools)
Temperature control unit (Temperature controller & alarm, TCAT-1A)
Thermometer (Physitemp, BAT-10)

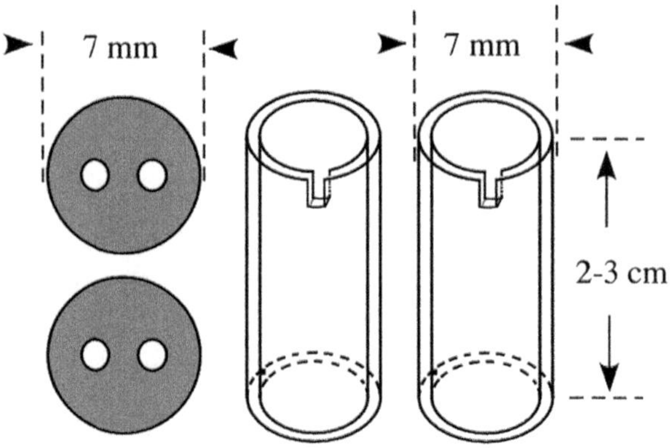

Fig. 2 Drawings of buttons and plastic tube for occlusion

Temperature probe: (Harvard) Rectal: BS4 52-1526; brain: BS4 52-1732 (OD, 0.64 mm)

Recording of Ischemic Depolarization

Amplifier (Neuroprobe, model 1600)
Electrode puller (KOPF model 750)
Electrode (Thin wall glass capillary with filament, OD 2.0 mm, ID 1.5 mm)

Procedures

Original 4-VO Model

Rats are anesthetized with isoflurane (a mixture of 1–2% isoflurane in 33% O_2, 66% N_2), and placed on an operating table. Anesthesia is maintained with a nasal mask.

Placement of Occluding Device Around Carotid Artery

1. Make a ventral midline cervical incision to open the skin.
2. Separate the muscles covering the ventral neck with two small curve forceps to expose the common carotid arteries.
3. Isolate the common carotid arteries free from the vagas nerves and the cervical sympathetic chains.
4. Place the occluding device, a piece of silicone tubing (0.025″ I.D., 0.047″ O.D., 10–15 cm in length), loosely around each common carotid artery and passes through the two holes in a small Teflon button (Fig. 3a).

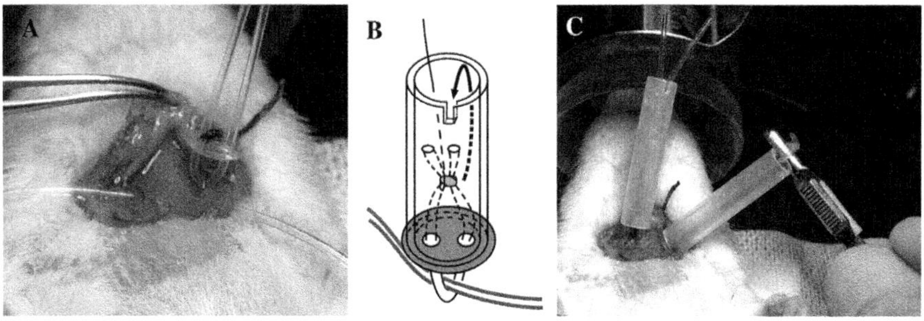

Fig. 3 (**a**, **c**) Photomicrograph of preparation of occluding device. (**b**) Schematic drawing of occluding devices for 4-VO

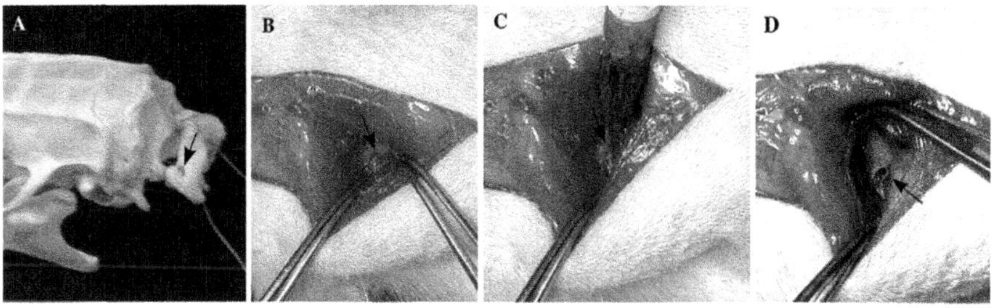

Fig. 4 (**a**) Skeletal specimen showing the alar foramen and vessels (arrow). (**b**) Exposure of alar foramen (arrow). (**c**) Electrocauterization of vertebrate artery. (**d**) Photomicrograph showing the alar foramen after electrocauterization

5. Pass two ends of the silicone tubing through the plastic tube and tie the tubing with a knot to form a loop. The length of the loop should be shorter than the length of the plastic tube that will be used as the supporting apparatus for occlusion. So when the silicone-tubing loop is pulled through the plastic tube, it will press the common carotid artery against the button and stop the blood flow (Fig. 3c).
6. Close the wound with 1–2 suture.

Electrocauterization of Vertebral Artery

1. Place rat on a stereotaxic frame. A dorsal neck incision is made from the occipital bone to the second cervical vertebrae (about 1 cm in length). The paraspinal muscles are separated to expose the alar foramina of the first cervical vertebrae. To improve the exposure of the alar foramina, the cervical spine should be gently extended with a rubber band tied on the rat's tail.
2. Insert the tip of the solder ion into the alar foramen to permanently cauterize the vertebral arteries (Fig. 4). It should be mentioned that the cauterization is crucial for successful ischemia. Heating too long will damage the pons underneath

the cervical vertebrate and increase the mortality, but heating too short will not coagulate the vertebral arteries, resulting incomplete ischemia. There is no perfect method to detect where the vertebral arteries are permanently occluded. In our experience, a small probe with sharp tip, such as syringe needle, is inserted into alar foramina and gently twisted. If bleeding occurs, another cauterization is needed.

3. Close the wounds with surgical clips and return the animal to cage for recovery.

Occlusion of Bilateral Common Carotid Arteries

1. On day 2, the awaked rat (fast over night to provide uniform plasma glucose levels) is gently held by one experimenter and put the animal upside down to expose the ventral neck (rat adapter could be used to secure the head and better expose the operating area).
2. Remove the sutures on ventral neck by another experimenter and pull the silicone tubing loop through the plastic tube and secure the knot to the edge of the tube to completely occlude carotid blood flow (Fig. 5). After occlusion of both common carotid arteries, the rat should become unresponsive and loss its righting reflex within 1 min if the ischemia is completed. The core body temperature should be maintained at 37 °C during occlusion, via a rectal probe coupled to a heating lamp.
3. Release the knot to restore blood flow after occlusion for certain period of time, which depends on the brain region of interest. The animal should recover shortly.

Fig. 5 Schematic drawing of occlusion of common carotid artery

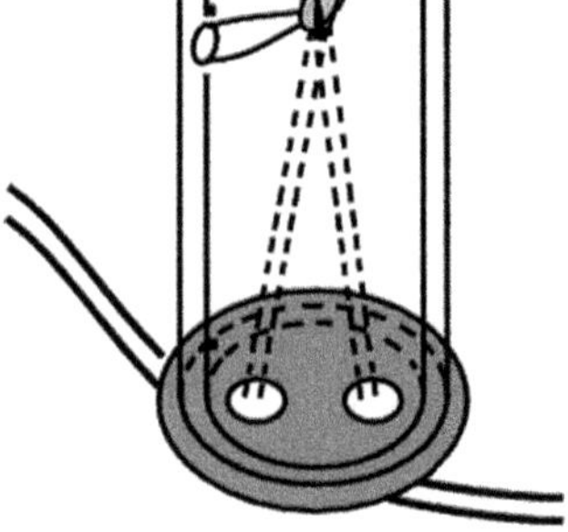

One-Day 4-VO Model (with ID Recording)

Brain temperature is the most important factor influencing the severity of cell death after ischemia. Another potential problem is the collaterals of blood vessels from the skull and neck muscles resulting in incomplete ischemia. This modified 4-VO model resolves these problems by maintain the brain temperature at 37 °C and use ID as an indication of complete ischemia [6].

Surgical Preparation

1. The procedures for placement of occluding devices on common carotid arteries and electrocauterization of vertebral arteries are the same as described above except a surgical suture thread is tied to the knot of the tubing.
2. Drill a burr hole on one side of the skull for placement of temperature probe.
3. Gently insert a tiny temperature probe (0.025″ in diameter) underneath the skull in the extradural space, and the brain temperature is maintained at 37 °C with a heating lamp using a temperature control system (Fig. 6).
4. Drill another hole on the other side of the skull above the brain region of interest. The coordinates for hippocampus are: AP 5.0 mm, ML 2.5 mm; the coordinates for striatum are: AP 9.5 mm, ML 3.0 mm (Interaural).
5. Peal off the dura mater with a sharp tip forceps and keep the brain surface moisture with saline.
6. Put a reference electrode (silver wire) underneath the skin at the opening of the dorsal neck.

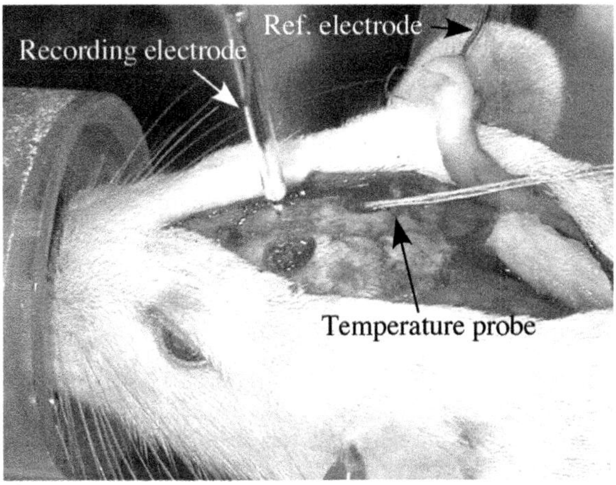

Fig. 6 The operation site of ID recording and temperature monitoring

ID Recording

1. Prepare a glass microelectrode with electrode puller and break the tip of the electrode to yield a tip diameter of 5–10 μm. Fill the electrode with 2 M NaCl.
2. Slowly advance the microelectrode to the region (Hippocampus DV 2.5 mm; Striatum DV 3.0 mm).
3. Connect the recording electrode and reference electrode to the amplifier. Adjust the DC potential to 0 mV.
4. Following baseline recording for 1–2 min, occlude the common carotid arteries bilaterally by pulling the thread and secure the knot to the notch of the plastic tube. A sudden drop of DC potential from 0 mV to ~20 mV should occur 2–3 min after occlusion if a complete ischemia is achieved (Fig. 7).
5. Release the occlusion when ID last for 10 min for studying hippocampus and 20 min for studying striatum. The DC potential should return to 0 mV in 1–2 min.
6. Remove the electrodes and close the wound with surgical clips.

Outcome Evaluation

Several endpoints, including behavioral, physiological and histological features, could be used to assess the brain damage and validate the success of the animal model. Among which, histological changes in ischemia vulnerable regions are the most reliable measurement. Hematoxylin-Eosin (HE) staining of paraffin section is a standard pathological technique and has been widely used to evaluate cell death after transient global ischemia.

Ischemic Cell Death in the Hippocampus

CA1 pyramidal neurons in the hippocampus are the most vulnerable neurons to ischemia. Neurons in the dorsal hippocampus are more sensitive to ischemia than those in the ventral hippocampus. Therefore only sections from the dorsal hippocampus are used for evaluation (Fig. 8). CA1 neurons start to show morphological signs of cell death, i.e. cell body swollen or shrinkage, and nuclear condensation, 2–3 days after ischemia. The cell death begins from the medial part of CA1 zone, gradually progresses towards the lateral part of CA1 zone and stops at the junction of CA3 zone. If the ischemia is relatively mild, the cell death will stop somewhere in the middle of CA1 zone. In contrast, the CA3 pyramidal neurons, dentate granule cells and interneurons in the CA1 region survive the ischemic insult. Another intriguing phenomenon in hippocampus is the delay cell death after ischemia. Approximately 2 days after 10–15 min global ischemia, CA1 neurons start to die and will reach the maximal cell death in 7 days.

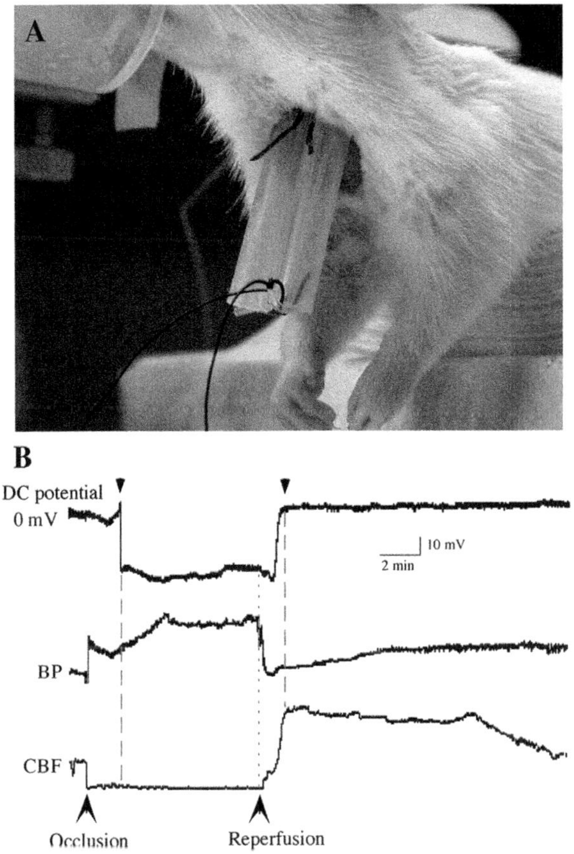

Fig. 7 (**a**) Occlusion of common carotid arteries. (**b**) Traces showing the changes of cerebral blood flow (CBF), arterial blood pressure (BP) and DC potential after ischemia. After occlusion, the blood pressure immediately increases and the DC potential drops to −20 mV (ID). After reperfusion, the blood flow significantly increases (hyperperfusion), the blood pressure and DC potential return to control levels. The ID duration is defined as the distance between the onset of potential drop and the return to 0 mV (arrow heads)

There are many ways to quantify the ischemic cell death. The grading method [7] is the one relatively simple and the results are well correlated to the severity of ischemic insult.

1. Perfuse animal with 4% paraformaldehyde 7 days after ischemia.
2. Embed brain blot containing hippocampus into paraffin.
3. Cut coronal sections of 8 μm thickness of hippocampus and stain with Hematoxylin-Eosin.

   (The details of the above procedure can refer to any histology manual.)

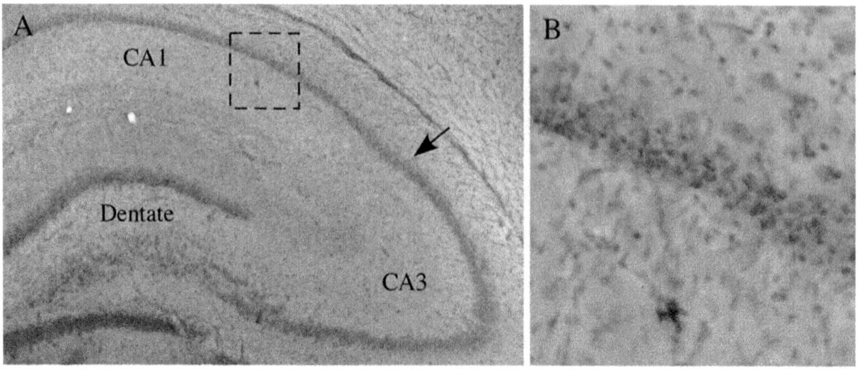

Fig. 8 Photomicrograph of paraffin sections showing the HE staining of the hippocampus 7 days after 10 min forebrain ischemia. (**a**) Most of CA1 pyramidal neurons have degenerated at this time. The arrow indicates the boundary of CA1 and CA3 zone where the cell death stops. (**b**) Higher magnification picture of square area in (**a**) showing the morphological features of cell death in CA1 region after ischemia

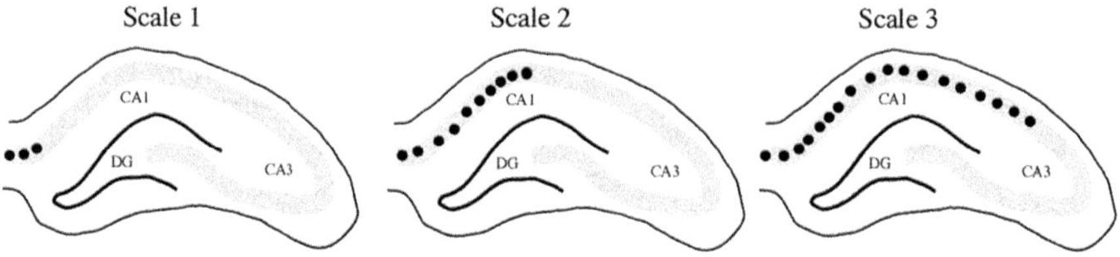

Fig. 9 Schematic drawing showing the grade scale method to quantify the cell death in the hippocampus after transient global ischemia

4. Select sections from the dorsal hippocampus (between AP 4.8–5.8 mm, interaural or AP -3.3~–4.3 mm, Bregma) and grade the damage with the following scale:

   No damage (control) = 0;
   A few neurons damage = 1;
   Many neurons damage = 2;
   Majority of neurons damage = 3.

According to the natural curves in the coronal plane of the CA1 cell layer, the cell death within the medial portion (subiculum) is defined as scale 1. The cell death reaches the middle part of CA1 zone (the peak of the dorsal curve) is defined as scale 2. The cell death reaches the junction of CA3 zone is defined as scale 3 (Fig. 9). Cell death progresses to between these landmarks could be defined as 1.5 or 2.5 respectively. This method could obtain reliable results across the experimental groups [8].

5. Evaluate at least four sections from each hemisphere. Get average grading from two hemispheres of each animal and use it as the score of that animal.
6. Compare the data between groups using nonparametric Mann-Whitney U-test.

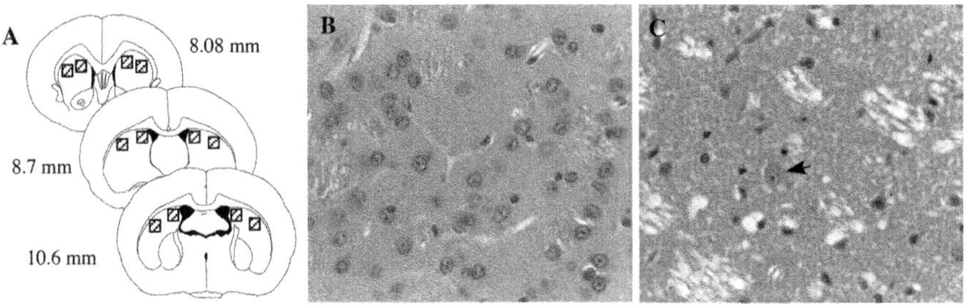

Fig. 10 (**a**) Schematic drawing showing the section plane and selected region (square) for cell counting. (**b**) HE staining of striatal section of a control rat. (**c**) HE staining of section 1 day after 20 min ischemia. Moat of the cells have died but some interneurons (arrow) remain viable

Ischemic Cell Death in the Striatum

In the striatum, small to medium-sized spiny neurons are highly sensitive to transient global ischemia. It requires ~25 min ischemia to induce cell death in the dorsal striatum. The cell death starts about 6–8 h after reperfusion and most of them die in 24 h. The neurons in the dorsolateral striatum are more vulnerable than those in the ventrolateral striatum. Interneurons in the striatum are resistant to ischemia (Fig. 10b, c). Quantification of cell death in the striatum is achieved by counting the number of survived neurons in a given area.

1. Prepare paraffin sections as described above.
2. Select four sections at the anterior, middle and posterial plans (AP 10.6 mm, 8.7 mm and 8.8 mm, interaural). Identify a region in the dorsomedial and a region in the dorsolateral striatum with the same size and approximately the similar location (Fig. 10a).
3. Count the number of survived neurons in the selected region.
4. Compare the data between groups using nonparametric Mann-Whitney U-test.
5. Get average grading from two hemispheres of each animal and use it as the score of that animal.

Advantages, Complications and Limitations

Advantages

1. Rat is widely used in neuroscience research so the results can be compared to the previous studies.
2. 4-VO model in rat is well established and yields reproducible neurophathological results.

3. The species is large enough to allow relatively easy surgery, physiological monitoring and maintenance.
4. Low costs in purchasing and maintenance.

Complications

Developing seizure is a common complication after cerebral ischemia. The incidence of seizure is positively correlated to the severity and duration of ischemic insult. In Wistar rats, a 20 min ischemia might induce seizure in 20–30% of the animals. The brain damage caused by seizure is different from that caused by ischemia. Therefore, the animals with seizure after ischemia must be identified and excluded from the study. One of the signs of seizure is the spreading of the flooring materials around the cage because these materials are kicked out of the cage due to the seizure activities.

Limitations

Technical Difficulties

In some animals, the alar foramina could deform or disappear, making the electrocauterization impossible. The more difficult task is the successful electrocauterization of the vertebral arteries that are hidden underneath the alar foramina. Unsuccessful electrocauterization is the major cause of incomplete ischemia. Therefore it is essential to make sure that the vertebral arteries are cauterized and the blood flow is completely stopped using the methods described above.

Recording of DC potential change during ischemia might be another difficulty for someone doesn't have experience in electrophysiology. Actually it is not as difficult as it sounds. Recording of ID only involves very basic electrophysiological technique and can be easily learn and master.

Mortality

Approximately 10% of the rats will die either during ischemia or shortly after reperfusion even in experienced hands. The respiratory failure during the first 2–3 min of occlusion is a major complication that contributes to animal death during ischemia. Immediate resuscitation normally rescues the animals. Death after reperfusion may stem from severe ischemia, cerebral vesicular blood clot, combined with surgical trauma in the neck and head. The mortality of stains other than Wistar rats might be even higher.

Variation

Despite the reproducible outcomes, the results of 4-VO might vary among animals. Even in animals with successful ischemia, the severity of cell death between two hemispheres might be different. Many factors, including cerebrovascular collaterals, brain temperature, brain glucose, anesthesia, oxygen and carbon dioxide tension etc., influence the ischemic outcome [9]. It is important to strictly control the variables and optimize the experimental conditions in order to obtain consistent results.

References

1. Pulsinelli WA, Brierley JB. A new model of bilateral hemispheric ischemia in the unanesthetized rat. Stroke. 1979;10(3):267–72.
2. Schmidt-Kastner R, Paschen W, Ophoff BG, Hossmann KA. A modified four-vessel occlusion model for inducing incomplete forebrain ischemia in rats. Stroke. 1989;20(7):938–46.
3. Xu ZC, Pulsinelli WA. Responses of CA1 pyramidal neurons in rat hippocampus to transient forebrain ischemia: an in vivo intracellular recording study. Neurosci Lett. 1994;171(1–2):187–91.
4. Yamaguchi M, Calvert JW, Kusaka G, Zhang JH. One-stage anterior approach for four-vessel occlusion in rat. Stroke. 2005;36(10):2212–4.
5. Ginsberg MD, Busto R. Rodent models of cerebral ischemia. Stroke. 1989;20(12):1627–42.
6. Xu ZC. Neurophysiological changes of spiny neurons in rat neostriatum after transient forebrain ischemia: an in vivo intracellular recording and staining study. Neuroscience. 1995;67(4):823–36.
7. Pulsinelli WA, Brierley JB, Plum F. Temporal profile of neuronal damage in a model of transient forebrain ischemia. Ann Neurol. 1982;11(5):491–8.
8. Xu ZC, Gao TM, Ren Y. Neurophysiological changes associated with selective neuronal damage in hippocampus following transient forebrain ischemia. Biol Signals Recept. 1999;8(4–5):294–308.
9. Pulsinelli WA, Jacewicz M. Animal models of brain ischemia. In: Barnett HJM, Stein BM, Mohr JP, Yatsu FM, editors. Stroke: pathophysiology, diagnosis, and management. New York: Churchill Livingstone; 1992. p. 49–64.

Rat Model of Global Cerebral Ischemia: The Two-Vessel Occlusion (2VO) Model of Forebrain Ischemia

Ami P. Raval, Chunli Liu, and Bingren R. Hu

Abstract Animal models of brain ischemia can be divided into focal and global brain ischemia. Among global (often referred to as forebrain) ischemia animal models, the 2-vessel occlusion (2VO) and 4-vessel occlusion (4VO) models in rats are most frequently employed in the studies of molecular mechanisms of neuronal damage after brain ischemia followed by reperfusion. This chapter mainly discusses the rat 2VO model. In this model, reversible forebrain ischemia is produced by bilateral common carotid artery (CCA) occlusion combined with systemic hypotension (50 mmHg) sufficient to induce brain ischemia. This ischemia and reperfusion model affects widespread areas of the forebrain brain. There are two chief histopathological changes in this 2VO ischemia model: (1) selective neuronal vulnerability; typically CA1 pyramidal neurons of the hippocampus are most vulnerable, followed by dorsoventral striatal small- and medium-sized neurons, and pyramidal neurons in the layers 3–4 of the neocortex; and (2) delayed neuronal death, i.e., neuronal death does not occur immediately after transient ischemic episode, but takes place after 2–3 days of reperfusion. This model during the surgical procedure requires artificially maintenance of: (1) rectal and brain temperatures at 37 °C; (2) the mean arterial pressure (MABP); and (3) blood gases. This 2VO ischemia model has the advantage of a one-step surgical procedure that produces high-grade forebrain ischemia. The reproducibility of the ischemic

---

A. P. Raval
Department of Neurology, Cerebral Vascular Disease Research Center, University of Miami Miller School of Medicine, Miami, FL, USA
e-mail: Araval@med.miami.edu

C. Liu · B. R. Hu (✉)
Tissue Ischemia-Reperfusion Protection Program, Departments of Anesthesiology and Surgery, Shock Trauma and Anesthesiology Research Center, University of Maryland School of Medicine, Baltimore, MD, USA
e-mail: Cliu@med.miami.edu; bhu@som.umaryland.edu

J. Chen et al. (eds.), *Animal Models of Acute Neurological Injury*, Springer Series in Translational Stroke Research, https://doi.org/10.1007/978-3-030-16082-1_7

histopathological damage of this model is more than 90%. This animal model is also suitable for molecular, biochemical and physiological studies, as well as for evaluation of neuroprotective procedures and agents.

Keywords Transient cerebral ischemia · Two-vessel occlusion · Reperfusion · Delayed neuronal death · Selective vulnerability · Hypotension · Survival studies · Hippocampus

Introduction

Human brain ischemia is enormously diverse in causes, manifestations and anatomic region. For instance, ischemic stroke is often caused by focal brain ischemia, whereas global brain ischemia likely results from cardiac arrest or heart surgeries. For a scientific study of molecular mechanisms, treatments and prevention of brain ischemia, it is essential to employ physiologically controlled and reproducible in vivo animal models [1, 2]. In this context, rodent models of cerebral ischemia have been widely employed owing to small animal size convenient for handling, low cost of animals, simple surgical procedures, and relative homogeneity within strains owing to inbreeding.

The intent of this chapter is to describe the two-vessel occlusion (2VO) model of global ischemia. Because global ischemic insult does not damage cerebellum, this model is also referred to as the forebrain ischemia or cerebral ischemia model. This model was first developed in Dr. Siesjö's laboratory in Lund University, Sweden, more than 30 years ago [3]. During the ischemic period, mean arterial blood pressure (MABP) is reduced to 50 mmHg by withdrawing blood from either jugular vein or femoral arterial catheter, immediately followed by occlusion of bilateral common carotid arteries, in order to produce brain ischemia to a degree that cause neuronal damage. The withdrawn blood is returned to the animals at the end of the ischemic period. This model produces a "square-wave" ischemic insult in which the transition from non-ischemia to ischemia and its subsequent reversal are abrupt [3, 4]. This model gives rise to delayed neuronal injury selectively in vulnerable brain regions, including the CA1 pyramidal neurons of the hippocampus, the dorsoventral striatum, and neocortex [3]. In this model, ischemic duration of more than 4 min in Wistar rats is required to produce consistent neuronal damage, and animals can survive an ischemic duration for as long as 20–30 min.

Materials, Equipment and Supplies

Animals

Male Wistar rats at 280–300 g body weight [3] are mostly employed in studies using this 2VO model. Although challenging technically, mice are also used for this 2VO model [5]. Selection of appropriate species is important owing to genetic variants which affects the ischemic outcome. Because of the known effects of estrogen on the ischemic outcome, male rats are most frequently used in this 2VO model [6]. Similarly, animal age also affects the ischemic outcome. Therefore, when using this 2VO model, animal strain and species, sex and age must be taken into consideration.

Equipment

Figure 1 illustrates typical surgical equipment required for performing the rat 2VO ischemia model.

(a) Anesthesia box (Harward Apparatus, Holliston, MA, USA)
(b) Anesthesia machine (Matrx Medical Inc., Orchard Park, NY, USA)
(c) Respiratory ventilator (Columbus Instruments, Columbus, OH, USA)
(d, e) Computerized Powerlab/400 for mean arterial blood pressure (MABP) and electroencephalogram (EEG) measurements (AD Instruments, Colorado Springs, CO, USA)
(f) Gas regulators (VWR, West Chester, PA, USA)
(g) Temperature controller (Omega, Stamford, CT, USA)
(h) Oxygen (O_2) tank
(i) Lamp (Luxo Corporation, New York, NY, USA)
(j) Nitrous oxide (NO) tank

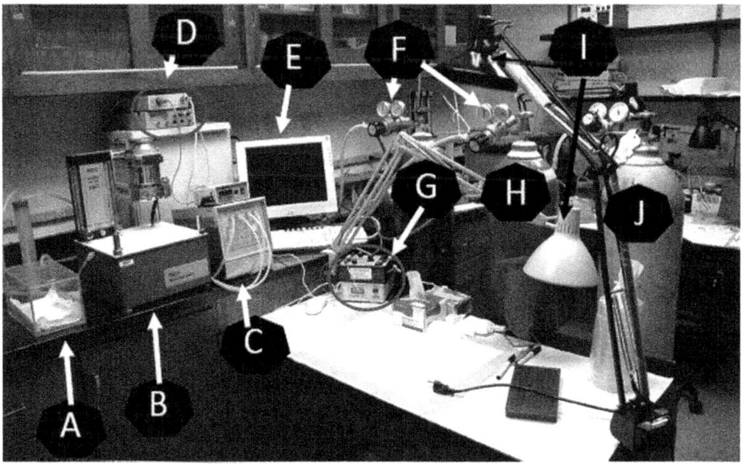

Fig. 1 A surgical equipment table for performing the rat 2VO ischemic model. (**a**) Anesthesia box, (**b**) anesthesia machine, (**c**) respiration ventilator, (**d**, **e**) Computerized Powerlab/400, (**f**) gas regulators, (**g**) temperature controller, (**h**) oxygen (O_2) tank, (**i**) lamps, (**j**) nitrous oxide (NO) tank

Anesthesia Agent

Isoflurane and halothane are commonly used to anesthetize animals during surgical procedures of the 2VO forebrain ischemic model. Because global brain ischemia may affect respiratory system function, ventilator-controlled respiration with a rodent ventilator is required [3].

Surgery Instruments and Supplies

These include a pair of serrated slightly curved forceps (Roboze Surgical Instrument Company, MD, USA; Cat #RS-8254), microdissecting spring scissors (Roboze, Cat #RS-5658), operating scissors (Roboze; Cat #RS-6951), a pair of microdissecting forceps (Roboze; Cat #RS-5244), a pair of delicate hemostatic forceps (Roboze; Cat #RS-7111), microdissecting tweezers (Roboze; Cat #RS-5058), and a bulldog-type serrated fine clamp (Fine Science Tools, Foster City, CA, USA; Cat #18374-43).

Syringes (1 and 10 ml; BD Biosciences, San Jose, CA, USA), syringe adaptors (23 and 20 gauge; BD Biosciences), intubation tube of 14GA, 1.75 in., 2.1 × 45 mm (BD Biosciences; Cat #381467), polyethylene tubing (PE-50, BD Biosciences), heparinized capillaries (Radiometer Copenhagen, Cat #D941G-240-85), silastic tubing (inner diameter 0.76 mm and outer diameter 1.65 mm; Dow Corning Corporation, Midland, MI, USA), rectal temperature probe (Omega, Stamford, CT, USA; Cat #HYPO-33-1-T-G-60-SMP-M), and head temperature probe (YSI, Dayton, OH, USA).

Surgical Procedure

All procedures are performed under aseptic conditions for survival studies. The person performing the surgery should wear a mask, clean gown and sterile gloves. All surgical instruments are sterilized before surgical procedures and maintained aseptic in a solution, such as hexazol solution (Burns Veterinary Supply Inc., Westbury, NY, USA), between the animal procedures.

A Male Wistar rat weighing 250–300 g is fasted overnight. The rat is anesthetized by placing it into an anesthesia box (Fig. 1a) supplied with 4% isoflurane in a 70/30% nitrous oxide/oxygen (NO/O_2) gas mixture for 5–8 min until the rat becomes numb and reaches a state of apathy. Subsequently, the rat is endotracheally intubated with a plastic tube by means of an otoscope (Fig. 2). In this technique, an otoscope is fixed on a frame (Fig. 2a). Anesthetized rat is positioned such that the rat's mouth can be introduced over otoscope tip. Through the otoscope, the surgeon can see the vocal cords in the trachea (Fig. 2b). A flexible guide wire with a blunt and smooth end (Fig. 2a) is gently introduced between the vocal cords into

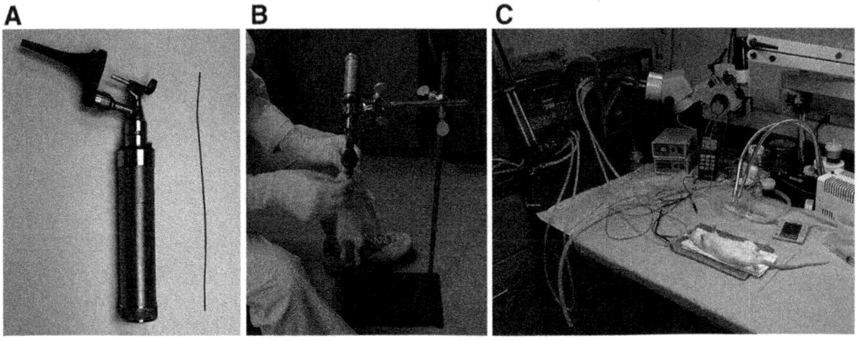

Fig. 2 Otoscope intubation method. (**a**) A guide wire and otoscope. (**b**) Endotracheal intubation of rat with otoscope assistance. (**c**) Connecting endotracheal intubated rat with a ventilator

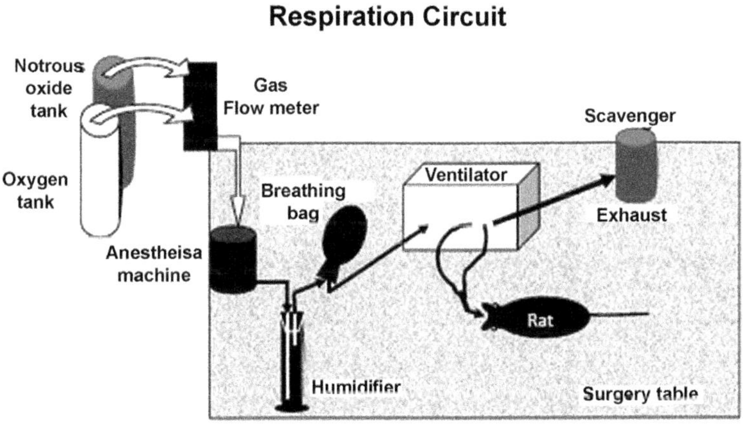

Fig. 3 Schematic drawing of a respiratory circuit

the trachea through the otoscope. The rat, together with the guide wire, is then pulled from the otoscope. A plastic intubation tube (BD Biosciences) can now be inserted via the guide wire into the trachea. There is usually a slight resistance as the intubation tube passes the vocal cords. After removing the guide wire, the intubation tube is fixed with suture onto the mouth skin and then connected to the ventilator. The rat is artificially ventilated at a respiration rate of 60/min with 1.5% ± 0.5 of isoflurane and 70/30% nitrous oxide/oxygen (NO/O_2). The circuit to connect ventilator to rat is illustrated in Fig. 3.

The rats skin is shaved and scrubbed with povidine iodine (10% solution, Purdue Products, Stamford, CT, USA) before any incision is made. Incisions are made to expose a tail artery and the right femoral artery and vein. They are then catheterized with a single lumen polyethylene tubes (PE-50; BD Biosciences). The tail artery catheter is for monitoring mean arterial blood pressure (MABP) and arterial sampling of blood gases and plasma glucose. Two needle electrodes are inserted through

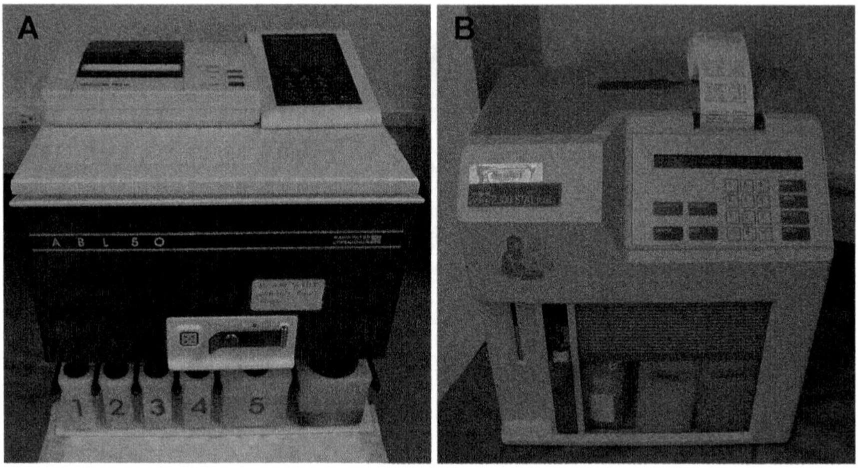

Fig. 4 Photographs of (**a**) blood gas analyzer and (**b**) blood glucose analyzer

Table 1 Physiological parameters

Variables	Before ischemia	During ischemia	After ischemia
Body weight (g)	300 ± 20	–	–
pH	7.4 ± 0.04	7.4 ± 0.08	7.4 ± 0.04
pCO_2 (mmHg)	37 ± 3	39 ± 3	37 ± 3
pO_2 (mmHg)	120 ± 10	110 ± 10	120 ± 10
MABP (mmHg)	120 ± 20	50 ± 2	120 ± 20
Plasma glucose (mg/ml)	120 ± 10	–	120 ± 10

small skin incisions into the temporalis muscle bilaterally to obtain an inter-hemispheric electroencephalogram EEG. MABP and EEG are monitored with a Powerlab system (AD Instruments, Inc., Colorado Springs, CO, USA) throughout the experiment. The femoral artery and vein catheters are for withdrawing of blood during ischemia and intravenous infusion of drugs such as heparin and pancuronium bromide, respectively (see below). Heparin at a dose of 0.2 ml (1000 USP units/ml) is slowly injected through the femoral vein catheter to heparinize the animal. Arterial blood (0.1 ml) is sampled with a heparinized glass tube (Radiometer Copenhagen, Brønshøj, Denmark) through the tail artery catheter for measuring arterial blood gases and plasma glucose levels. An arterial blood gas analyzer (Fig. 4a; Model: ABL 50; Radiometer Copenhagen, Brønshøj, Denmark), and plasma glucose level detector (Fig. 4b; Model: YSI-2300 STAT Plus; YSI, Dayton, OH, USA) are used (Fig. 4). Arterial blood gases and plasma glucose levels are measured 15 min before and 15 min after the ischemic period, as well as intermittently throughout the experiment whenever required. Blood gases are maintained in a physiological range throughout the experiment (Table 1).

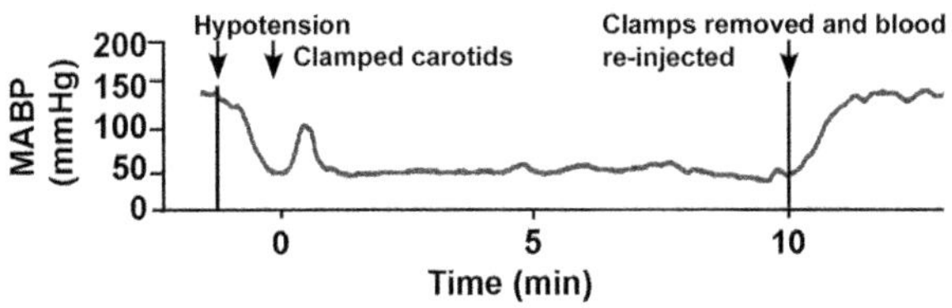

Fig. 5 Representative curve of MABP measurement before, during, and after ischemia period

An incision is made in the ventral neck skin of the rat. Both common carotid arteries are gently dissected free of surrounding tissues and nerve fibers. Thick sutures are loosely placed under the arteries for pulling them out when clamping during induction of ischemia (see below).

A rectal temperature probe (Omega, Stamford, CT, USA) is gently inserted into the rectum to monitor body temperature. A thermocouple 33-gauge needle temperature probe (YSI, Dayton, OH, USA) is implanted into the temporalis muscle for monitoring head temperature. Body and head temperatures are maintained at 37 °C ± 0.2 throughout the experiment with assistance of lamps placed above the animal's body and head. When the surgical preparation is ready for induction of ischemia, pancuronium bromide (2 mg/kg) is injected (intravenously) to immobilize the rat while the isoflurane concentration is maintained at 1.0%.

Induction of Cerebral Ischemia

Fifteen minutes before induction of ischemia, blood gases, MABP, EEG and rectal and head temperatures are monitored and maintained within the physiological range (see Table 1). Immediately before induction of ischemia, blood is gradually withdrawn from the femoral artery with a heparinized syringe (10 ml; adaptor 20 gauge, BD Biosciences) to reduce systemic blood pressure to 50 mmHg. Cerebral ischemia is then induced by clamping both common carotid arteries with surgical clamps (bulldog type serrated fine clamp; Fine Science Tools, Foster, CA, USA). After bilateral occlusion of the common carotid arteries, mean arterial blood pressure usually increases sharply (Fig. 5) which must be reduced instantly by withdrawal of blood and controlled at 50 mmHg by reinfusion and withdrawal of blood through the femoral artery catheter during the ischemic period. The onset of cerebral ischemia is defined as the time of cessation of EEG activity, usually within 1 min after clamping of arteries. Cerebral ischemic durations of 10–15 min are mostly used according to the literature.

To end the ischemic episode, the common carotid artery clamps are removed first to allow post-ischemic reperfusion, and then the withdrawn blood is returned immediately through the femoral artery at a slow rate. The reinfusion of blood usually restores mean arterial blood pressure to a range of 120–140 mmHg. After about

15–30 min of a stabilizing period and measurement of blood gases (see above), the catheters can now be removed, arteries and vein closed with sutures, and anesthesia can be stopped. However, the rat still must be continuously ventilated with 70/30% nitrous oxide/oxygen (NO/O_2) and head and rectal temperatures are maintained at 37 °C ± 0.2 until the onset of spontaneous respiration during recovery from anesthesia, which usually takes about 30–60 min depending on ischemic duration. The ventilator can now be disconnected while the endotracheal tube is kept in place. After removing the head and rectal temperature probes, the rat can be returned to its cage. The rat is monitored closely until spontaneous respiration smoothly recovers before removing the endotracheal tube. After 1–2 h monitoring, the animal can be transferred to the housing room with food and water supplied.

Outcome Evaluation

Ischemia duration of 10–15 min in this 2VO model leads to histopathological damage within selectively vulnerable structures, including neurons in the hippocampal CA1, striatum and neocortical regions [3]. Histopathological changes are thus reliable indicators of ischemic damage in this model. Because neuronal death takes place in a delayed fashion, within 2–3 days of reperfusion, animals are usually perfusion-fixed at 7 days for histopathological examination (see below).

Histopathology

After 7 days of reperfusion following an episode of cerebral ischemia, the rat is anesthetized in an anesthesia box supplied with 4% isoflurane in 70/30% nitrous oxide/oxygen, intubated through tracheotomy and ventilated with 2% isoflurane in 70/30% nitrous oxide/oxygen. The rat chest case is opened surgically, and a catheter is inserted into the ascending aorta through the left ventricle. After clamping the descending aorta with a surgical clump, the rat upper body is immediately perfused through an ascending aorta catheter initially with saline for 1–2 min, and then with FAM (a mixture of 40% formaldehyde, glacial acetic acid, and methanol, 1:1:8 by volume) for 10 min at a constant pressure of 110–120 mmHg. The rat is decapitated. The brain is removed from the skull, and immersed in FAM at 4 °C for 1 day. Coronal brain blocks of about 3–4 mm are cut and embedded in paraffin; coronal sections of 5–10 μm are sectioned with a microtome (Leica Microsystems GmbH, Ernst-Leitz-Strasse 17-3735578 Wetzlar), and stained with hematoxylin and eosin. Brain sections at different levels can be examined.

Delayed neuronal death after transient cerebral ischemia under light microscopy has the following changes: (1) cells shrunken to about half or 1/3 of normal neuronal sizes; (2) shrunken, darkly stained and polygonal-shaped nuclei; and (3) eosinophilic or acidophilic cytoplasm (Fig. 6).

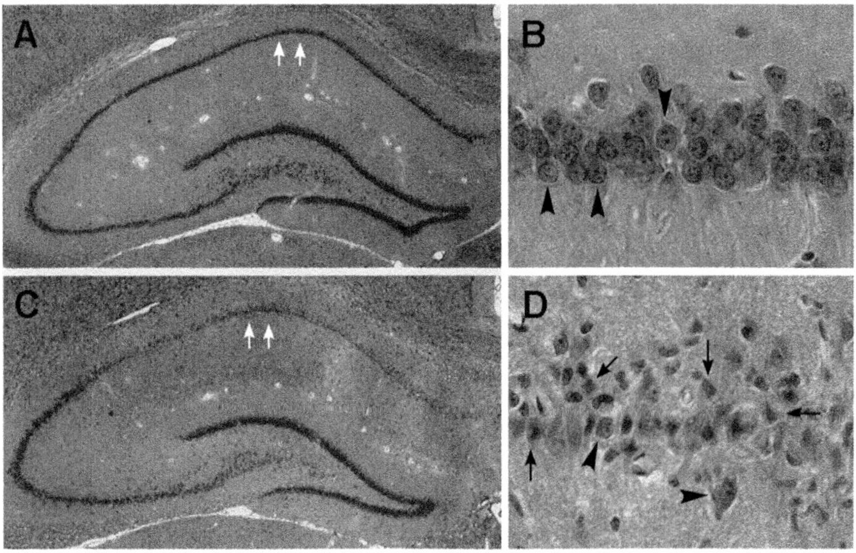

Fig. 6 Histopathology changes in the hippocampus after 10 min of cerebral ischemia in the 2VO model. (**a**) A hippocampal section from a sham-operated rat. (**b**) A higher magnification of the CA1 region indicated by white arrows in the sham-operated rat (**a**). (**c**) A hippocampal section from a rat subjected to 10 min of ischemia followed by 24 h of reperfusion. (**d**) A higher magnification of the CA1 region indicated by white arrows in the post-ischemic rat (**c**). Arrowheads in (**b**) indicate normal neurons. Black arrows in (**d**) point to schemic dead neurons that are identified by cells and nuclear shrunken as well as acidophilically stained cytoplasm

Quantification of Neuronal Death

The number of dead neurons in defined brain regions after transient cerebral ischemia in this 2VO model can be quantified. Quantification is commonly done by counting all the normal and dead neurons (see above) at the same anatomic levels of neocortical, striatal and hippocampal brain sections from both sham-operated control and post-ischemic subjects under light microscopy [3]. The extent of injury is usually expressed as the percent of dead neurons of the total neurons in the defined regions examined. For statistical comparisons of neuronal death between experimental groups, the Kruskal-Wallis test followed by the Mann-Whitney U test is commonly used. Typically, in this 2VO model, more than 4 min of transient ischemia is required to produce consistent neuronal death. Ten minutes of ischemia induces delayed neuronal death selectively in about 80% of CA1 pyramidal neurons, and a few neocortical neurons. Fifteen min of ischemia damages more than 90% CA1 neurons, the dorsolateral striatal neurons, and about 5–10% of neocortical neurons, whereas 20 min of ischemia leads to delayed neuronal death in about 30% of neocortical neurons and also extend delayed neuronal death to neurons in the CA3 and DG regions.

Figure 6 illustrates typical histopathological changes in the hippocampus after transient cerebral ischemia in this 2VO model. Male Wistar rats were subjected to either sham-surgery without ischemia, or 10 min of ischemia followed by 7 days of reperfusion, respectively. Paraffin-embedded coronal sections (10 μm) were photographed by light microscopy. Selective neuronal death is evident in more than 80% of pyramidal neurons in the CA1 region of the hippocampus after 10 min of cerebral ischemia in the 2VO model.

Advantages and Limitations

The advantages of this 2VO transient cerebral ischemia model include: (1) One-stage surgical procedure; (2) highly reproducible ischemic damage; (3) high animal survival rate for studies of reperfusion injury; and (4) suitability for studies of physiological, molecular, biochemical, and behavioral alterations during the post-ischemic phase and for testing efficacy of neuroprotective agents.

The limitations of this 2VO transient cerebral ischemia model are that: (1) brain ischemia cannot be induced in awake rats; and (2) the requirement of systemic control of hypotension.

Potential Technical Difficulties and Pitfalls

Because of the global ischemia may affect animal respiration in this 2VO model, mechanical maintenance of animal respiration and blood gases with a ventilator is essential during and after ischemia [3]. Endotracheal intubation of rats and mice for connecting a ventilator can be technically challenging and often requires personal training and practice for several weeks. There are two methods commonly used for endotracheal intubation: (1) with direct vision of the trachea through the animal mouth under local illumination [3]; and (2) with otoscopic vision of the tracheal vocal cords as described in the present chapter. In comparison, the otoscopic method is easier but requires an otoscope.

The following is vital for reproducing the same degree of histopathological damage in this 2VO transient cerebral ischemia model: (1) both rectal and brain temperatures must be controlled constantly at 37° ± 0.2 during the procedure. The brain temperature is particularly important because cutoff of blood supply to brain during the ischemic period can lead to reduction of brain temperature. Small changes in brain temperature alter markedly ischemic histopathological outcome; (2) it is critical to withdraw food (fast) overnight to maintain constant blood glucose level. The higher blood glucose level before ischemia often leads to more sever ischemic neuronal damage; and (3) it is also essential to control mean arterial blood pressure as close as possible to 50 mmHg in order to induce the same degree of brain ischemia during the ischemic period.

Mortality

In our hands, the success rate of this model is more than 90%. The mortality is mostly due to post-operative complications. Inappropriate intubation can damage lung and lead to trachea swell, causing respiratory difficulties during the post-ischemic period. A long duration of surgery period and careless surgical damage can lead to poor recovery. In addition, an excessive dose of anesthesia can causes animal death during surgery.

References

1. Dietrich WD, Busto R, Alonso O, Globus MY, Ginsberg MD. Intraischemic but not post-ischemic brain hypothermia protects chronically following global forebrain ischemia in rats. J Cereb Blood Flow Metab. 1993;13:541–9.
2. Hu BR, Martone ME, Jones YZ, Liu CL. Protein aggregation after transient cerebral ischemia. J Neurosci. 2000;20:3191–9.
3. Smith ML, Bendek G, Dahlgren N, Rosen I, Wieloch T, Siesjo BK. Models for studying long-term recovery following forebrain ischemia in the rat. 2. A 2-vessel occlusion model. Acta Neurol Scand. 1984a;69:385–401.
4. Smith ML, Auer RN, Siesjo BK. The density and distribution of ischemic brain injury in the rat following 2-10 min of forebrain ischemia. Acta Neuropathol (Berl). 1984b;64:319–32.
5. Wellons JC 3rd, Sheng H, Laskowitz DT, Burkhard Mackensen G, Pearlstein RD, Warner DS. A comparison of strain-related susceptibility in two murine recovery models of global cerebral ischemia. Brain Res. 2000;868:14–21.
6. Bramlett HM. Sex differences and the effect of hormonal therapy on ischemic brain injury. Pathophysiology. 2005;12:17–27.

Rodent Model of Pediatric Asphyxial Cardiac Arrest

Mioara D. Manole, Henry L. Alexander, Robert W. Hickey, and Robert S. B. Clark

Abstract A model of asphyxial cardiac arrest in post-natal day 17 rats is described. This clinically relevant model includes a period of hypoxemia followed by ischemia and resuscitation. Continuous physiologic monitoring is performed before, during and after the insult. Graded insults produce consistent and dose dependent brain injury with histological damage and behavioral impairment. The details of the procedures, along with outcome assessments and applications are discussed.

Keywords Asphyxia · Cardiac arrest · Cerebral blood flow · Neurodegeneration · Sprague Dawley rat

M. D. Manole (✉)
Division of Pediatric Emergency Medicine, Department of Pediatrics, Safar Center for Resuscitation Research, Children's Hospital of Pittsburgh, Pittsburgh, PA, USA
e-mail: mioara.manole@chp.edu

H. L. Alexander
Department of Critical Care Medicine, Safar Center for Resuscitation Research, Pittsburgh, PA, USA
e-mail: alexanderhl@anes.upmc.edu

R. W. Hickey
Division of Pediatric Emergency Medicine, Department of Pediatrics, Children's Hospital of Pittsburgh, Pittsburgh, PA, USA
e-mail: robert.hickey@chp.edu

R. S. B. Clark
Department of Critical Care Medicine, Safar Center for Resuscitation Research, Children's Hospital of Pittsburgh, Pittsburgh, PA, USA

Department of Pediatrics, Safar Center for Resuscitation Research, Children's Hospital of Pittsburgh, Pittsburgh, PA, USA
e-mail: clarkrs@ccm.upmc.edu

J. Chen et al. (eds.), *Animal Models of Acute Neurological Injury*, Springer Series in Translational Stroke Research, https://doi.org/10.1007/978-3-030-16082-1_8

Model Selection

Cardiac arrest (CA) in infants and children is secondary to asphyxia in a majority of cases [1]. This differs from adults where CA is typically caused by cardiac arrhythmias such as ventricular fibrillation (VF) or tachycardia (VT). Asphyxial CA differs in physiology as compared with VF- or potassium chloride-induced cardiac arrest because a period of hypoxemia, hypercarbia, and hypotension precedes cardiovascular collapse and circulatory arrest. Regardless of the underlying etiology, all CA have a period of total body ischemia that can cause impaired and interrelated post-ischemic organ dysfunction. This multisystem organ dysfunction is reproduced in animal models of CA (VF or asphyxial), but not in models of isolated brain ischemia, mimicking the clinical condition.

To model pediatric asphyxial CA, our laboratory modified an established protocol of CA in adult rats [2] and applied it to immature, postnatal day (PND) 17 rats [3]. PND 17 rats resemble a 1–4 year old human child in terms of a number of developmental parameters [4]. The model of pediatric asphyxial CA described in this chapter is thus especially relevant to scientists whose research interest involves CA in children.

Materials

Animal Selection

The use of all photographs was pre-approved by the Institutional Animal Care and Use Committee at the University of Pittsburgh.

We use Sprague-Dawley rat pups, PND 16–18. Pups are breastfed at this age and are thus housed with a post-partum female rat, allowing them full access to nutrition prior to the experiment. Litters of one mother with eight pups are ordered to obtain rats with similar weights. It is our experience that pups below 30 g have lower survival rates post-resuscitation. Thus, we typically use PND 16–18 rats weighing 30–40 g on the day of the experiment. Normal cardiovascular physiologic parameters of PND 17 rats under anesthesia with fentanyl or isoflurane are a heart rate (HR) of 350–450 beats/min and a mean arterial pressure (MAP) or 65–80 mmHg.

Anesthesia

In this model, we anesthetize the animals with either isoflurane or fentanyl. The choice of anesthetic depends on the parameter studied. Isoflurane provides an excellent plane of anesthesia and stable HR and MAP. Anesthesia with isoflurane has the advantages of being easier to titrate and more predictable. The recovery of the animal is rapid after isoflurane anesthesia. However isoflurane markedly increases cerebral blood

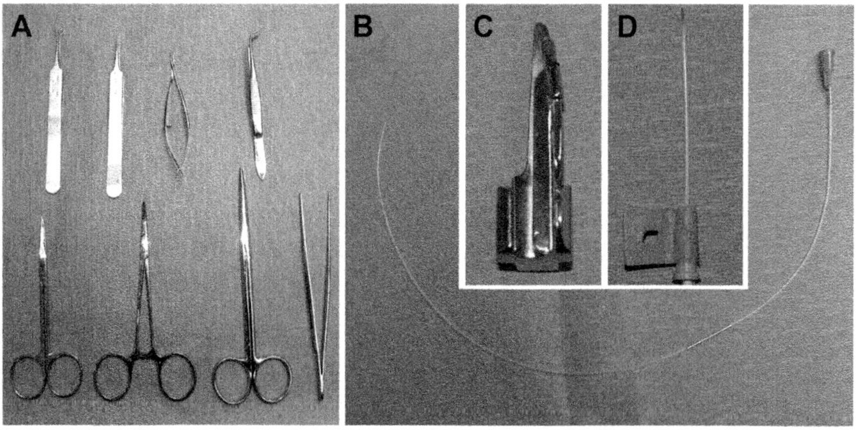

Fig. 1 Examples of equipment necessary for the pediatric asphyxial CA model. (**a**) Surgical instruments (clockwise from top left) Roboz by Dumont 45° angle forceps, straight Dumont forceps, Iris scissors (Roboz RS-5602), 45° angle forceps (Roboz RS-5137), various scissors, straight forceps. (**b**) Vascular catheter, observe the natural curvature of the catheter, which is placed in the vessel with the concavity cephalad. (**c**) Modified Miller 0 straight blade used for endotracheal intubation. The straight Miller 0 blade is filed to from the base of the blade to the tip, resulting in a narrower, triangular shape. (**d**) Endotracheal tube with stylet. The endotracheal tube is an 18 gauge, 2 in. angiocatheter with the needle removed. The stylet is a blunt vascular catheter introducer, and is advanced 2 mm past the tip of the angiocatheter. The stylet permits intubation before glottic spasm occurs, and reduces trauma to the vocal cords from the sharper angiocatheter. The hub of the angiocatheter has a piece of tape with an orifice to secure the catheter after placement

flow (CBF) as compared with fentanyl [5]. Thus, in experiments where measurement of CBF is important, fentanyl is a more desirable anesthetic choice than isoflurane.

Equipment

Surgery

- Surgical instruments (Fig. 1a): Roboz by Dumont 45° angle forceps; Straight Dumont Forceps; Iris scissor Roboz RS-5602; Roboz RS-5137 forceps 45° angle; Scissors; Curved hemostat; Straight forceps
- Dissecting microscope
- Femoral arterial and venous catheters: PE 10 and PE 50 polyethylene tubing (Becton Dickinson; Fig. 1b)
- Glue: 3M Scotch Weld Instant Adhesive CA 4OH
- Syringes
- 6.0 silk sutures
- 23 Gauge short blunted needle
- Heparinized saline 1 unit/ml
- Heating pad

- Cotton tip applicators
- Iodine
- Lidocaine solution

Anesthesia and Intubation

- Isoflurane
- Plexiglas chamber
- Laryngoscope with a modified Miller 0 Blade machine-filed from base to tip to reduce the width of the blade (Fig. 1c)
- 18 Gauge, 2-in. long Teflon angiocatheter modified with the hub of the catheter surrounded by tape with a 0.5 cm edge and a central orifice, necessary for securing the angiocatheter after placement (Fig. 1d)
- Stylet custom made from a central line introducer passed through the angiocatheter and advanced 2 mm beyond the tip. This facilitates advancement of the angiocatheter in the trachea with less trauma. The stylet can be stabilized at the hub of the catheter by taping the stylet to a cotton tip applicator
- Forceps 45° angle, Roboz RS-5137
- Rubber band
- Rodent mechanical ventilator

Physiologic Monitoring

- Arterial pressure transducer
- Electroencephalogram (EEG) recorder with scalp electrodes
- Electrocardiogram (ECG) electrodes
- Temperature probe

Procedures

Intubation

The rat is anesthetized with isoflurane (mixture of 3% in 50% N_2O/balance O_2) in a Plexiglas chamber until unconscious. The rat is then placed supine on the anesthesia table and isoflurane 3%/50% N_2O/balance O_2 is provided via a nasal cone made from the barrel of a 20 ml syringe. The front legs and tail are secured with tape to prevent lifting the rat off the table during intubation. A rubber band secured to the anesthesia table is placed over the upper incisors of the rat to stabilize the head during intubation, as shown in Fig. 2a. The tongue is lifted outside the mouth with forceps, then held between the index and thumb of the operator's left hand and pulled towards the left as shown in Fig. 2a. The laryngoscope blade is then introduced with

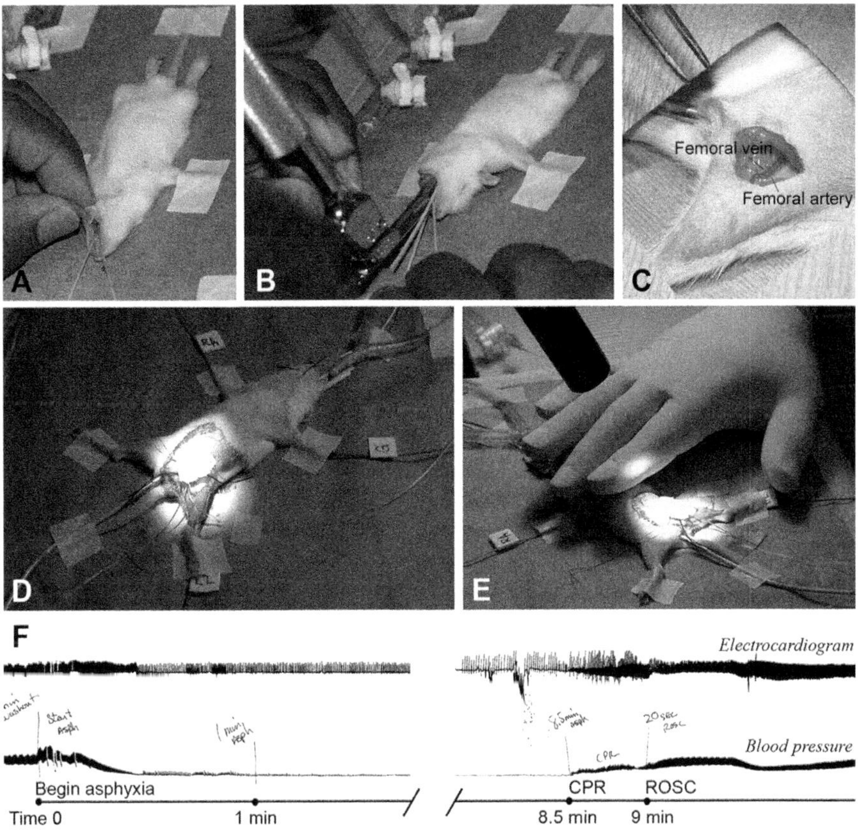

Fig. 2 Asphyxial CA in PND 17 rats. (**a**) The limbs are taped to the operating table and the upper incisors are secured with a rubber band to prevent inadvertent lifting of the head during endotracheal intubation. (**b**) Endotracheal intubation. The tongue is pulled to the left side and the custom-modified laryngoscope blade is used to facilitate placement of the 18 gauge angiocatheter into the trachea. (**c**) Isolation and identification of femoral vein and femoral artery. The artery lies laterally to the vein. (**d**) Proper positioning of the rat and securing of invasive lines, monitors, and endotracheal tube. (**e**) Cardiopulmonary resuscitation (CPR) includes manual chest compressions. (**f**) Physiologic monitoring includes ECG and MAP tracings. Note pulseless electrical activity followed by return of spontaneous circulation (ROSC) after CPR

the right hand until resistance is felt (approximately 2 cm in a 30–40 g rat). The tongue is then released and the operator switches the laryngoscope handle to the left hand. The operator places the endotracheal tube (18 gauge angiocatheter) in the right hand. The laryngoscope is then retracted until the glottic opening is seen. The glottic opening is seen as a small (1 mm) orifice that opens and closes with breathing. The 18 gauge catheter with an introducer is passed into the trachea as shown in Fig. 2b. The introducer is then removed, the ventilator is connected and the depth of the tube is adjusted to assure equal chest expansion. The endotracheal tube is secured in place with a suture to the skin of the anterior neck of the rat.

It is our experience that multiple intubation attempts in these small rats will produce edema of the airway and decreased survival after extubation. When more than one attempt is necessary for intubation of the trachea we stop the procedure and return the rat to the dam after recovery from anesthesia, and a different rat is used. The rat unsuccessfully intubated can be used on the following day.

Catheterization of Vessels

- Prepare the intravascular catheters: The catheters have a natural curvature (Fig. 1b). Maintain the normal curvature of the catheters, and orient the curvature towards the rats' head. Cut 25 cm of PE 50 catheter. Connect a 23 g blunted needle to one end. Cut 4 cm of PE 10 catheter. Thread the PE 10 into the PE 50 catheter and glue the catheters together. Cut the tip of the PE 10 catheter at a 45° angle bevel, ensuring that a clean cut with no burrs is obtained. Fill catheter with heparinized saline solution and attach a 1 ml syringe to the needle hub.
- Shave the femoral area and clean the area with iodine using aseptic technique.
- While holding the skin with small smooth tissue forceps, make an oblique incision over the femoral area with Iris scissors.
- Under the magnifying microscope, bluntly dissect the tissues to expose the femoral vessels with a cotton tip applicator (Fig. 2c).
- Identify the femoral artery and vein (the vein is medial to the artery).
- With Dumont 45° angle forceps and straight Dumont forceps, separate the femoral vessels.
- Due to manipulation and drying, the vessels will contract. To prevent this, place two drops of lidocaine on the surgical area to dilate the vessels.
- Place one silk suture proximally and one silk suture distally on the isolated vessel to cannulate.
- Tie the distal suture permanently, the extremity will get blood supply through collateral vessels.
- Make a loose knot with the proximal suture, which will be tightened after the catheter is in place.
- Place a hemostat over the proximal suture ends and apply gently tension for temporary vessel occlusion.
- Make a small incision in the lower portion of the vessel with the iris scissors (Roboz RS 5602).
- With the left hand, stabilize the proximal side of the incision in the vessel with a Roboz RS 5137 90° forceps.
- With the right hand, pick up the catheter a few mm from the tip with a 45° angle forceps. The catheter should be held with the curvature toward the head of the rat, and with the bevel cephalad.
- Introduce the catheter up to the level of the proximal ligature, which is held under tension with the suture and the hemostat. Release the tension from the hemostat, and advance the catheter no more than 5 mm. At this point, the tip of the catheter is not visible. Advancing the catheter further may damage the vessel wall.

- Tie the proximal suture. Align the catheter with the vessel and tie the catheter with the ends of the distal suture to secure the femoral line.
- Repeat the procedure for the other vessel to obtain both arterial and venous access.
- Connect the arterial transducer to the arterial line.
- Close the skin incision.

Asphyxial Cardiac Arrest

The rats are ventilated with a tidal volume of 0.6 ml and a rate of 35–45 breaths/min using a Harvard Model 683 Rodent Ventilator (Holliston, MA). Arterial blood gas parameters are measured before the asphyxial arrest and the ventilator is adjusted if necessary to achieve a $PaCO_2$ between 35 and 45 mmHg. Pre-arrest, the temperature is monitored and is maintained at 37 °C with a heating pad or a warming light. Vecuronium at a dose of 1 mg/kg is administered intravenously 10 min before CA to prevent spontaneous respirations and allow for apnea [6]. Two minutes before asphyxia the isoflurane is discontinued, to allow for anesthetic washout to reduce the potential confounding effects of inhaled anesthetics on neurological injury [7]. Also, anesthetics can cause myocardial depression and hypotension making it more difficult to successfully resuscitate the rats. Previous pilot experiments testing the anesthetic regimen in uninjured rats show that EEG activity begins to recover in amplitude and frequency after 2 min of anesthetic washout. The FiO_2 is reduced to 0.21 one minutes before asphyxia by disconnecting the oxygen tube from the ventilator, to avoid hyperoxygenation pre-arrest. After the anesthetic washout, the ventilator is disconnected at the endotracheal tube for the desired duration of asphyxia, typically 8–12 min. During CA, the ECG tracing progresses from normal sinus rhythm, to bradycardia, then to pulseless electrical activity and rarely to asystole or ventricular fibrillation. Typically, CA occurs 2–4 min after disconnecting the ventilator, and is defined as occurring when there are no pulsations in the arterial line tracing, i.e. pulseless electrical activity (Fig. 2f).

After the desired duration of asphyxia, epinephrine 0.005 mg/kg and sodium bicarbonate 1 mEq/kg are administered intravenously. The sequence of events is as follows: at time −1 min before resuscitation the arterial line is flushed to assure patency (an accurate arterial pressure wave is essential for monitoring successful return of circulation) and then the medications are administered in quick succession via the venous line, followed by a 10 ml/kg intravenous fluid bolus of normal saline. At time −30 s before resuscitation the endotracheal tube is re-connected to the ventilator and the rats are mechanically ventilated at a rate 10 breaths/min above baseline using an FiO_2 of 1.0. At time 0 rapid manual chest compressions are performed with two fingers on the mid chest at a rate of approximately 300/min, until return of spontaneous circulation, which typically occurs in less than 1 min. During the CA the rat is kept on a heating pad. Usually the temperature will spontaneously decrease during CA and the rat will spontaneously re-warm after return of spontaneous circulation. Figure 2f depicts a typical tracing of the HR and MAP at the start of asphyxia through return of spontaneous circulation.

Post-resuscitation Care

Arterial blood gas values are measured 15–30 min after CA and the ventilator rate is adjusted if necessary to maintain a $PaCO_2$ of 35–45 mmHg. The FiO_2 is maintained at 1.0, though this could be reduced if desired. The rat is kept on the ventilator until 1 h after return of spontaneous circulation, at which time the arterial and venous lines are removed and the incision is closed with 6.0 silk sutures. The endotracheal tube is disconnected from the ventilator and the ability of the rat to breath spontaneously through the endotracheal tube is observed for 5 min prior to extubation. The rat is placed on supplemental O_2 (100%) via nose cone for 5 min after extubation. Afterwards, the rat is given a fluid bolus of D5W ½ NS, 20 cc/kg subcutaneously in the posterior mid thoracic region because typically the rats are comatose between 12–48 h depending upon the duration of asphyxia, and thus do not eat or drink. The rat is then placed in an incubator with supplemental oxygen for one hour, and then returned to the dam and littermates. The rats typically emerge from coma and are able to feed between 12–48 h depending upon the duration of asphyxia. For longer insults (>10 min), the rats are observed overnight in the incubator, reexamined in 24 h, and given another fluid bolus of D5W ½ NS, 20 cc/kg subcutaneously. Longer insults are possible; however, it is our experience that after 12 min of asphyxia, the rats cannot be weaned from the ventilator and would require prolonged ventilation and post-resuscitative care.

Outcome Evaluation

Several endpoints can be used to evaluate neurological outcome. These include both functional outcome and histological assessment. The duration of coma and spontaneous activity can be used as gross functional assessments, as the duration of coma is directly proportional to the duration of asphyxia.

Vestibulomotor function can be evaluated using a number of quantitative tests. The beam balance test employs a narrow wooden beam (1.5 cm wide) and the duration the rat remains on the beam is measured (for up to 60 s). Prominent beam balance deficits are detected after asphyxia for up to 3 days after 8 min of asphyxia [3, 8]. Gross motor function can also be evaluated using an inclined plane test. This test involves a flat rubber covered surface that is increased at 5° angle increments until the rat can no longer maintain its position on the surface. Prominent inclined plane deficits are detected after asphyxia for up to 2 days after 8 min of asphyxia [3, 8].

Cognitive deficits can be evaluated using the Morris water maze, although there are unique considerations for younger rats. The Morris water maze employs a 180 cm diameter and 60 cm high pool filled with water to a depth of 28 cm kept at 20–22 °C. A platform 10 cm in diameter and positioned 2 cm below the water's surface is used as the hidden goal platform in a room with constant extra-maze cues. Since spatial memory acquisition is not felt to be developed in rats until PND 24, testing begins at least 7 days after asphyxia in PND 17 rats. Because only moderate

Morris water maze deficits are seen after 8 min of asphyxia [3], increasing durations of asphyxia were tested. However, injury effects remained difficult to detect, primarily because of the poor performance of sham rats (relative to adult rats). The Morris water maze paradigm has since been adjusted to include a 3 day acclimation period, using a visible platform testing, followed by a 4 day hidden platform testing period. Using this modified paradigm robust injury effects are observed.

Histological assessment has been performed at different time points after different durations of asphyxial CA. Hematoxylin and eosin staining shows eosinophilic, pyknotic neurons in CA1 hippocampal neurons and layer 5 cortical neurons after 8 min of asphyxial CA [3]. These cells are also Flurojade B positive. Fluorojade B staining identifies neurons that are undergoing degeneration [9]. Terminal deoxynucleotidyl transferase-mediated biotin dUTP nick end labeling (TUNEL) identifies neurons with DNA damage including those undergoing apoptosis [10], and is detected in CA1 hippocampal but not cortical neurons after 8 min of asphyxia [3]. CA1 hippocampal neurons are readily quantifiable, and a 40–60% reduction in CA1 hippocampal neuron survival is seen after 8 min of asphyxia [3]. Remarkably, neurodegeneration as assessed using Fluorojade B is detectable up to 5 weeks after asphyxial CA [8].

While after 8 min of asphyxia neuronal damage appears to be restricted to layer 5 cortical and CA1 hippocampal neurons, longer durations of asphyxia result in extension of neuronal damage to other brain regions. Figure 3 shows Fluorojade B staining 48 h after 8.5 and 9 min of asphyxial CA. After 8.5 min of asphyxia, neurodegeneration extends to the subiculum and CA2 hippocampal regions (Fig. 3b), and beyond cortical layer 5 (Fig. 3e). After 9 min of asphyxia, neurodegeneration extends to CA3, CA4, dentate gyrus, and hilar hippocampal regions, throughout the cerebral cortex and thalamus, and Purkinje cells in the cerebellum. Thus, asphyxia-dose dependent increases in neurodegeneration are observed in this model.

Advantages and Applications

This model has many advantages, including the capability to: (1) perform invasive physiologic monitoring and acute resuscitation that closely mimic guidelines used in pediatric patients; (2) assess acute and long term functional outcome using standardized behavioral tools; (3) assess acute and long term biochemical and cellular markers of cell death and molecular mechanisms; and (4) test clinically relevant and novel interventions and treatments for efficacy after asphyxial CA in the developing brain. One example of the later is therapeutic hypothermia. While hypothermia is now recommended after VF-induced cardiac arrest in adult patients [11], its efficacy after pediatric asphyxial arrest is less clear. Using this model, brief induced post-resuscitative hypothermia—32 °C for 1 h followed by a 1 h rewarming period—improved motor function and provided long lasting neuroprotection compared with normothermic controls [8].

This model can also be used to evaluate CBF patterns after CA in immature rats, information that is currently extrapolated from models of CA in adult animals. To

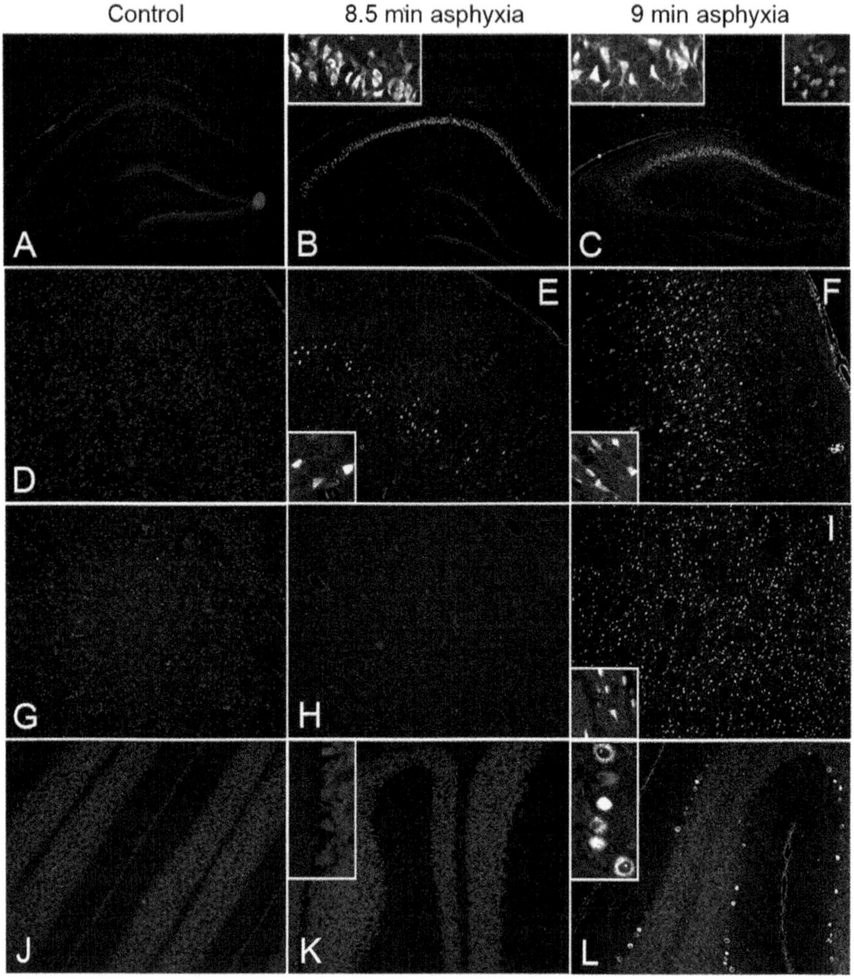

Fig. 3 Asphyxia duration-dependent increase in neurodegeneration as assessed by Fluorojade B staining 48 h after 8.5 (panels **b**, **e**, **h**, **k**) and 9 (panels **c**, **f**, **i**, **l**) min vs. naïve control (panels **a**, **d**, **g**, **j**). Fluorojade positive cells are bright white in this black and white image. Note expansion of Flurojade B staining in regions of the hippocampus (panels **a**–**c**), cortex (panels **d**–**f**), thalamus (panels **g**–**i**), and cerebellum (panels **j**–**l**) with increasing durations of asphyxia

measure CBF, arterial spin labeling (ASL) magnetic resonance imaging (MRI) was used to allow for non-invasive, serial measurement of regional CBF. This method uses the magnetic signal generated by protons in arterial blood as a tracer. To facilitate this in this model, the anesthetic regimen was modified. Anesthesia is initially achieved with isoflurane until arterial and venous lines are placed. Afterwards, the isoflurane is discontinued and fentanyl is administered at a dose of 50 μg/kg/h for anesthesia. A continuous vecuronium infusion at a dose of 5 mg/kg/h is used to prevent motion during MRI. Figure 4 shows serial CBF measurements in a sham rat

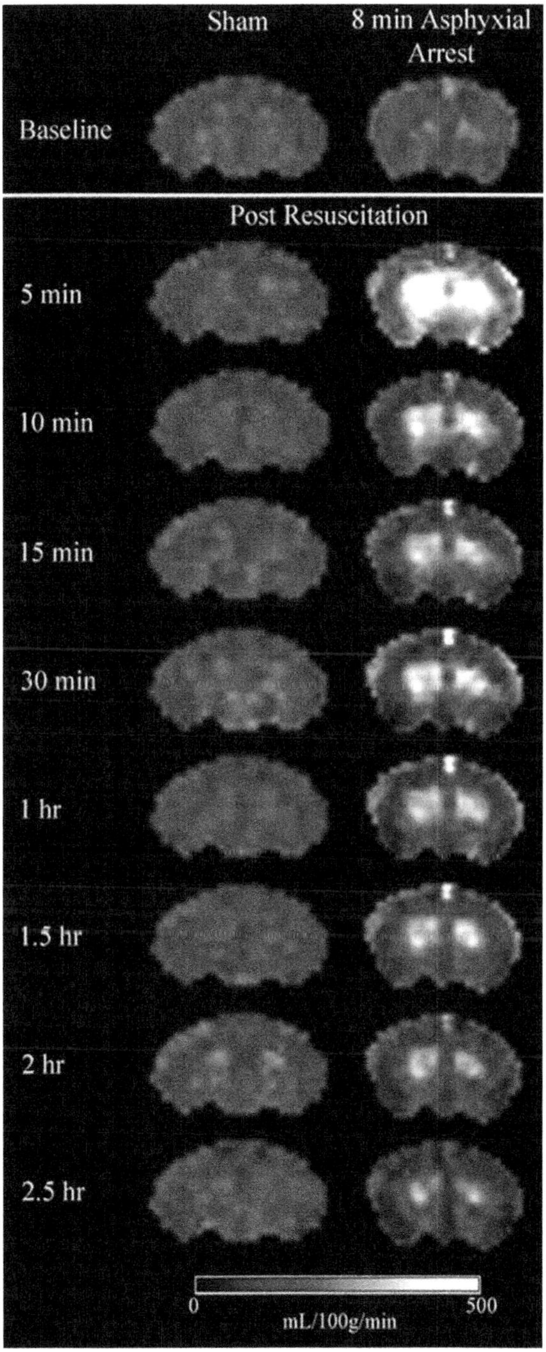

Fig. 4 Regional CBF maps measured serially at baseline and after 8 min of asphyxia. CBF was measured using ASL in a 7 T Bruker Biospec MRI. Note post-arrest hyperemia (bright areas in this black and white image) after asphyxia vs. sham

and a rat after 8 min of asphyxial CA. CBF was measured using the ASL method and a 7 T Bruker Biospec MRI. After 8 min of asphyxia prominent hyperemia is observed in subcortical brain structures. Thus, this model coupled with ASL-MRI will permit a comprehensive evaluation of cerebrovascular disturbances after asphyxial CA in the developing brain.

Conclusions

We describe a clinically relevant model of pediatric asphyxial CA. This model can be used to characterize cerebrovascular disturbances and molecular mechanisms relevant to infant and childhood victims of respiratory arrest, for example, those due to respiratory infections or near-drowning accidents. This model also has the capacity to evaluate potential therapies aimed at improving outcome and reducing the devastating effects of hypoxic-ischemic encephalopathy after CA in infants and children.

References

1. Young KD, Gausche-Hill M, McClung CD, Lewis RJ. A prospective, population-based study of the epidemiology and outcome of out-of-hospital pediatric cardiopulmonary arrest. Pediatrics. 2004;114(1):157–64.
2. Katz L, Ebmeyer U, Safar P, Radovsky A, Neumar R. Outcome model of asphyxial cardiac arrest in rats. J Cereb Blood Flow Metab. 1995;15(6):1032–9.
3. Fink EL, Alexander H, Marco CD, et al. Experimental model of pediatric asphyxial cardiopulmonary arrest in rats. Pediatr Crit Care Med. 2004;5(2):139–44.
4. Rice D, Barone S Jr. Critical periods of vulnerability for the developing nervous system: evidence from humans and animal models. Environ Health Perspect. 2000;108(Suppl 3):511–33.
5. Hendrich KS, Kochanek PM, Melick JA, et al. Cerebral perfusion during anesthesia with fentanyl, isoflurane, or pentobarbital in normal rats studied by arterial spin-labeled MRI. Magn Reson Med. 2001;46(1):202–6.
6. Manole MD, Hickey RW, Momoi N, et al. Preterminal gasping during hypoxic cardiac arrest increases cardiac function in immature rats. Pediatr Res. 2006;60(2):174–9.
7. Statler KD, Kochanek PM, Dixon CE, et al. Isoflurane improves long-term neurologic outcome versus fentanyl after traumatic brain injury in rats. J Neurotrauma. 2000;17(12):1179–89.
8. Fink EL, Marco CD, Donovan HA, et al. Brief induced hypothermia improves outcome after asphyxial cardiopulmonary arrest in juvenile rats. Dev Neurosci. 2005;27(2–4):191–9.
9. Schmued LC, Hopkins KJ. Fluoro-Jade B: a high affinity fluorescent marker for the localization of neuronal degeneration. Brain Res. 2000;874(2):123–30.
10. Loo DT, Rillema JR. Measurement of cell death. Methods Cell Biol. 1998;57:251–64.
11. 2005 International Consensus on Cardiopulmonary Resuscitation and Emergency Cardiovascular Care Science with Treatment Recommendations. Part 6: Paediatric basic and advanced life support. Resuscitation. 2005;67(2–3):271–91.

Transient Middle Cerebral Artery Occlusion Model in Rodents

Ying Mao, Wei Zhu, and Guo-Yuan Yang

Abstract Rodent transient middle cerebral artery occlusion (MCAO) is one of the widely used focal ischemia models in the world. This model mimics human cerebral ischemic injury and provides a unique tool for the mechanistic studies of ischemia in brain tissues *in vivo*. Rodent ischemia and reperfusion models are commonly used because of a greater understanding of rodent genetics, the availability of specific antibodies to rodent epitopes and molecular probes in rodents, and the ability to study transgenic mouse strains. In this chapter, we discuss: (1) the preparation of MCAO in rodents; (2) the intra-luminal MCAO techniques and its modifications; (3) the quantitative evaluation of model success; and (4) the advantages and limitations of MCAO models.

Keywords Blood flow · Infarct volume · Middle cerebral artery occlusion · Rodent · Transient focal ischemia

Introduction

Ischemic stroke is the third most common disease-related killer in the United States. Developing reliable and reproducible transient focal cerebral ischemia models, which mimic features of human ischemic stroke syndromes, is necessary for the

Y. Mao · W. Zhu
Department of Neurosurgery and Institute of Neurosurgery, Huashan Hospital, Fudan University School of Medicine, Shanghai, China
e-mail: yingmaoc@online.sh.cn

G.-Y. Yang (✉)
Department of Anesthesia and Perioperative Care, Center for Cerebrovascular Research, University of California San Francisco, San Francisco, CA, USA

Department of Neurological Surgery, University of California San Francisco, San Francisco, CA, USA
e-mail: gyyang@anesthesia.ucsf.edu

J. Chen et al. (eds.), *Animal Models of Acute Neurological Injury*, Springer Series in Translational Stroke Research, https://doi.org/10.1007/978-3-030-16082-1_9

systematic study of pathophysiology and treatment of stroke. Rodent models of focal cerebral ischemia has been increasingly gaining acceptance in recent years because they are relevant to human clinical settings. The first commonly used technique of rat middle cerebral artery occlusion (MCAO) involves coagulation of the middle cerebral artery via a craniotomy [1, 2]. The focal ischemia model has an advantage in comparing the difference in ischemic changes between ischemic and non-ischemic hemispheres. However, this permanent MCAO model is invasive and does not permit reperfusion. Post-ischemic blood restoration occurs frequently in human ischemic stroke, especially in cerebral embolism. To mimic this clinical condition, Koizumi and Zea-Longa developed a novel and a relatively noninvasive transient MCAO model using an intraluminal suture [3, 4]. The advantage of this model is its ability to induce focal reperfusion injury. Since then, several groups have modified this model to make it more feasible and reproducible. The greatest improvement is the coating of the sutures using poly-L-lysine, a polycationic polymerized amino acid [5].

Yang et al. further developed a mouse transient MCAO model using the intraluminal suture technique [6, 7]. The lesion induced by suture MCAO is involved both in the cortex and striatum. Working with rodents in ischemia or reperfusion research is advantageous because of our greater understanding of animal genetics, the availability of specific antibodies and molecular probes in mice, and our ability to study transgenic strains [8]. This mouse transient focal ischemic model is widely used for the study of molecular mechanisms in ischemic brain injuries. To minimize the lesion, a modification of Tamura's model has been developed. This model induces less injury and localizes only in the cortex region.

Several other focal cerebral ischemia models in rodents have also been developed, such as the rat or mice embolic focal cerebral ischemia model [9–11], and the mini-stroke model [12]. Since these models are not as commonly used as the suture and distal MCAO models, we did not include them in this chapter.

Preparation of MCAO in Rodents

Anesthesia and Physiological Parameter Monitoring

Rats

Adult male Sprague-Dawley (or other strains) rats weighing 260–290 g are optimal, although the use of animals that are not within this range has also been reported. The most commonly used anesthesia for MCAO is isoflurane inhalation. Other inhalants include halothane (1–3% in a mixture of 70% nitrous oxide and 30% oxygen), ketamine (80–100 mg/kg) and xylazine (5–10 mg/kg). A PE-50 catheter is inserted into the femoral artery for continuous monitoring of arterial blood pressure, and sampling for analysis of blood gases and blood pH. The temperature of both the temporal muscle and the rectum is measured by thermocouples (E type, OMEGA

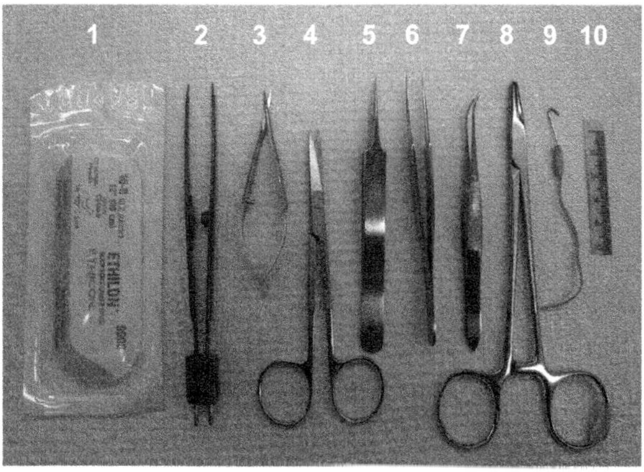

Fig. 1 Photograph shows a set of basic microsurgical equipment for the MCAO model. All items must be sterilized before use: 1. nylon suture; 2. bipolar forceps; 3. microscissors for small vessels; 4. surgical scissors for animal skin and tissue use; 5 and 6. microforceps; 7. tissue forceps; 8. needle holder; 9. skin hook for the exposure; 10. ruler

Engineering, Inc.). Brain temperature is regulated by using a heating lamp and a heating pad to maintain normothermia. Blood pressure is maintained at 100 mmHg or above, which is regulated by the intensity of the anesthetic. Blood gases are analyzed once during the experiment to maintain an arterial PaO_2 level of 100 mmHg or higher (Fig. 1).

Mice

Adult male CD-1 (or other strains) mice weighing 30–35 g would be ideal for the procedure. The anesthesia for MCAO in mice is the same as for rats. A PE-10 catheter is inserted into the femoral artery for continuous monitoring of arterial blood pressure and sampling for the analysis of blood gases and blood pH. Measuring temporal muscle and rectum temperature and monitoring blood pressure and blood gas are the same as for rats.

Equipment

The optimal small animal surgery room needs to be kept quiet and pathogen-free, and the room temperature maintained at 25 ± 3. The following items are recommended for small animal surgery:

1. Operation microscope (OPMI 6-SH, Zeiss, Germany)
2. Bipolar coagulator (Malis™ Precision-Control bipolar coagulator, CMC™-II-PC, Codman and Shurtleff Inc., Randolph, MA)
3. Exercise Physiological System (AD Instruments, PowerLab/4SP, Castle Hill, Australia)
4. Temperature controller (Homeothermic Blanket Control Unit, Harvard Apparatus, Cambridge, MA)

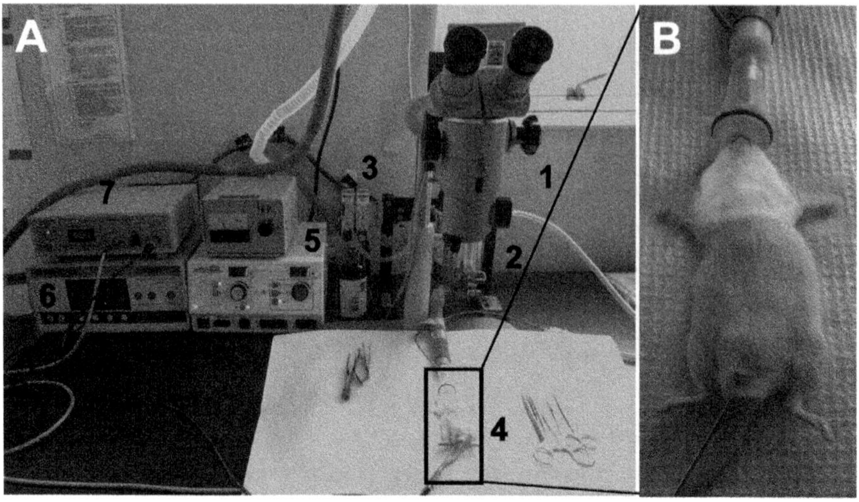

Fig. 2 (**A**) The animal surgical set for MCAO procedure. The basic set includes: 1. Surgical microscope; 2. Vaporizer in anesthesia; 3. Flowmeter for air gas control; 4. Animal surgical table; 5. Bipolar coagulator; 6. Laser Doppler Flowmeter; 7. Temperature controller. (**B**) Photograph shows a mouse under inherent anesthesia using isoflurane

5. Laser Doppler Flowmeter (LASERFLO® Blood perfusion monitor BPM^2, VASAMEDICS, St. Paul, MN)
6. Dry sterilizer (Simon Keller AG, Fine Science Tools, CH 3400, Foster City, CA)
7. Vaporizer (Summit Anesthesia Solutions, Dragerwerk AG Lubeck, Isofluorane Vapor 19.1, Bend, OR)
8. Stereotaxic frame (Model 900, DAVID KOPF INSTRUMENTS, Tujunga, CA)
9. pH/blood gas analyzer (Bayer, Radiolab 248, Tarrytown, NY)
10. High-speed micro drill (Fine Science Tools, Foster City, CA)
11. Surgical equipment (Fig. 2).

Transient MCAO Techniques and Their Modifications

Intraluminal Suture MCAO Model

Rat

Under the operating microscope, the left common carotid artery (CCA) is exposed through a midline incision. All other branches are isolated and coagulated along with the terminal lingual maxillary artery branches. The internal carotid artery (ICA) is then isolated and its extracranial branch, the pterygopalatine artery, is ligated close to its origin. Thus, the ICA, which is the only extracranial branch of the CCA, remains open. A 3-cm length of 3–0 nylon suture with a slightly enlarged and rounded tip is inserted into the transected lumen of the external carotid artery (ECA), and gently advanced from the ICA across to the opening of the MCA. The distance from the tip of the suture to the bifurcation of CCA is 17–19 mm. For reperfusion, the suture is withdrawn into the ECA to restore ICA-MCA blood flow. The skin is sutured and the animal is returned to the cage when it recovers from the anesthesia [13].

Mouse

The method for mouse MCAO is similar to the rat model described above. The only difference is the suture size (2-cm length of 5–0 nylon). The distance from the tip of the suture to the bifurcation of CCA is 10–11 mm (Fig. 3) [6, 7].

Distal MCAO Model [14]

Rat

The most commonly used anesthetic for the distal MCAO model is isoflurane inhalation. Under a surgical microscope, a 2-cm incision is made between the right orbit and the tragus. The temporal muscle is retracted laterally and a 3-mm diameter craniotomy is made just rostral to the foramen ovale. The dura is incised with a tiny hook and the MCA is exposed. The arachnoid is then opened and the MCA ligated with a 8–0 suture for the permanent MCAO. Interruption of the blood flow is confirmed visually under the microscope. For the transient distal MCAO, the MCA is temporarily occluded with a microclip or a 8–0 suture ligation. Immediately after, both CCAs are occluded with microclips, and the rats undergo occlusion for 60 min. This model produces an infarct restricted to the cerebral cortex (Fig. 4). Reperfusion is achieved by removing the microclips or suture. Restoration of MCA blood flow can be verified visually. Blood flow is monitored during the MCAO using laser Doppler flowmetry. The temporal muscle is repositioned and the skin is closed.

Mouse

The method for the mouse distal MCAO is similar to the rat distal MCAO model above. Under a surgical microscope, a 1-cm incision is made between the right orbit and tragus. The arachnoid is then opened and the MCA ligated with a 10–0 suture for the permanent MCAO. For the transient distal MCAO, the MCA is temporarily occluded with a microclip or a 10–0 suture ligation. Reperfusion is achieved by removing the microclips or the suture. Restoration of MCA blood flow can be verified visually. Blood flow is monitored during the MCAO using Laser Doppler Flowmetry. The temporal muscle is repositioned and the skin is closed.

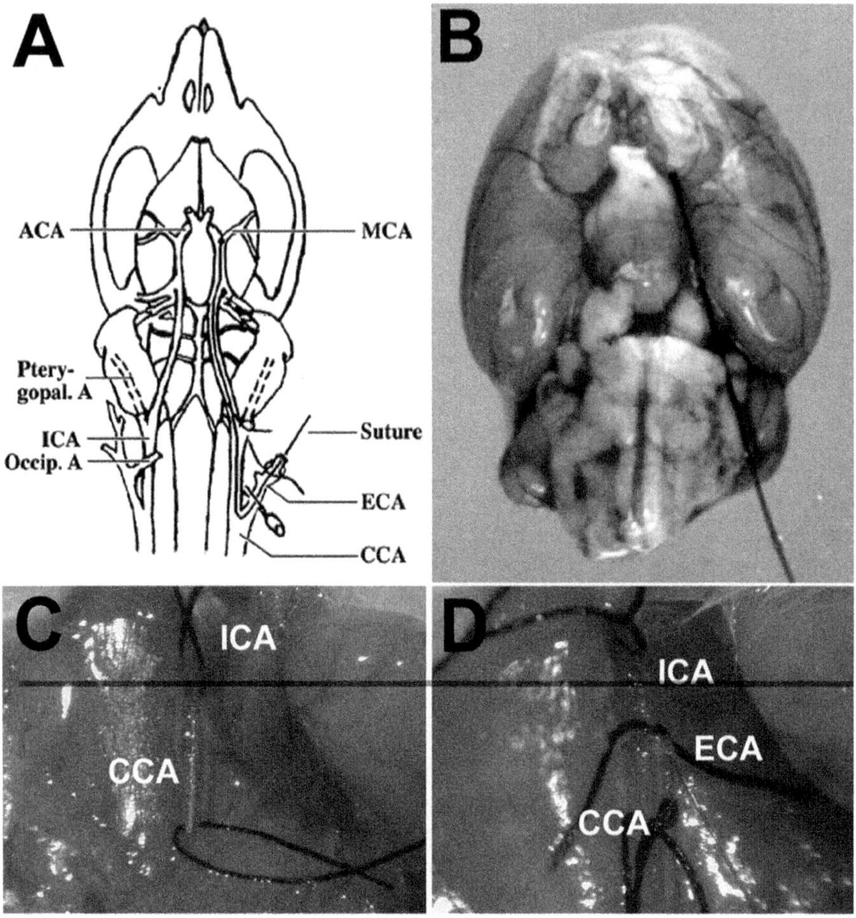

Fig. 3 (**a**) Diagram of cerebral anatomy in mice illustrates extracranial and intracranial vascular relations. Anterior communicate artery (ACA), middle carotid artery (MCA), internal carotid artery (ICA), external carotid artery (ECA), common carotid artery (CCA) and their branches are shown. (**b**) Photograph of mouse brain at autopsy showing intravascular suture, inserted through ICA, within lumen of ECA. Suture occludes origin of MCA from ICA. (**c**) Photograph shows the anatomy of CCA and ICA in the neck; ICA is looped by suture. (**d**) Photograph shows a right blue suture inserted into the opening of the ECA and advanced into ICA

Quantitative Evaluation of Model Success

Focal Cerebral Blood Flow Measurement

Measurement of Local CBF Using Laser Doppler Flowmetry (LDF)

LDF can be used to verify that MCAO has occurred and to follows the flow over time. A BPM2 Laser Doppler Flowmeter is equipped with a small caliber probe with a 0.7-mm diameter. The mice are placed in a stereotactic frame with a mouse

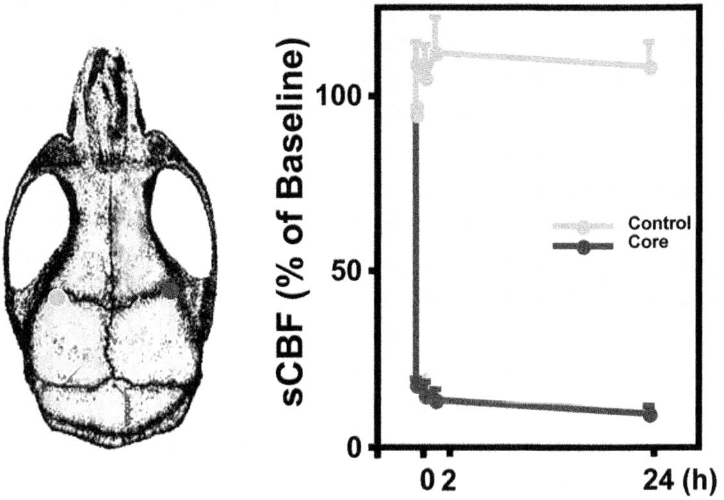

Fig. 4 The left side shows a diagram of a mouse skull. Two points are marked to indicate the same area each time: yellow dot is 3.5 mm lateral to the sagittal suture which indicates the control region; red dot is 3.5 mm lateral to the sagittal suture on the right side which indicates the ischemic core region. Graph on the right shows the changes of regional CBF at 5 min, 1 and 2 h, and 1 day after MCAO, in the ischemic core and control point in the contralateral hemisphere

holder, a midline incision is made, and a thermocouple is placed in the right temporal muscle and secured with cyanoacrylate glue. The pericranium over the coronal and sagittal sutures is exposed and kept moist with normal saline for LDF measurement [15]. The probe, held in a micromanipulator, is lowered until it comes in contact with the surface of the dura. Warmed saline is used during the experiment to slowly rinse around the probe and maintain a clear medium between the probe and the dura [16]. Stable baseline LDF readings are obtained for 20 min prior to occlusion. Continuous digital display of LDF values is averaged over 5-s intervals and recorded every 20 min during ischemia and early reperfusion. The CBF values are calculated and expressed as percentages of baseline values (Fig. 5).

Infarct Volume Measurement

2,3,5-Triphenyltetrazolium Chloride (TTC) Staining

This method is simple and fast; but not as accurate as the Cresyl violet staining method [17, 18]. After MCAO, animal brains are removed immediately. From rats, six coronal slices of 1, 3, 5, 7, 9 and 11 mm distal from the frontal pole are dissected using a brain slicer, and from mice, four coronal slices of 1, 3, 5, and 7. All brain slices are stained with 2% (wt/vol) TTC in Dulbecco's phosphate buffer (pH 7.4) at 37 °C for 20 min. The volume of infarction is calculated by multiplying the distance between sections (Fig. 6) [19].

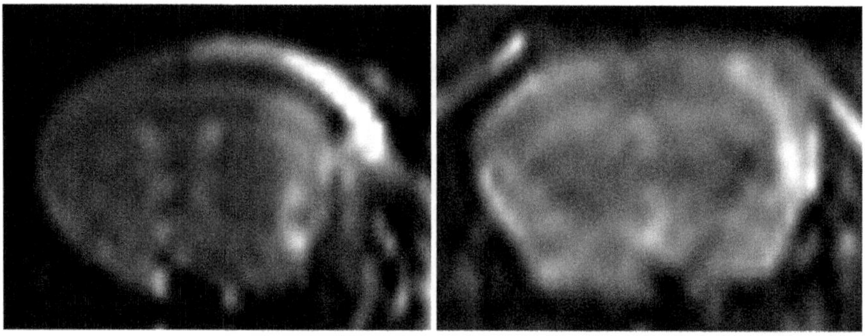

Fig. 5 MRI image shows the higher density signal in the right cortical region following 24 h of distal MCAO

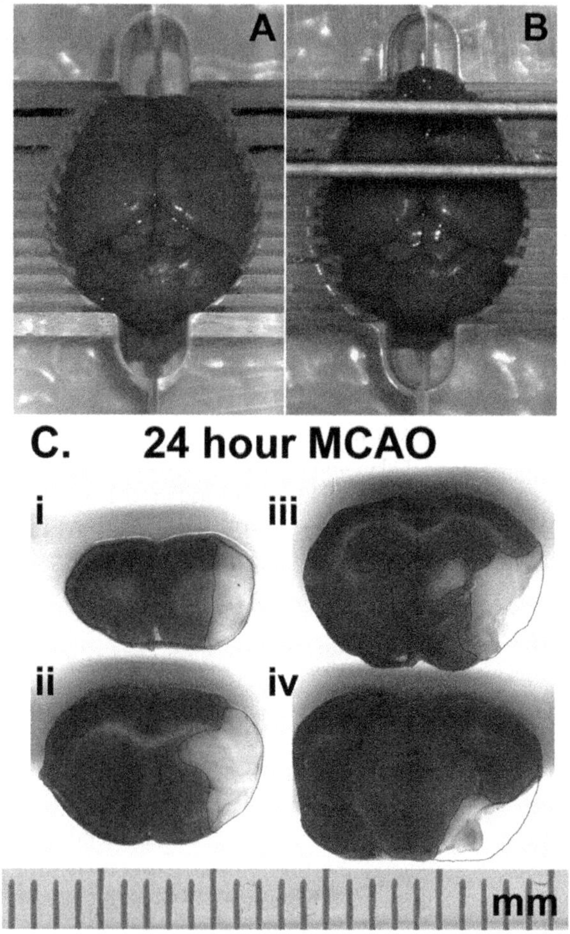

Fig. 6 Experimental sampling after ischemic injury. (**a**) Mouse brain matrix, which can be used for 1 mm coronal sectioning. (**b**) A mouse brain under the sectioning. (**c**) A set of TTC stained coronal sections, which is sectioned by using this mouse brain matrix

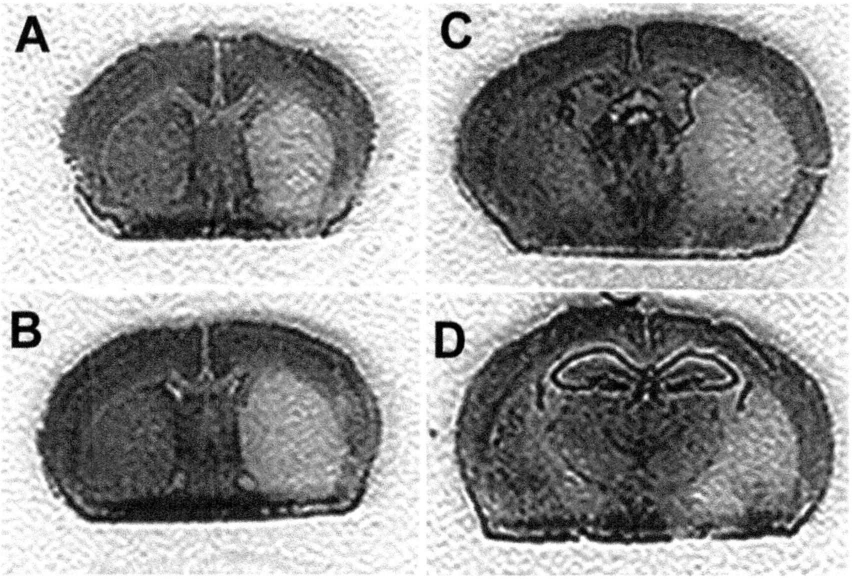

Fig. 7 A set of Cresyl violet staining mouse brain coronal sections. The mouse underwent 45 min of MCAO and followed by 24 h of reperfusion. The infarct areas are usually limited in the basal ganglia region

Cresyl Violet Staining Section

Following MCAO, the animals are killed and the brains are removed and frozen immediately in dry ice, in 2-methylbutane at −42 °C for a 5-min period. Twenty coronal sections from the frontal pole, 20 μm in thickness and in 200 μm intervals, are cut with a cryostat and mounted on slides. The sections are dried and then stained with Cresyl violet [20]. Non-Cresyl violet staining is used to identify the infarct area. Using NIH Image J software, the ischemic lesion area is calculated as the difference in the area of the non-ischemic hemisphere and the normal area in the ischemic hemisphere. The infarct volume is calculated by multiplying the infarct areas by the thickness of the sections (Fig. 7).

Neurobehavioral Evaluation

Body Asymmetry Test

This test is a simple and fast method to estimate the degree of ischemic brain injury. To measure motor asymmetry, animals are examined by the elevated body swing test as described [21]. Mice are examined for head swing while being suspended by their tails. The direction of the swing, either right or left, is recorded when the mouse turns its head sideways by an approximate 10° angle to the body's midline. The mouse is allowed to move freely for at least 30 s before the next swing is counted, and the trials are repeated 20 times for each animal. The frequency of head swings toward the non-injured side is counted and normalized as follows: (number of lateral swings in 20 tests − 10)/10 × 100%.

Water-Maze Learning

In the water maze test, the ability of the animal to locate a hidden submerged platform in a circular pool (142 cm in diameter, 50 cm deep) filled with warm opaque water is assessed. Animals are trained first to locate a visible platform, and then to locate a hidden platform; in a subsequent trial the hidden platform is removed. Animals receive two training sessions per day for 5 consecutive days. Each session consists of three trials with a 10-min inter-trial interval. The interval between the two daily sessions is 3 h. Once the animals locate the platform they are allowed to remain on it for 10 s. The time to reach the platform (latency), path length, and swimming speed are recorded with a Noldus Instruments EthoVision video tracking system set to analyze two samples per second.

Rotor-Rod Test

The rotor-rod test is used to screen for motor deficits that might influence performance in the water maze testing. Rotor-rod balancing requires a variety of proprioceptive, vestibular, and fine-tuned motor abilities. The task requires the animal to balance on a rotating rod. Performance can be evaluated in trials of fixed speed; however, a slow acceleration during the test session limits inter-individual variability in performance. After a 1-min adaptation period on the rod at rest, the rod is accelerated at 5 rpm every 15 s, and the length of time when the animal remains on the rod (fall latency) is recorded.

Novel Open Field Activity

To quantitate open field activity, animals are placed in brightly lit, individually automated infrared photocell activity cages interfaced with a computer. The open field is divided into nine zones, in which the center zones contain 6 × 6 photocells and the side zones, 5 × 5 photocells. The data are converted into a zone map file, and the following parameters are calculated per interval: active time, distance moved, rearing times and events, corner entries, center entries, and percentage of time spent in the center. The open field activity is recorded for three consecutive 10-min intervals after a 1-min adaptation period. The number of fecal boli deposited during the 30-min period is recorded.

Novel Object Recognition Task

Animals are individually habituated to the open field for 3 days. During the training session, two novel objects are placed in the open field and each animal is allowed to explore for 15 min. During retention tests, the animals are placed back in the open field, in which one of the familiar objects is replaced by a novel object, and are allowed to explore for 5 min. The time spent exploring each object during the training and retention sessions is recorded. The ratio of the amount of time spent exploring any of the two objects (training session) or the novel one (retention session), to the total time spent exploring both objects is calculated and used to measure object recognition memory.

Advantages and Limitations

Advantages

The MCAO model produces cerebral ischemia in the vascular territory of the MCA and gives rise to focal metabolic disturbance resulting in brain edema, selective neuronal necrosis, and infarction. The suture MCAO model is relatively simple to operate because no craniotomy needs to be performed; therefore, other possible effects, such as craniotomy, focal inflammation, and focal facial muscle damage, can be excluded. Suture MCAO producing ischemic brain injury involves both cortex and striatum; this injury further produces obvious sensori-motor dysfunction, which can be assessed by neurobehavioral evaluation.

The distal MCAO model produces ischemic brain injury that is restricted in the cortex of the ipsilateral hemisphere. The overall mortality rate remains low, although craniotomy is needed to expose and clip the MCA branch. Focal ischemia model producing injury is limited in the ipsilateral hemisphere, and therefore, the same region opposite to the injury in the contralateral hemisphere can be used for the control.

Limitations

Mortality

Approximately 10% of the rats die either during ischemia or shortly after reperfusion even in experienced hands. The main cause of the mortality is subarachnoid hemorrhage, which is caused by inappropriate insertion of the suture. Improving the top of the suture and practicing surgery more often can attenuate this mortality rate. Animal deaths also increase gradually following the time course of MCAO, especially after 5 days, caused by the animals' inability to eat and drink. Injecting saline and feeding them with milk-chocolate may help reduce mortality.

Variation

Despite the reproducible outcomes, the transient suture MCAO model has a relatively higher variation of infarct volume. Zea-Longa has reported that the success rate in achieving infarction is only 56% [4]. Other studies have also reported that approximately 30% of the experimental animals have to be excluded. To reduce variability of infarct volume, performing a pilot study to optimize the experiment (e.g., choosing appropriate suture to fit the diameter of animal ICA) is necessary. Furthermore, carefully controlling brain temperature, blood pressure, cerebral blood flow during and immediately after MCAO, as well as the length of occlusion period, is extremely important. In addition, a suture coated with poly-L-lysine solution (0.1% [wt/vol], in deionized water; Sigma) may help reduce variability [5].

References

1. Tamura A, Graham DI, McCulloch J, Teasdale GM. Focal cerebral ischaemia in the rat: 2. Regional cerebral blood flow determined by [14C]iodoantipyrine autoradiography following middle cerebral artery occlusion. J Cereb Blood Flow Metab. 1981;1:61–9.
2. Tamura A, Graham DI, McCulloch J, Teasdale GM. Focal cerebral ischaemia in the rat: 1. Description of technique and early neuropathological consequences following middle cerebral artery occlusion. J Cereb Blood Flow Metab. 1981;1:53–60.
3. Koizumi J, Yoshida Y, Nakazawa T, Ooneda G. Experimental studies of ischemic brain edema. I: A new experimental model of cerebral embolism in rats in which recirculation can be introduced in the ischemic area. Jpn J Stroke. 1986;8:1–8.
4. Zea-Longa E, Weinstein PR, Carlson S, Cummins R. Reversible middle cerebral artery occlusion without craniectomy in rats. Stroke. 1989;20:84–91.
5. Belayev L, Alonso OF, Busto R, Zhao W, Ginsberg MD. Middle cerebral artery occlusion in the rat by intraluminal suture. Neurological and pathological evaluation of an improved model. Stroke. 1996;27:1616–22; discussion 1623.
6. Yang GY, Chan PH, Chen J, et al. Human copper-zinc superoxide dismutase transgenic mice are highly resistant to reperfusion injury after focal cerebral ischemia. Stroke. 1994;25:165–70.
7. Yang GY, Liu XH, Kadoya C, Zhao YJ, Betz AL. Attenuation of ischemic inflammatory response in mouse brain using an adenoviral vector to induce overexpression of interleukin-1 receptor antagonist. J Cereb Blood Flow Metab. 1998;18:840–7.
8. Mao Y, Yang GY, Zhou LF, Stern JD, Betz AL. Focal cerebral ischemia in the mouse: description of a model and effects of permanent and temporary occlusion. Brain Res Mol Brain Res. 1999;63:366–70.
9. Zhang ZG, Zhang L, Ding G, et al. A model of mini-embolic stroke offers measurements of the neurovascular unit response in the living mouse. Stroke. 2005;36:2701–4.
10. Zhang Z, Chopp M, Zhang RL, Goussev A. A mouse model of embolic focal cerebral ischemia. J Cereb Blood Flow Metab. 1997;17:1081–8.
11. Zhang RL, Chopp M, Zhang ZG, Jiang Q, Ewing JR. A rat model of focal embolic cerebral ischemia. Brain Res. 1997;766:83–92.
12. Wei L, Rovainen CM, Woolsey TA. Ministrokes in rat barrel cortex. Stroke. 1995;26:1459–62.
13. Zhao YJ, Yang GY, Domino EF. Zinc protoporphyrin, zinc ion, and protoporphyrin reduce focal cerebral ischemia. Stroke. 1996;27:2299–303.
14. Won SJ, Xie L, Kim SH, et al. Influence of age on the response to fibroblast growth factor-2 treatment in a rat model of stroke. Brain Res. 2006;1123:237–44.

15. Dirnagl U, Pulsinelli W. Autoregulation of cerebral blood flow in experimental focal brain ischemia. J Cereb Blood Flow Metab. 1990;10(3):327–36.
16. Dirnagl U, Kaplan B, Jacewicz M, Pulsinelli W. Continuous measurement of cerebral cortical blood flow by laser-Doppler flowmetry in a rat stroke model. J Cereb Blood Flow Metab. 1989;9:589–96.
17. Bederson JB, Pitts LH, Germano SM, Nishimura MC, Davis RL, Bartkowski HM. Evaluation of 2,3,5-triphenylterazolium chloride as a stain for detection and quantification of experimental cerebral infarction in rats. Stroke. 1986;17(6):1304–8.
18. Lundy EF, Solik BS, Frank RS, et al. Morphometric evaluation of brain infarcts in rats and gerbils. J Pharmacol Methods. 1986;16:201–14.
19. Swanson RA, Morton MT, Tsao WG, Savalos RA, Davidson C, Sharp FR. A semiautomated method for measuring brain infarct volume. J Cereb Blood Flow Metab. 1990;10:290–3.
20. Yang GY, Pang L, Ge HL, et al. Attenuation of ischemia-induced mouse brain injury by SAG, a redox-inducible antioxidant protein. J Cereb Blood Flow Metab. 2001;21:722–33.
21. Borlongan CV, Saporta S, Poulos SG, Othberg A, Sanberg PR. Viability and survival of hNT neurons determine degree of functional recovery in grafted ischemic rats. Neuroreport. 1998;9:2837–42.

Targeted Occlusion to Surface and Deep Vessels in Neocortex Via Linear and Nonlinear Optical Absorption

David Kleinfeld, Beth Friedman, Patrick D. Lyden, and Andy Y. Shih

Abstract We discuss two complementary methods for the study of cerebral blood flow and brain function in response to the occlusion of individual, targeted blood vessels. These bear on the study of microstroke and vascular dysfunction in cortex. One method makes use of linear optical absorption by a photosensitizer, transiently circulated in the blood stream, to induce an occlusion in a surface or near-surface vessel. The second method makes use of nonlinear optical interactions, without the need to introduce an exogenous absorber, to induce an occlusion in a subsurface microvessel. A feature of both methods is that the dynamics of blood flow and functional aspects of the vasculature and underlying neurons in the neighborhood of the occluded vessel may be monitored before, during, and after the occlusion. We present details of both methods and associated surgical procedures, along with example data from published studies.

Keywords Ablation · Clot · Hypoxia · Ischemia · Microstroke · Photosensitizer · Plasma · Rodent · Two-photon

D. Kleinfeld (✉)
Department of Physics, University of California at San Diego, La Jolla, CA, USA

Graduate Program in Neurosciences, University of California at San Diego, La Jolla, CA, USA
e-mail: dk@physics.ucsd.edu; dk@ucsd.edu

B. Friedman
Department of Neuroscience, University of California at San Diego, La Jolla, CA, USA
e-mail: befriedman@ucsd.edu

P. D. Lyden
Graduate Program in Neurosciences, University of California at San Diego, La Jolla, CA, USA

Department of Neuroscience, University of California at San Diego, La Jolla, CA, USA
e-mail: plyden@ucsd.edu

A. Y. Shih
Department of Physics, University of California at San Diego, La Jolla, CA, USA
e-mail: a2shih@ucsd.edu

J. Chen et al. (eds.), *Animal Models of Acute Neurological Injury*, Springer Series in Translational Stroke Research, https://doi.org/10.1007/978-3-030-16082-1_10

Introduction

There are several forms of vascular dysfunction that can contribute to the onset of blood flow disturbances and neurodegeneration [1]. We consider dysfunction that is caused by vessel occlusion, which leads to ischemia. Our focus is on techniques that can occlude individual, targeted single vessels. The importance of targeting is three-fold. First, it provides a means to select a particular type of vessel, e.g., capillary versus surface artery. Second, It provides a means to select a vessel at a particular location, e.g., within a selected cortical column. Lastly, it provides a means to make both pre- and post-measurements of hemodynamic and neuronal state variables in the region that neighbors the occluded vessel.

The diameter of vessels in neocortex range over an order of magnitude in size. Surface communicating arterioles, which form a supply network for the underlying parenchyma, span 20–50 μm in diameter, penetrating arterioles, which transfer blood from the surface to the parenchyma, span 10–25 μm in diameter, and the microvessels and capillaries that supply the parenchyma span 5–10 μm in diameter [2]. A variety of techniques have been developed to induce occlusions in all of these vessels (Table 1). One class of procedures involves the introduction of small particles to the blood stream, such as microspheres or small thrombo-emboli, that can be tailored to lodge in arterioles of a particular size and produce occlusions. However, the location of the occlusion cannot be targeted. A second class of procedures involves the local injection of a powerful vasoconstrictor, particularly endothelin-1, that can be targeted to specific arteries and arterioles but typically spreads within the extracellular space to impact many vessels rather than just the desired target. A final

Table 1 Methods for occlusion of individual small vessels in cortex

Method	Vessel(s)	Targeted	Reperfusion	Representative references
Intravascular injection of microspheres (25–50 μm diameter)	Arterioles	No	No	[3, 4]
Intravascular injection of macrospheres (300–400 μm diameter)	Arteries	No	No	[5]
Intravascular injection of thrombo-emboli	Arteries Arterioles	No	Yes	[6]
Focal injection of vasoconstrictor (endothelin-1)	Arteries Arterioles	Partially	Yes	[7–10]
Suture ligation Micro-clip	Surface arteries	Yes	Yes	[11–13]
Thermo-coagulation Electro-coagulation	Surface arteries	Yes	No	[14, 15]
Photosensitizer (Rose Bengal)	Surface arteries Penetrating arterioles	Yes	Incompletely tested	[10, 16–18]
Plasma-mediated ablation (amplified 100 fs pulses)	Subsurface microvessels	Yes	No	[17]

class makes use of physical blockages, including ligation by a fine suture or a fine cautery. This approach is appropriate for relatively large arteries and veins on the cortical surface. Our goal has been to exploit the physical localization of highly focused laser light to form occlusions.

We consider two complementary techniques to form occlusions, both of which require the concurrent use of *in vivo* confocal laser scanning microscopy or *in vivo* two-photon laser scanning microscopy (TPLSM) as a means to image the underlying vasculature and select a vessel for a targeted occlusion. *In vivo* TPLSM is particularly advantageous as it allows fluorescently labeled structures to be visualized at depths of 500–1000 μm below the cortical surface [19]. The first technique makes use of near-threshold optical activation of the photosensitizer Rose Bengal and is suitable for vessels at the surface of the brain [16, 17]. These include the communicating arteries and the initial segment of the penetrating arterioles that branch off of the communicating arteries. The second makes use of plasma-mediated ablation through the nonlinear absorption of 100-fs, ultrashort laser pulses by blood plasma or the vascular wall [20]. This technique is suitable for subsurface vessels, such as arterioles, capillaries, and venules, as vessels above and below the target vessel are unaffected. However, this technique appears not to be suitable for surface vessels, as the energy deposited can lead to rupture of the free surface.

Common Procedures

Our procedures are applicable to any laboratory animal, although the majority of our experiments to date involve the adult rat with an observation window placed over parietal cortex. The underlying vasculature is imaged with *in vivo* TPLSM [21], with contrast generated by labeling of the blood plasma with high molecular weight fluorescein isothiocyanate-conjugated dextran [22–24] (fluorescein-dextran) (2 MDa, Sigma). Vessels throughout the cranial window are first mapped at low-magnification, typically 4-X (e.g., UB950, Olympus), to form a single map of an area of the exposed cortex. Target vessels for photodisruption are typically selected using the criteria that a portion of the vessel courses parallel to the cortical surface. We then map the local vascular architecture in the vicinity of the target vessel with a three-dimensional stack of images at high-magnification, typically 40-X (e.g., UM568, Olympus).

Photosensitizer-Mediated Occlusion of Surface Vessels

This technique involves the introduction of a photosensitizer, in our case Rose Bengal, into the blood stream and the subsequent irradiation of the target vessel with actinic light, green laser-light in our case (Fig. 1a). The intensity of the light is adjusted to be at the threshold for photo-thrombosis in the plane of the target surface vessel. Light that propagates into the parenchyma will be lower in intensity and thus below the threshold level to induce photo-thrombotic damage. This allows formation of the

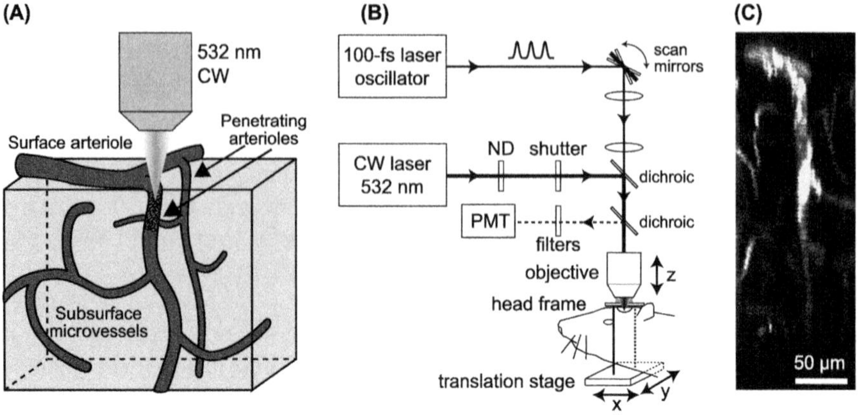

Fig. 1 Overview for forming intravascular occlusions in surface vessels using focal illumination of Rose Bengal. (**a**) Green laser light is tightly focused through an objective and directed into the lumen of a target vessel. The circulating Rose Bengal dye is photo-activated and initiates a clotting cascade. (**b**) Schematic of a two-photon laser scanning microscope system modified to include a continuous wave (CW) green-light laser for photo-thrombosis. The green laser is directed onto the optical axis of the near-infrared beam with dichroic mirror 1 (625 DRLP; Chroma). An approximately 3.5-mm hole was etched in the coating of dichroic mirror 2 (700 DCXRU; Chroma) to allow transmission of the green laser beam; this results in at most a 10% loss in collection efficiency. *CW* continuous wave, *ND* neutral density filter. (**c**) X–Z maximal projection of TPLSM image stack showing a penetrating arteriole selectively occluded by photo-thrombosis. All panels adapted from Schaffer et al. [16]

occlusion that is restricted to the surface vessel based on three physical mechanisms: (1) absorption of the incident light by the target vessel; (2) divergence of the beam beyond the target vessel as a result of the strong focus; and (3) scattering of light by brain tissue. However, small thrombi may be formed during the period of illumination that travel downstream rather than stick at the site of the irradiated region.

The persistence of Rose Bengal within the blood plasma may be estimated from *in vivo* measurements of its fluorescence, which peaks near 580 nm (omlc.ogi.edu/spectra/PhotochemCAD/html/), in the plasma. In mouse, which has approximately a tenfold smaller blood volume than rat, Rose Bengal is cleared from circulation within 5 min [25] (Fig. 2).

Equipment

Formation of an occlusion with a photosensitizer is readily combined with TPLSM to enable the targeting and real-time monitoring of blood vessels. The typical set-up makes use of a standard two-photon microscope, either home-built [26, 27] or a commercial instrument (e.g., Ultima with optional high speed two-photon optics set and external detector; Prairie Technologies), in which a 20 mW or greater green

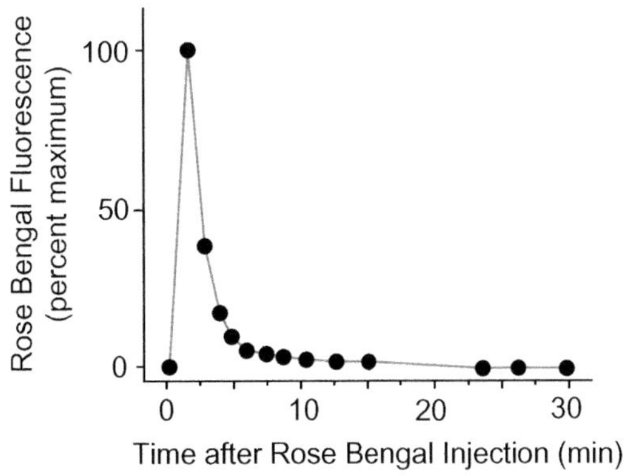

Fig. 2 Time-scale for Rose Bengal clearance from circulation. Rose Bengal fluorescence in a cerebral arteriole was repeatedly measured after tail vein injection using TPLSM planar scans. Rose Bengal is largely cleared from circulation within 5 min. Data was collected from mouse, courtesy of Zhang and Murphy [25]

(λ = 532 nm) laser light (e.g., GCL-100-M, CrystaLaser) is introduced into the beam pathway with a dichroic mirror (Fig. 1b). We used a fixed location of the beam, centered along the optical axis. The diameter of the beam is typically adjusted to pass through a hole in the emission dichroic (2 in Fig. 1b) and underfills the back aperture of the objective. The final spot size is approximately 3 μm diameter and is brought to the same focus as the pulsed near-infrared beam for TPLSM using an appropriate telescope. In this case, the motorized/encoding X–Y stage can be used to reposition the animal to change the location of photoactivation. Alternatively, the green actinic light may be introduced through a second set of scanners so that the position may be shifted electromechanically (e.g., Ultima with optional second-set of scanners). In all cases, the intensity of the laser light at the focus should initially be held at 0.1 mW, using a tunable neutral density filter in the beam path, and increased as needed up to a maximum of 5 mW.

Procedure

1. Prior to surgery:
   (a) Choice of anesthetic:
      - Isoflurane in a 30% O_2–70% N_2O gas mixture, i.e., 4% (v/v) induction and 1–2% (v/v) maintenance, provides adjustable anesthesia and the option of animal recovery.
      - Urethane provides long-lasting and stable anesthesia, i.e., 1000 mg per kg rat initial dose with 100 mg per kg supplements as required, but is difficult to adjust and is inappropriate for recovery. When using urethane, a 30% O_2–70% N_2O gas mixture is delivered to maintain normal blood gas levels.
      - Other anesthetic agents are also suitable for this procedure [28].

   (b) Check for lack of toe pinch reflex to ensure adequate level of anesthesia.
   (c) Secure animal in stereotaxic frame.

(d) Inject atropine, 0.05 mg per kg rat, intraperitoneal (i.p.) to reduce respiratory distress.
(e) Apply ophthalmic ointment to eyes to retain moisture.
(f) Inject 2% (v/v) Lidocaine™ into the scalp subcutaneously before incision for local anesthesia.

2. Throughout surgery and recording (Fig. 3d):

(a) Ensure heart and breathing rates are within a normal range, i.e., 300–400 and 60–120 events per minute, respectively, using a pulse oximeter (8600V; Nonin).

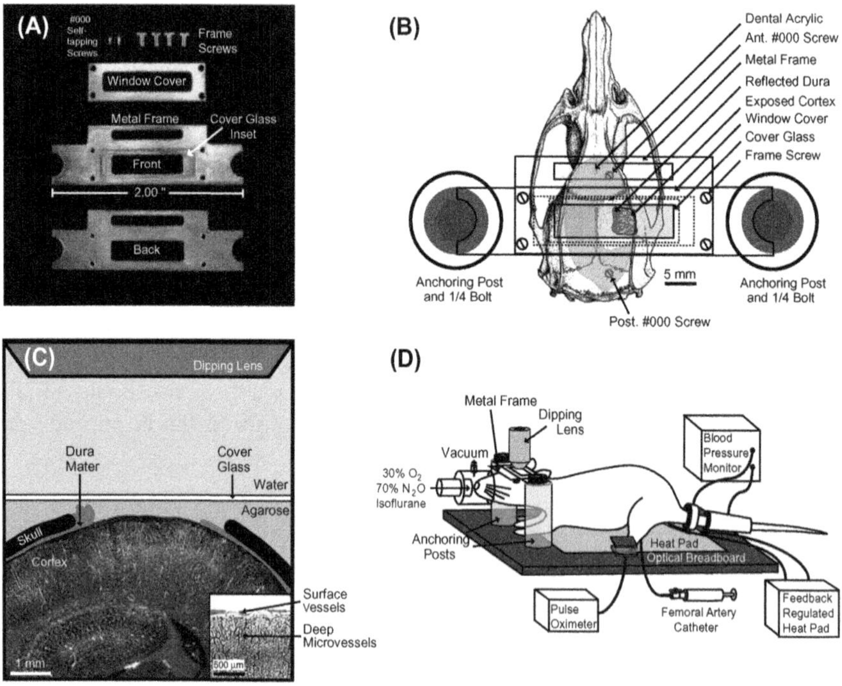

Fig. 3 Methods for *in vivo* imaging of blood flow through a cranial window. (**a**) To immobilize the head of the animal during imaging, we designed a metal frame that could be cemented to the skull, and then anchored to an optical set-up. The frame is constructed of type 410 stainless steel with dimensions of 2.00 in long, 0.61 in wide and 0.029 in thick, and can be secured between two posts in a standard optical breadboard. An inset region, 0.015 in deep, borders the frame window to hold a cover glass over the craniotomy. The window cover, 0.015 in thick type 301 stainless steel, is then secured to the frame with four screws, sandwiching the no. 1 cover glass in place. (**b**) Surgical procedure for acute imaging experiments. The animal is placed in a stereotaxic frame. Lidocaine™ is subcutaneously injected into the scalp for local anesthesia. An approximately 5 cm midline incision is made in the scalp, and the periostium is scraped to the sides with a bone curette. The scalp is retracted using hemostats. The right temporalis muscle is separated from the bone, without severing, and retracted. The skull surface is cleaned with cotton swabs and bleeding from the skull is controlled with

(b) Maintain body temperature at 37 °C using a feedback regulated rectal probe and heat pad (Harvard).
(c) Inject (i.p.) saline with 5% (w/v) dextrose, 3 mL per kg rat every 2 h to maintain body fluids and energy requirements.
(d) Install a femoral artery catheter to allow blood samples to be collected. Blood gas is monitored once every 2 h (Rapid Lab 248; Bayer). The normal range is 80–100 mmHg for pO_2, 35–45 mmHg for pCO_2, and 7.35–7.45 for pH.
(e) Monitor arterial blood pressure (XBP1000; Kent Scientific). The normal range is 80–130 mmHg for systolic pressure and 60–100 mmHg for diastolic pressure. We use tail-cuff device to measure blood pressure, but measurement through an intra-arterial catheter is also possible.

Note that fluctuations in physiological parameters, often caused by variation in the depth of anesthesia, can greatly influence cerebral blood flow. It is important to maintain these variables in a normal range and to take note of abrupt changes during the experiment.

Fig. 3 (continued) maintained pressure from the swab, or superficial abrasion with a high-speed drill. A thin layer of VetBond™, is applied to the skull surface to aid attachment of the dental acrylic and frame. The location of the cranial window is marked in pencil. We typically place the window above the somatosensory cortex by centering a 4 × 4 mm craniotomy at −3.0 mm anterior-posterior and +4.5 mm medial-lateral. The edges of the marked cranial window are slowly thinned with a high-speed drill. The bone is flushed regularly with physiological saline to avoid over-heating. Drilling is stopped just as the bone begins to crackle and cerebral spinal fluid begins to leak through. The resulting island of bone is gently lifted away horizontal to the skull surface using two forceps on opposite corners of the bone flap. The underlying dura is flushed with ACSF and bleeding is controlled with small pieces of Gelfoam™ soaked in ACSF and KimWipes™, twisted to a fine point with the fingers. The cranial window is kept moist with a piece of moist Gelfoam™. Two burr holes are made with a 1/4 bit, anterior and posterior to the intended frame placement, self-tapping #000-3/32 screws are placed in the burr holes, and the threads secured with a small dab of VetBond™. The metal frame is fixed tangentially to the clean, dry skull surface with dental acrylic. Extra dental acrylic is built up around the screws to provide further anchoring support. A small nick is made in the dura with a bent 27 gauge syringe needle. The dura is carefully teased apart with fine forceps (Dumont #5), starting at the nick, and reflected to one side of the cranial window. The interior of the window chamber is filled with 37 °C, 1.5% (w/v) low-melting-point agarose, dissolved in a modified version of ACSF (125 mM NaCl, 10 mM glucose, 10 mM HEPES, 3.1 mM $CaCl_2$, 1.3 mM $MgCl_2$, pH 7.4). The chamber is immediately sealed using a cover glass (no. 1, cut to size), and the window cover is secured over the cover glass with four screws. The skin is sutured together around the frame, and agarose is trailed around the cover glass to hold water for the dipping lens. (**c**) A cross-sectional view of the cranial window. The inset diagram shows an inverted coronal view of surface vessels and deep microvessels that are targeted for occlusion. (**d**) A typical experimental set-up used in our laboratory. The metal frame attached to the skull is immobilized between two anchoring posts inserted into an optical breadboard. Anesthesia is maintained with isoflurane in a 30% O_2–70% N_2O gas mixture. Blood pressure is measured with a tail cuff device. The femoral artery catheter is used to collect blood samples for blood gas measurements, and is also the delivery route for fluorescein-dextran and Rose Bengal. Heart rate and blood O_2 saturation is continuously monitored using a pulse oximeter, which is useful in the assessing the depth of anesthesia. Rat skull diagram in panel **b** adapted from Paxinos and Watson [29]

3. Surgery:
   (a) Generate a craniotomy above the brain region of interest using a high-speed drill (Fig. 3b). We typically consider measurements over primary somatosensory cortex, which is the part of parietal cortex that nominally lies between −1 and −5 mm relative to the Bregma point and between 1 and 7 mm from the midline on the medial-lateral axis [29].
   (b) Glue the metal frame to the skull with dental acrylic (Grip Cement; Dentsply). The frame rigidly holds the head of the animal to the optical apparatus (Fig. 3a, b). To achieve a reliable connection between acrylic and bone, the contact regions on the bone are cleaned of soft tissue, a thin layer of VetBond™ is applied, #000-3/32 self-tapping screws (Small Parts) are introduced to the anterior and posterior aspects of the skull, and screws, one of which passes through an opening in the frame, are then mechanically linked to the frame with dental acrylic (Fig. 3b).
   (c) Remove the dura after the dental acrylic has cured. It is crucial to avoid any bleeding, and thus not cut or tear across any large vessels in the dura. Removal of the dura greatly improves the optical field and is recommended.
   (d) Fill the interior of the chamber with 1.5% (w/v) low melting point agarose (A-9793; Sigma) dissolved in artificial cerebral spinal fluid (ACSF) that contains neither carbonate nor phosphate which cause precipitation when the solution is boiled to dissolve the agarose [30] (Fig. 3b, c). The temperature of the agarose must not exceed approximately 37 °C before it is applied to the brain.
   (e) Immediately seal the chamber using a cover glass as a window (Fig. 3c). With the addition of a cover glass, the frame serves to provide a chamber that seals and protects the cortex. Resealing the craniotomy is crucial for the suppression of excessive motion that otherwise ensues as a result of cranial-pressure fluctuations that are caused by heart beat and breathing.[1]
   (f) Stabilize the animals on a platform for imaging, using the frame as a head support (Fig. 3d). We prepare a separate plate that can be transported between surgical and imaging suites with the animal and all physiological monitoring devices assembled as one unit.
   (g) Label the blood plasma with 2 MDa fluorescein-dextran through the femoral artery catheter as a means to generate contrast between the red blood cells (RBCs) and the plasma.[2] We inject a 1.0 mL per kg rat bolus of 5% (w/v) solution of fluorescein-dextran in physiological saline. This high molecular

[1] The agarose between the cover glass and the cortex allows the cranial opening to be effectively resealed even if a narrow opening is left on the sides as a means to insert glass microelectrodes into cortex [21].

[2] The blood plasma may also be labeled through a tail-vein injection. Submerge the tail in warm water, i.e., 37 °C, for approximately 2 min and place a soft clamp at the base of the tail to dilate the veins. Starting as close to the tip of the tail as possible, insert a 24-gauge catheter and inject the bolus of fluorescein-dextran in physiological saline.

weight dye provides stable labeling, typically 3 h, and further acts to stain the damage in occluded vessels [16]. The dye is supplemented as needed.

4. Imaging and occlusion:

   (a) Collect baseline vascular images at successive depths to construct a three dimensional stack of images for maximal projection or volume rendered views of the vasculature. Flow parameters, such as the speed of RBCs, are derived from line-scan data (Fig. 4a). Structural parameters, such as the cross-section of vessels, are derived from planar scans.
   (b) Select a vessel to target for occlusion.
   (c) Inject a 1.0 mL per kg rat bolus of 1% (w/v) solution of Rose Bengal in physiological saline through the catheter. Rose Bengal should be injected immediately prior to formation of the occlusion.
   (d) To initiate photo-thrombosis, aim the focal point of the green-light laser within the lumen of the target vessel but near the vessel wall. This will nucleate the clotting cascade and create an attachment point for the growing clot. The irradiation is periodically interrupted for brief epochs of imaging, typically 1 frame (0.2 s per frame) every 1 s for an 80% duty cycle, so that formation of the clot can be observed in real time. If a clot does not occur close to the wall within 5–10 s of irradiation, increase the power by factors of two until clot begins to form.
   (e) Oscillate the relative position between the animal and the beam, at 0.1–0.5 Hz, to traverse the diameter of the blood vessel. Occlusion requires 10–300 s or irradiation. Larger vessels with relatively fast RBC flow, i.e., surface vessels as opposed to penetrating arterioles, require more time to fully occlude; the exact time must be determined empirically.
   (f) A successful occlusion appears densely packed with immobile non-fluorescent RBCs often surrounded by a brighter region of stagnant fluorescent plasma (Figs. 1c, 4a, and 5a).
   (g) Collect post-occlusion measurements, as desired.
   (h) The hypoxia marker pimonidazole hydrochloride (Hypoxyprobe™; www.hypoxyprobe.com) forms stable protein-adducts in viable tissues with low oxygen content that can be detected by *post hoc* immunohistology. In particular, Hypoxyprobe™ can be used to locate relatively small zones of hypoxia caused by single vessel occlusions (Fig. 5b). We typically inject Hypoxyprobe™ intravenously (i.v.) just after occlusion and 1 h before sacrifice.

5. Histological examination.

   (a) It is often desirable to examine the consequences of single vessel occlusion using *post hoc* histology. The animal is trans-cardially perfused with saline followed by 4% (w/v) paraformaldehyde in phosphate buffered saline, the brain is removed, equilibrated with 30% (w/v) sucrose, and frozen-sectioned. Extravasated fluorescein-dextran is trapped within the vessel wall at or near the site of the occlusion (Figs. 4b and 5b).

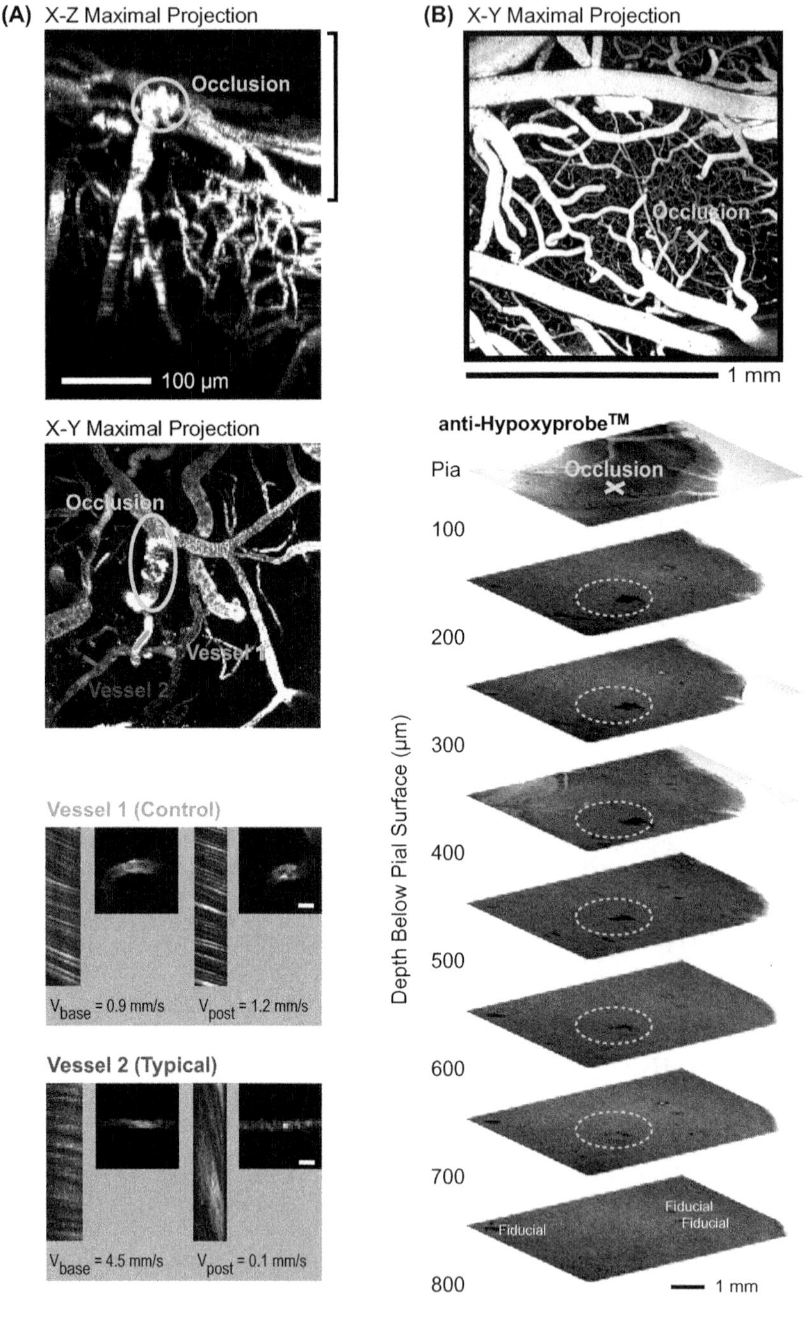

Fig. 4 Photo-thrombotic occlusion of penetrating arterioles using Rose Bengal. (**a**) Projected TPLSM image of a single penetrating arteriole occluded by photo-thrombosis (yellow circle) shown in the coronal plane (X–Z maximal projection), with surface arterioles near the top of the image. For the same field of interest, the surface vascular network is visible in the plane tangential to the cortical

Example Data

The pial arterioles of the cortical surface arteries form a communicating network, from which penetrating arterioles branch and dive to perfuse underlying capillary beds. In the first experiment, we selectively occluded a penetrating arteriole (yellow circle in Fig. 4a). Baseline and post-occlusion values of RBC velocities were measured in the target vessel, as well as neighboring microvessels, using continuous line-scans [31] of a segment of the vessel in the plane of the cortical surface. The location of the occlusion appeared as a bright region of stagnant serum within the vessel lumen and extravasated fluorescein-dextran indicated damage to the vessel wall. This occlusion was specific to the target vessel since the velocity of RBCs in a neighboring penetrating arteriole was unaffected (Vessel 1 in Fig. 4a). However, a downstream microvessel directly perfused by the target vessel experienced severe loss of blood flow (Vessel 2 in Fig. 4a). In a second animal, the hypoxic volume that resulted from the occlusion of a single penetrating arteriole was examined using *post hoc* histology (Fig. 4b). Immunostaining for Hypoxyprobe™ revealed a column of hypoxic tissue that extended 700 μm beneath the pial surface. This result is consistent with the idea that penetrating arterioles, which are largely devoid of collateral flow, act as bottlenecks in the supply of blood to the cortex [17].

Occlusions placed within the redundant surface network [32] lead to a fundamentally different outcome. A typical example is shown in Fig. 5a, where a single vessel divides into downstream branches (first frame). Green laser light was focused in the lumen of the main arteriole (yellow circle). In the absence of circulating Rose Bengal, no occlusion was formed (second frame). However, immediately after injection of Rose Bengal, the irradiation initiates a clotting cascade against the vessel wall (third frame). Within 3 min of continuous irradiation, the vessel is completely occluded by a dense mass of clotted RBC with bright stagnant plasma directly upstream (fifth frame). As with the example given above, surface vessels that neighbor the target vessel are not affected by photo-thrombosis, which demonstrates the specificity imparted by this technique.

Fig. 4 (continued) surface (X–Y maximal projection). The velocity of RBCs in two neighboring surface vessels (Vessel 1 and 2) was measured using repeated line-scans. Vessel 1 shows no change in post-occlusion RBC velocity (v_{post}) compared to baseline (v_{base}), whereas Vessel 2 shows essentially no flow after the occlusion. (**b**) Examination of the hypoxic volume generated after a single penetrating arteriole was occluded (yellow cross). Hypoxyprobe™ was injected (i.v.) after clot formation and 1 h before sacrifice. The animal was trans-cardially perfused with 100 mL of saline followed by 100 mL of 4% (w/v) paraformaldehyde. To locate the occluded vessel after brain extraction, fiducials markers were electrolytically placed near the target vessel before the brain was extracted from the skull. The entire ipsilateral cortex was removed and sandwiched between two glass slides separated by a distance of 2.5 mm. The tissue was post-fixed for 24 h, equilibrated with 30% (w/v) sucrose, and sectioned on a freezing-sliding microtome into 100-μm sections tangential to the cortical surface. Immunostaining for Hypoxyprobe™ revealed a column of hypoxia, within the dashed yellow circles, that reached 700 μm below the pial surface. Panel **a** adapted from Nishimura et al. [17]

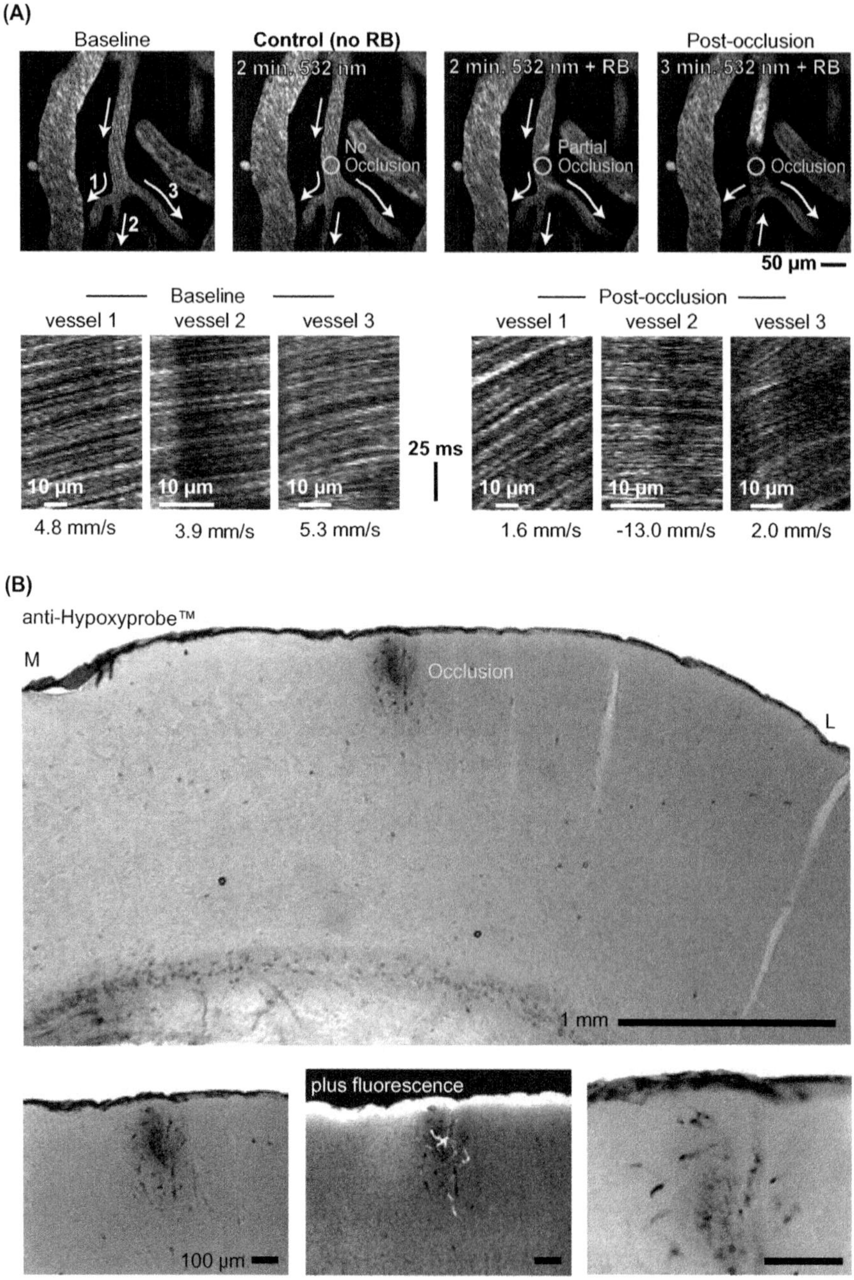

Fig. 5 Photo-thrombotic occlusion of surface arterioles using Rose Bengal. (**a**) Planar TPLSM images showing the time-course for photo-thrombotic occlusion of a single surface arteriole. The first frame was taken at baseline, and white arrows indicate the directions of RBC flow as determined by line-scans. Three vessels branching from the main arteriole, numbered 1–3, were examined for flow direction and RBC velocity before and after the occlusion. In the second frame, the yellow circle indicates the location of green laser irradiation. In the absence of circulating Rose Bengal, a 2-min

One feature of disrupted flow in the surface arteriole network is the occurrence of robust flow reversals in downstream arterioles, which compensate for loss of flow caused by the occlusion. White arrows indicate baseline flow directions (Fig. 5a). Line-scans show that vessel 2 reverses after the occlusion, which allows the flow in vessels 1 and 3 to be maintained at near-baseline levels. *Post hoc* Hypoxyprobe™ staining revealed a much smaller hypoxic region compared to penetrating arteriole occlusion (Fig. 5b), showing that re-routing of blood flow offers protection against the formation of a large damage zone. A small field of fluorescein-dextran extravasation is still visible in the vicinity of the occlusion and overlaps with the hypoxic region. High-magnification views of Hypoxyprobe™ stained tissue reveals the staining of individual neurons (Fig. 5b).

Plasma-Mediated Occlusion of Deep Vessels

This technique makes use of the dissociation of matter by high-fluence, 100–300 fs pulses of near infrared light. These pulses lead to ionization of the material, such as blood plasma or vascular lumen, within a femtoliter-sized focal-volume of the incident laser pulse. The interaction of light towards the end of the pulse with the initial light-induced ionization results in a spatially-limited release of mechanical energy in the form of a shock wave (Fig. 6a). This method allows occlusions to be induced below the surface of cortex, typically down to 500 μm but in principle down to 1000 μm below the pia, without disruption of the neighboring tissue.

Equipment

Plasma-mediated occlusion is readily combined with TPLSM, as described above, to enable the targeting and real-time monitoring of blood vessels. The typical set-up makes use of a standard two-photon microscope in which an amplified 100 fs light source is introduced to the beam path through a polarizing beam splitter (Fig. 6b). The imaging and photodisruption beams are focused to the same focal plane and the

Fig. 5 (continued) period of irradiation did not form a clot. After the intravenous injection of Rose Bengal, irradiation initiated clot formation within 2 min and completely occluded the target vessel by 3 min. Surrounding vessels were unaffected by targeted photo-thrombosis. The occlusion appears as a dark mass of clotted RBCs with brightly fluorescent blood plasma directly upstream from the occlusion. Linescans show that the occlusion resulted in a re-routing of flow in downstream vessels. Notably, a flow reversal occurred in vessel 2 allowing vessels 1 and 3 to maintain near baseline levels of RBC velocity. (**b**) Hypoxyprobe™ was injected (i.v.) 1 h after occlusion. The animal was transcardially perfused and the brain was prepared for histology (Fig. 4). The brain was cryo-sectioned along the coronal plane and immuno-stained for Hypoxyprobe™. Hypoxia was restricted to a zone immediately beneath the site of occlusion. Lateral and medial directions are labeled by M and L, respectively. Below, a mixed brightfield-fluorescent image, plus fluorescence, shows that the extravasated fluorescein-dextran overlaps with the hypoxic region. High-magnification images show specific Hypoxyprobe™ labeling of individual neurons. All panels adapted from Schaffer et al. [16]

amplified beam is centered on the optical axis so that photodisruption occurred at the center of the imaged field. The energy per pulse of the photodisruption beam is varied with neutral density filters and the number of pulses is controlled by a mechanical shutter. There are multiple suppliers of "single-box" amplifiers, which consist of four devices in one housing, i.e., a 100-fs Ti:Sapphire oscillator with associated pump to serve as the seed for the optical amplifier, together with a regenerative Ti:Sapphire-based amplifier and it's associated pump laser (*e.g.*, Libra, Coherent Inc.).

The diameter of the amplified 100-fs beam is adjusted to overfill the back aperture of the objective, using an appropriate telescope, to form the smallest possible spot size. The energy per pulse of the amplified beam should initially be set at 0.03 μJ, which corresponds to a near-threshold fluence for damage of approximately 1 J/cm^2 at the focus with a 40× dipping objective, consistent with literature values for damage.

Procedure

The surgical and animal maintenance procedures are identical to those described for the photosensitizer-based occlusions. Similarly, histological examination of the tissue proceeds as described above. It is imperative to perform occlusions with concurrent imaging as a means to continuously monitor the state of the vasculature.

6. Imaging and occlusion:
   (a) Collect baseline vascular images at successive depths to construct a three dimensional stack of images for maximal projection or volume rendered views of the vasculature (Fig. 6c). Flow parameters, such as the speed of RBCs, are derived from line-scan data. Structural parameters, such as the cross-section of vessels, are derived from planar scans.
   (b) Select a subsurface microvessel to target for occlusion.
   (c) Aim the focal point of the amplified 100 fs beam near the center of the lumen and irradiate for approximately 10 pulses at the minimum pulse energy. The first indication of photodisruption is visualized as the accumulation of fluorescein-dextran outside the vessel lumen; this corresponds to extravasation. If extravasation fails to occur, increase the energy by approximately a factor of 2 and irradiate again; extravasation typically occurs by 2- to 3-times times the threshold energy and in the range of 0.1–0.5 μJ per pulse.
   (d) Once extravasation occurs, the irradiation is repeated with increased pulse energy, typically by less than a factor of 2, and increasing number of pulses, typically 10–100, until there is a complete cessation of RBC movement in the target vessel. Increasing the energy further may result in the rupture of the vessel and a hemorrhage.[3]

---

[3] Preliminary findings suggest that irradiation of the wall of a penetrating arteriole with amplified 100-fs laser pulses, at a level deep to the surface, may be used to form an occlusion [33]. This builds on work that used unamplified pulses with subthreshold energies to irritate the vasculature [34], possibly by direct damage to RBCs through two-photon absorption of the incident 100-fs pulses [35].

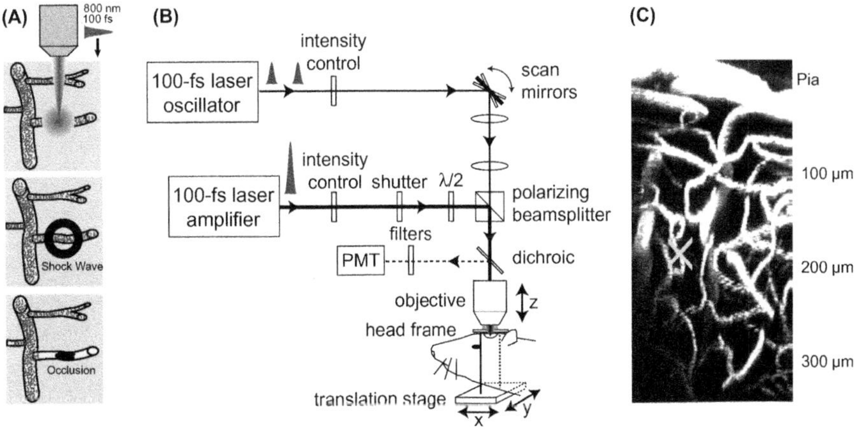

Fig. 6 Overview for forming intravascular occlusions in deep microvessels using amplified 100-fs laser pulses. (**a**) Laser pulses directed on a deep microvessel (top image) leads to a release of mechanical energy in the form of a shock wave (middle image). Repeated delivery of pulses, starting at low energies and gradually increasing, forms an occlusion within the target vessel (bottom image). (**b**) Schematic of a two-photon laser-scanning microscope system modified for delivery of amplified 100-fs pulses. The dichroic is a 700 DCXRU (Chroma). (**c**) A projected X–Z TPLSM image stack showing deep microvessels (yellow cross) that are routinely targeted for selective occlusion using this technique. All panels adapted from Nishimura et al. [20]

(e) A successful occlusion appears densely packed with immobile non-fluorescent RBCs, abutted by a brighter region of stagnant fluorescent plasma. The occlusion is further delimited by the uptake of fluorescein-dextran in the endothelial cells of the vessels.

Example Data

We consider a microvessel, approximately 10 μm in diameter, that lay approximately 250 μm below the pia (Fig. 7a). Uniform flow is seen in the vessel prior to irradiation, as evidenced by streaks in the raster-scanned imaged (panel 1 in Fig. 7a). Labeling of the vessel wall with trapped fluorescein-dextran is observed, but flow is still present, subsequent to the first burst of pulses (panels 2 and 3 in Fig. 7a). A second burst of pulses leads to cessation of flow and a stagnant pool of fluorescent plasma behind the occlusion (panels 4 and 5 in Fig. 7a). Lastly, measurements of the speed and direction of RBC flow before and after formation of the occlusion indicate that the occlusion effectively halts flow in vessels that lie immediately downstream (Fig. 7a).

A histological examination of a plasma-mediated occlusion in a different preparation shows that the hypoxia marker Hypoxyprobe™ stains tissue in a diffuse manner near the site of the occlusion (Fig. 7b). This staining is highly overlapped with the extravasated fluorescein-dextran in this section. Clotting activity is observed in the

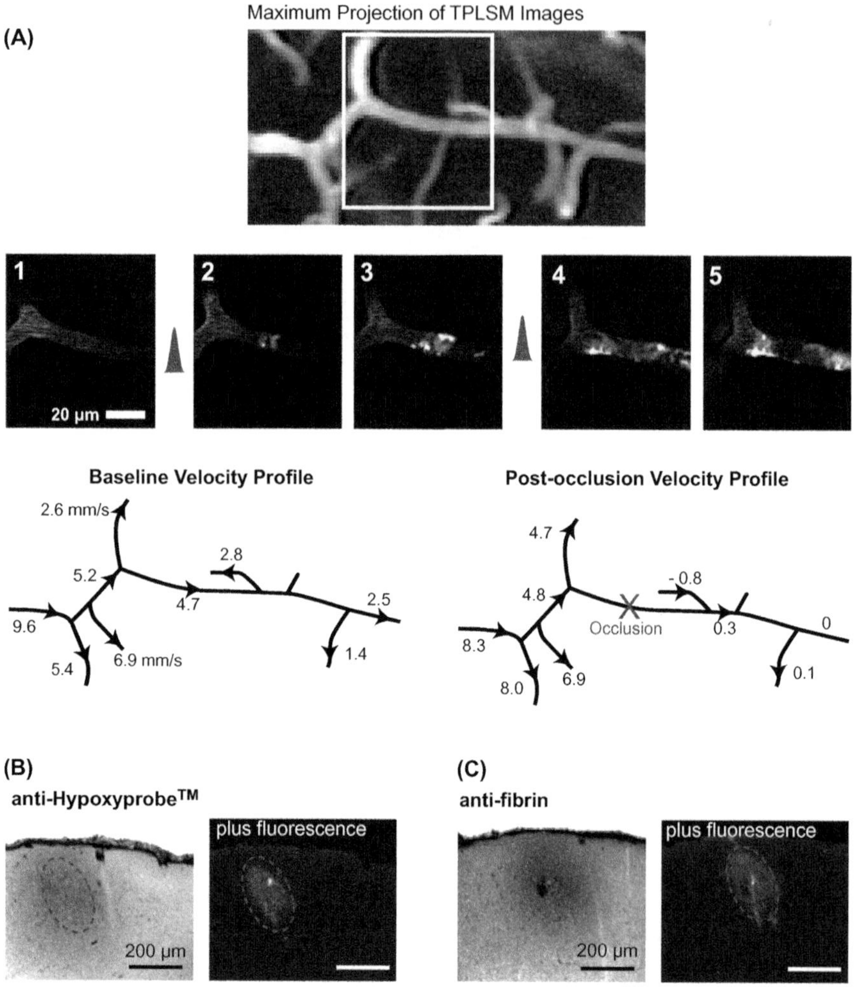

Fig. 7 Example of targeted deep microvessel occlusion using amplified 100-fs laser pulses. (**a**) Maximal projection of a TPLSM stack showing a tortuous network of sub-surface microvessels. The second row shows planar images taken from a region of interest (white rectangle) depicting the time-course for intravascular clot formation (frames 1–5). The red pulses indicate irradiation with multiple trains of 0.03-µJ pulses delivered at 1 kHz. After the first pulse, there was temporary cessation of RBC motion and swelling of the target vessel (frames 2 and 3). A second pulse led to limited extravasation and finally occlusion of the vessel lumen (frames 4 and 5). The third row shows vascular traces with baseline and post-occlusion RBC velocity profiles, in mm/s, of the vascular network. Arrow-heads denote the direction of RBC movement and the red cross marks the occluded microvessel. (**b**) Immunolabeling of tissue after deep microvessel occlusion. The red ellipses delineate a common area in the vicinity of the occlusion. Diffuse Hypoxyprobe™ immunostaining overlaps with affected region surrounding the occlusion, as marked by the territory of fluorescein-dextran extravasation. (**c**) Anti-fibrin immunostaining reveals localization of fibrin/fibrinogen clots, which appears as a mixed diffuse zone that surrounds a small dense region of intense staining. A mixed brightfield-fluorescence image (plus fluorescence) shows that fibrin immunoreactivity is also confined to the immediate vicinity of the occlusion. All panels adapted from Nishimura et al. [20]

targeted vessel through anti-fibrin immunoreactivity, a marker of fibrin/fibrinogen clots. The fibrin forms a dense core deposit with a more diffuse surround that is highly overlapped with the extravasated fluorescein-dextran in the section (Fig. 7c).

Synopsis

We have described two methods that, collectively, allow occlusions to be targeted to single vessels that lie anywhere within the upper 500–1000 μm of cortex. These occlusions provide a means to address the consequences of vascular disease and microstrokes on a microscopic level. While past investigations have employed these relatively new methods to only acute studies, it is anticipated that the same occlusion and concurrent imaging techniques will be exploited for chronic investigations as well.

Acknowledgements The authors are pleased to thank the Canadian Institutes of Health Research, the National Institutes of Health (NCRR, NIBIB and NINDS), the National Science Foundation, and the Veterans Medical Research Foundation for support.

References

1. Iadecola C. Neurovascular regulation in the normal brain and in Alzheimer's disease. Nat Rev Neurosci. 2004;5:347–60.
2. Scremin OU. Cerebral vascular system. In: Paxinos G, editor. The rat nervous system. 2nd ed. San Diego: Academic Press, Inc.; 1995.
3. Miyake K, Takeo S, Kaijihara H. Sustained decrease in brain regional blood flow after microsphere embolism in rats. Stroke. 1993;24:415–20.
4. Sugi T, Schuier FJ, Hossmann KA, Zulch KJ. The effect of mild microembolic injury on the energy metabolism of the cat brain. J Neurol. 1980;223(4):285–92.
5. Gerriets T, Li F, Silva MD, et al. The macrosphere model: evaluation of a new stroke model for permanent middle cerebral artery occlusion in rats. J Neurosci Methods. 2003;122(2):201–11.
6. Zhang Z, Zhang RL, Jiang Q, Raman SB, Cantwell L, Chopp M. A new rat model of thrombotic focal cerebral ischemia. J Cereb Blood Flow Metab. 1997;17(2):123–35.
7. Fuxe K, Bjelke B, Andbjer B, Grahn H, Rimondini R, Agnati LF. Endothelin-1 induced lesions of the frontoparietal cortex of the rat. A possible model of focal cortical ischemia. Neuroreport. 1997;8(11):2623–9.
8. Macrae IM, Robinson MJ, Graham DI, Reid JL, McCulloch J. Endothelin-1-induced reductions in cerebral blood flow: dose dependency, time course, and neuropathological consequences. J Cereb Blood Flow Metab. 1993;13(2):276–84.
9. Sharkey J, Ritchie IM, Kelly PA. Perivascular microapplication of endothelin-1: a new model of focal cerebral ischaemia in the rat. J Cereb Blood Flow Metab. 1993;13(5):865–71.
10. Zhang S, Boyd J, Delaney KR, Murphy TH. Rapid reversible changes in dendritic spine structure *in vivo* gated by the degree of ischemia. J Neurosci. 2005;25:5333–228.
11. Wei L, Rovainen CM, Woolsey TA. Ministrokes in rat barrel cortex. Stroke. 1995;26:1459–62.
12. Chen ST, Hsu CY, Hogan EL, Maricq H, Balentine JD. A model of focal ischemic stroke in the rat: reproducible extensive cortical infarction. Stroke. 1986;17(4):738–43.
13. Buchan AM, Xue D, Slivka A. A new model of temporary focal neocortical ischemia in the rat. Stroke. 1992;23(2):273–9.

14. Tamura A, Asano T, Sano K. Correlation between rCBF and histological changes following temporary middle cerebral artery occlusion. Stroke. 1980;11:487–93.
15. Katsman D, Zheng J, Spinelli K, Carmichael ST. Tissue microenvironments within functional cortical subdivisions adjacent to focal stroke. J Cereb Blood Flow Metab. 2003;23(9):997–1009.
16. Schaffer CB, Friedman B, Nishimura N, et al. Two-photon imaging of cortical surface microvessels reveals a robust redistribution in blood flow after vascular occlusion. PLoS Biol. 2006;4:258–70.
17. Nishimura B, Schaffer CB, Friedman B, Lyden PD, Kleinfeld D. Penetrating arterioles are a bottleneck in the perfusion of neocortex. Proc Natl Acad Sci U S A. 2007;104:365–70.
18. Markgraf CG, Kraydieh S, Prado R, Watson BD, Dietrich WD, Ginsberg MD. Comparative histopathologic consequences of photothrombotic occlusion of the distal middle cerebral artery in Sprague-Dawley and Wistar rats. Stroke. 1993;24:286–93.
19. Helmchen F, Denk W. Deep tissue two-photon microscopy. Nat Methods. 2005;2:932–40.
20. Nishimura N, Schaffer CB, Friedman B, Tsai PS, Lyden PD, Kleinfeld D. Targeted insult to individual subsurface cortical blood vessels using ultrashort laser pulses: three models of stroke. Nat Methods. 2006;3:99–108.
21. Svoboda K, Denk W, Kleinfeld D, Tank DW. *In vivo* dendritic calcium dynamics in neocortical pyramidal neurons. Nature. 1997;385:161–5.
22. Kleinfeld D, Mitra PP, Helmchen F, Denk W. Fluctuations and stimulus-induced changes in blood flow observed in individual capillaries in layers 2 through 4 of rat neocortex. Proc Natl Acad Sci U S A. 1998;95:15741–6.
23. Kleinfeld D, Denk W. Two-photon imaging of neocortical microcirculation. In: Yuste R, Lanni F, Konnerth A, editors. Imaging neurons: a laboratory manual. Cold Spring Harbor: Cold Spring Harbor Laboratory Press; 2000. p. 23.1–15.
24. Kleinfeld D, Denk W. Two-photon imaging of cortical microcirculation. In: Yuste R, Konnerth A, editors. Imaging in neuroscience and development. Cold Spring Harbor: Cold Spring Harbor Laboratory Press; 2005. p. 701–5.
25. Zhang S, Murphy TH. Imaging the impact of cortical microcirculation on synaptic structure and sensory-evoked hemodynamic responses in vivo. PLoS Biol. 2007;5:e119.
26. Tsai PS, Nishimura N, Yoder EJ, Dolnick EM, White GA, Kleinfeld D. Principles, design, and construction of a two photon laser scanning microscope for *in vitro* and *in vivo* brain imaging. In: Frostig RD, editor. In vivo optical imaging of brain function. Boca Raton: CRC Press; 2002. p. 113–71.
27. Majewska A, Yiu G, Yuste R. A custom-made two-photon microscope and deconvolution system. Pflugers Arch. 2000;441:398–408.
28. Short CE. Principles and practice of veterinary anesthesia. Baltimore: Williams and Williams; 1987.
29. Paxinos G, Watson C. The rat brain in stereotaxic coordinates. San Diego: Academic Press; 1986.
30. Kleinfeld D, Delaney KR. Distributed representation of vibrissa movement in the upper layers of somatosensory cortex revealed with voltage sensitive dyes. J Comp Neurol. 1996;375:89–108.
31. Dirnagl U, Villringer A, Einhaupl KM. *In-vivo* confocal scanning laser microscopy of the cerebral microcirculation. J Microsc. 1992;165:147–57.
32. Brozici M, van der Zwain A, Hillen B. Anatomy and functionality of leptomeningeal anastomoses: a review. Stroke. 2003;34:2750–62.
33. Nishimura N, Rosidi NL, Mandell J, Iadecola C, Schaffer CB. Neighboring arterioles do not dilate to increase collateral flow after the occlusion of a cortical penetrating arteriole. In: Abstracts of the society for neuroscience annual meeting, neuroscience. 2007.
34. Nimmerjahn A, Kirchhoff F, Helmchen F. Resting microglial cells are highly dynamic surveillants of brain parenchyma in vivo. Science. 2005;308(5726):1314–8.
35. Clay GO, Schaffer CB, Kleinfeld D. Large two-photon absorptivity of hemoglobin in the infrared range of 780–880 nm. J Chem Phys. 2007;126:025102.

Intraluminal Middle Cerebral Artery Occlusion Model in Mice

Masaya Oda, Taku Sugawara, Kohei Kokubun, Kazuo Mizoi, and Hiroyuki Kinouchi

Abstract Intraluminal middle cerebral artery occlusion (MCAO) model in rodents has been widely used for cerebral ischemia studies. The procedure is relatively simple and non-invasive compared to the other focal cerebral ischemia models, and the ischemic insult is highly reproducible. The use of genetically modified mice has been attracting attention because it allows us to dissect molecular and cellular mechanisms of various pathological conditions in the brain and therefore, the mouse intraluminal MCAO model has become more prevalent. The authors describe here our method of mouse intraluminal MCAO and characterized its effects on infarct areas, local cerebral blood flow, extracellular amino acids concentration and physiological parameters.

Keywords C57BL/6 mouse · Focal cerebral ischemia · Intraluminal occlusion · Cerebral infarct · Amino acids · Microdialysis · High-performance liquid chromatography

Model Selection

Rat intraluminal MCAO model was established more than two decades ago and utilized to investigate the mechanisms of ischemia-induced brain damage. More recently, genetic modifications using transgenic techniques in mice had been widely used to elucidate molecular mechanisms of various pathological conditions and therefore,

M. Oda · T. Sugawara · K. Kokubun · K. Mizoi
Division of Neurosurgery, Department of Neuro and Locomotor Science, Akita University School of Medicine, Akita, Japan
e-mail: taku@nsg.med.akita-u.ac.jp

H. Kinouchi (✉)
Department of Neurosurgery, Faculty of Medicine, University of Yamanashi, Tamaho, Yamanashi, Japan
e-mail: hkinouchi@yamanashi.ac.jp

J. Chen et al. (eds.), *Animal Models of Acute Neurological Injury*, Springer Series in Translational Stroke Research, https://doi.org/10.1007/978-3-030-16082-1_11

mouse ischemia models have become prevalent. Compared to the other focal ischemia models, such as direct middle cerebral artery (MCA) coagulation model [1, 2], the intraluminal MCAO model reported here is technically easier and less invasive because of the simple procedures without craniectomy [3, 4]. This ischemia model has been found to be useful in various investigations on cerebral ischemia.

Materials

Animal

The choice of animal for MCAO model is male C57BL/6 mice obtained from Charles River Inc. (Japan). The C57BL/6 strain from this breeder is widely used as a background strain for genetic alternations and shows little variation in intracranial collaterals and lacks the patent posterior communicating artery (PComA) on both sides [5]. For successful induction of the ischemia, body weight of the animals has been shown to be important. The mice in 25–30 g body weight range have been used in many ischemia studies and we also found them to have more consistent outcomes than the ones outside of the range. The mice are allowed to have free access to water and food before and after surgery.

Anesthesia

Anesthesia is induced with 2% halothane in a closed chamber and maintained with 1.0–1.5% halothane in 30% O_2 and 70% N_2O using a face mask. The rectal temperature is monitored and maintained at 37 ± 0.5 °C with a thermal blanket throughout the surgical procedure.

Equipment

Anesthesia

Facemask.
Halothane vaporizer
Flowmeter of O_2 and N_2O

Surgery

Surgical instruments (microscissors, forceps, knives, bipolar coagulator, etc.)
Microvascular clip

Stereotaxic device
Dissecting microscope
High-speed drill for implantation of vertical microdialysis probes
Temperature control unit (thermometer, temperature probe and thermal blanket)
6–0 silk suture
5–0 nylon monofilament suture with a round tip, blunted by heat (Dermaron 1756-41, Devis & Geck, Manati, Puetro Rico)

Recording of Regional Cerebral Blood Flow (rCBF)

Laser Doppler flowmeter (LDF, Laserflo BPM2, Vasamedics, St. Paul, MN, USA)

In Vivo Microdialysis

Microdialysis System (Eicom Corp., Tokyo, Japan)
Vertical microdialysis probes (0.22 mm outer diameter, 2 mm membrane length: Eicom Corp., Tokyo, Japan)
High-performance liquid chromatography (Eicom Corp., Tokyo, Japan)

Measurement of Infarct Size

Brain slicer (Mouse Brain Matrix, Harvard Apparatus, MA, USA)
Image scanner
NIH image analysis software

Procedures

Focal cerebral ischemia is induced by occlusion of the MCA using the intraluminal filament technique [6, 7] (Figs. 1 and 2). Anesthesia is induced with 2% halothane in a closed chamber and maintained with 1.0–1.5% halothane in 30% O_2 and 70% N_2O delivered via facemask. The animal is put on the operating table in supine position and the neck is extended with the use of a small pillow. Rectal temperature is monitored and maintained at 37 ± 0.5 °C with a thermal blanket throughout the surgical procedure.

1. Through a ventral midline incision, the left external carotid artery (ECA), superior thyroid artery (STA) and occipital artery (OA) are exposed and the STA and OA are coagulated.
2. The left common carotid artery (CCA) is carefully freed from the adjacent vagus nerve and the internal carotid artery (ICA) and pterygopalatine artery (PPA) are identified under the OA.

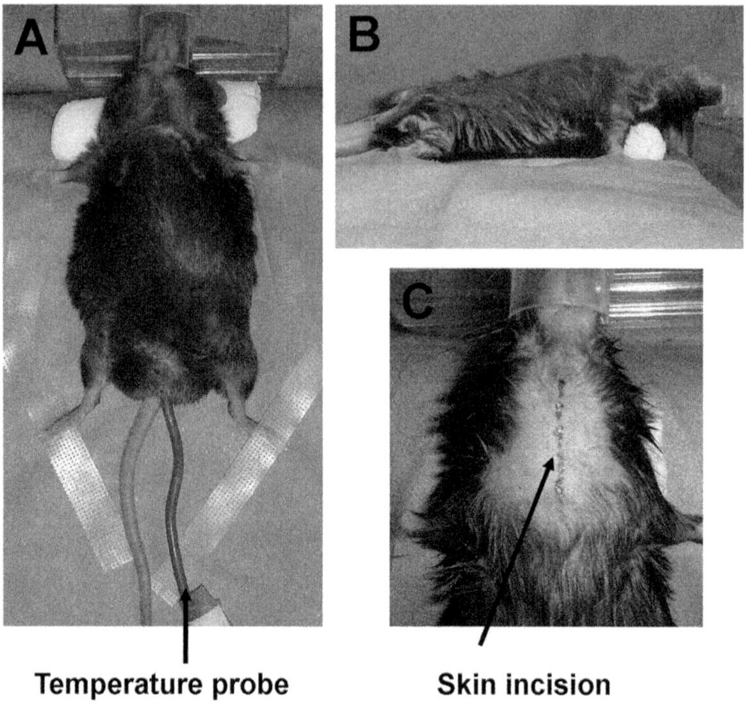

Fig. 1 Surgical setup. Mice are anesthetized with 2% halothane in a closed chamber and placed on an operating table in supine position (**a**). The neck is extended with the use of a small pillow (**b**). Rectal temperature is monitored and maintained at 37 ± 0.5 °C with a thermal blanket throughout the surgical procedure. Midline skin incision is to be performed as in panel **c**

3. The distal ECA is ligated with 6–0 silk suture just proximal to the lingual and maxillary artery bifurcation.
4. A microvascular clip is temporarily applied to the proximal ECA.
5. A 6–0 silk suture is loosely tied around the ECA between the clip and the distal ligature of the ECA.
6. An arteriotomy is carried out in the ECA between the distal and proximal ligatures and a 5–0 monofilament suture with a round tip is advanced into the ECA. After confirming that the suture tip is in the proximal ECA and adjacent to the clip, the proximal ligature is tightened enough to allow the suture to advance while preventing leakage of arterial blood from the ECA stump.
7. The clip is removed and the distal ECA is transected. The proximal ECA stump is retracted caudally.
8. A 5–0 nylon monofilament suture with a round tip is inserted into ICA 11 ± 0.5 mm from the bifurcation of the CCA until the LDF signal abruptly dropped.
9. After 30 or 60 min of MCAO, the nylon suture is removed and the blood flow restoration is confirmed by the LDF signal. Mice, in which the LDF signal

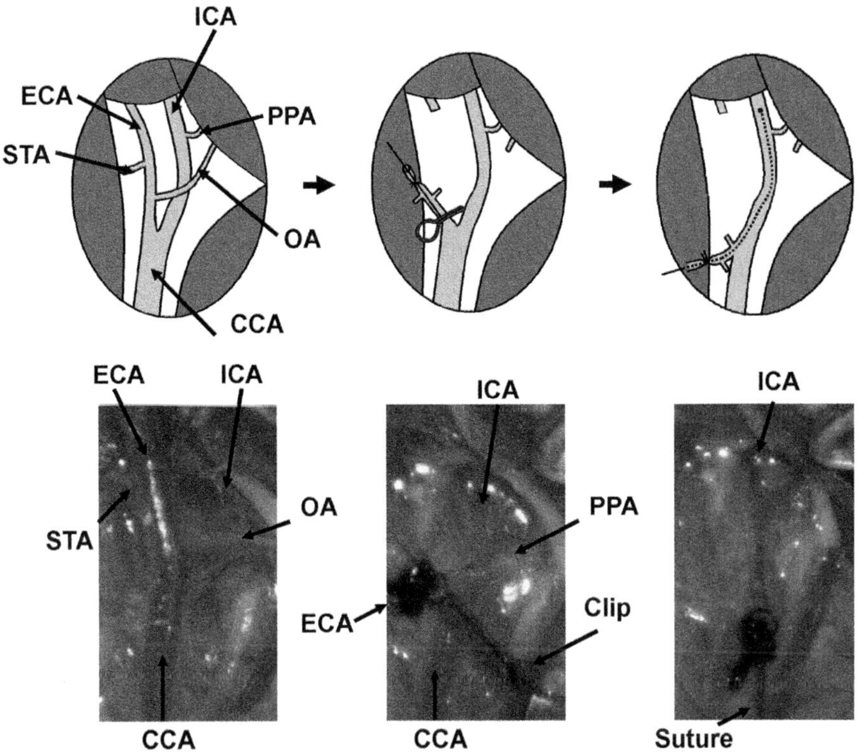

Fig. 2 Diagram of operative procedure (above) and corresponding views under a microscope (below). Through a ventral midline incision, the left external carotid artery (ECA), superior thyroid artery (STA) and occipital artery (OA) are exposed and the STA and OA are coagulated. The left common and internal carotid artery (CCA and ICA), and pterygopalatine artery (PPA) are identified. A microvascular clip is temporarily applied to the proximal ECA and a 5–0 monofilament suture with a round tip is inserted in ECA and advanced into the ICA

during ischemia exceeded 10% of the pre-ischemic signal, are excluded from this study.

10. The wound is sutured, the animal is returned to its cage and allowed free access to water and food.

Outcome Evaluation

Infarct Size

Twenty four hours after MCAO, the mice were deeply anesthetized and decapitated. The brain was removed and sectioned coronally into four 2-mm slices with a brain slicer. The slices were placed in 2% 2,3,5-triphenyltetrazolium chloride (TTC) solution at 37 °C for 10 min, and fixed in a 10% buffered formalin solution. The sections were then placed on the image scanner and the color images were obtained. The infarction area, appeared in white, was measured using the NIH image analysis software, and infarction volume was calculated by summing the infarct volume of sequential 2-mm-thick sections. Infarct areas were seen in the striatum and the dorsolateral cortex in 60 min MCAO mice, while 30 min MCAO mice demonstrated them only in the striatum, with a small portion of cortex (Fig. 3a). The size of the infarct area in each slice was shown in Fig. 3b. The infarct volume in 60 min MCAO mice was 54.5% larger than that of 30 min animals (Fig. 3c).

Regional Cerebral Blood Flow (rCBF) Measurement

rCBF was measured by the LDF. The flow probe (0.5 mm diameter) was placed on the cranial bone above the MCA territory (0.5 mm posterior and 4 mm lateral from the bregma) and away from large surface vessels. Steady-state baseline values were recorded before MCAO, and blood flow data were then expressed as percentage of the pre-occlusion baseline. rCBF immediately dropped after a suture insertion by approximately 90% of pre-occlusion value and no significant difference in the percent change during ischemia. After removal of the suture, rCBF returned to the pre-occlusion level (Fig. 4).

Profiles of Extracellular Aminoacid Concentrations in MCAO

Twenty-four hours before MCAO, vertical microdialysis probes were stereotaxically implanted in the left striatum of mice under anesthesia. The probe was coordinately implanted at 0.6 mm anterior and 2.0 mm lateral to the bregma and 2 mm ventral from the brain surface, according to the atlas of Franclin and Paxinos [8]. The external portion of the probes was fixed to the skull with dental cement. Throughout ischemia, dialysis probes were perfused with Ringer's solution (147 mmol/L NaCl, 2.3 mmol/L $CaCl_2$, 4.0 mmol/L KCl; pH 7.0) at a rate of 2 μl/min. The microdialysate (20 μl) were collected every 10 min. After stabilization period of 60 min, the samples were collected from 60 min before MCAO to 90/120 min after MCAO in 30/60 min MCAO model, respectively. Amino acid concentrations in the samples were measured by high-performance liquid chromatography [9, 10]. The aminoacid concentrations in all groups increased significantly at 20 min after MCAO and reached their peak at 10 min after reperfusion. The peak concentrations of all aminoacids were higher in 60 min MCAO mice compared with in 30 min MCAO mice (Fig. 5).

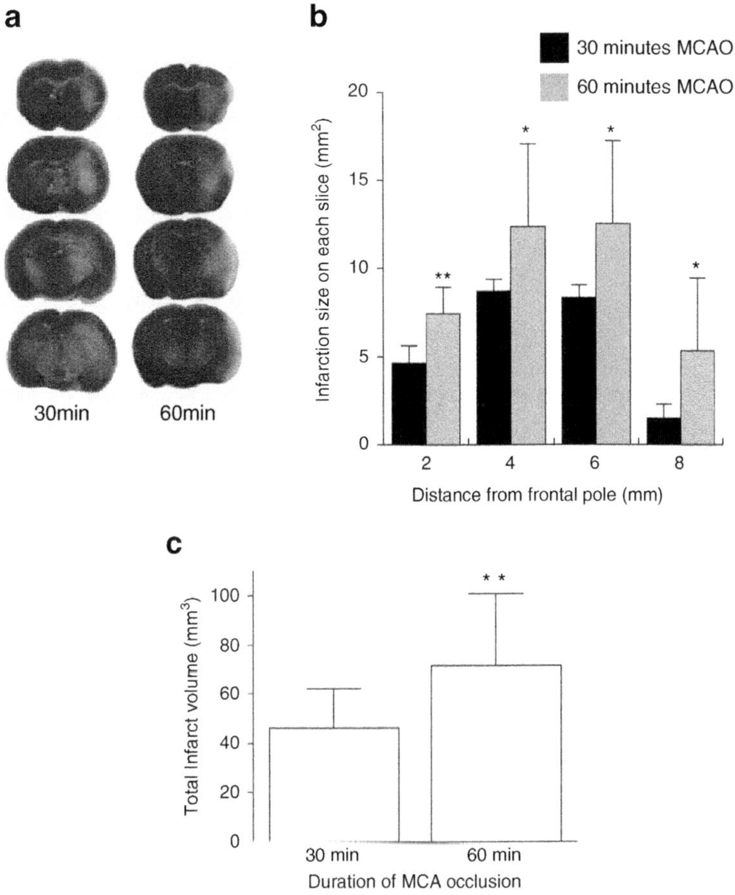

Fig. 3 Evaluation of brain injury after MCAO in 30 min MCAO mice and 60 min MCAO mice. A 2-mm slice of brain sample was stained with triphenyltetrazolium chloride for visualization of the infarct region (white). Representative photographs of each mouse group (n = 10) are shown (**a**). Infarct area (in mm^2) in each slice was measured (**b**). (**c**) The infarct volume (in mm^3) in 60 min MCAO mice was 54.5% larger than that of 30 min MCAO mice. Values are mean ± SD. $^{*}P < 0.05$, $^{**}P < 0.01$

Physiological Parameters

The femoral arteries of the animals in this experiment were cannulated using a polyethylene catheter (PE-10, Clay Adams, NJ, USA) for blood pressure monitoring and serial blood gas sampling. Mean arterial blood pressure was monitored using a transducer amplifier (Nihon Koden, Tokyo, Japan), and PaO_2, $PaCO_2$ and pH are measured with pH and blood gas analyzer. Rectal temperatures were measured and maintained at 37 ± 0.5 °C by using temperature control unit during all surgical

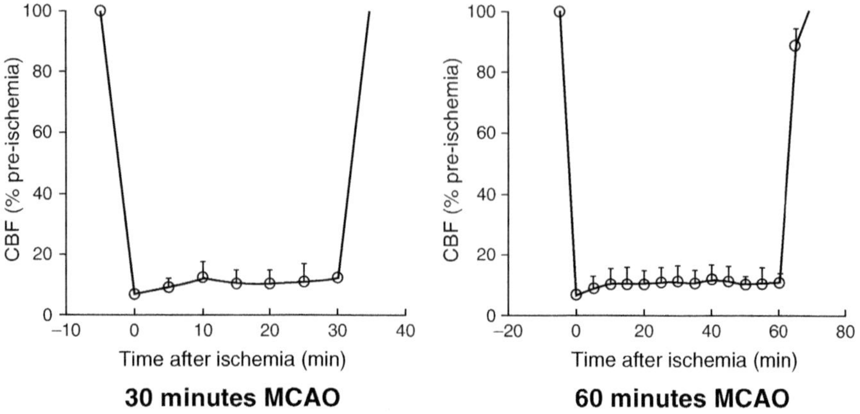

Fig. 4 rCBF by LDF during MCAO in 30 min MCAO mice and 60 min MCAO mice. The blood flow declined reproducibly after thread insertion and returned to the control levels within 5 min after thread retraction. There was no significant changes in rCBF during ischemia in both models. Values are mean ± SD (n = 10 in each group)

procedures and during recovery from anesthesia. These physiological variables were measured before and 60 min after MCAO in 60 min MCAO model mice in this study and showed no significant differences (Table 1).

Advantages and Complication

Advantages

This mouse intraluminal MCAO model is less invasive than direct MCAO methods such as the procedures requiring craniectomy or electro-coagulation of the brain surface. And MCAO of various durations can be obtained by restoring MCA blood flow by suture removal at adequate time points without substantial technical difficulties. In our transient MCAO models, both 30 and 60 min MCAO induce consistent cerebral infarcts with a significant difference between the groups.

Complications

Hypothalamic damage following insertion of a monofilament suture into the ICA to occlude the MCA has been documented [11, 12]. The hypothalamus is fed by the ACA, ICA, and the PComA. Because the suture head potentially occludes the ACA and ICA, rather than the MCA alone, the risk of hypothalamic injury should be considered. And subarachnoid hemorrhage is also a common complication after

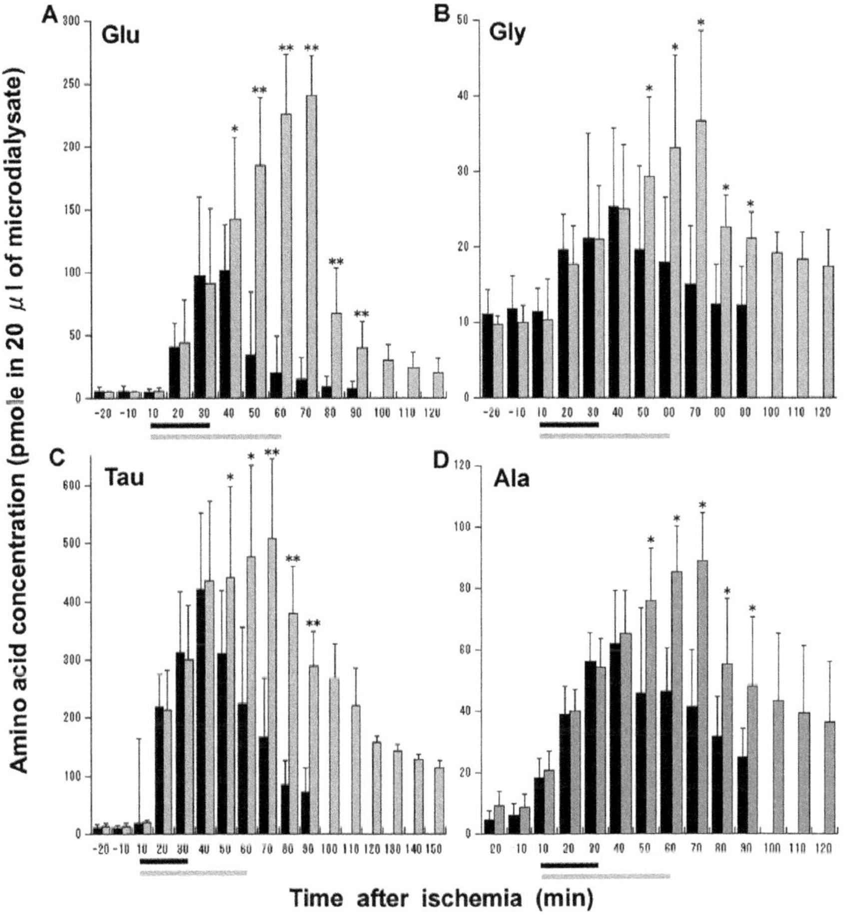

Fig. 5 Extracellular concentrations of aminoacid neurotransmitters in focal ischemia. The horizontal bar stands for the MCAO period (30 and 60 min). Values are mean ± SD. $^{*}P < 0.05$, $^{**}P < 0.01$. The aminoacid concentrations in all groups increased significantly at 20 min after MCAO and reached their peak at 10 min after reperfusion. The peak concentrations of all aminoacids were higher in 60 min MCAO mice compared with in 30 min MCAO mice (**A**–**D**). *Glu* Glutamate, *Gly* Glycine, *Tau* Taurine, *Ala* Alanine

Table 1 Physiological parameters in 60 min MCAO

	Preischemia	Postischemia
MABP (mmHg)	102.1 ± 1.7	104.3 ± 3.5
pH	7.27 ± 0.05	7.28 ± 0.01
PaO_2 (mmHg)	163.9 ± 12.7	169.2 ± 18.8
$PaCO_2$ (mmHg)	42.6 ± 2.6	45.2 ± 2.3
Rectal temperature (°C)	37.1 ± 0.09	37.0 ± 0.06

Values are mean ± SD; n = 10. *MABP* mean arterial blood pressure insertion of monofilament suture into ICA. Approximately 10% of the mice die during or after ischemia. These complications occur often after excessive insertion of the monofilament suture into ICA. Therefore, it is important that monofilament suture is inserted into ICA until the LDF signal abruptly dropped, not until resistance is felt.

References

1. Tamura A, Graham DI, McCulloch J, Teasdale GM. Focal cerebral ischaemia in the rat: 1. Description of technique and early neuropathological consequences following middle cerebral artery occlusion. J Cereb Blood Flow Metab. 1981;1(1):53–60.
2. Yamamoto M, Tamura A, Kirino T, Shimizu M, Sano K. Behavioral changes after focal cerebral ischemia by left middle cerebral artery occlusion in rats. Brain Res. 1988;452(1–2):323–8.
3. Kamii H, Kinouchi H, Sharp FR, Koistinaho J, Epstein CJ, Chan PH. Prolonged expression of hsp70 mRNA following transient focal cerebral ischemia in transgenic mice overexpressing CuZn-superoxide dismutase. J Cereb Blood Flow Metab. 1994;14:478–86.
4. Kinouchi H, Epstein CJ, Mizui T, Carlson E, Chen SF, Chan PH. Attenuation of focal cerebral ischemic injury in transgenic mice overexpressing CuZn superoxide dismutase. Proc Natl Acad Sci U S A. 1991;88:11158–62.
5. Maeda K, Hata R, Hossmann KA. Differences in the cerebrovascular anatomy of C57black/6 and SV129 mice. Neuroreport. 1998;9(7):1317–9.
6. Longa EZ, Weinstein PR, Carlson S, Cummins R. Reversible middle cerebral artery occlusion without craniectomy in rats. Stroke. 1989;20(1):84–91.
7. Belayev L, Alonso OF, Busto R, Zhao W, Ginsberg MD. Middle cerebral artery occlusion in the rat by intraluminal suture. Neurological and pathological evaluation of an improved model. Stroke. 1996;27(9):1616–22.
8. Franclin and Paxinos (1997). Book, California, USA. International standard book number: 0-12-547623-X.
9. Oda M, Kure S, Sugawara T, Yamaguchi S, Kojima K, Shinka T, Sato K, Narisawa A, Aoki Y, Matsubara Y, Omae T, Mizoi K, Kinouchi H. Direct correlation between ischemic injury and extracellular glycine concentration in mice with genetically altered activities of the glycine cleavage multienzyme system. Stroke. 2007;38(7):2157–64.
10. Shimizu-Sasamata M, Bosque-Hamilton P, Huang PL, Moskowitz MA, Lo EH. Attenuated neurotransmitter release and spreading depression-like depolarizations after focal ischemia in mutant mice with disrupted type I nitric oxide synthase gene. J Neurosci. 1998;18:9564–71.
11. Chen H, Chopp M, Zhang ZG, Garcia JH. The effect of hypothermia on transient middle cerebral artery occlusion in the rat. J Cereb Blood Flow Metab. 1992;12(4):621–8.
12. Garcia JH, Yoshida Y, Chen H, Li Y, Zhang ZG, Lian J, Chen S, Chopp M. Progression from ischemic injury to infarct following middle cerebral artery occlusion in the rat. Am J Pathol. 1993;142(2):623–35.

Inducing Hemorrhagic Transformation Following Middle Cerebral Artery Occlusion via Acute Hyperglycemia in Rats

Devin W. McBride, Derek Nowrangi, Wing Mann Ho, Jiping Tang, and John H. Zhang

Abstract Herein, the model for inducing hemorrhagic transformation following middle cerebral artery occlusion (MCAO) in acutely hyperglycemic rats is presented. Fifty percent dextrose, administered 15 min before MCAO, causes reproducible hemorrhagic transformation at a rate of 100% in rats. This chapter details the procedure for creating this model, as well as the outcomes, evaluation, limitations, and alternatives.

Keywords Middle cerebral artery occlusion · Hemorrhagic transformation · Hyperglycemia · Rat

Introduction

Stroke disproportionally affects patients with diabetes and/or hyperglycemia which results in higher incidence, increased morbidity and mortality, and worse outcome compared to non-diabetic and non-hyperglycemic stroke patients [1]. Hyperglycemia is associated with larger infarct volume and more cerebral edema

Electronic supplementary material: The online version of this chapter (https://doi.org/10.1007/978-3-030-16082-1_12) contains supplementary material, which is available to authorized users.

D. W. McBride · D. Nowrangi · J. Tang
Department of Physiology & Pharmacology, Loma Linda University School of Medicine, Loma Linda, CA, USA

W. M. Ho
Department of Neurosurgery, Medical University Innsbruck, Innsbruck, Austria

J. H. Zhang (✉)
Department of Physiology & Pharmacology, Loma Linda University School of Medicine, Loma Linda, CA, USA

Department of Anesthesiology, Loma Linda University School of Medicine, Loma Linda, CA, USA
e-mail: jhzhang@llu.edu

J. Chen et al. (eds.), *Animal Models of Acute Neurological Injury*, Springer Series in Translational Stroke Research, https://doi.org/10.1007/978-3-030-16082-1_12

in stroke patients [2, 3]. Following the primary insult, secondary injury, including delayed neuronal apoptosis, brain edema, and inflammation begins [3–7].

Acute hyperglycemia, typically induced by dextrose injection, and chronic hyperglycemia, typically streptozotocin-induced diabetes, have been evaluated in various species (rat, mouse, primate, rabbit) and utilizing a number of different ischemic models (i.e. global ischemia or focal ischemia, permanent or transient occlusion, thromboemboli or intraluminal suture) [8, 9] (good reviews are available, see [10, 11]).

Several rodent models of focal ischemia are available which can be used to investigate the effects of hyperglycemia on stroke and for developing therapies which reduce, either directly or indirectly, hyperglycemic consequences after stroke [8, 9, 12, 13]. The most common agents which induce hyperglycemia are dextrose, for acute hyperglycemia, and streptozotocin, for chronic hyperglycemia and diabetes [11]. Each method for creating hyperglycemia allows various stroke pathophysiologies to be studied. Herein, we describe the procedures for the intraluminal suture middle cerebral artery occlusion (MCAO) in acutely hyperglycemic rats. Acute hyperglycemia is induced by an intraperitoneal injection of dextrose.

Materials

Adult male Sprague-Dawley rats weighing between 260 and 290 g (Harlan, Indianapolis, IN, USA) were used for this model. The following equipment was utilized to complete all surgical procedures (Fig. 1):

1. Standard surgical equipment: scalpel, fine forceps, skin retractors (Fine Science Tools, Foster City, CA)
2. Surgical microscope (Zeiss, West Germany)
3. Hair clippers and skin disinfectant: alcohol prep pads, betadine
4. Cotton tipped applicator or absorbent surgical spears
5. 2-0 Perma-Hand Silk Suture, Black Braided

   Note: These sutures are unbraided to create the thin sutures used for vessel ligation (external carotid artery and pterygopalatine artery)
6. 3-0 Suture silk for wound closure
7. Electrosurgical system for cauterizing (Macan, Chicago, IL) with smooth straight bipolar forceps (Macan, Chicago, IL)
8. Vessel clips (00398-01 and 00398-02, Fine Science Tools, Foster City, CA)
9. 4-0 Nylon suture (Ethilon, Ethicon) and silicon for creating the intraluminal suture with a rounded tip for occlusion

   Note: See Steps 1–4 for the procedure on creating the intraluminal suture with a rounded tip
10. Rectal thermometer-controlled heat pad (Fine Science Tools, Foster City, CA)
11. Blood glucose meter with strips
12. A mixture of ketamine (80 mg/kg) and xylazine (20 mg/kg) for anesthesia
13. 50% Dextrose solution (6 mL/kg)

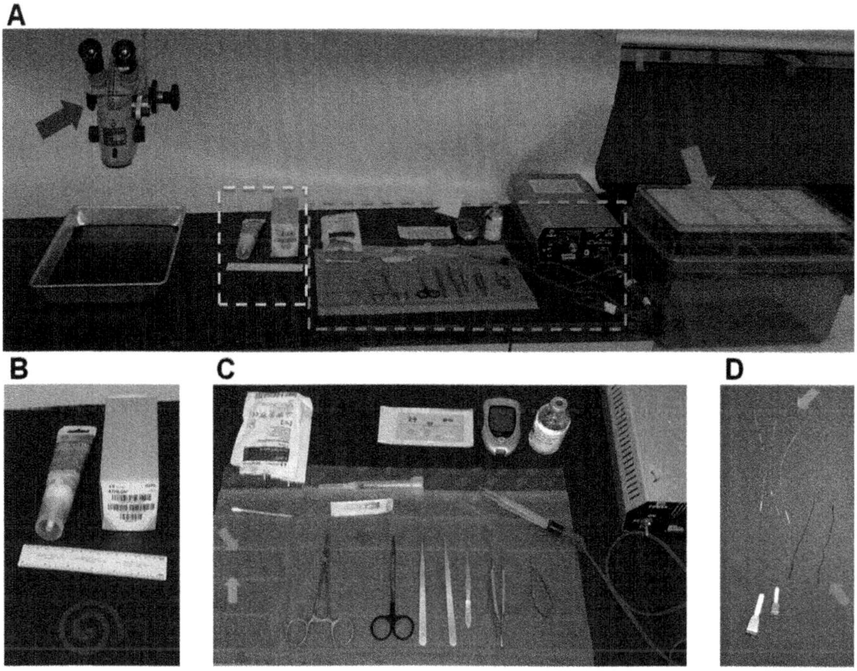

Fig. 1 Preparation for the MCAO Model. (**a**) Surgical Station. The surgical tools needed to perform the model (blue dashed box), equipment needed to make the intraluminal sutures (yellow dashed box), surgical microscope (brown arrow), and the recovery chamber (red arrow) are displayed. Also shown is the glucose monitor (orange arrow). (**b**) Equipment for creating the intraluminal sutures. Shown is the silicone, a ruler, 4-0 nylon suture, and a steady low-heat source for blunting the tip of the nylon sutures (incense). (**c**) Surgical tools. Top row (from left to right): cotton tipped applicators, syringe and needle for anesthetic or dextrose injection, 3-0 silk suture for wound closure, glucose monitor, 50% dextrose, and electrocautery forceps. Bottom row (from left to right): sutures for vessel ligation (pink arrow), intraluminal sutures (green arrow), vessel clips (red arrow), Mayo-Hegar needle holder, scissors, two fine forceps, two forceps, and surgical scissors. (**d**) Sutures and vessel clips. Intraluminal sutures silicone coated and with round tips (green arrow), silk sutures for vessel ligation (pink arrow), and vessel clips (red arrow)

14. Atropine (0.1 mg/kg)
15. Buprenorphine (0.02 mg/kg)
16. Yohimbine (1 mg/kg) for anesthesia reversal

Procedures

All procedures were approved by the Loma Linda University IACUC. Rats were given access to food and water ad libitum. Rats were housed on a 12 h light-dark cycle in a temperature controlled room.

Preparation of the 4.0 Monofilament Suture with Rounded Tip for Occlusion

1. Cut the 4-0 Nylon suture into 30 mm lengths.
2. On one end of the suture, slowly move it close to a heat source until the tip begins to soften. As the tip softens, it becomes rounded. Allow the tip to round to the appropriate size based on your animal weights (about 300–350 μm in diameter for rats weighing between 260 and 290 g).

   Note: The heat source used may be an open flame (such as from a candle), but we recommend using a smoldering heat source (such as a burning incense).

   Note: If the suture tip is not perpendicularly advanced close to the heat source the tip may bend or the bulb may not be equally rounded all around the suture.
3. Allow the suture to cool. Then dip the rounded part of the suture into silicon up to a length of about 3–4 mm.

   Note: We recommend wiping the silicon off the rounded tip. Not doing so may result in rounded tips which have variable diameters.

   Note: The silicon sheath should be about 300–350 μm in diameter for rats weighing between 260 and 290 g. It may be slightly smaller or larger than the rounded tip.

   Note: We recommend marking the suture at the lengths of 18 and 22 mm from the tip. These marks indict the typical length of insertion for blocking the middle cerebral artery.
4. Allow the silicon to air dry at room temperature overnight before using.

Anesthetizing Animals and Inducing Acute Hyperglycemia

5. Animals are anesthetized with an intraperitoneal injection of a ketamine (80 mg/kg) and xylazine (20 mg/kg) mixture.
6. Rats then receive a subcutaneous dose of atropine (0.1 mg/kg).
7. Upon reaching an adequate level of anesthesia (i.e. loss of paw pinch reflex), administer an intraperitoneal injection of 50% dextrose (6 mL/kg) 15 min before surgery.

   Note: Monitor blood glucose to ensure that all animals have a similar glucose concentration within the blood at designated time points (such as pre-dextrose, before beginning surgery, immediately after beginning occlusion, at the start of reperfusion, and every 1–2 h post-reperfusion). Unequal blood glucose levels may cause differences in the rate of hemorrhagic transformation and differences in hematoma volume, ultimately confounding the study. Blood glucose can be monitored with a standard blood glucose monitor (such as TRUEresult).
8. Place the rat onto an operating table on top of a heating pad. Place the rectal probe to maintain a body temperature of 37 °C ± 0.5 °C.

Intraluminal Suture Middle Cerebral Artery Occlusion

9. Make a 2–3 cm vertical, midline incision in the skin from the sternum jugular notch to the hyoid bone (Fig. 2a).
10. Separate the thyroid gland lobes to expose the underlying trachea which is covered by the sternohyoid muscle.
11. Bluntly dissect the fascia to expose the anatomical triangle created by sternohyoid muscle (medially), the omohyoid muscle (superiorly), and the sternomastoid muscle (inferolateral border) (Fig. 2a).

    Note: The sternohyoid muscle can be cauterized at its origin to the hyoid bone. Performing this step allows for better exposure of the internal carotid and pterygopalatine arteries (see Step 18).
12. Use retractors to separate the omohyoid and sternomastoid muscles, exposing the carotid sheath (Fig. 2a, b).
13. Continue blunt dissection of the fascia to expose the common carotid (CCA) and external carotid arteries (ECA).

    Note: Take care when dissecting the fascia in this step. There are several blood vessels branching from the external carotid artery. While damaging these vessels may not cause major blood loss, reframe from puncturing these vessels.

    Note: It is recommended to either start the dissection proximally or distally to find either the common carotid or external carotid arteries, respectively. This will aid in reducing the chance of damaging the branching arteries. Minor bleeds can be stopped by applying pressure using a cotton tipped applicator or absorbent surgical spear. If needed, electrocauterization can be used to stop bleeds.
14. Dissect the vessels, taking extreme care not to damage the vagus nerve (Fig. 2b).
15. Find the superior thyroid artery (branches off medially from the ECA) and occipital artery (branches off laterally from the ECA near the carotid bifurcation) (Fig. 2b). Ligate both of these arteries using electrocauterization.

    Note: The occipital artery lies on top of the internal carotid artery (ICA) and is attached to the ICA via fascia. Using the fine forceps, the occipital artery can be separated from the ICA by carefully inserting the forceps underneath the occipital artery and separating the forceps tips.
16. Next, completely separate the ECA from the surrounding tissue. Place 2 silk sutures around the ECA about 3 and 5 mm from the carotid bifurcation. Tie both sutures, cutting off blood flow (Fig. 2c).
17. Use a pair of fine surgical scissors to ligate the ECA between the two sutures, leaving an ECA stump of about 3 mm in length.
18. Continue to dissect the fascia, exposing the ICA and pterygopalatine artery.
19. Next, stop the blood flow to the pterygopalatine artery by tying a suture around it (Fig. 2c).

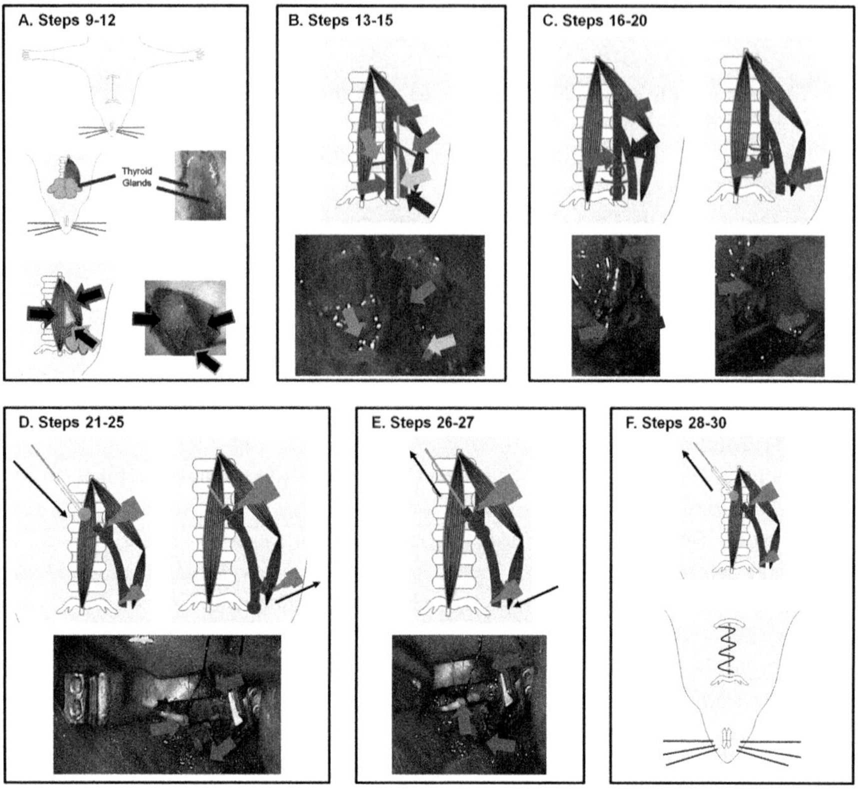

Fig. 2 Surgical Procedure. Schematics and intraoperative images of the steps are provided. (**a**) Steps 9–12. The sternohyoid (black arrow with blue outline), omohyoid (black arrow with orange outline), and sternomastoid (black arrow with red outline) muscles are separated to expose the carotid sheath. (**b**) Steps 13–15. The fascia is dissected to expose the common carotid (blue arrow) and external carotid (brown arrow) arteries, and the vagus nerve (yellow arrow). The superior thyroid artery (orange arrow) and occipital artery (green arrow) branch off from the external artery. These small arteries need to be ligated. (**c**) Steps 16–20. Once the occipital artery is ligated, the common carotid (blue arrow), internal carotid (purple arrow), and external carotid (brown arrow) artery bifurcation can be seen. Two sutures are placed around the external artery to create the external carotid artery stump. An additional suture is placed around the pterygopalatine artery (pink arrow). (**d**) Steps 21–25. Two vessel clips are placed on the artery (one on the common carotid artery and the other on the internal carotid artery) to stop blood flow and allow the intraluminal suture to be inserted into the external carotid artery stump. (**e**) Steps 26–27. The intraluminal suture is withdrawn to the external carotid artery stump and a vessel clip is placed on the internal carotid artery to allow for complete removal of the intraluminal suture. (**f**) Steps 28–30. After complete removal of the intraluminal suture, the animal is sutured and allowed to recover

20. Place loosely tied suture around the ECA stump at the base of the carotid bifurcation.
21. Use a vessel clip to interrupt the blood flow through the CCA and a second vessel clip on the ICA (proximal to the pterygopalatine artery branch).

22. Use surgical scissors to make a small cut in the ECA stump.
    Note: 1–2 drops of blood may exit the lesion on the ECA stump. If blood begins to pool, quickly pull the loose suture on the ECA stump tight. If this happens, typically one of the vessel clips are not completely stopping flow. Reposition the vessel clips and loosen the ECA stump suture.
23. Insert a 4.0 monofilament nylon suture with a rounded tip into the ECA stump through the lesion. Advance the suture up through the carotid bifurcation and into the ICA (Fig. 2d).
24. Pull the suture on the ECA stump tight and unclamp the vessel clip on the ICA.
25. Continue to advance the intraluminal suture with a rounded tip into the ICA up towards the middle cerebral artery. Once resistance is felt (approximately 18–22 mm from the ECA stump), stop advancing the intraluminal suture. Pull on the ECA stump suture tightly to make sure the intraluminal suture will not move. Leave the intraluminal suture in place for the desired amount of time (1.5 h in this procedure).
    Note: If resistance is not felt and the intraluminal suture advances more than 25 mm, pull the intraluminal suture back to 15 mm and try advancing it again. If no resistance is felt a second time, completely remove the intraluminal suture (following Steps 26–28) and repeat Steps 23–25 again with another intraluminal suture (preferably with a larger rounded tip). The likely cause of this is that the rounded tip of the intraluminal suture is not large enough to occlude the blood vessel.

Restoring Cerebral Blood Flow

26. After the allotted time for middle cerebral artery occlusion, gently withdraw the intraluminal suture back through the ICA and then the carotid bifurcation to the ECA stump (Fig. 2e).
27. Place a vessel clip into the ICA proximal to the pterygopalatine artery branch.
28. Loosen the suture on the ECA stump and completely withdraw the intraluminal suture.
29. Tighten the suture on the ECA stump, then remove the two vessel clips, completely restoring blood flow (Fig. 2f).
30. Remove the retractors. Then suture the skin closed. Remove the rectal probe.
31. Administer buprenorphine (0.02 mg/kg, subcutaneously) and yohimbine (1 mg/kg, intraperitoneally). Place the rat into a heated recovery cage until fully awake.

Note: Post-operative normal saline (1 mL, subcutaneously) should be administered to aid in recovery and minimize possible dehydration.

Outcomes and Evaluations

The MCAO rat model allows for several pathological outcomes to be examined, including infarction size [14–18], neuroinflammation [6, 17], blood-brain barrier disruption [3, 5, 6, 14–16, 18–20], and neuronal apoptosis. Furthermore, hyperglycemia-exacerbated MCAO in rats have the additional deleterious outcomes of hemorrhagic transformation [3–6, 14–16, 18, 19], severe brain swelling [3, 5, 14, 18], and elevated brain water content [4–6, 18, 19, 21, 22], and oxidative stress [4, 23]. Finally, MCAO in hyperglycemic rats results in significant short- and long-term neurological deficits [14, 19].

Physiological Outcomes

Administration of dextrose 15 min before beginning MCAO surgery rapidly elevates blood glucose concentration (via blood glucose strips and monitor). At onset of reperfusion, 2 h after dextrose injection, blood glucose is 194 ± 55 mg/dL for MCAO animals and 218 ± 89 mg/dL in Sham animals; this concentration is maintained for another 1.5 h (3.5 h post-dextrose administration) (Fig. 3).

Hemorrhagic transformation caused by MCAO in acutely hyperglycemic rats induced significant weight loss 24 h post-reperfusion. Rectal temperature was not different between Sham or rats subjected to MCAO (Table 1).

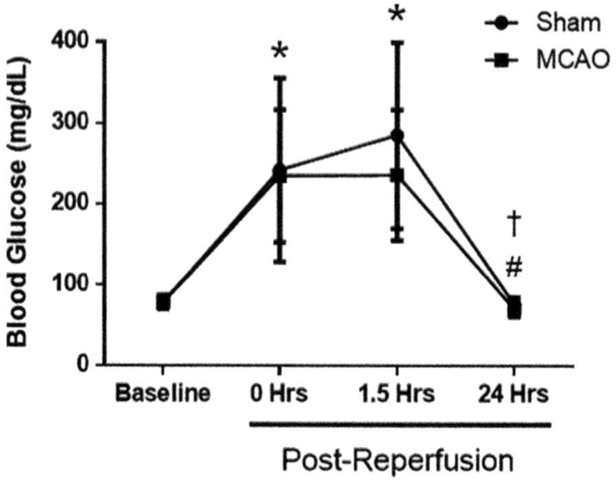

Fig. 3 Blood Glucose Concentration. Baseline blood glucose concentrations were measured before injection of the 50% dextrose solution. Post-reperfusion blood glucose concentrations were measured at the indicated time after removal of the intraluminal suture. n = 6–12/group/time point. Mean ± SD. Two-way ANOVA with Tukey post-hoc test. $^{*}p < 0.05$ vs Baseline, $^{\#}p < 0.05$ vs 0 h post-reperfusion, and $^{\dagger}p < 0.05$ vs 1.5 h post-reperfusion

Pathological Outcomes

Infarct Volume

Hyperglycemic rats subjected to 90 min of MCAO have a hemorrhagic transformation rate of 100% (Fig. 4a) and presented with infarction volumes of 26.5 ± 7.3% 24 h after reperfusion; the infarction developed and reached its maximum within 15 min post-reperfusion (Fig. 4b). The infarction is typically observed using histological staining: either TTC (seen in Fig. 4a) or Nissl staining. In TTC-stained brains, the infarction appears as the unstained, white tissue. Here, infarction volume is measured as the difference in the infarction area (white, un-stained tissue) and the whole ipsilateral hemisphere, then normalized to either the contralateral hemisphere or the whole brain. A number of different algorithms are available for quantifying the infarction size, each with there limitations [24–31]. The algorithm used here is

$$\text{Infarct Volume }(\%) = \left(\frac{\sum_i \left(\left(\frac{I_i - N_i}{I_i} \right) C_i \right)}{2 \sum_i C_i} \right) 100,$$

where I_i is the ipsilesional area of slice i, N_i is the non-infarcted ipsilesional hemisphere tissue for slice i, and C_i is the contralesional area of slice i.

Hemoglobin Volume

Contrary to the infarction size, the hemoglobin volume within the ipsilateral, infarcted hemisphere (measured via the hemoglobin assay) develops within 1 h after beginning reperfusion and continues to increase over 24 h (Fig. 4c). Briefly, the hemoglobin assay consists of homogenizing and sonicating the brain tissue on PBS, centrifugation to remove cell debris, then measuring the absorbance (at 540 nm) of the supernatent using Drabkin's reagent.

Table 1 Physiological parameters

		Temperature (°C)	
	Weight change (g)	Pre-MCAO	Post-MCAO
Sham	−16.5 ± 6.55	35.9 ± 0.35	35.7 ± 0.72
MCAO	−25.2 ± 2.99*	35.5 ± 0.75	35.6 ± 0.89

Weight change 24 h post-reperfusion (g). Temperature is measured with a rectal probe 1 h before MCAO (Pre-MCAO) and 24 h after reperfusion (Post-MCAO). n = 6–9/group. *$p < 0.05$ vs Sham

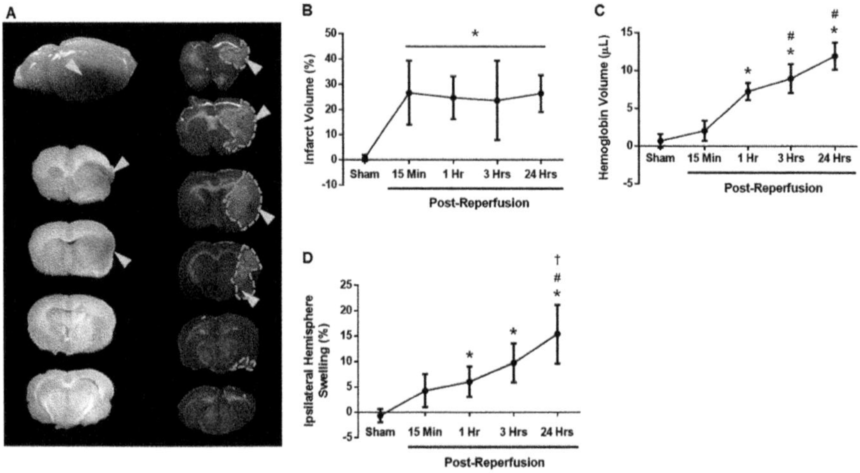

Fig. 4 Development of the Infarction, Hemorrhagic Transformation, and Cerebral Edema after MCAO in Acutely Hyperglycemic Rats. (**a**) Representative images of the whole brain (top left) and slice brains (bottom left and right) showing hemorrhagic transformation (yellow arrow heads). The left brain slices are not stained. The right brain slices are stained with 2% TTC to show the infarcted area (white tissue, outlined with green dashed lines). In the TTC-stained brain slices, the hemorrhage is colored brown. (**b**) Time course of infarct volume (%). (**c**) Time course of hemoglobin volume in the ipsilateral hemisphere. (**d**) Time course of ipsilateral hemisphere swelling (%). n = 6–9/group/time point. Mean ± SD. One-way ANOVA with Tukey post-hoc test. *$p < 0.05$ vs Sham, #$p < 0.05$ vs 15 min post-reperfusion, and †$p < 0.05$ vs 1 h post-reperfusion

Ipsilesional Swelling

Similar to hemoglobin volume, swelling of the ipsilesional hemisphere develops within 1 h after MCAO and continues to increase over 24 h (Fig. 4d). Ipsilesional hemispheric swelling is quantified using the difference in the sum of the ipsilesional hemisphere slices, areas (no stain necessary) and that of the contralateral hemisphere. This difference is normalized to the contralateral hemisphere volume [30]. The algorithm used herein is

$$\text{Ipsilesional Hemispheric Volume}\left(\%\right)=\left(\frac{\sum_i \left(I_i-C_i\right)}{\left(\sum_i C_i\right)}\right)100,$$

where I_i is the ipsilesional area of slice i and C_i is the contralesional area of slice i.

Functional Outcomes

Acute hyperglycemia during MCAO results in significant neurological and behavioral deficits. Three common and simple functional tests is the modified Garcia neuroscore, the Corner Turn Test, and the Forelimb placement test. The modified Garcia neuroscore is a composite score of 7 subtests evaluating various aspects of sensori-motor function, including forelimb and hindlimb use, grip strength and climbing ability, reflex, and spontaneous activity. Each subtest has a minimum score of 0 and a maximum score of 3, with a total composite neuroscore having a minimum of 3 and a maximum of 21 (Table 2).

Table 2 Modified Garcia neuroscore criteria

	Score			
Sub-test	0	1	2	3
Spontaneous activity	Animal is akinesitic	Animal moves slowly or minimally	Animal approaches one to two walls	Animal approaches at least three walls of the cage or raises on hindlimbs to explore the top of the cage on two or more sides
Body proprioception	–	Animal has a unilateral response	Animal has either a weak bilateral response or weak left response and brisk right response	Animal has a brisk bilateral response
Vibrissae touch	–	Animal has a unilateral response	Animal has either a weak bilateral response or weak left response and brisk right response	Animal has a brisk bilateral response
Limb symmetry	Hemiparesis	Left forelimb or left hindlimb flexed	Asymmetric extension	All limbs extend symmetrically
Lateral turning	Animal does not turn to one side	Animal has unequal turning between the two sides	Animal turns bilaterally less than 45° on both sides	Animal turns bilaterally more than 45° on both sides
Forelimb outstretching	Animal has a paretic forelimb	Animal walks in circles	Animal walks asymmetrically or to one side	Animal briskly walks symmetrically on forepaws
Climbing	–	Animal fails to climb or circles instead of climbing	Animal climbs to the top and has a weak grip or animal does not climb to the top but has a strong grip	Animal climbs to the top and has a strong grip

The total composite neuroscore is the sum of all seven sub-tests: minimum composite neuroscore is 3 and the maximum is 21

The Corner Turn Test evaluates the animals tendency to rear and turn when it reaches a corner. The animal is allowed to walk towards a 30° corner (created using two sheets of plexi-glass). The animal will rear and turn around to walk out of the corner. The choice of turn is recorded. Test each animal for 20 trials. The number of left turns is reported as a percentage of the total number of turns. Non-injured animals usually have no preference for turning direction (i.e. 50% of turns are to the right and 50% of the turns are to the left). This test is very sensitive and can measure functional deficits in animals with even a minimal injury [32].

The Forelimb Placement Test examines the sensitivity of the vibrissae to stimulation; when stroked, the animal should reach its forelimb out to reach a stable surface. Hold the rat in a single hand; ensure the grip on the rat is comfortable yet strong. Allow one of the animal's forelimbs to be freely moveable; hold the other forelimb so that it is restrained against the animal's torso. Bring the animal towards a flat surface, such as a table top, to stimulate the vibrissae which should elicit a forelimb stretching response towards the surface. Record the number of positive responses to the vibrissae stimulations for 20 trials. Repeat this process for the other forelimb. Report the number of positive vibrissae-evoked responses as a percentage of the total number of stimulations for each forelimb [32].

Here, acutely hyperglycemic rats subjected to 90 min of MCAO have functional deficits in each of these three behavioral tests. The composite neuroscore is significantly lower in MCAO animals than in Sham animals 24 h after MCAO (Fig. 5a). Hyperglycemic rats subjected to right MCA occlusion preferentially turn right 24 h post-ictus (Fig. 5b). Finally, the forelimb placement test displays functional deficiencies in the left forepaw but not the right forepaw (Fig. 5c).

A number of other functional tests can be utilized for evaluation of behavioral deficits following MCAO in acutely hyperglycemic rats, including grip strength tests [33], as well as long term behavioral deficits, such as open field and novel objection recognition [34].

Caveats, Limitations, and Alternatives

The middle cerebral artery occlusion model has a number of limitations and alternatives which are discussed elsewhere. Here, only the limitations and alternatives for the model of hemorrhagic transformation by acute hyperglycemia following MCAO are described below.

Mortality and Weight-Loss

One drawback of this model is the significant mortality associated with hemorrhagic transformation; a morality of 25–35% is common with this model. The mortality rate increases if the occlusion time is lengthened. For example, a 2 h occlusion can result in a mortality of 50–60%.

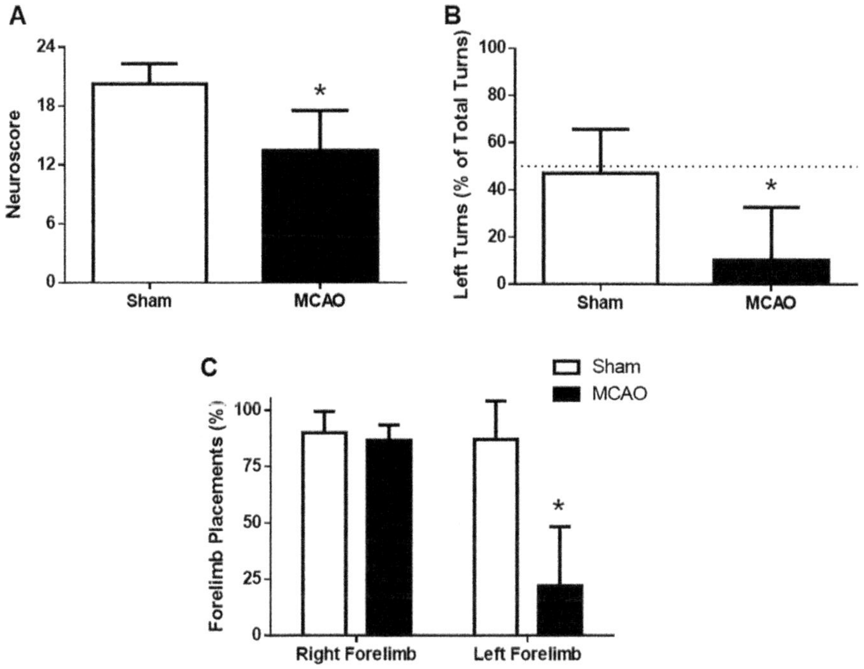

Fig. 5 Neurological Deficits 24 h after MCAO in Acutely Hyperglycemic Rats. (**a**) Modified Garcia Neuroscore. Maximum score is 21, indicating no neurological deficits. (**b**) Corner turn test. The amount of left turns (% of total turns, 20 total turns). A dotted line indicates the expected number of left turns. (**c**) Forelimb placement test. The percent of forelimb placements in response to vibrissae stimulation for each forelimb (20 stimulations for each side). n = 6–9/group. Mean ± SD. T test. $^{*}p < 0.05$ vs Sham

Another concern is the weight-loss which occurs post-MCAO due to a decreased ability and/or desire to eat and drink. Animals chosen to survive for longer than 2–3 days post-ictus should be observed for normal feeding and drinking. If needed, normal saline can be administered and food given by oral gavage.

Model Variations

Acute hyperglycemia can be induced by intraperitoneal or intravenous dextrose injection, as well as orally administered dextrose. Here we described the model for hemorrhagic transformation induced by pre-injury dextrose injection. Additional variations for creating hemorrhagic transformation are the time of dextrose administration and other factors. Hemorrhagic transformation following MCAO can also be caused by post-MCAO dextrose injection, tPA administration, long-term MCAO (4 h or greater of occlusion increases the rate of hemorrhagic transformation), or diabetic/chronic hyperglycemic rats.

References

1. Peng G, Yuan Y, Wu S, He F, Hu Y, Luo B. MicroRNA let-7e is a potential circulating biomarker of acute stage ischemic stroke. Transl Stroke Res. 2015;6(6):437–45.
2. Bruno A, Biller J, Adams HP, et al. Acute blood glucose level and outcome from ischemic stroke. Neurology. 1999;52(2):280–4.
3. Chen C, Hu Q, Yan J, et al. Early inhibition of HIF-1alpha with small interfering RNA reduces ischemic-reperfused brain injury in rats. Neurobiol Dis. 2009;33(3):509–17.
4. Hu Q, Ma Q, Zhan Y, et al. Isoflurane enhanced hemorrhagic transformation by impairing antioxidant enzymes in hyperglycemic rats with middle cerebral artery occlusion. Stroke. 2011;42(6):1750–6.
5. Yura S. Effects of hyperglycemia on ischemic brain damage, local cerebral blood flow and ischemic cerebral edema. Hokkaido Igaku Zasshi. 1991;66(1):1–15.
6. Bian H, Hu Q, Liang X, et al. Hyperbaric oxygen preconditioning attenuates hemorrhagic transformation through increaing PPARg in hyperglycemic MCAO rats. Exp Neurol. 2015;265:22–9.
7. Dock H, Theodorsson A, Theodorsson E. DNA methylation inhibitor zebularine confers stroke protection in ischemic rats. Transl Stroke Res. 2015;6(4):296–300.
8. Ergul A, Li W, Elgebaly MM, Bruno A, Fagan SC. Hyperglycemia, diabetes and stroke: focus on the cerebrovasculature. Vasc Pharmacol. 2009;51(1):44–9.
9. Hafez S, Coucha M, Bruno A, Fagan SC, Ergul A. Hyperglycemia, acute ischemic stroke, and thrombolytic therapy. Transl Stroke Res. 2014;5(4):442–53.
10. Howells DW, Porritt MJ, Rewell SS, et al. Different strokes for different folks: the rich diversity of animal models of focal cerebral ischemia. J Cereb Blood Flow Metab. 2010;30(8):1412–31.
11. MacDougall NJ, Muir KW. Hyperglycaemia and infarct size in animal models of middle cerebral artery occlusion: systematic review and meta-analysis. J Cereb Blood Flow Metab. 2011;31(3):807–18.
12. Fan X, Qiu J, Yu Z, et al. A rat model of studying tissue-type plasminogen activator thrombolysis in ischemic stroke with diabetes. Stroke. 2012;43(2):567–70.
13. Hung L-M, Huang L-P, Liao JM, et al. Insulin renders diabetic rats resistant to acute ischemic stroke by arresting nitric oxide reaction with superoxide to form peroxynitrite. J Biomed Sci. 2014;21:92.
14. Li W, Prakash R, Kelly-Cobbs AI, et al. Adaptive cerebral neovascularization in a model of type 2 diabetes. Relevance to focal cerebral ischemia. Diabetes. 2010;59(1):228–35.
15. Soejima Y, Ostrowski RP, Manaenko A, Fujii M, Tang J, Zhang JH. Hyperbaric oxygen preconditioning attenuates hyperglycemia enhanced hemorrhagic transformation after transient MCAO in rats. Med Gas Res. 2012;2:9.
16. Tan F, Fu W, Cheng N, Meng DI, Gu Y. Ligustrazine reduces blood-brain barrier permeability in a rat model of focal cerebral ischemia and reperfusion. Exp Ther Med. 2015;9(5):1757–62.
17. Wu MH, Huang CC, Chio CC, et al. Inhibition of peripheral TNF-alpha and downregulation of microglial activation by alpha-lipoic acid and etanercept protect rat brain against ischemic stroke. Mol Neurobiol. 2016;53:4961–71.
18. Xing Y, Hua Y, Keep RF, Xi G. Effects of deferoxamine on brain injury after transient focal cerebral ischemia in rats with hyperglycemia. Brain Res. 2009;1291:113–21.
19. Chen C, Ostrowski RP, Zhou C, Tang J, Zhang JH. Suppression of hypoxia-inducible factor-1 alpha and its downstream genes reduces acute hyperglycemia-enhanced hemorrhagic transformation in a rat model of cerebral ischemia. J Neurosci Res. 2010;88(9):2046–55.
20. Merali Z, Leung J, Mikulis D, Silver F, Kassner A. Longitudinal assessment of imatinib's effect on the blood-brain barrier after ischemia/reperfusion injury with permeability MRI. Transl Stroke Res. 2015;6(1):39–49.
21. Cipolla MJ, Huang Q, Sweet JG. Inhibition of protein kinase Cbeta reverses increased blood-brain barrier permeability during hyperglycemic stroke and prevents edema formation in vivo. Stroke. 2011;42(11):3252–7.

22. Kusaka I, Kusaka G, Zhou C, et al. Role of AT1 receptors and NAD(P)H oxidase in diabetes-aggravated ischemic brain injury. Am J Physiol Heart Circ Physiol. 2004;286(6):H2442–51.
23. Qi Z, Dong W, Shi W, et al. Bcl-2 phosphorylation triggers autophagy switch and reduces mitochondrial damage in limb remote ischemic conditioned rats after ischemic stroke. Transl Stroke Res. 2015;6(3):198–206.
24. Bederson JB, Pitts LH, Germano SM, Nishimura MC, Davis RL, Bartkowski HM. Evaluation of 2,3,5-triphenyltetrazolium chloride as a stain for detection and quantification of experimental cerebral infarction in rats. Stroke. 1986;17(6):1304–8.
25. Belayev L, Khoutorova L, Deisher TA, et al. Neuroprotective effect of SolCD39, a novel platelet aggregation inhibitor, on transient middle cerebral artery occlusion in rats. Stroke. 2003;34(3):758–63.
26. Lee J, Lee JK, Han K. InfarctSizer: computing infarct volume from brain images of a stroke animal model. Comput Methods Biomech Biomed Engin. 2011;14(6):497–504.
27. Lin TN, He YY, Wu G, Khan M, Hsu CY. Effect of brain edema on infarct volume in a focal cerebral ischemia model in rats. Stroke. 1993;24(1):117–21.
28. Swanson RA, Morton MT, Tsao-Wu G, Savalos RA, Davidson C, Sharp FR. A semiautomated method for measuring brain infarct volume. J Cereb Blood Flow Metab. 1990 Mar;10(2):290–3.
29. Swanson RA, Sharp FR. Infarct measurement methodology. J Cereb Blood Flow Metab. 1994;14(4):697–8.
30. McBride DW, Klebe D, Tang J, Zhang JH. Correcting for brain swelling's effects on infarct volume calculation after middle cerebral artery occlusion in rats. Transl Stroke Res. 2015;6(4):323–38.
31. McBride DW, Tang J, Zhang JH. Development of an infarct volume algorithm to correct for brain swelling after ischemic stroke in rats. Acta Neurochir Suppl. 2016;121:103–9.
32. Schallert T. Behavioral tests for preclinical intervention assessment. NeuroRx. 2006;3:497–504.
33. Hafez S, Hoda MN, Guo X, Johnson MH, Fagan SC, Ergul A. Comparative analysis of different methods of ischemia/reperfusion in hyperglycemic stroke outcomes: interaction with tPA. Transl Stroke Res. 2015;6(3):171–80.
34. Zhu W, Casper A, Libal NL, et al. Preclinical evaluation of recombinant T cell receptor ligand RTL1000 as a therapeutic agent in ischemic stroke. Transl Stroke Res. 2015;6(1):60–8.

Cerebral Palsy Model of Uterine Ischemia in Pregnant Rabbits

Zhongjie Shi, Kehuan Luo, and Sidhartha Tan

Abstract Rabbits, like humans, are perinatal brain developers. The model of uterine ischemia mimics acute placental insufficiency states in humans such as placental abruption. A detailed description of the fetal model is presented. Following fetal hypoxia-ischemia, movement disorders are observed in postnatal rabbit kits akin to those observed in cerebral palsy in humans.

Keywords Behavior · Brain · Dystonia · Fetus · Hypoxia · Infant, newborn · Infant, premature · Ischemia · Placental insufficiency · Uterus

Introduction

In selecting an animal model for human perinatal disease, one must take into account the normal brain development of that animal. In mammalian perinatal models, rodents are most often used but rodents are postnatal brain developers. White matter and motor development occurs primarily after delivery. Other perinatal models use ungulates and nonhuman primates. Both ungulates and nonhuman primates are prenatal brain developers. At birth most of the motor development has been completed. This probably is a protective strategy in the wild as the newborn needs to have some motor function in order to survive. Humans are unlike other primates in that for some reason, humans have become perinatal brain developers, just like rabbits [1]. Thus, for investigation of the etiology of motor deficits, the right model for perinatal diseases is probably not the closest animal relative to man, the chimpanzee or other great apes, but animals that are perinatal brain developers such as the rabbit.

Z. Shi · K. Luo
Department of Pediatrics, Wayne State University, Detroit, MI, USA

S. Tan (✉)
Department of Pediatrics, Wayne State University, Detroit, MI, USA

Department of Pediatrics, Children's Hospital of Michigan, Detroit, MI, USA

J. Chen et al. (eds.), *Animal Models of Acute Neurological Injury*, Springer Series in Translational Stroke Research, https://doi.org/10.1007/978-3-030-16082-1_13

Table 1 Similarity of rabbit to humans compared to other animal species

Advantages of the rabbit model	Comparison with other species
True fetal model	Postnatal models in rodents
Global hypoxia-ischemia	Unilateral, ischemic models in rodents
Workable size of brain from fetuses for MRI, biochemical and molecular biology analysis	Rodent fetuses and pups have very small brains
Rabbits, similar to humans, are perinatal brain developers especially for WM. Beginning myelination shortly after birth	Sheep and baboons are prenatal. Rodents are postnatal developers
Limited neurobehavioral development at birth, similar to humans	Sheep and great apes almost fully developed at birth
Oxidant systems, such as xanthine oxidase are similar to humans	Rodents have very high levels of xanthine oxidase in circulation
Relatively inexpensive, litter size ~8 and easy to get enough **n** for experiments	Sheep and nonhuman primates are expensive

Hypoxia-ischemia results in immediate fetal bradycardia, logarithmic decrease in blood pH with increasing hypoxia-ischemia, parenchymal hemorrhage in preterm fetuses, and after preterm hypoxia-ischemia, cysts, hydrocephalus and ventriculomegaly found at P1

The second important requirement for animal models is to mimic human disease conditions. One of the criticisms of the variations of the Levine model in rodents is that these are not reflective of any pathological conditions of humans in the perinatal period. For example, if one wants to establish a model of cerebral palsy, one has to look at the antecedents causing such a disease. Cerebral palsy is often believed to result from perinatal hypoxia-ischemia. One of the important conditions resulting in fetal hypoxia-ischemia is acute placental insufficiency that occurs in diseases such as placental abruption. Placental abruption has been shown to result in a higher incidence of cerebral palsy [2]. Modeling acute placental insufficiency states in animals causes a problem because this involves opening up the abdomen [3]. We were able to establish a model without entering the abdominal cavity that subjects the fetuses to hypoxia-ischemia [4, 5].

The major advantages of this model are listed in Table 1.

Methods

Experimental Animal

Species:	Rabbit
Breed/strain:	New Zealand white (NZW)
Sex:	Female timed pregnant
Weight range:	3.5–5.0 kg

Vendors:	Covance Research Products, Denver, PA or Charles River, Wilmington, MA
Requirement for vendors:	Need a constant time for visualized conception in order to ensure accurate timed pregnancies
Term gestation:	31.5 days
Ages studied:	70% gestation (22 days, E22); 79% gestation (25 days, E25) and 92% gestation (29 days, E29)

Pre-Operative Evaluation

1. NZW rabbits need some time to recover from the stress of ground transport. Plan for added recovery time if air transport. It is best to wait 2–5 days after transport before experiments.
2. Conduct thorough physical examination before any procedure and make sure there is not any condition that will compromise the planned experiment. It is advised to use ultrasound to confirm pregnancy.
3. Monitor the rabbit dam's eating, drinking, urination and defecation. Records must be kept and displayed on the cage. The normal values are [6]:

Water intake:	50–100 ml/kg/day
Food intake:	50 g/kg/day
Fecal production:	15–60 g/day
Urine volume:	50–75 ml/kg/day
Rectal temperature:	38.0–40.0 °C

4. When first starting it is advisable to draw blood for biochemical tests. This is also recommended if there is any doubt of the physical well-being of the animal. An ear vein can be used with a 23 G butterfly needle.

Pre-Anesthesia Preparation

1. Fasting rabbits before anesthesia is unnecessary. Rabbits are thought to lack the vomiting reflex. In addition, eating certain amount of high fiber diet (Teklad Global High Fiber Rabbit Diet-2031, Harlan Teklad, Madison, WI) before surgery can aid bowel movements after surgery, helping prevent postoperative constipation and possible ileus.
2. Weigh the rabbits before the procedure to accurately determine the dose of anesthetics and experimental compounds.

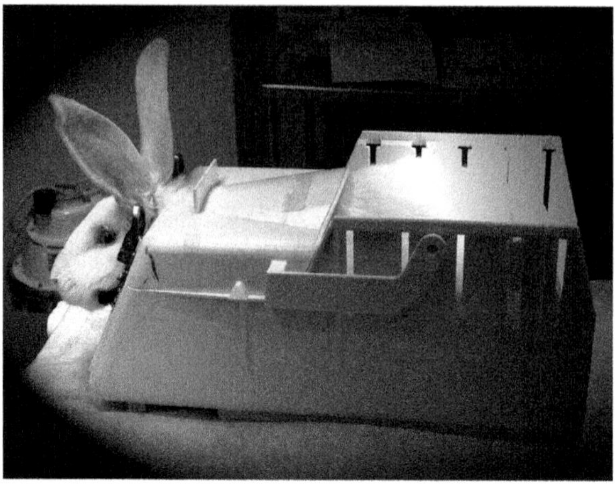

Fig. 1 The pregnant rabbit is restrained in a Nalgene Rabbit Restrainer with pans that is made up of polypropylene and is autoclavable (Lab4Less)

3. Check all records before operation to make sure there are no pre-existing contra-indications to surgery, such as fever, refusal to drink or feed.
4. Pre-anesthetic sedation is not necessary. The investigator must be calm and soothing even when the rabbits are overactive. A calming strategy is to hold the rabbits in the arms for a couple of minutes, and cover the rabbits' eyes.
5. Restrain the rabbit in a Nalgene Rabbit Restrainer with pans (polypropylene, autoclavable, from Lab4Less) (Fig. 1). A struggling rabbit in an oversized restrainer can result in many complications including vertebral fracture, stripped nails and internal bleeding, and even death. Stuffing the rabbits into the restrainer should also be avoided.
6. Shave the hair of the left ear (usually left). It is suggested to use a shop vacuum to suck away the hair. Less powerful vacuum suction devices leave rabbit hair floating in the air.
7. Cannulation of margin vein and middle artery of the ear.

   (a) Preparation of cannulation tray:

   - One 22 GA (0.9 × 25 mm) 1.00 IN sterile shield I.V. Catheter (BD, CE 0086).
   - One Resealable male adapter (Hospira, Lot 35195 5H).
   - 70% Alcohol Prep, Povidone Iodine Topical solution.
   - Three 1.0 × 12.0 cm adhesive tapes.
   - Sterile saline, 5 ml syringe, Heparin Lock Flush.
   - Sterile gloves and a tray.
   - One clean vessel Bulldog Clamp (Vascu-Statt 1001-574) or paperclip.

   (b) Use povidone iodine and 70% alcohol preps to disinfect the distal part of the vein.
   (c) Choose the lateral marginal vein of either ear (usually the left), put the vessel clamp or paperclip at its proximal end to block the vein, wait 1–2 min for the vein to dilate (Fig. 2). Use the left thumb and middle finger to pull the vein

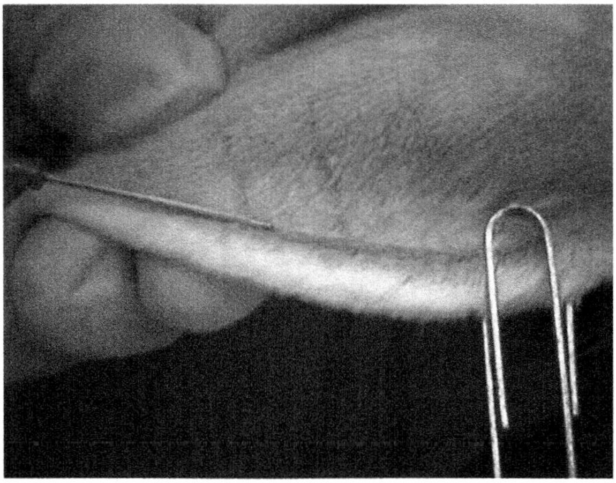

Fig. 2 A paperclip is used to block the lateral marginal vein of the left ear and the vein cannulated with a 22 GA (0.9 × 25 mm) sterile shield intravenous Catheter (left)

straight, use the index finger to push the middle part of the vein up from the underneath to form a 30° angle. Keep the catheter parallel to the proximal part of the vein and punch the vein right at the angle with the tip bevel of the needle facing upward, slide the needle in ~1.0 cm. Watch for blood return and then slide the catheter forward over the introducer and withdraw the introducer, keeping the catheter in the vein.

(d) Connect the heparin-filled adaptor to the catheter tightly.

(e) Use the three strips of tape, one as an L-shape to secure catheter to the ear.

(f) Flush the catheter with 2–3 ml of saline.

(g) The middle ear artery is accessed for an arterial line. To dilate the artery, tap the skin over the artery a few times. Cannulation is similar to the venous cannulation except that there is no need to clamp the vessel. The other difference is that once the catheter is in the artery, the operator must put pressure over the artery at the distal end of the catheter to prevent bleeding during the connection to the adaptor.

8. Vital signs: All vital signs should be monitored and recorded at a minimum of every 10–15 min. The normal values (without anesthesia) are [6]: (1) Respiratory rate: 32–60/min; (2) Heart rate: 200–300/min; (3) Arterial systolic pressure: 90–130 mmHg; (4) Arterial diastolic pressure: 80–90 mmHg; (5) Rectal temperature: 38.0–40.0 °C.

General Anesthesia

1. Equipment

An OMNI-FLOW 4000 (Abbott Laboratories) multiple channel I.V. pump through which fluids and anesthetics can be infused together with separate speeds.

A thermometer (Type K, Fisher Scientific) with accuracy 0.1 °C.
A pulse oximeter for monitoring oxygen saturation of the arterial blood and pulse.
Blood Pressure Monitor (Model 6000, Sensor Devices Inc.).
Bag and mask with oxygen supply. A premature baby mask offers the best fit for the rabbit's snout.
Anesthetics: (1) Fentanyl Citrate, 1.0 mg/20 ml/amp, (2) Droperidol, 5 mg/2 ml/amp.
Solutions: (1) Lactated Ringer's injection; (2) 0.9% Sodium Chloride injection; (3) 5% sodium bicarbonate solution.

2. Procedure of General Anesthesia

   (a) Induction of general anesthesia:

   - A dose of Fentanyl 0.075 mg/kg/h, and Droperidol 3.75 mg/kg/h is given, mixed with normal saline or Ringer's (100 ml/h) IV through the ear vein.
   - Induction lasts 5–10 min, during which the rabbit dam should be closely watched. There is a certain amount of variation between different rabbits in their response to anesthesia. Heart rate, respiration rate, the size of pupils and its reflex to light and touch and the color of the skin should be observed and recorded to monitor the effect of anesthesia.
   - Pupils will constrict and the respiration will become very shallow. This necessitates occasional use of bag and mask ventilation (initially 10 breaths with 60% mask oxygen). A 50 ml bolus injection of normal saline is routinely given in our laboratory as a precautionary measure against hypotension.
   - The dam will allow itself to be turned to its side by gentle handling at the end of induction. Attempts to abruptly move the animal or use too much force wakes the animal up and more anesthetic is eventually needed for induction.

   (b) Maintenance of general anesthesia

   - The rabbit is moved out of restrainer to receive spinal/epidural anesthesia, and the maintenance period of the general anesthesia is started either before or after spinal/epidural anesthesia at the dose of Fentanyl 0.025–0.05 mg/kg/h, Droperidol 1.25–2.5 mg/kg/h depending the rabbit's condition.
   - The spinal anesthesia or epidural anesthesia is dealt in a later section.

3. After spinal or epidural anesthesia, the rabbit is laid supine on an operating board (custom made), and all four extremities are restrained using umbilical cord tape. The dose of anesthesia is dropped so that the levels of fentanyl and droperidol maintain the animal in deep sedation. The dam is allowed to breathe spontaneously. If any sign of pain is noted, the dose of anesthesia should be increased by 25%.

(a) The following signs usually indicate inadequate sedation

- Increased respiratory and heart rate,
- Vocalization or limb movements,
- Hyper-reactivity to the pedal reflex, ear pinch, corneal reflex.

(b) Deep general anesthesia leading to respiratory failure and cardiovascular collapse needs to be avoided. An anesthesia assistant is needed to closely watch the rabbits during the entire procedure and who can make timely adjustments to the dose of anesthetics and prevent possible cardiorespiratory collapse. The following signs suggest too much anesthetic dose:

- Loss of muscle tone, including loss of resistance to opening of the mouth,
- Loss of reflexes which include the corneal reflex, pedal reflex, the ear pinch reflex,
- Dilatation of pupils,
- Slow heart rate, slow respiration rate, arrhythmias on ECG, and hypotension,
- Hypothermia,
- Lower SaO_2; abnormally low PaO_2, high $PaCO_2$; and low pH in arterial blood gas determinations.

4. Arterial blood gas should be measured every 20–30 min. Oximeter readings are also very useful. The normal values for NZW rabbits for blood gases are [6]:

   (a) Arterial blood pH: 7.2–7.5 (our laboratory limits are 7.25–7.45)
   (b) PaO_2: 85–102 mmHg
   (c) $PaCO_2$: 20–46 torr (our limits are 35–49)
   (d) HCO_3^- 12–24 mmol/l (our limits are 20–24)
   (e) SaO_2: 95–99%.

5. Some precautionary measures need to be used during the maintenance period:

   (a) A temperature-controlled water pad is laid underneath the rabbit.
   (b) Occasional breaths with bag and mask and 40–60% FiO_2 may be needed.
   (c) Periodic suctioning or cleaning of the mouth secretions.
   (d) The head and chest should be slightly elevated if blood pressure is normal to prevent excessive pressure on the diaphragm by abdominal organs which could affect the lung function.
   (e) In addition to IV Ringer's infusion of 100 ml/h, an extra dose of 10 ml/kg may need to be given to maintain hydration and blood pressure in case of excessive blood loss. The rabbit will present with hypotension and fast weak pulses. The rabbit can also be placed in the Trendelenburg position in case of hypotension.
   (f) Heat lamp is often necessary to maintain temperature. This is usually necessary during aortic occlusion. Heating the IV line may be useful if hypothermia is a problem.

Spinal Anesthesia

1. Advantages: Spinal anesthesia allows the dam to breathe spontaneously and thus avoid the use of a ventilator. In addition, the amount of medications used for sedation or general anesthesia is considerably reduced. Thus, the confounding effect of drugs is reduced for any experiment.
2. Disadvantages: Occasionally the procedure results in post-operative paralysis from possible damage to nerve roots. In our experience, we have had two dams not moving their hindlimbs out of 50 survival surgeries, with a total incidence of 4%.
3. Contraindications to doing spinal anesthesia:

   (a) Local infections (skin diseases) at the injection site;
   (b) Cerebrospinal fluid (CSF) is mixed with blood which does not clear;
   (c) Spinal deformity;
   (d) Operator is not familiar with the anatomy and procedure.

4. Preparation:

   (a) Shave the rabbit back over an area of 4 × 6 in. with electric clipper;
   (b) The rabbit is held by an assistant in a left lateral decubitus position when operator is right handed and a right lateral decubitus position if left handed;
   (c) The rabbit's back should be positioned at and parallel to the edge of the operating table or bed so that the rabbit is within easy reach of the operator;
   (d) The assistant stands in front of the rabbit, stretches the front and back legs and crosses them in front of the rabbit and holds them firmly with the left hand, covering the dam's eyes with the right hand;
   (e) The spine is slightly bent to allow optimal expansion of the intervertebral spaces;
   (f) The skin is disinfected with povidone-iodine starting at the point of insertion and in ever expanding circles. This is repeated twice. The povidone iodine is allowed to dry for 2–3 min, and then the skin is wiped with 70% alcohol. The alcohol is then allowed to dry. Remember all antiseptic solutions are neurotoxic, and care must be taken not to contaminate spinal needles or local anesthetics with the antiseptic solution;
   (g) The rest of the rabbit skin is covered with sterile drapes leaving a very small area exposed.

5. Preparation of spinal tray (sterile):

   (a) 25 G spinal needle,
   (b) Sterile gloves of different sizes,
   (c) Sterile syringes of different sizes (1, 3, 5, 10, 20 ml),
   (d) Sterile gauze,
   (e) Sterile drapes,
   (f) Local anesthetic: 1% Lidocaine or 0.125% Bupivacaine,

   (g) Spinal anesthetic: 2% Lidocaine or 0.25% Bupivacaine,
   (h) Adhesive tapes.

6. Anatomic landmarks: The injection of local anesthetic into subarachnoid space is usually carried out in the midline between L4–5 or L5–6 intervertebral spaces. A virtual line is drawn from one iliac crest to the other, which crosses the spinous process of L6. This is then used to determine the injection site.
7. Local anesthesia: 0.3–0.5 ml 1% Lidocaine is injected into the skin over the intervertebral space to produce a skin wheal. Another 0.2–0.3 ml is injected intradermally. This needle injection also functions as a "probing" needle to verify the intervertebral space between the two spines.
8. Procedure:

   Using the dominant hand to hold the spinal needle, pierce through the middle of the skin wheal, pointing cranially at a 10–30° angle. The operator can feel the resistance produced by supraspinous and interspinous ligaments when the needle is advanced through them. If the operator releases the needle at this moment, it maintains the proper direction. If it points laterally, it is likely that the needle is not in the interspinous space, and should be withdrawn. The bevel of the needle should face upward to avoid spinal nerve injury. A characteristic "pop" is felt when the subarachnoid space is entered. Needle advancement should stop as soon as the pop is felt. Advancing the needle further increases the risk of paresthesias and neuronal injury. Remove the stylet to allow CSF flow freely. CSF appears within 3–5 s. Sometimes gentle aspiration is necessary to get some CSF. If there is still no CSF, rotate the needle very slowly in all the directions before careful advancing or withdrawing the needle. Remember to put the stylet back before advancing the needle. Often CSF appears after slowly adjusting the needle's position. Attach the syringe containing the spinal anesthetic solution to the hub of the spinal needle while the other hand stabilized the spinal needle. The syringe should be free of air. The position of the needle in the subarachnoid space should be once more verified by gentle aspiration of CSF. Avoid careless movement of the needle during aspiration. Effort should be made to keep the syringe tightly attached to the spinal needle. For a surgical procedure less than 1.5 h, 0.3–0.5 ml of 2% Lidocaine or 0.25% Bupivacaine will produce satisfactory anesthesia; for a procedure between 1.5–2.5 h, we use 0.3–0.5 ml Marcaine Spinal (each ml contains 7.5 mg bupivacaine HCl and 82.5 mg dextrose). Procedures running longer than that limit need either leaving a catheter in subarachnoid space for epidural anesthesia, or requires additional general anesthesia.
9. Spinal anesthesia Test: Normally after injection of local anesthetic into subarachnoid space, the motor blockage appears about 20 s later. In some cases it appears 1–2 min later. The hindlimbs should be completely relaxed. Anesthesia can be evaluated by using skin pinch stimulus with a hemostat moving from lower abdomen to cephalad positions.

Epidural Anesthesia

Epidural anesthesia is an alternative method and is preferred. Its advantages are that it can provide prolonged anesthesia, it is safer for various procedures, and produces less severe complications than spinal anesthesia. The disadvantages are that technically it is more difficult to master, it can cause total spinal anesthesia, or incomplete anesthesia when it fails. The approaches to the lumbar epidural space are similar to those described for spinal anesthesia.

1. Preparation of the spinal tray (sterile):
   - (a) 9.0 cm 18-gauge epidural needle (standard Touhy needle),
   - (b) 20-Gauge epidural catheter with connector,
   - (c) Sterile gloves of different sizes,
   - (d) Sterile syringes of different sizes (1, 3, 5, 10, 20 ml),
   - (e) Alcohol wipes,
   - (f) Sterile gauze,
   - (g) Sterile drape, Adhesive tapes,
   - (h) Local anesthetic for the skin infiltration (1% Lidocaine or 0.125% Bupivacaine) and epidural space injection (2% Lidocaine or 0.25% Bupivacaine), all of which contain epinephrine 1:200,000,
   - (i) A 5 ml syringe filled with anesthetic.
2. Procedure: There are two methods used to identify entrance of the epidural needle into the epidural space: Loss-of-Resistance (LOR) and hanging-drop. The LOR technique is the most popular, and easier to learn.
   - (a) Skin incision: Using a large-diameter needle or a No.11 sharp surgical blade, make a tiny incision to facilitate the skin passage of a blunt epidural needle.
   - (b) Passage through the ligaments: Spread the index and middle fingers of the left hand to fix the injection site in between, a 9 cm epidural needle is held between the thumb, index and middle fingers of the right hand and is advanced through the skin incision. Its bevel tip should face cephalad when bilateral blockage is desired. When the needle passes the supraspinous ligament and intervertebral ligament, the operator feels consistent resistance. The distance from skin to the ligamentum flavum is ~1.5 cm. Stop and check if the needle stays in the midline before reaching epidural space. If the needle is deviated or can be moved freely, it is not in the midline and needs to be replaced.
   - (c) Take out the stylet. Then one has choice of two techniques.
   - (d) Loss-of-Resistance technique: The LOR syringe is filled with either saline or air in a 2–4 ml syringe. Apply certain pressure on the syringe plunger. The saline or air should not move since the needle tip is in the tough ligament tissue. Keep continuous pressure on the plunger of the syringe while the needle is advanced very slowly (1 mm each time). The needle shaft is

controlled near the skin by left thumb and index while it is advanced by the right hand. Stop after every advance of 2 mm and check whether the epidural space has been reached. The operator can feel as the needle tip advances into the ligamentum flavum because it is tougher than interspinous ligament. When the needle tip reaches epidural space, the resistance against plunger suddenly disappears (thus LOR), and saline or air partially disappears into the epidural space. The advancement of the needle should be stopped immediately to prevent the needle from punching through the dura mater to injure the spinal cord. One should immediately stop injecting any more saline or air as it may lead to partial anesthesia. The depth of the epidural space from skin in adult NZW White rabbit with 20° caudal-side angle is about 2.0 cm.

(e) "Hanging drop" technique: After the ligamentum flavum has been reached, take the stylet out and put a drop of saline into the hub of the epidural needle. When the ligamentum flavum is passed, and the epidural space reached, the drop will suddenly disappear.

(f) Aspiration test and injection of testing dose through the needle: Gentle aspiration is carried out in several directions to ensure there is no CSF coming out. The needle is secured by the thumb and index finger of the left hand. After a negative aspiration test, a test dose of anesthetic, 0.5–0.75 ml 0.25% Bupivacaine or 2% lidocaine with epinephrine, can be injected. It takes 5 min for test dose to take effect: loss of muscle tone of lower part of the rabbit's body and back legs and absent skin response to painful stimuli. You can also choose to give test dose through the catheter later.

(g) Insertion of the epidural catheter: After the epidural space is identified, an epidural catheter is threaded through the epidural needle. The catheter is threaded in short increments by grasping the catheter 1 cm from the hub of the needle and advancing it through the needle. Once at the 12 cm mark at the needle hub, the catheter has reached the tip of the epidural needle and is about to enter the epidural space. The catheter is usually advanced into the space 1.5–2.0 cm which means that the catheter should reach the 14–15 cm mark at the needle hub. The epidural needle can then be removed slowly with one hand while the other hand continues to thread the catheter through the needle. The length of the catheter inside the space should be rechecked and may be adjusted. Never attempt to pull back the catheter when the needle has not been withdrawn first. It will cut off the catheter tip.

(h) Securing the epidural catheter: Connect the catheter to an epidural connector, take the cap off, and lower it below to a position below the epidural space. If the catheter tip has been misplaced in the subarachnoid space, or in veins, CSF or blood would appear by gravity. This procedure should be used to discover any misplaced catheter before anesthetic administration. The catheter is secured at the site of puncture by sterile dressing and adhesive tapes.

(i) Epidural test dose: This tests whether the epidural catheter tip is misplaced in the subarachnoid space or a blood vessel. A negative test result does not guarantee the correct placement of the epidural catheter in the epidural

space, but decreases the likelihood that the catheter tip is in a blood vessel or the subarachnoid space. 0.5 ml 2% lidocaine is injected through the catheter. Wait 5 min for response.
(j) Injection of epidural anesthetic solution: Drugs should be double-checked before each injection. 2 ml 2% lidocaine or 0.25% bupivacaine both with 1:200,000 epinephrine is injected slowly and effect lasts 60–90 min.
(k) Removal of the catheter: Removal of the epidural catheter is a simple process.

Femoral Arterial Catheterization

In our rabbit model, an aortic balloon catheter is introduced into the distal aorta. The balloon is inflated and results in uterine ischemia. The uterine ischemia results in fetal hypoxia-ischemia. A single-lumen balloon catheter can be used. If a double-lumen catheter is employed, blood gas sampling, blood pressure monitoring and drug administration can be carried out simultaneously during the same hypoxia period.

1. Equipment:
   (a) A sterile arterial catheterization package (gauze, hemostat, needle holder, micro-needle holder, scalpel, drapes, and 2–0, 7–0 sutures),
   (b) A sterile instrument tray containing: single-use micro blood vessel clips, micro vascular introducer, 1.0, 5, 10 ml syringes one of each, one three-way-stopcock and 4 French arterial embolectomy catheter (usable length 40 cm—single lumen or double).
2. Preparation of the surgical area: The left femoral artery has been traditionally chosen as the site of arterial catheterization by our laboratory. The left groin area is shaved by an electric clipper 2–3 in. around the incision site. Care should be taken not to injure the local skin. It may later become the source of infection which sometimes leads to postoperative laceration of the groin wound. Use Povidone Iodine and 70% alcohol to disinfect the area The disinfected area is covered with sterile drapes, and exposing only the area overlying left femoral artery in the center.
3. 3–5 ml 1% lidocaine or 0.125% bupivacaine is injected subcutaneously along the planned incision site.
4. Exposure of femoral artery: In most instances a 3–5 cm longitudinal incision along the course of left femoral artery provides adequate exposure to femoral artery. Incision is made smoothly by a surgical scalpel (#10). Some subcutaneous bleeding spots will be seen and can be cauterized by an electro-cautery. When the deep fascia is reached, use two fingers to separate subcutaneous fat tissue and a femoral neurovascular bundle will be identified right underneath the deep fascia. The femoral artery is located right below the inner part of the

inguinal ligament with its course running downwards in a longitudinal shallow ditch between two muscle groups. Palpation of the artery helps determine the position of the incision. At the lateral middle point of the neurovascular bundle, make a small (3–5 mm) incision on the fascia over the bundle with a sharp scalpel, and then pick the fascia up with a tissue forceps, extend the fascia incision up and down with tissue scissors along the fiber direction of the bundle to about the length of the incision. This will usually give enough exposure to the neurovascular bundle. Pick up both edges of the incision site and use the handle of the scalpel to separate the fascia from the underneath structures. The femoral neurovascular bundle is sometimes covered with a thin layer of muscles which can be easily removed by careful dissection with fine scissors. The three structures of the bundle are: (1) femoral vein—front-median with larger diameter and dark blue color, (2) femoral artery retro-lateral to the femoral vein with characteristic pulsations, fresh red color and smaller diameter, and (3) femoral nerve in the lateral groove, a thin silver-color bundle. The mnemonic is VAN for vein, artery and nerve. Using fine tweezers pick up the outer membrane of the artery and cut open with micro-scissors, and then a hemostat is inserted to separate the membrane from femoral artery wall along its course. During this process some small branches of the artery may be met and should be tied with 4–0 silk sutures before cutting. Do not cut these branches without tying them because then the bloody field will make visualization difficult. The femoral artery should be exposed at least 2.0 cm to allow enough exposure for catheterization (one can use hemostat only to do this dissection step if one is leery of using scissors at this juncture). A poorly exposed artery is the most common reason for failed catheterization. The isolated artery is held by a micro-clip at its proximal end and looped with two 2–0 silk sutures both proximally and distally. The two loop sutures then will be secured to control unexpected hemorrhage and help facilitate the incision of the artery and introduction of the catheter into the femoral artery.

5. Catheterization: With left hand holding up the distal loop to raise the middle part of the artery, a 1–2 mm incision is made in the longitudinal axis of the artery. The incision is only carefully made in the front wall of the artery. Usually, a little bit blood can be seen where the incision is made to help identify the incision. Releasing the proximal clip shortly and letting the blood come through the incision can also help identify its location. (Watch out! Blood can spatter into your eyes and face! Did you wear goggles?) Hanging up the proximal loop with the left hand to stop the blood flow and releasing the clip near it, an assistant could help open the incision with a tiny L-shaped retractor. The operator can then insert a 4 French Fogarty embolectomy balloon catheter at the distal opening. The catheter is inserted 10 cm from the incision. This distance will place the balloon in the descending aorta just above the uterine arteries and below the renal arteries. The balloon is inflated with 0.2–0.3 ml of saline. The looped sutures are used to secure the catheter to the artery. This is important to prevent accidental removal of the catheter with its hemorrhaging consequences.
6. Closure of vessel and incision: After the end of uterine ischemia the balloon is deflated and the catheter withdrawn. Just before final removal a clip is placed

proximal to the incision in the artery. Then the catheter is completely withdrawn. There are two more precautionary measures to implement at this point. The incision site is flooded with heparin lock flush solution and normal saline. Then, the clip is released for an instant to allow for the arterial blood to flush away a possible blood clot formed during procedure. The bloody fluid in the site is then cleared and the artery visualized again. 1.0 ml 1% lidocaine is dropped around the vessel to vasodilate the femoral artery. The opening in the femoral artery is then closed with 4–6 interrupted over-and-over stitches using 7–0 silk sutures. This suturing is made easier by visualizing with magnifying loupes. Attention should be taken not to narrow the vessel lumen and prevent blood flow. After the suturing the vessel, pressure should be applied on the suturing site with warm-saline-wet gauze, the clips removed to let blood flow through, and to check for obvious leakage through the sutures. Minor leakage often can be stopped by application of pressure for 3–5 min. Interrupted over-and-over suture with 4–0 silk sutures are used to close fascias and soft tissue. The skin incision site is prepped with 70% alcohol before closing. An interrupted vertical mattress suture is used to close the skin, which gives more exact apposition, approximation and more strength to the sutures and reduces the chance of postoperative re-opening.

7. Trouble shooting:

   (a) Difficulty distinguishing between femoral artery and vein. 2% lidocaine can be used to dilate the artery and visualize the pulsations.
   (b) The femoral artery has an outer membrane that needs to be stripped gently. This will help in the insertion of the catheter. Care must be taken not to tear the artery in this process.
   (c) Arterial spasms can prevent insertion of the catheter. One must be careful not to use force because the artery can tear or the catheter can break through the arterial wall. Dripping 1.0 ml 1% lidocaine or warm saline can help dilate the artery and make the insertion easier.
   (d) The catheter egresses the artery through another hole at the back side. Withdraw the catheter. Reintroduce the L-shaped introducer and separate the walls of the artery at a more proximal location. Then insert the catheter again.

Laparotomy

This procedure is provided for the investigator if there is need for direct manipulation of the fetus or direct administration of drugs to the fetus.

1. Equipment:

   (a) A sterile laparotomy package (gauzes, hemostats, needle holder, micro-needle holder, scalpel, drapes, retractors, steel bowls);

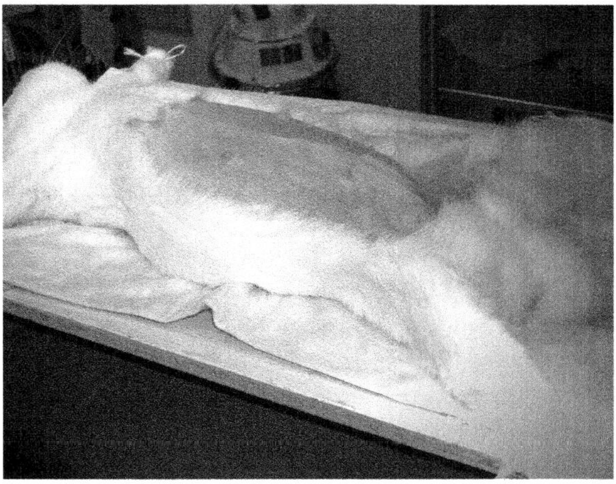

Fig. 3 After spinal or epidural anesthesia, the sedated pregnant rabbit is laid on her back. The left groin area and abdomen is shaved of all fur using clippers and the fur removed by a shop vacuum (background) in preparation for catheterization of the left femoral artery and laparotomy

(b) Different size sutures (2–0, 4–0, 7–0);
(c) Different size syringes (1.0, 5, 10 ml, one of each);
(d) Sterile gloves, tray, 2% lidocaine or 0.25% bupivacaine and normal saline.

2. Preparation of the surgical area: The hair on the abdomen is shaved carefully using an electric clipper from the xiphoid process to the pubic symphysis, including two groin areas, and extending to the flank (Fig. 3). Use 70% alcohol to clean the area three times before povidone iodine soaked gauzes. The operator should start wiping the area from upper middle downwards and from the midline to the periphery. The pubic area should be wiped with separate gauzes should be applied. Allow 2–3 min to dry, and then apply alcohol with the same method to wipe off the povidone iodine. The disinfected area is then covered with sterile drapes, leaving the midline incision area in the center exposed. If local anesthetic infiltration is needed, 1% lidocaine or 0.125% bupivacaine can be given along the incision.
3. Incision and exploration: An incision is made in the lower midline area, starting 2–3 cm above the umbilicus and extending in a cephalo-caudal direction downwards for 10–15 cm. The mammary tissue full with milk can be seen after the skin is incised. Attention should be paid not to damage the mammary tissue. Some small veins cross the midline. They should be cauterized or tied and cut with tissue scissors. Then the midline of the abdominal fascia is exposed. Together with an assistant, two toothed forceps are used to pick up the abdominal muscle and fascia layer making a small tent and ensuring no abdominal contents are caught in the tent. A small longitudinal incision is made in the middle with a scalpel, fine tissue scissors is then used to extend the incision and cut the peritoneum. As the incision progresses downwards a fat-covered bladder will appear underneath, which can be mistaken for fat tissue and accidentally incised by an inexperienced operator. The bladder can be pushed gently downward with gauze over a right index finger. The peritoneum can then be picked up to avoid the possible intestinal damage and incised with tissue scissors. A few bleeding

points are often met in the midline and must be clamped, tied or controlled by electro-coagulation. While protecting both edges of the incision with saline soaked gauzes, hold the edges with retractors to expose the abdominal cavity. Exploration should be very gentle. The intestines are pushed aside carefully and packed away with warm, moist gauze pads, or placed in a plastic bag and moistened with warm saline to minimize heat and evaporative water loss. The uterus filled with fetuses is usually inspected inside abdominal cavity, and can only be taken outside for a short period. It is easy to cause intrauterine hemorrhage and placental separation.

4. Injection of experimental drugs: After the number and position of the fetuses are determined, an experimental drug solution can be injected with a 1 ml TB syringe into the fetal peritoneal cavity, subarachnoid space or nasal cavity through the wall of the uterus. Sometimes an ultrasound is used to guide the injection. Be sure to avoid the placenta and inject through other side of the placenta.

Hysterotomy

1. Following laparotomy the uterus is isolated. The incision is made in the avascular part of the uterus in a longitudinal plane. Fetuses can be safely brought out of uterus, the umbilical cord ligated and uterus can be sutured up with 3–0 sutures.
2. Closure of abdominal incision: After finishing all the steps of operation, use warm saline to wash the area. The peritoneum is closed by a running over-and-over suture with 2–0 silk in a cephalo-caudal direction while the intestines and uterus are protected by a saline-wet gauze pad. Be sure not to suture the internal organs with peritoneum, which could lead to postoperative ileus. Check the leakage and use interrupted over-and-over suture with 2–0 silk to close fascia. Prep the skin with alcohol and close subcutaneous soft tissue with 4–0 silk interrupted over-and-over suture. Be sure to check and have all bleeding points tied or cauterized to prevent subcutaneous hematoma that could be a reason for postoperative wound infection or laceration. An interrupted vertical mattress suture is used to close the skin.

Neurobehavioral Abnormalities at Postnatal Day 1 (P1) and 11 (P11)

A range of neurobehavioral deficits is observed in the P1 newborn kits following fetal hypoxia-ischemia at E22 [4]. There is increasing stillbirths with increasing hypoxia-ischemia. There are increasing neurobehavioral deficits following greater than 30 min of sustained fetal hypoxia-ischemia at E22. What is striking is the observation of hypertonia and resulting postural abnormalities in ~75% of the survivors. Some of the stillbirths also have postural abnormalities. We are able to keep alive some of the severely affected kits, born following hypoxia-ischemia at E22, by careful nursing and orogastric feeding of rabbit milk. In all cases with hypertonia and postural deficits, the hypertonia persists and the motor deficits at P11 are a stark contrast to the motor capabilities of normal controls [5].

Acknowledgements Xinhai Ji for his input.
Supported by grants from NIH, HD01138, NS41476, NS43285, NS051402, NS081936, NS100088 (S.T.).

References

1. Harel S, Shapira Y, Hartzler J, Teng EL, Quiligan E, Van Der Meulen JP. Neuromotor development in relation to birth weight in rabbits. Biol Neonate. 1978;33(1–2):1–7.
2. Matsuda Y, Maeda T, Kouno S. Comparison of neonatal outcome including cerebral palsy between abruptio placentae and placenta previa. Eur J Obstet Gynecol Reprod Biol. 2003;106(2):125–9.
3. Iwasa H, Aono T, Fukuzawa K. Protective effect of vitamin E on fetal distress induced by ischemia of the uteroplacental system in pregnant rats. Free Radic Biol Med. 1990;8(4):393–400.
4. Derrick M, Luo NL, Bregman JC, et al. Preterm fetal hypoxia-ischemia causes hypertonia and motor deficits in the neonatal rabbit: a model for human cerebral palsy? J Neurosci. 2004;24:24–34.
5. Tan S, Drobyshevsky A, Jilling T, et al. Model of cerebral palsy in the perinatal rabbit. J Child Neurol. 2005;20(12):972–9.
6. Suckow MA, Douglas FA. The laboratory rabbit. Boca Raton, FL: CRC Press; 1997.

A Model of Neonatal Focal Cerebral Ischemia-Reperfusion

Zinaida S. Vexler and Donna M. Ferriero

Abstract It has become clear that susceptibility to hypoxic-ischemic, excitotoxic and oxidative insults is increased at specific stages of postnatal brain maturation, affecting both the acute and lasting injury patterns seen after the insult. For decades, a model of hypoxia-ischemia in postnatal days 7–9 (P7–P9) was the only rodent model relevant to ischemic brain injury in the neonate. This model is more likely to mimic global hypoxic-ischemic encephalopathy rather that focal transient ischemia of a single artery, such as the middle cerebral artery (MCA). To study the mechanisms of arterial pediatric stroke, a transient MCA occlusion model was first developed in juvenile rats. Since focal occlusion of the MCA is more commonly seen in term human babies than children and brain maturation affects the response to cerebral ischemia, we developed a transient MCA occlusion in P7 rats to satisfy the need for age-appropriate stroke models. We produced a series of models of different stroke severity by varying the duration of MCA occlusion and the age of the animals. In this chapter we will describe in detail surgical procedures to induce MCA occlusion, ways to ascertain that the procedure is successful, and discuss factors and limitations that can affect short and longer-term injury outcomes.

Keywords Arterial stroke · Middle cerebral artery occlusion · Neonate · Neonatal · Age · Reperfusion

Z. S. Vexler (✉)
Department of Neurology, University of California San Francisco, San Francisco, CA, USA
e-mail: Zena.Vexler@ucsf.edu

D. M. Ferriero
Department of Neurology, University of California San Francisco, San Francisco, CA, USA

Department of Pediatrics, University of California San Francisco, San Francisco, CA, USA

J. Chen et al. (eds.), *Animal Models of Acute Neurological Injury*, Springer Series in Translational Stroke Research, https://doi.org/10.1007/978-3-030-16082-1_14

Introduction

The incidence of arterial stroke in newborns—at least 1 in 4000 live term babies [1]—is similar to that in the elderly. The traditional view that neonates have greater resistance to CNS injury than adults has not been confirmed. Rather, it has become clear that at specific early stages of brain maturation susceptibility to hypoxic-ischemic, excitotoxic and oxidative insults may be increased [2–4]. The exact reasons for differences in response are yet to be understood but few known major contributing factors include differences in energy metabolism [5, 6], the status of excitatory and inhibitory receptors [2, 7–9], oxidative defenses [4, 10, 11], inflammatory responses [12] and a substantially more prominent role of caspase-3-mediated apoptosis in the immature post-hypoxic-ischemic brain than in that of the adult [13–15]. These data have been obtained in a model of hypoxia-ischemia (HI), a model that consists of a combined ligation of the common carotid artery of one hemisphere and systemic hypoxia for various periods of time [16]. Due to the presence of a systemic hypoxic exposure, HI model, which was essentially the only available model related to cerebral ischemia in neonatal rodents for many years, mimics hypoxic ischemic encephalopathy (HIE) more closely than it does focal stroke.

To better understand mechanisms of pediatric arterial stroke, age-appropriate animal stroke models have been developed by modifying the monofilament MCA occlusion model established by Longa et al. in 1989 [17] for studying ischemic brain injury in adult rats. Ashwal et al. [18] first developed a model of transient middle cerebral artery (MCA) occlusion in juvenile, postnatal day 14–18 (P14–P18) spontaneously hypertensive rats. Considering that P14 is the age that corresponds to an age of a toddler and, based on data in adult rats, reperfusion is poorer and infarct is more extensive in spontaneously hypertensive as compared to other rat strains, we developed a monofilament MCA occlusion model in normotensive P7 rat pups [19]. The MCA occlusion model developed by our group produces severe disruption of cerebral blood flow in the region of the occluded MCA and confirmed partial reperfusion upon suture removal, as is evident from contrast-enhanced MRI studies [20]. Developing a model of transient suture occlusion is important because reperfusion is a common situation seen in human neonates with arterial stroke [1, 21]. Another age-appropriate rat stroke model was developed by Renolleau et al. [22] with the use of a combined permanent MCA occlusion and transient occlusion of the common carotid artery in P7 rats. The transient monofilament MCA occlusion model in P7 rat proved useful for short-term stroke studies [19, 20, 23]. This surgical procedure was, however, associated with limited ability of neonates to thrive over longer post-reperfusion periods [24]. To overcome problems in gaining weight by injured pups over time and therefore make the model suitable for long-term studies, we have recently modified the surgical procedure to produce a less invasive surgical procedure [25], a procedure that does not require the use of vessel electocoagulation and use of aneurysm clips.

In this chapter we will describe in detail the modified surgical procedure to induce MCA occlusion, ways to ascertain that the procedure is successful, and discuss factors and limitations that can affect short and longer-term injury outcomes. We will also discuss the effects of different duration of MCA occlusion in P7 rats, including permanent occlusion, on injury and the development of transient MCA occlusion in P10 rats.

Detailed Methods

Animals

All animal research should be approved by institutional Committee of Animal Research. The described experimental procedures were performed in accordance with NIH guidelines for humane handling of animals with prior approval from the Committee of Animal Research at University of California San Francisco. Female Sprague Dawley rats with a 6 day-old litter (10–11 pups per litter) were obtained from Simonson Labs (Gillroy, CA). The mother and her litter were given food and water ad libitum and housed in a temperature/light controlled animal care facility until the pups are 7 days old.

Suture Preparation

Proper suture coating is critical to achieve complete blockage of cerebral blood flow to the MCA and to prevent intracerebral hemorrhage. The coating is accomplished as follows:A 6–0 dermalon monofilament suture (United States Surgical, Norwalk, CT) is cut into 12 mm segments. A 2 mm end of each suture is lightly scratched on all sides with fine, 400 silicone carbide paper, and the tip rounded by heat [19]. To prepare a silicone based polysiloxane impression material (CutterSil, Miles Inc. Dental Products, South Bend, IN), 0.6 ml of the base is mixed for 20–30 s with one drop of the liquid activator on a clean non-porous surface. Under the microscope, the 2 mm scratched end of the suture is dipped in the silicone, lightly wiped against paper to remove excess of silicone and immediately taped upside down. When executed properly, this results in a finely coated, smooth, tapered suture, resembling a baseball bat. The adhesion and hardening process is quick, and the sutures may be used within 10 min of preparation. Each suture is examined under a microscope prior to use and the diameter measured with a micrometer. The ideal dimensions are a distal width of 0.18 ± 0.02 mm. Sutures that do not have a smooth continuous taper or those with a distal tip width of >0.2 mm or <0.14 mm are rejected. The silicone is stripped and the suture is recoated later.

Animal Surgery

1. The P7 pups are anesthetized utilizing 3% isoflurane in a mixture of 70% N_2O/30% O_2, and then the isoflurane is reduced to 1–1.5%, as needed. Throughout the 15–20 min procedure, the animal temperature is maintained at 38 °C by a heating pad and an overhead lamp.
2. The cervical incision is made directly over the right common carotid (CCA), external carotid (EC) and internal carotid (IC) bifurcation. Beginning at its origin, a 1 mm segment of IC is carefully dissected and tied off with a single strand of silk at its origin [25]. A second silk suture is looped around the IC, just above the pterygopalatine artery (PPA), and gently pulled laterally to prevent retrograde blood-flow. A small incision is made in the proximal isolated IC vessel segment.
3. The coated 6–0 occluding suture is inserted and advanced 7.5–9.0 mm, depending on the pup weight. In smaller pups, 12–14 g, the suture is advanced 7.5–8.5 mm; pups over 15 g have the suture filament advanced 8.5–9 mm (see Fig. 1a).
4. The lower knot is gently tightened, securing the suture, and the skin incision is closed with 3 interrupted 7–0 silk sutures.
5. The pups are transferred in a container placed on a heating pad under the lamp until animals are recovered from anesthesia. Hyperthermia must be avoided.
6. The pups are returned to dams after they are fully recovered from anesthesia.
7. For reperfusion, following anesthesia (described in step 1 of Section "Animal Surgery") the middle skin suture is removed, both knots are removed, and the suture filament together with the coating is gently pulled out and a small piece of surgical with gentle pressure is applied to the arteriotomy. Typically, there is no blood loss during either occlusion or reperfusion.

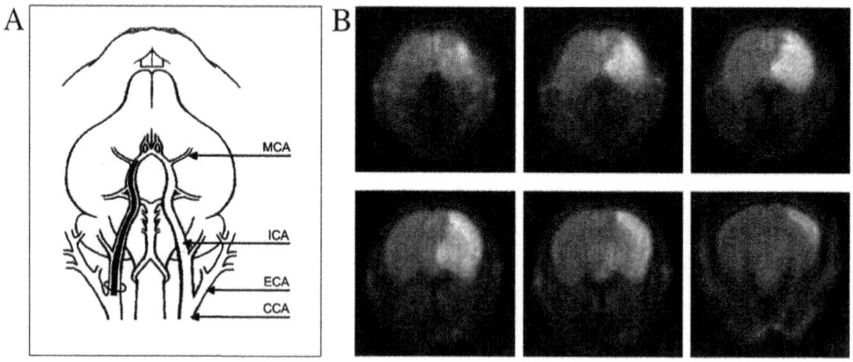

Fig. 1 (**a**) A schematic diagram of the suture insertion and advancement to occlude the MCA. *CCA* common carotid artery, *ECA* external carotid artery, *ICA* internal carotid artery. (**b**) An example of anterior-to-posterior injury distribution pattern on DW-MRI 2 h post-MCA occlusion. DW hyperintensity is observed in brain regions with slowed water diffusion

8. The pups are transferred in a container placed on a heating pad under the lamp until animals are recovered from anesthesia, as described in step 5 of Section "Animal Surgery".
9. Pups are returned to dams after they are fully recovered from anesthesia.
10. Animal weights are recorded daily. If no weight gain is seen within 2 days after surgical procedure upon the development of surgical techniques, animals are to be gavage-fed with 2% condensed milk until signs of weight gain are evident [24].

Non-invasive Delineation of Injury Patterns

Non-invasive injury delineation by MRI is not required but is helpful during the development of the procedure to confirm that the desired injury pattern has been achieved in individual animals. Upon development of the surgical procedure, MRI can also be used as a criterion for excluding uninjured animals in drug-studies.

1. Diffusion-weighted magnetic resonance imaging (DW-MRI) is a non-invasive method which is sensitive to reduced water diffusivity in the tissue following occlusion-induced disruption of CBF. DW-MRI provides information on the presence of injury in minutes; animals need to be anesthetized for imaging. Therefore, the technique can be used for delineation of the spatial pattern of early injury during the occlusion [20, 25]. An example of the anterior-to-posterior injury distribution on DW-MRI 2 h after MCA occlusion is shown in Fig. 1b.
2. Contrast-enhanced MRI can ascertain the completeness of CBF reduction in the MCA vascular territory during the occlusion and restoration of CBF upon retraction of the suture [20]. This method requires intra-jugular injection of the magnetic susceptibility contrast.

Caveats, Limitations and Concerns

1. Inadequate occlusion of the MCA. As described in Section "Suture Preparation", a combination of factors can affect the successful occlusion of the MCA. The weight of animals can vary substantially even within the same litter thus affecting the length of the suture advancement and the thickness of the suture needed for proper placement of the tip of the suture in proximity to MCA. The use of a shorter than necessary suture will not result in blockage of the MCA. In contrary, the use of a longer than needed suture may produce partial rather than nearly complete disruption of cerebral blood flow through the MCA as the thickest portion of the suture, its coated tip, will be far from the MCA. Unusually high branching of the MCA, which is occasionally seen, can provide sufficient residual blood flow, thus precluding from complete occlusion.

2. Insufficient suture coating. An uncoated suture may perforate the vessel. Thick or uneven coating will make the suture more rigid, making it more difficult to advance or retract. Excessive coating may affect the ability to reperfuse due to vasospasm and affected responsiveness of the vessel. Reuse of the coated suture may affect consistency of results, as the suture will become more rigid and/or the coating may slide off.
3. Excessive anesthesia. Multiple lengthy anesthesia episodes may affect the ability of animals to thrive and therefore the duration of anesthesia should be kept to a necessary minimum, with frequent checking of pinch and other reflexes. Due to the small size of P7 rats, physiological monitoring is limited as MABP monitoring or blood gas analysis are currently not available. The type of anesthetic may affect time necessary for pups to recover form anesthesia. Isoflurane is preferable as animals recover from this agent rapidly. Animals should be put back to their dam for nursing as soon as they are fully awake. If surgery and recovery from anesthesia takes longer then what we described in Section "Detailed Methods", animals may become dehydrated and saline is to be injected (i.p., 0.1–0.2 ml).
4. Lack/delay in weight gain.

   (a) Like with stroke models in adults and HI model in neonates, essentially all animals lose weight over the first 24 h period after the insult. However, weight loss can persist over time, as we described in our previous publications [24, 25], especially following the originally described procedure [19]. The exact reasons for lack of weight gain after MCAO are not known but the coagulation and/or ligation of several arteries, including the external carotid, occipital, superior thyroid, lingual and maxillary arteries may reduce or completely cut off blood flow to the ipsilateral neck muscles, tongue, salivary glands. This will and the face is the likely possibilities. Compared to original procedure, the modified procedure described in this chapter is less invasive due to elimination of bipolar coagulation, electrocautery and use of aneurysm clips from the surgical procedure.
   (b) Reduced nursing can occur in case of frequent moving cages from facility to facility. Normal feeding can be reduced due to frequent removal of pups from the cage.
   (c) Gavage feeding with milk may be necessary if weight gain of injured pups is diminished [24, 25].

5. The incidence of injury. Intra-litter variations in the incidence may occur.
6. The route of suture insertion. Insertion of suture through the ECA is the common way of occlusion of the MCA, which produces consistent injury acutely. The limitation is that this surgical procedure interferes substantially with animal grow and long-term survival. The surgical procedure described in this chapter that involves suture insertion through the ICA results in an overall similar injury pattern as the later surgical procedure [25].
7. Variable extent of recovery over time. With the use of combined multiple MRI sessions during MCA occlusion, 24 h and 7 days after reperfusion following a 3 h.

MCA occlusion we showed that individual pups with nearly identical spatial injury pattern within 24 h may recover to a different extent by 7 days post-MCA occlusion [25], from no to a nearly complete recovery based on histology.
8. The effect of gender. Gender may affect injury evolution and outcome, as has been shown for HI [26, 27].

Advantages and Unknowns

1. Varying the time of MCA occlusion in animals of the same age results in lesions of different size. The use of a 3 h MCA occlusion [24, 25] produced a larger lesion than a 2 h [28]. MCA occlusion in P7 rats. Lack of reperfusion is associated with even larger injury [23].
2. Permanent occlusion of the MCA favors a necrotic type of cell death whereas occlusion-reperfusion induces both apoptosis and necrosis [23]. Following transient MCA occlusion, apoptotic neuronal death is substantial both in the ischemic core and in penumbra [23]; no data on the presence of continuum cell death which has been reported after HI in P7 rats [29] are available.
3. Age of neonatal rats and injury outcome. MCA occlusion in neonatal rats of different ages may require different duration of occlusion to achieve similar injury severity histologically but no systematic comparative studies have been performed. In addition to focal stroke model in P7 rats we have developed a similar model in P10 rats [30]. The latter model produces consistent injury pattern sub-chronically [31] and animals survive for months [32].

Summary and Future Directions

Several focal stroke models in neonatal rats have become available over the last decade, making it possible to study mechanisms of neonatal stroke experimentally. These models are based on occlusion of MCA in P7 [19, 20, 22–25, 28, 33] and P10 [30–32, 34, 35] rats. Reperfusion following monofilament occlusion is induced in some but not all models. The completeness of reperfusion may vary depending on the way MCA occlusion is induced (transient placement of a monofilament to block the MCA compared to transient CCA ligation combined with permanent MCA ligation) and duration of the occlusion (transient compared to permanent). Another important aspect of availability of models in rats of different ages is the ability to determine maturational differences in the response to stroke within the neonatal period (stroke in P7 [19, 22, 25] compared to P10 rats [30, 35]) as well as comparison to stroke effects in juvenile rats (P14–P17 [18]). While much is to be learned about the role of gender and strain on short- and long-term outcomes, availability of several models of focal cerebral ischemia provides ways to advance the field of neonatal stroke.

Other common conditions in human newborns are venous thrombosis and intracerebral hemorrhage. So far, there have been no age-appropriate models developed for those diseases.

Acknowledgements This study was funded by NIH grants NIH NS44025 to Z.S.V. and NS35902 to D.M.F. and Z.S.V. and AHA GIA grant 0655236Y to Z.S.V. The authors thank Nikita Derugin for critical reading of the manuscript and Joel Faustino for technical assistance.

References

1. deVeber G, Roach ES, Riela AR, Wiznitzer M. Stroke in children: recognition, treatment, and future directions. Semin Pediatr Neurol. 2000;7:309–17.
2. Ikonomidou C, Bosch F, Miksa M, et al. Blockade of NMDA receptors and apoptotic neurodegeneration in the developing brain. Science. 1999;283:70–4.
3. Olney JW, Wozniak DF, Jevtovic-Todorovic V, Farber NB, Bittigau P, Ikonomidou C. Drug-induced apoptotic neurodegeneration in the developing brain. Brain Pathol. 2002;12:488–98.
4. Sheldon RA, Jiang X, Francisco C, et al. Manipulation of antioxidant pathways in neonatal murine brain. Pediatr Res. 2004;56:656–62.
5. Vannucci RC, Vannucci SJ. Glucose metabolism in the developing brain. Semin Perinatol. 2000;24:107–15.
6. Vannucci SJ, Simpson IA. Developmental switch in brain nutrient transporter expression in the rat. Am J Physiol Endocrinol Metab. 2003;285:E1127–34.
7. Ikonomidou C, Mosinger JL, Salles KS, Labruyere J, Olney JW. Sensitivity of the developing rat brain to hypobaric/ischemic damage parallels sensitivity to N-methyl-aspartate neurotoxicity. J Neurosci. 1989;9:2809–18.
8. McDonald JW, Johnston MV. Physiological and pathophysiological roles of excitatory amino acids during central nervous system development. Brain Res Brain Res Rev. 1990;15:41–70.
9. Bickler PE, Fahlman CS, Ferriero DM. Hypoxia increases calcium flux through cortical neuron glutamate receptors via protein kinase C. J Neurochem. 2004;88:878–84.
10. Aspberg A, Tottmar O. Development of antioxidant enzymes in rat brain and in reaggregation culture of fetal brain cells. Brain Res Dev Brain Res. 1992;66:55–8.
11. Fullerton HJ, Ditelberg JS, Chen SF, et al. Copper/zinc superoxide dismutase transgenic brain accumulates hydrogen peroxide after perinatal hypoxia ischemia. Ann Neurol. 1998;44:357–64.
12. Hedtjarn M, Mallard C, Hagberg H. Inflammatory gene profiling in the developing mouse brain after hypoxia-ischemia. J Cereb Blood Flow Metab. 2004;24:1333–51.
13. Cheng Y, Deshmukh M, D'Costa A, et al. Caspase inhibitor affords neuroprotection with delayed administration in a rat model of neonatal hypoxic-ischemic brain injury [see comments]. J Clin Invest. 1998;101:1992–9.
14. Hu BR, Liu CL, Ouyang Y, Blomgren K, Siesjo BK. Involvement of caspase-3 in cell death after hypoxia-ischemia declines during brain maturation. J Cereb Blood Flow Metab. 2000;20:1294–300.
15. Zhu C, Wang X, Xu F, et al. The influence of age on apoptotic and other mechanisms of cell death after cerebral hypoxia-ischemia. Cell Death Differ. 2005;12:162–76.
16. Rice JE, Vannucci RC, Brierley JB. The influence of immaturity on hypoxic-ischemic brain damage in the rat. Ann Neurol. 1981;9:131–41.
17. Longa EZ, Weinstein PR, Carlson S, Cummins R. Reversible middle cerebral artery occlusion without craniectomy in rats. Stroke. 1989;20:84–91.
18. Ashwal S, Cole DJ, Osborne S, Osborne TN, Pearce WJ. A new model of neonatal stroke: reversible middle cerebral artery occlusion in the rat pup. Pediatr Neurol. 1995;12:191–6.

19. Derugin N, Ferriero DM, Vexler ZS. Neonatal reversible focal cerebral ischemia: a new model. Neurosci Res. 1998;32:349–53.
20. Derugin N, Wendland M, Muramatsu K, et al. Evolution of brain injury after transient middle cerebral artery occlusion in neonatal rat. Stroke. 2000;31:1752–61.
21. Ferriero DM. Neonatal brain injury. N Engl J Med. 2004;351:1985–95.
22. Renolleau S, Aggoun-Zouaoui D, Ben-Ari Y, Charriaut-Marlangue C. A model of transient unilateral focal ischemia with reperfusion in the P7 neonatal rat: morphological changes indicative of apoptosis. Stroke. 1998;29:1454–60; discussion 1461.
23. Manabat C, Han BH, Wendland M, et al. Reperfusion differentially induces caspase-3 activation in ischemic core and penumbra after stroke in immature brain. Stroke. 2003;34:207–13.
24. Fox C, Dingman A, Derugin N, et al. Minocycline confers early but transient protection in the immature brain following focal cerebral ischemia-reperfusion. J Cereb Blood Flow Metab. 2005;25:1138–49.
25. Derugin N, Dingman A, Wendland M, Fox C, Vexler ZS. Magnetic resonance imaging as a surrogate measure for histological sub-chronic endpoint in a neonatal rat stroke model. Brain Res. 2005;1066:49–56.
26. Hagberg H, Wilson MA, Matsushita H, et al. PARP-1 gene disruption in mice preferentially protects males from perinatal brain injury. J Neurochem. 2004;90:1068–75.
27. van den Tweel ER, van Bel F, Kavelaars A, et al. Long-term neuroprotection with 2-iminobiotin, an inhibitor of neuronal and inducible nitric oxide synthase, after cerebral hypoxia-ischemia in neonatal rats. J Cereb Blood Flow Metab. 2005;25:67–74.
28. Dingman A, Lee SY, Derugin N, Wendland MF, Vexler ZS. Aminoguanidine inhibits caspase-3 and calpain activation without affecting microglial activation following neonatal transient ischemia. J Neurochem. 2006;96:1467–79.
29. Northington FJ, Graham EM, Martin LJ. Apoptosis in perinatal hypoxic-ischemic brain injury: how important is it and should it be inhibited? Brain Res Brain Res Rev. 2005;50:244–57.
30. Mu D, Jiang X, Sheldon RA, et al. Regulation of hypoxia-inducible factor 1alpha and induction of vascular endothelial growth factor in a rat neonatal stroke model. Neurobiol Dis. 2003;14:524–34.
31. Mu D, Chang YS, Vexler ZS, Ferriero DM. Hypoxia-inducible factor 1alpha and erythropoietin upregulation with deferoxamine salvage after neonatal stroke. Exp Neurol. 2005;195(2):407–15.
32. Chang YS, Mu D, Wendland M, et al. Erythropoietin improves functional and histological outcome in neonatal stroke. Pediatr Res. 2005;58:106–11.
33. Denker S, Ji S, Lee SY, et al. Macrophages are comprised of resident brain microglia not infiltrating peripheral monocytes acutely after neonatal stroke. J Neurochem. 2007;100:893–904.
34. Ashwal S, Tone B, Tian HR, Chong S, Obenaus A. Comparison of two neonatal ischemic injury models using magnetic resonance imaging. Pediatr Res. 2007;61:9–14.
35. Ashwal S, Tone B, Tian HR, Chong S, Obenaus A. Serial magnetic resonance imaging in a rat pup filament stroke model. Exp Neurol. 2006;202:294–301.

Part III
Cerebral Hemorrhage Animal Models

Monofilament Perforation Subarachnoid Hemorrhage Rat Model

Wing Mann Ho, Cesar Reis, Onat Akyol, Julian Cahill, Devin McBride, and John H. Zhang

Abstract Since subarachnoid hemorrhage is a devastating condition with high mortality and poor outcome, an appropriate animal model for research is needed. Although the monofilament perforation model in rats is a challenging procedure to learn, the induction of subarachnoid hemorrhage in this way is by far the most suited model for investigating the pathophysiological mechanisms of subarachnoid hemorrhage. Additionally, this surgical technique shares similarities with stroke models. The following chapter offers a step-by-step manual of how to induce subarachnoid hemorrhage in rats with ease.

Keywords Subarachnoid hemorrhage · Rat model · Endovascular perforation · Monofilament puncture · Vasospasm

W. M. Ho
Department of Physiology and Pharmacology, Loma Linda University School of Medicine, Loma Linda, CA, USA

Department of Neurosurgery, Medical University Innsbruck, Innsbruck, Austria

C. Reis · O. Akyol · D. McBride
Department of Physiology and Pharmacology, Loma Linda University School of Medicine, Loma Linda, CA, USA

J. Cahill
Department of Neurosurgery, Southampton General Hospital, Southampton, Hampshire, England

J. H. Zhang (✉)
Department of Physiology and Pharmacology, Loma Linda University School of Medicine, Loma Linda, CA, USA

Departments of Anesthesiology and Neurosurgery, Loma Linda University School of Medicine, Loma Linda, CA, USA
e-mail: jhzhang@llu.edu

J. Chen et al. (eds.), *Animal Models of Acute Neurological Injury*, Springer Series in Translational Stroke Research, https://doi.org/10.1007/978-3-030-16082-1_15

SAH Animal Models

There are various techniques to induce subarachnoid hemorrhage (SAH) in the rodents, including the most popular methods of the monofilament perforation model and the single or double hemorrhage injection model. While blood injection is more often used in large animals such as the canines, the perforation model is difficult to perform in larger animals. Additionally, the puncture model is associated with a much higher mortality rate compared with the injection model, which can be very expensive, especially when you are using canine or primate subjects. In the injection model, autologous blood is injected into the cisterna magna or into the prechiasmatic cistern. The superior method inducing SAH in the rodent model is currently unclear. However, the injection model does not damage to the vessel, and in addition it is well recognized that the injury and subsequent degree of vasospasm is not as severe in the injection model compared with the monofilament model [1]. Since the etiology of vasospasm is currently unknown and may lie within the vessel wall, the cause would be missed in the injection model. Therefore, the perforation induced SAH matches the human conditions more appropriately in all parameters including morbidity, mortality, and severity.

Materials

1. Animals: Adult Sprague-Dawley (SD) rats, 280–300 g body weight
2. Surgical equipment
   (a) Hair clippers
   (b) Skin disinfectant (idophors or chlorhexidine)
   (c) Surgical drapes, gown, cap, face mask, and gloves
   (d) Surgical microscope and/or fiber optic illuminator recommended
   (e) Electrosurgical system for cauterization
   (f) Surgical instruments: scalpel/scissors for skin incision, bipolar, fine forceps, microscissors, needle holder, retractors, and 2–3 micro vessel clips recommended (e.g. type: Acland, Biemer, or Mehdorn). Instruments should be sterilized before each procedure
   (g) Sterile cotton swabs
   (h) Sterile sutures:
      - for puncture: adjusted 4/0 nylon monofilament (straightened, cut at 45° angle 3 cm long)
      - for vessel ligature: 10/0 silk
      - for skin closure: 3/0 silk

   (i) Well lit and aired solid working place
3. Anesthesia and perioperative care

(a) Ophthalmic ointment
(b) Isoflurane, oxygen, medical air
(c) Intubation equipment
(d) Animal ventilator
(e) Heating pad with feedback regulator, rectal temperature sonde
(f) Pain medication (e.g. buprenorphine 0.03 mg/kg s.c. or carprofen 4 mg/kg s.c.)
(g) Optional: ICP-monitoring, flowmetry, blood-pressure-catheter

Procedure

Preparation and Anesthesia

1. Anesthesia should be induced in a small chamber with 4% isoflurane in 60% medical air and 40% oxygen mixture for 2–3 min, until the animal loses consciousness and toe pinch reflex is negative. Next, the animal is intubated orotracheally and ventilated with 1.5–2% isoflurane. All following tasks should be done carefully to maintain the correct endotracheal tube position during the rest of the procedure.
2. The animal is gently placed supine on a heating pad with the neck slightly extended. For feedback regulation the temperature sonde is inserted rectally.
3. All surgical fields should be shaved with a hair clipper, and skin-disinfectant should be applied on the exposed skin.
4. Invasive monitoring devices are placed at this step.

Monofilament Procedure

1. An approximately 2–3 cm vertical skin incision is made in the midline extending from the jugular notch of the sternum to the hyoid bone.
2. The thyroid gland lobes are separated exposing the trachea underneath covered by the sternohyoid muscle.
3. Subsequently, the left gland including the connective tissue is moved laterally. The cervical fascia is incised showing the anatomical triangle confined by the sternohyoid muscle medially, the omohyoid muscle superiorly, and the sternocleidomastoid muscle inferiorly.
4. Further dissection is performed bluntly shifting the omohyoid and sternomastoid muscle laterally to uncover the carotid sheath, which contains the common carotid artery (CCA), the vagus nerve, and the internal jugular vein. We recommend to start dissecting proximally until the CCA is identified, since less arteries branch off of the proximal CCA.

5. Once this has been achieved, the proximal carotid vessels are exposed following the exposed CCA and carefully freed of the carotid sheath. Prioritize prevention of accidental tearing and bleeding of the superior thyroid artery or the occipital artery. Other minor bleeding usually stops by applying gentle pressure with a cotton swab. Also, coagulation with a bipolar forceps can be used.
6. The superior thyroid artery branches off medially from the external carotid artery (ECA). First, it is coagulated, and then cut.
7. The occipital artery, which exits the external artery laterally near the carotid bifurcation, is also coagulated and cut. Care should be taken, since the occipital artery lies over the internal carotid artery (ICA), and they may be tightly attached together.
8. The CCA, the ECA and the ICA are detached from the surrounding connective tissue.
9. A 10/0 silk suture is placed around the ECA and loosely tied around the base close to the carotid bifurcation.
10. The ECA is either ligated twice or thoroughly coagulated, and then cut leaving an ECA stump of 2–3 mm, which is reflected caudally.
11. Next proximal and distal blood flow are controlled, first by placing a small vessel clip on the pterygopalatine artery, which branches off the ICA laterally. Attention should be paid not to block the distal ICA entering the skull base, otherwise in the following steps the monofilament will not be able to pass through. Alternatively, the pterygopalatine artery can be ligated with a 10/0 suture.
12. The CCA is clamped as proximally as possible to avoid blocking the surgical field.
13. Finally, the ICA is clipped. Placing the clip proximal from the pterygopalatine artery branch avoids mechanical manipulation on the distal delicate ICA and reduces ICA vasospasm.
14. The isoflurane can be reduced to 1.5% at this point.
15. There should be no blood flow through the ECA vessel stump or the CCA. To confirm that blood flow has been interrupted, pulsations should not be observable in the ECA stump. Then a small incision in the tip of the ECA stump can be made without tremendous bleeding, preferably using microscissors.
16. The stump of the ECA is deflected forming a continuous channel with the ICA. The prepared 4/0 nylon monofilament is inserted into the ECA incision. Thereafter, it is carefully forwarded until resistance is felt, where the filament tip touches the vessel clip on the ICA. That coincidental movement of the clip signals the correct position of the filament in the ICA. Ideally, the marks on the filament are noticeable through the thin vessel wall during the whole process.
17. The loose ligature around the base of the ECA is gently tightened, and the ICA clip is removed. The filament is advanced furthermore observing its marks, to reassure the tip is following the ICA into the skull. No resistance should be felt until approximately 1.8–2.5 cm, where the bifurcation of the middle cerebral artery and the anterior common artery is located.

18. When this point is reached, the filament should be pushed a little further forward to puncture the vessel. Then, the position of the filament should be maintained for a few seconds, since immediate withdrawal may lead to insufficient SAH induction.
19. Release the blood flow by removing the clip on the CCA. The nylon filament is withdrawn slowly under awareness to quickly tighten the ligature occluding the ECA anytime, since strong bleed may occur while removing the monofilament.
20. The stump of the ECA is ligated and coagulated. The vessel clip or ligature on the pterygopalatine artery and the retractors are removed.
21. Finally after meticulous hemostasis, the skin incision is closed using a 3/0 silk suture.

Evaluation

SAH Grading

It is recommended to score the extent of subarachnoid bleeding for a representable analysis of brain injury, outcome, and treatment effects after SAH induction. Suzuki and colleagues introduced the broadly established method for SAH grading after euthanasia, which divides the basal cistern into six segments and scores the amount of visible blood [2]. Alternatively, MRI is able to grade non-euthanized animals by evaluating the amount and localization of SAH [3]. This technique requires the possibility of MRI and the animal needs to be under anesthesia.

Neurological Behaviour Testing

The neurological score should be evaluated by a blinded examiner prior to surgery as the baseline and repeated after surgery. Several scoring methods in rodent models of experimental SAH have been introduced [4]. The most frequently used tests are the modified-Garcia-score [5], beam-balance [6], and the Morris-Water-Maze-test [7]. For short term outcome (24 h and 72 h) modified-Garcia-score and beam-balance are eligible. While for long term outcome (up to 4 weeks), the Morris-Water-Maze-test should be added to evaluate spatial learning and memory (Tables 1 and 2).

Brain-Water-Content

After euthanasia, brains are quickly removed and divided into four parts, such as the left, the right hemispheres, the cerebellum, and the brain stem. These segments are weighed (wet weight) quickly after removal and dried in an oven at 105 °C for 72 h

Table 1 Modified Garcia score: /18

	0	1	2	3
Spontaneous activity (in cage for 5 min)	No movement	Barely moves position	Moves but does not approach at least three sides	Moves and approaches at least three sides
Spontaneous movements of all limbs	No movement	Slight movement	Moves all limbs but slowly	Move all limbs same as pre-SAH
Movements of forelimbs (outstretching)	No out-stretching	Slight out-stretching	Outstretching limited and less than pre-SAH	Outstretch same as pre-SAH
Climbing wall of wire cage	–	Fails to climb	Climbs weakly	Normal climbing
Touch of trunk	–	No response	Weak response	Normal response
Vibrissae touch	–	No response	Weak response	Normal response

Table 2 Beam balance:/4 (Able to walk: defined as moving more than 10 cm)

	0	1	2	3	4
Beam Walking (60 s)	Not walk & fall	Not walk, remains on beam	Walk, but fall	Walk less than 20 cm	Walk beyond 20 cm

and weighed again (dry weight). The percentage of water content was calculated with the following equation [8]:

$$\left(\frac{\text{wet weight} - \text{dry weight}}{\text{wet weight}}\right) \times 100\%$$

Evans-Blue-Extravasation

For the measurement of blood-brain-barrier permeability, Evans-blue-dye is injected intravenously or subcutaneously following the protocol of Satin and Lundberg [9], and then the Evans-blue in the brain samples is quantified using a spectrophotometer.

Molecular and Morphological Analysis

For molecular analysis protein levels is quantified with the method of Western-Blot, and mRNA expression is analyzed via PCR. Depending on the study protocol, Western-Blot samples are extracted from the whole brain, the ipsilateral hemisphere, or the basal brain parenchyma around the circle of Willis [8].

Immunohistochemistry enables analysis of vessel diameters for vasospasm evaluation, and morphological visualization of protein expression and cellular apoptosis.

The presence and severity of hydrocephalus can be evaluated with ventricular volumetry using either MRI [10] or histological slices [11].

Comments

Any surgical procedure should be performed under sterile environment, and the instruments should be therefore sterilized before each procedure. Intubation is recommended, since it reduces mortality of this model, but alternatively the anesthetics can be inhaled by mask or given systemically in a intravenous or intraperitoneal way. The surgery and the neurological testing should be planned preferably according to the rodents circadian rhythm. Blunt dissection to separate arteries completely is helpful before coagulation and cutting.

Bleeding

- Skin incision and dissection of the superficial cervical fascia bleed less, if performed at the midline.
- When positioning the vessel clips, try to visualize the clip endings reassuring the correct positioning of the vessel in the branches. Best occlusion is achieved in the middle of the branches.
- Before removing the vessel clips completely, first loosen them briefly and be prepared to reclip in case of bleeding.
- The source of bleeding is usually where the last dissection or manipulation was done.
- Coagulation with the bipolar forceps is more sufficient with clean branches and wet branches. This prevents the vessel to stick to the forceps, when being coagulated.

Filament Insertion

- If the filament is stuck before reaching the point of puncture, withdraw it, and try reinserting the slightly rotated monofilament.
- Check the animal's neck position for any dislocation. Sometimes, slightly turning the head a little to the right side is beneficial.
- Retract the filament, but keep the tip in the vessellumen, and then open the CCA for reperfusion for a short time. Now reclamp the CCA or proceed insertion under continuous blood flow.

- Try inserting a new monofilament.
- Continuous observation of the marks on the filament through the thin vessel wall indicates the inserted length and correct position in the ICA. If not visible, careful bending and repositioning of the filament towards the vessel wall can improve the visualization.
- Looping the suture around the ECA base too tight at step 9, makes feeling resistance while perforating the vessel wall difficult.

Postoperative Signs for Severe SAH induction

- Prolonged weaning phase
- Abnormal breathing pattern and other vital signs, cerebral blood flow, ICP
- Seizures
- Abnormal decorticate, decerebrate posturing, and ophistotonus

Advantages, Limitations, and Comparison with Other Models

The animal models to study SAH are more frequently performed in mice and rats, than rabbits, dogs, cats, or non-human primates. Several techniques are available to induce experimental SAH. Well established methods are the single-, the double-blood injection or the endovascular perforation model.

The endovascular perforation model produces more profound pathophysiological and histological changes to the brain and cerebral vessels. It is the only model with a vascular injury and direct hemorrhagic brain lesion under arterial blood pressure. It induces rapid increase of relatively long-lasting, high intracranial pressure and subsequent global ischemia in the acute stages of SAH. Especially in vasospasm research, the perforation model is well suited, but this technique can easily be adapted to stroke research. Another advantage is the lack of potential damage to the surrounding brain tissue and structures, since the perforation model does not require a craniotomy (Table 3).

Weaknesses and limitations of the perforation model are considered the possible failure of SAH induction and the range of variable SAH grades. Combination with grading methods like based on MRI helps avoiding heterogeneous study groups without euthanizing the animals [12]. The relatively high mortality rate compared to injection models and more challenging surgery lead to the predominant number of blood injection SAH studies. Thorough modification and improvements develop the perforation model continuously for broader application in experimental SAH research. Since it matches the morbidity, mortality, and severity of SAH in the clinical setting, it is well suited for evaluation and research of SAH (Figs. 1, 2, 3, and 4).

Despite its limitations, the endovascular perforation is suggested the superior model in SAH induction in several aspects.

Table 3 Comparison of the three most frequent rodent SAH models

	Rat injection model	Rat perforation model	Mice perforation model
SAH grading	+ (mild)	+++ (severe)	++ (moderate)
Subarachnoid blood volume	Equal in every animal	Variable	Variable
Neurological deficits	+	+++	++
Mortality	+ (<10%)	+++ (<20–45%)	+++ (<30%) Bühler protocol SAH mice
Possible rebleeding	–	+	+
Vascular injury	–	+	+
Craniotomy	+	–	–
Surgical procedure	Easy to learn and reproducible	Prolonged learning curve, surgeon-dependant	Prolonged learning curve, surgeon dependant

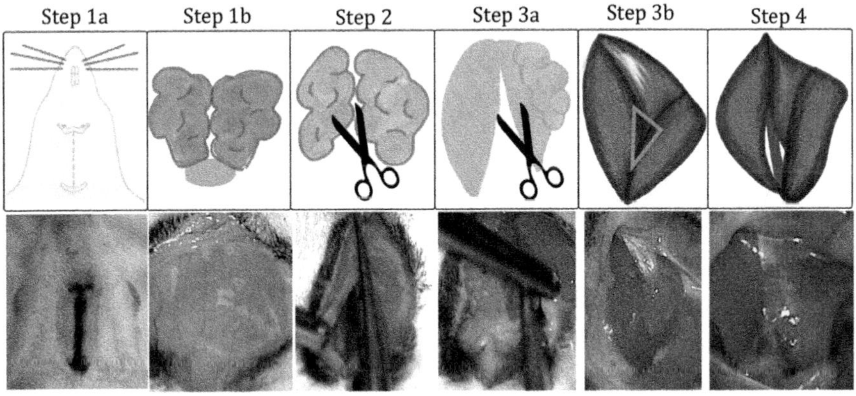

Fig. 1 Makroscopical surgical steps 1–5

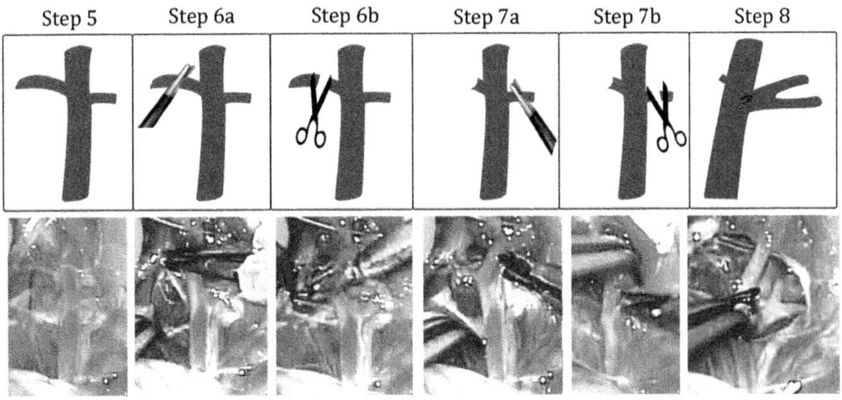

Fig. 2 Mikroscopical surgical steps, preparating the vessels, step 5–8

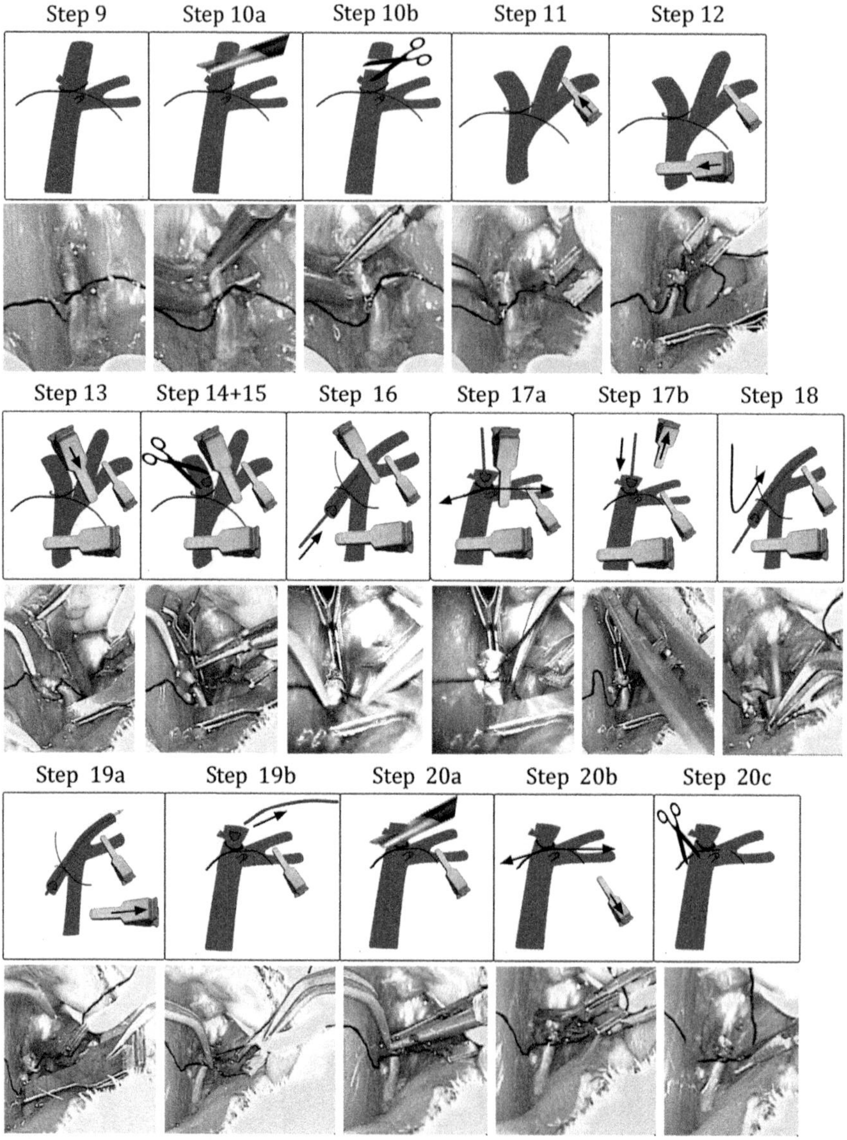

Fig. 3 Mikroscopical surgical steps, step 9–20

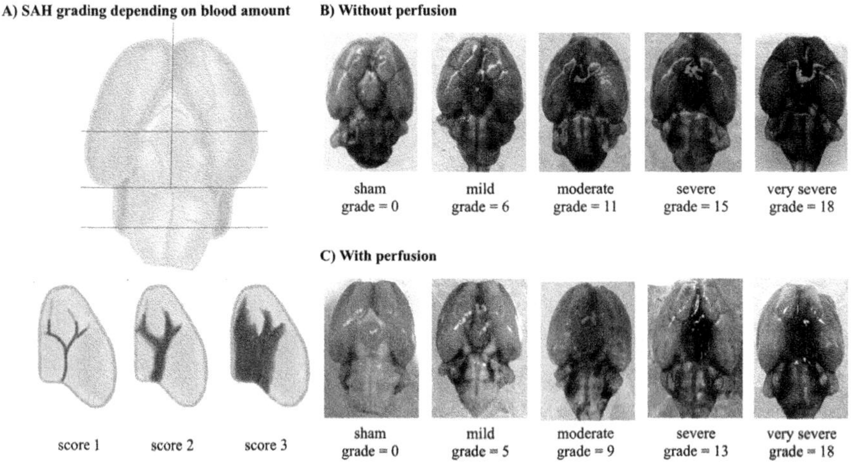

Fig. 4 Examples of different SAH grades using the method by Sugawara et al. [2]. (**a**) Basal brain surface divided into six sections and scored by blood amount visible (**b**) Brain samples collected without systemic perfusion. (**c**) Samples were collected after systemic perfusion with 0.01 M phosphate-buffered saline (PBS)

References

1. Gules I, Satoh M, Clower BR, Nanda A, Zhang JH. Comparison of three rat models of cerebral vasospasm. Am J Physiol Heart Circ Physiol. 2002;283(6):H2551–9.
2. Sugawara T, Ayer R, Jadhav V, Zhang JH. A new grading system evaluating bleeding scale in filament perforation subarachnoid hemorrhage rat model. J Neurosci Methods. 2008;167(2):327–34.
3. Egashira Y, Shishido H, Hua Y, Keep RF, Xi G. New grading system based on magnetic resonance imaging in a mouse model of subarachnoid hemorrhage. Stroke. 2015;46(2):582–4.
4. Jeon H, Ai J, Sabri M, Tariq A, Shang X, Chen G, et al. Neurological and neurobehavioral assessment of experimental subarachnoid hemorrhage. BMC Neurosci. 2009;10(1):103.
5. Garcia JH, Wagner S, Liu KF, Hu XJ. Neurological deficit and extent of neuronal necrosis attributable to middle cerebral artery occlusion in rats. Statistical validation. Stroke. 1995;26(4):627–34; discussion 635.
6. Germanò AF, Dixon CE, d' Avella D, Hayes RL, Tomasello F. Behavioral deficits following experimental subarachnoid hemorrhage in the rat. J Neurotrauma. 1994;11(3):345–53.
7. Morris R. Developments of a water-maze procedure for studying spatial learning in the rat. J Neurosci Methods. 1984;11(1):47–60.
8. Yatsushige H, Calvert JW, Cahill J, Zhang JH. Limited role of inducible nitric oxide synthase in blood-brain barrier function after experimental subarachnoid hemorrhage. J Neurotrauma. 2006;23(12):1874–82.
9. Saria A, Lundberg JM. Evans blue fluorescence: quantitative and morphological evaluation of vascular permeability in animal tissues. J Neurosci Methods. 1983;8(1):41–9.
10. Okubo S, Strahle J, Keep RF, Hua Y, Xi G. Subarachnoid hemorrhage-induced hydrocephalus in rats. Stroke. 2013;44(2):547–50.

11. Lackner P, Vahmjanin A, Hu Q, Krafft PR, Rolland W, Zhang JH. Chronic hydrocephalus after experimental subarachnoid hemorrhage. PLoS One. 2013;8(7):e69571.
12. Shishido H, Egashira Y, Okubo S, Zhang H, Hua Y, Keep RF, et al. A magnetic resonance imaging grading system for subarachnoid hemorrhage severity in a rat model. J Neurosci Methods. 2015;243:115–9.

Filament Perforation Subarachnoid Hemorrhage Mouse Model

Hideyuki Kamii and Teiji Tominaga

Abstract The filament perforation subarachnoid hemorrhage (SAH) mouse model mimics a clinical situation with aneurysmal SAH in humans. In this model, subarachnoid clot induces vasospasm which has a peak at 3 days after SAH. Moreover, a sudden increase in intracranial pressure immediately after the arterial perforation induces cortical neuronal damage similar to aneurysmal SAH in humans. Therefore, this model is suitable for researches on the various pathophysiologies of SAH. Although uncontrollable volume of subarachnoid hemorrhage basically causes high mortality rate in this model, our techniques perforating the anterior cerebral artery can lead to less mortality and relatively reproducible state after SAH. The complications and limitations in this model are also discussed.

Keywords Subarachnoid hemorrhage · Mouse · Model · Filament · Perforation

Model Selection

The filament perforation subarachnoid hemorrhage (SAH) mouse model has recently become popular to investigate the mechanisms underlying vasospasm and neuronal injury after SAH [1–4], since our laboratory has developed the procedure in mice [5–7]. This model has modified rat SAH model on the basis of endovascular arterial rupture [8, 9]. As in the rat SAH model [10], the mortality rate in the filament perforation SAH mouse model is higher than that in the blood injection SAH mouse model, which induces SAH by autologous blood injected into the cisterna

H. Kamii (✉)
Department of Neurosurgery, Aoba Neurosurgical Clinic, Sendai, Japan

Department of Neurosurgery, Tohoku University Graduate School of Medicine, Sendai, Japan
e-mail: h-kamii@fa2.so-net.ne.jp

T. Tominaga
Department of Neurosurgery, Tohoku University Graduate School of Medicine, Sendai, Japan
e-mail: tomi@nsg.med.tohoku.ac.jp

J. Chen et al. (eds.), *Animal Models of Acute Neurological Injury*, Springer Series in Translational Stroke Research, https://doi.org/10.1007/978-3-030-16082-1_16

magna [11]. When only vasospasm after SAH is investigated in the mouse, the blood injection SAH model might be suitable for the study because of the low mortality rate. In the filament perforation SAH model, however, SAH causes brain parenchymal damage as well as vasospasm [12], that can mimic the pathological conditions of SAH in humans. With the recent successful development of transgenic and knockout mice, the studies on SAH using a mouse model have become more prevalent. The filament perforation SAH mouse model can be used to elucidate molecular mechanisms underlying the pathogenesis of SAH.

Material

Animal

Adult male CD-1 mice, 10–12 weeks of age (35–40 g), are selected in our laboratory. Also, C57BL/6J mice are used in several reports [11, 13, 14]. Although the strain of mice is not so an important factor in SAH model as in ischemic model, it seems easier to make SAH model in bigger mice. Since CD-1 mice (35–40 g) are a little bigger than C57BL/6J mice (30–35 g) in 10–12 week of age, we usually use male CD-1 mice for SAH model.

Anesthesia

The mice are housed under a 12-h light/dark cycle with free access to food and water. Anesthesia is induced in a chamber with 2% halothane in 30% oxygen and 70% nitrous oxide after an intraperitoneal injection of xylazine (10 mg/kg) [15]. The mice are placed in the supine position on an operating table, and the rectal temperature of the animals is maintained at 37 °C by a Homeothermic Blanket Control Unit (Harvard Apparatus). Mice are allowed to respire spontaneously, and the anesthesia is maintained with 0.5% halothane in 30% oxygen and 70% nitrous oxide using a face mask during the operation. The left femoral artery is cannulated with a PE-10 catheter for measurement of mean arterial blood pressure, PaO_2, $PaCO_2$, and pH.

Procedures

Induction of SAH

1. Make a midline incision of the neck skin, and expose the left common carotid artery (CCA) with careful conservation of the vagus nerve. Then, isolate the external carotid artery (ECA), superior thyroid artery (STA), and occipital artery

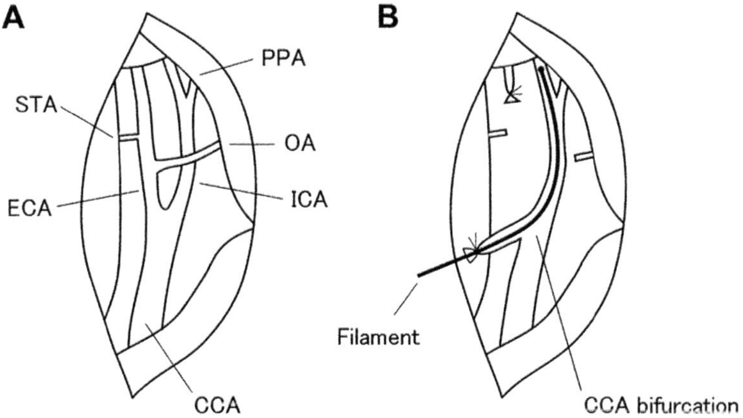

Fig. 1 Schematic representations showing procedures to introduce the filament into the left ICA. (**a**) After a midline incision of the neck skin, the left common carotid artery (CCA) is exposed, and then the external carotid artery (ECA), superior thyroid artery (STA), and occipital artery (OA) are isolated. The internal carotid artery (ICA) and pterygopalatine artery (PPA) are observed under the OA. (**b**) The nylon filament is introduced into the ICA through the ECA stump. Take care to avoid introducing the filament into the PPA

(OA). The internal carotid artery (ICA) and pterygopalatine artery (PPA) are observed under the OA (Fig. 1a).

2. Electrocoagulate and divide the STA and OA. After placing a microsurgical clip across the ECA origin, ligate the distal end of the ECA with 6-0 silk suture and cut it. Tie another 6-0 suture loosely around the ECA stump. Then, introduce a 2-cm length of 5-0 nylon monofilament, blunted at the tip by heat, into the ECA stump, and tighten the silk suture around it. After removing the microsurgical clip, advance the nylon filament into the ICA through the ECA stump. Take care to avoid introducing the filament into the PPA (Fig. 1b).
3. Advance the filament gently until encountering a resistance. At this point, the tip of the filament is in the left anterior cerebral artery (ACA) near the confluence with the right ACA (Fig. 2a). After pulling back the filament 2 mm (Fig. 2b), advance the filament 5 mm further to perforate the left ACA (Fig. 2c), and withdraw immediately through the ICA into the ECA stump, allowing reperfusion and SAH (Fig. 3a).
4. When endovascular SAH occurs, mice show a respiratory failure for several seconds, followed by recovering regular respiration without any treatments. The distance from the CCA bifurcation to the confluence with the right ACA (the origin of the azygous ACA) measures approximately 11 mm. Sham-operated control mice undergo identical procedures except that the filament is withdrawn just after encountering the resistance.
5. After cutting the nylon filament protruding over the ECA stump, close the incision with silk sutures. The entire procedure is completed within approximately 20 min.

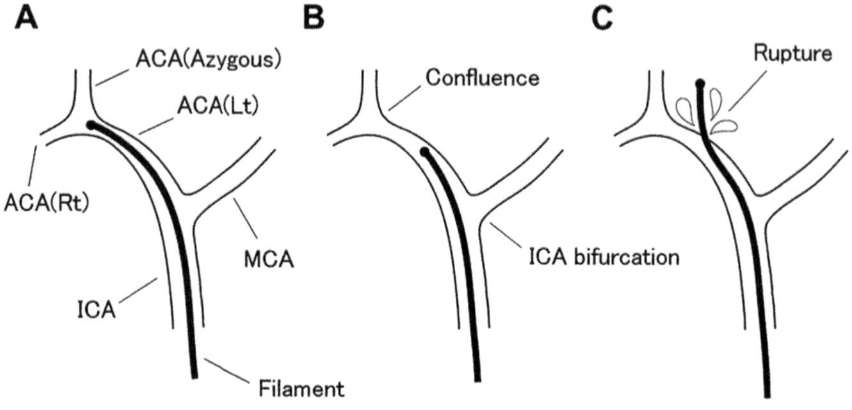

Fig. 2 Schematic representations showing procedures to perforate the left ACA. (**a**) The tip of the filament is in the left anterior cerebral artery (ACA) near the confluence with the right ACA, when a resistance is encountered. (**b**) Pull back the filament 2 mm. (**c**) Then, advance the filament 5 mm further to perforate the left ACA

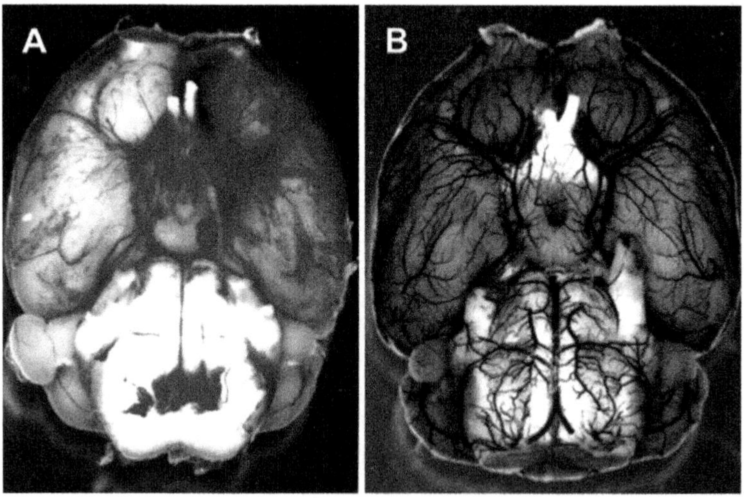

Fig. 3 Representative photographs of the brain. (**a**) At 4 h after SAH, blood clots are diffusely observed on the brain surface, especially around the left ACA. (**b**) All vessels of the brain are casted with carbon ink and 10% gelatin

Perfusion-Fixation and Casting Method

1. Anesthetize the mice with an intraperitoneal injection of 20 mg/kg pentobarbital. Open the chest, and cannulate the aorta through the left ventricle with a 23-gauge needle. Just after making an incision in the right atrium, begin perfusion with

200 mL of normal saline including heparin, followed by 200 mL of 10% formalin or 4% paraformaldehyde. Perfusion pressure is controlled by setting a container of the solution at 90 cm height from the mice.

2. In case of evaluating the arterial diameter in the mice, casting method is performed following the perfusion-fixation. Gelatin powder (10 g) is dissolved in 100 mL of saline and mixed with 100 mL of carbon ink, and the solution is maintained at 40 °C. After the perfusion with 10% formalin, perfuse all vessels with 100 mL of the carbon ink/gelatin solution. Then, refrigerate the mice for 2 h to allow gelatin solidification. Finally, remove the brains, and store them in 10% formalin (Fig. 3b).

Outcome Evaluation

Neurological Examination

It is important to confirm a mouse without hemiparesis after the surgery. When the mice showed right hemiparesis upon the recovery, our preliminary study revealed that the left MCA was often occluded completely, and that a big cerebral infarction in the left MCA area was observed in the brain. Therefore, when a mouse shows right hemiparesis after the surgery, the mouse should be excluded from the study.

Measurement of the Arterial Diameter

Vasospasm is evaluated by measuring the diameter of proximal portion of the left MCA by a stereomicroscope (Fig. 4). We do not use the ICA and ACA for evaluation of vasospasm, since the filament passes through them, resulting in mechanical injuries to the arteries. In our series, vasospasm peaked at 3 days after SAH (Fig. 5).

Immunohistochemistry in the Cerebral Artery and Cortex

The mouse brains are embedded in paraffin. Coronal 5-μm-thick sections near the ICA bifurcation are cut and mounted on slide. The sections can be advanced to the ordinary or immunohistochemical staining for the artery as well as the cortex. In our filament perforation SAH mouse model, iNOS protein was induced in the cerebral arteries as well as in the cerebral cortices (Fig. 6).

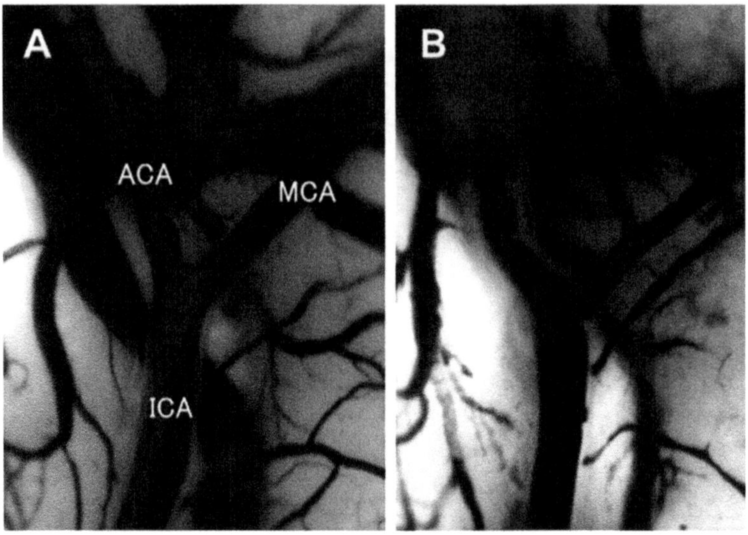

Fig. 4 Representative photographs of the cerebral arteries casted with carbon ink and 10% gelatin. (**a**) The left ICA, ACA, and MCA are clearly demonstrated by a stereomicroscope in a sham-operated control mouse. (**b**) At 3 days after SAH, the diameter of the MCA is reduced compared with that in a sham-operated control mouse. Note that the left ACA is perforated between the ICA bifurcation and the confluence with the right ACA

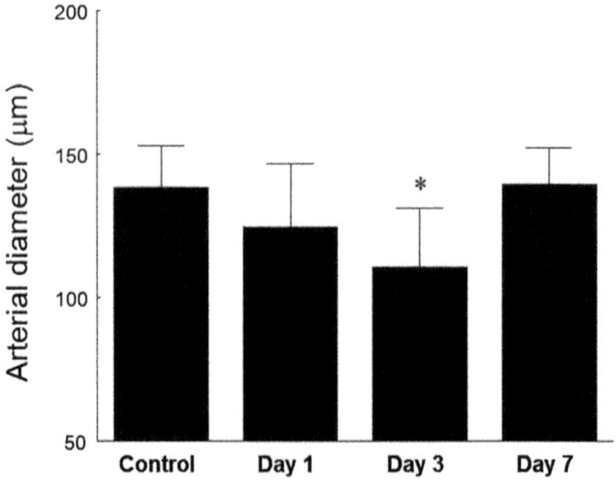

Fig. 5 Sequential changes in the diameter of the MCA after SAH in our series. At 3 days after SAH, the MCA diameter was significantly reduced compared with that in sham-operated control mice (*$p < 0.01$)

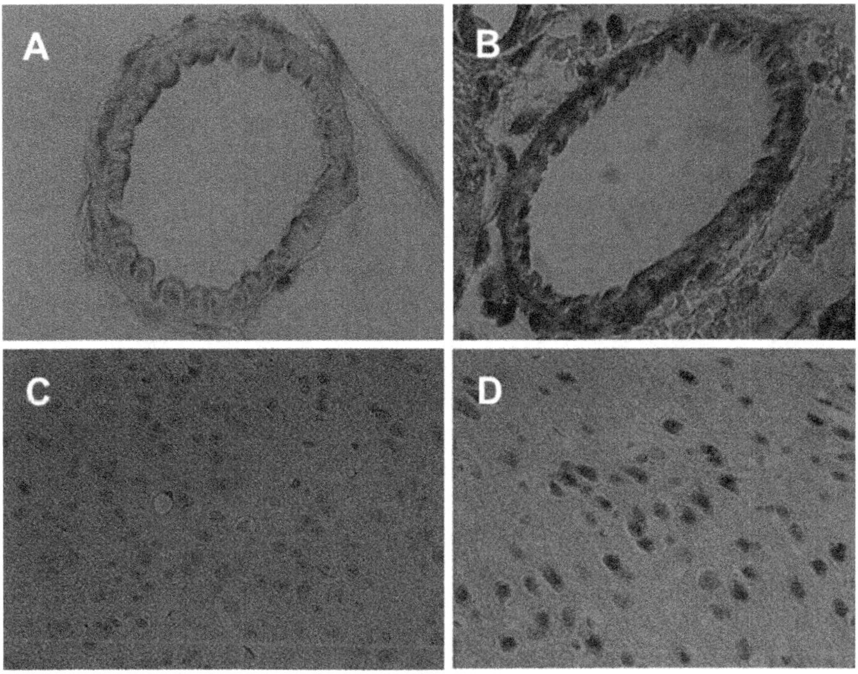

Fig. 6 Immunohistochemical demonstration of iNOS. At 24 h after SAH, iNOS immunoreactivity is observed in the cerebral artery (**b**) and the cortex (**d**). In sham-operated control mice, iNOS immunoreactivity is negative in the cerebral artery (**a**) and the cortex (**c**)

Advantages, Complications, and Limitations

Advantages

1. The filament perforation SAH mouse model mimics a clinical patient with aneurysmal SAH in humans, compared with the blood injection SAH model. A sudden increase and the prolongation of intracranial pressure (ICP) cannot been reproduced in the blood injection SAH model [10]. Moreover, the blood injection SAH model may not distribute the blood in the subarachnoid space all over the brain. Therefore, the filament perforation SAH model is the most suitable for the researches to investigate vasospasm as well as neuronal injury and neurogenesis following SAH [6, 7, 12, 16].
2. In our filament perforation SAH model, the left ACA is perforated by a nylon filament. That is, since the perforation site is between the ICA bifurcation and the confluence with the right ACA (the origin of the Azygous ACA) (Fig. 2), the filament does not injure the MCA directly. Thus, we can evaluate the vasospasm by SAH itself, not by the mechanical injury, in the proximal portion of the MCA. Although some reports showed that the perforation site was near the

ICA bifurcation [13, 14], the arterial rupture near the ICA bifurcation often causes the MCA occlusion resulting in brain infarction of the MCA area in our preliminary studies. Perforating the ACA is a very important technique to avoid ischemic brain damage, and it is a big advantage in our filament perforation SAH mouse model [5].
3. The filament perforation SAH model is less invasive for mice because of no craniotomy. The techniques are simple, and the entire procedure, skin incision to skin closure, is completed within approximately 20 min in our laboratory.
4. The costs are low to purchase and maintain mice, compared with other species. In addition, the recent successful development of transgenic and knockout mice has made mice more prevalent in numerous researches on stroke including SAH [17, 18].

Complications

1. In our SAH model, when typical SAH occurs, mice show respiratory failure for several seconds, while severe SAH often causes long respiratory arrest in the mice. Although the mice with long respiratory arrest can be resuscitated, they usually die within 24 h due to the high ICP with hypoxic brain injury.
2. When the perforation site is the ICA bifurcation, not the left ACA, the ICA/MCA occlusion often occurs, resulting in brain infarction in the MCA area. Therefore, when the mice show right hemiparesis after the surgery, they should be excluded from the study.
3. At 7 days after SAH, the mice showed hydrocephalus in the brain at approximately 30% in our laboratory.
4. Small intracerebral hematoma was seen in the left frontal cortex at approximately 3% in our series.

Limitations

1. In the filament perforation SAH model, the blood volume in the subarachnoid space cannot be controlled basically. The blood volume is believed to correlate well with ICP and mortality rate [10]. Recent studies on the blood volume using MRI [19–21] can help us evaluate SAH severity.
2. The mortality rate in our SAH model is approximately 20% within 72 h after SAH. Main cause of the death is respiratory failure due to severe SAH. Although the high mortality rate is inevitable in the filament perforation SAH model even in rats [8–10], our techniques perforating the left ACA can lead to less mortality rate. In our laboratory, the mortality rate on perforation of the ICA bifurcation or the left ACA was approximately 45% or 20%, respectively.

3. At 7 days after SAH, approximately 30% of the mice demonstrated hydrocephalus in the brain in our series. The mice with hydrocephalus could not survive for a long time. Thus, the filament perforation SAH mouse model might not be suitable for researches on a chronic phase over a month following SAH.

References

1. Muroi C, Fujioka M, Okuchi K, Fandino J, Keller E, Sakamoto Y, Mishima K, Iwasaki K, Fujiwara M. Filament perforation model for mouse subarachnoid hemorrhage: surgical-technical considerations. Br J Neurosurg. 2014;28:722–32.
2. Buhler D, Schuller K, Plesnila N. Protocol for the induction of subarachnoid hemorrhage in mice by perforation of the circle of Willis with an endovascular filament. Transl Stroke Res. 2014;5:653–9.
3. Du GJ, Lu G, Zheng ZY, Poon WS, Wong KC. Endovascular perforation murine model of subarachnoid hemorrhage. Acta Neurochir Suppl. 2016;121:83–8.
4. Kamp MA, Lieshout JHV, Dibue-Adjei M, Weber JK, Schneider T, Restin T, Fischer I, Steiger HJ. A systematic and meta-analysis of mortality in experimental mouse models analyzing delayed cerebral ischemia after subarachnoid hemorrhage. Transl Stroke Res. 2017;8:206–19.
5. Kamii H, Kato I, Kinouchi H, Chan PH, Epstein CJ, Akabane A, Okamoto H, Yoshimoto T. Amelioration of vasospasm after subarachnoid hemorrhage in transgenic mice overexpressing CuZn-superoxide dismutase. Stroke. 1999;30:867–72.
6. Saito A, Kamii H, Kato I, Takasawa S, Kondo T, Chan PH, Okamoto H, Yoshimoto T. Transgenic CuZn-superoxide dismutase inhibits NO synthase induction in experimental subarachnoid hemorrhage. Stroke. 2001;32:1652–7.
7. Mino M, Kamii H, Fujimura M, Kondo T, Takasawa S, Okamoto H, Yoshimoto T. Temporal changes of neurogenesis in the mouse hippocampus after experimental subarachnoid hemorrhage. Neurol Res. 2003;25:839–45.
8. Bederson JB, Germano IM, Guarino L. Cortical blood flow and cerebral perfusion pressure in a new noncraniotomy model of subarachnoid hemorrhage in the rat. Stroke. 1995;26:1086–92.
9. Veelken JA, Laing RJ, Jakubowski J. The Sheffield model of subarachnoid hemorrhage in rats. Stroke. 1995;26:1279–84.
10. Prunell GF, Mathiesen T, Diemer NH, Svendgaad N-A. Experimental subarachnoid hemorrhage: subarachnoid blood volume, mortality rate, neuronal death, cerebral blood flow, and perfusion pressure in three different rat models. Neurosurgery. 2003;52:165–75.
11. Lin CL, Calisaneller T, Ukita N, Dumont AS, Kassell NF, Lee KS. A murine model of subarachnoid hemorrhage-induced cerebral vasospasm. J Neurosci Methods. 2003;123:89–97.
12. Li M, Wang W, Mai H, Zhang X, Wang J, Gao Y, Wang Y, Deng G, Gao L, Zhou S, Chen O, Wang X. Methazolamide improves neurological behavior by inhibition of neuron apoptosis in subarachnoid hemorrhage mice. Sci Rep. 2016;6:35055.
13. Parra A, McGirt MJ, Sheng H, Laskowitz DT, Pearlstein RD, Warner DS. Mouse model of subarachnoid hemorrhage associated cerebral vasospasm: methodological analysis. Neurol Res. 2002;24:510–6.
14. McGirt MJ, Parra A, Sheng H, Higuchi Y, Oury TD, Laskowitz DT, Pearlstein RD, Warner DS. Attenuation of cerebral vasospasm after subarachnoid hemorrhage in mice overexpressing extracellular superoxide dismutase. Stroke. 2002;33:2317–23.
15. Kamii H, Kinouchi H, Sharp FR, Chan PH. A model of transient focal cerebral ischemia in the mouse. In: Ohnishi ST, Ohnishi T, editors. Central nervous system trauma: research techniques. Boca Raton: CRC Press; 1995. p. 139–46.

16. Feiler S, Friedrich B, Scholler K, Thal SC, Plesnila N. Standardized induction of subarachnoid hemorrhage in mice by intracranial pressure monitoring. J Neurosci Methods. 2010;190:164–70.
17. Cahill J, Calvert JW, Zhang JH. Mechanisms of early brain injury after subarachnoid hemorrhage. J Cereb Blood Flow Metab. 2006;26:1341–53.
18. Okano H, Kishi N. Investigation of brain science and neurological/psychiatric disorders using genetically modified non-human primates. Curr Opin Neurobiol. 2017;50:1–6.
19. Egashira Y, Shishido H, Hua Y, Keep RF, Xi G. New grading system based on magnetic resonance imaging in a mouse model of subarachnoid hemorrhage. Stroke. 2015;46:582–4.
20. Muroi C, Kashiwagi Y, Rokugawa T, Tonomura M, Obata A, Nevzati E, Tsuboi A, Okuchi K, Mishima K, Abe K, Fujioka M. Evaluation of a filament perforation model for mouse subarachnoid hemorrhage using 7.0 tesla MRI. J Clin Neurosci. 2016;28:141–7.
21. Sasaki K, Mutoh T, Nakamura K, Kojima I, Taki Y, Suarez JI, Ishikawa I. MRI-based in vivo assessment of early cerebral infarction in a mouse filament perforation model of subarachnoid hemorrhage. Neurosci Lett. 2017;653:173–6.

The Non-human Primate Model of Cerebral Vasospasm

R. Loch Macdonald

Abstract Cerebral vasospasm classically refers to narrowing of the large subarachnoid cerebral arteries occurring 3–14 days after blood clots are deposited in the subarachnoid space after subarachnoid hemorrhage (SAH) from any cause. It is most common after aneurysmal SAH because this process deposits the most dense and voluminous blood in the subarachnoid space and it is this blood clot that causes vasospasm. It is recommended that cerebral vasospasm be called angiographic vasospasm and to use this term to refer to the angiographic appearance of the cerebral arteries. The clinical consequences of angiographic vasospasm should be referred to as delayed cerebral ischemia (DCI). DCI is an important cause of morbidity and mortality after SAH and is usually associated with severe angiographic vasospasm. Animal models of angiographic vasospasm have been developed because of the difficulties modeling delayed vasospasm *in vitro*. These models use one of three techniques to simulate SAH: (1) an intracranial artery is punctured allowing blood to surround the artery; (2) an artery is surgically exposed and clotted autologous blood obtained from another site is placed around the artery; or (3) autologous blood from another site is injected into the subarachnoid space and allowed to surround the intracranial arteries. Each technique has advantages and disadvantages. In this chapter, we discuss the advantages and disadvantages of and methods for creation of SAH and angiographic vasospasm in non-human primates. Arterial diameters are easily quantified by angiography and SAH is created by frontal craniectomy, exposure of the intracranial arteries of the anterior circle of Willis and placement of autologous clotted blood around these arteries.

Keywords Angiographic vasospasm · Cerebral angiography · Non-human primate · Subarachnoid hemorrhage

---

R. L. Macdonald (✉)
Division of Neurosurgery, St. Michael's Hospital, Toronto, ON, Canada

Labatt Family Centre of Excellence in Brain Injury and Trauma Research, Keenan Research Centre for Biomedical Science, Li Ka Shing Knowledge Institute, St. Michael's Hospital, Toronto, ON, Canada

Department of Surgery, University of Toronto, Toronto, ON, Canada

J. Chen et al. (eds.), *Animal Models of Acute Neurological Injury*, Springer Series in Translational Stroke Research, https://doi.org/10.1007/978-3-030-16082-1_17

Model Selection

The animal model to use depends in part on what aspect(s) of the pathophysiology or management of SAH is being studied. The pathophysiology of SAH includes an initial early brain injury that is secondary to acute cerebral ischemia and effects of intracranial and subarachnoid blood itself [1]. DCI occurs 3–14 days after SAH and is believed to be due to a combination of angiographic vasospasm, cortical spreading ischemia, microthrombosis, autoregulatory failure and capillary transit time heterogeneity [2–6]. Early understanding of DCI was that angiographic vasospasm was the most important cause, leading to extensive studies modeling it *in vitro* and *in vivo*. These models have led to new understanding of vascular biology but tend to be used now in conjunction with models *in vivo* because it has been difficult to generalize results of acute studies of arterial segments studied *in vitro* with the complex situation *in vivo* after SAH [7, 8]. Arteries *in vitro* are denervated, not subject to pulsative pressure and flow and may not have normal endothelial function. Changes in gene expression can occur within hours and may influence the responses [9]. In most but not all monitoring systems *in vitro* the intra- and extraluminal surfaces are exposed to vasoactive substances which probably does not occur *in vivo*. Pathological and pharmacologic changes occur in vasospastic arteries but are less well documented in experiments *in vitro* which may be important [10]. Improved technology for studying the cerebral circulation *in vivo* also has contributed to this change [11].

Mice, rats, rabbits, cats, dogs and monkeys have been subjected to SAH in order to create angiographic vasospasm [12, 13]. In general, injection of blood into the cisterna magna is used in rodents and results in mild, short-lived vasospasm. Vasospasm can be prolonged by repeating the injection a day or 2 later. An endovascular perforation model also is used in mice and rats [14–16]. The most well documented models that produce prolonged, severe vasospasm are in dogs and monkeys [12, 17]. Cisternal injections are typically done in dogs. The monkey model described here was invented by Bryce Weir in Edmonton, Alberta in the 1980s. His experiments began with transcranial blood injections into the Rhesus monkey subarachnoid space but these produced only short-lived and mild vasospasm [18]. Injections of large volumes of blood did worsen the vasospasm but he conceived the intracranial clot placement model which was adopted because it was more reliable. The use of an operating microscope to perform a frontal craniectomy in the cynomolgus monkey, open the arachnoid and expose the internal carotid, posterior communicating, middle and anterior cerebral arteries and pack clotted autologous arterial blood around them was first reported in 1984 [19]. A modification of this model was described by Pluta and colleagues [20, 21]. He provides an excellent, detailed description of Weir's model, as well as of the one modification in the procedure to remove the arachnoid around the Sylvian fissure.

Advantages of this model include [13, 22]

1. The model reliably creates severe, prolonged angiographic vasospasm most like that occurring in humans [7]. The only other model that causes severe, prolonged angiographic vasospasm is the dog double-hemorrhage model. The double-hemorrhage dog model differs from humans in that it involves two injections of autologous blood into the cisterna magna several days apart, which complicates time course studies and has been likened to a rebleeding model. The time course is longer than that occurring in man [23]. Rodents do not develop angiographic vasospasm that is severe or long-lasting and this prolonged, severe course may be associated with processes that are pathophysiologically important and that are not even detected in these other models.
2. There may be more evidence of translation of results from animal model to human. Treatments that reduce angiographic vasospasm in this model do the same thing in humans and those that do not work in this model are very unlikely to affect angiographic vasospasm in humans. This cannot be said of rodent models. The non-human primate model demonstrated that sodium nitroprusside [24], nimodipine [25] and tirilazad [26] had no or very modest effects on vasospasm whereas clot lysis with tissue plasminogen activator [27], nitric oxide donors [28] and endothelin antagonists [29] were effective, corresponding to results in humans [30–32]. One of the many recommendations about stroke research is to investigate treatments in non-human primates before doing so in humans but these recommendations have not yet been demonstrated to ensure that a treatment for stroke in something other than humans is actually effective in humans in adequate, well-controlled clinical trials [33].
3. There are clot-side arteries exposed to SAH that can be compared within animals to the contralateral, non clot-side arteries.
4. Some animals develop cerebral ischemia and infarction, which are the clinically-important consequences of angiographic vasospasm.
5. Intracranial arteries and a subarachnoid space similar to that in humans are present so experiments with clot placement, replacement and removal as well as detailed time-course studies and systemic and intracranial drug delivery can be done easily.
6. Non-human primates are more like us than rats, mice and other species. This includes many aspects from their genome to molecular biology, anatomy, physiology, psychology, behavior and such. Toxicity of some drugs, the immune and other body systems, gene expression and cognitive/neuropsychiatric/behavior/social function are often unique to primate [34–36].

Disadvantages of the non-human primate model are

1. There is relatively limited data on gene sequences of cynomolgus monkeys and paucity of specific molecular biological supplies such as antibodies for monkeys.
2. Variability is greater. The animals are not an inbred strain like rodents used in biological research. Therefore, interanimal variations occur that affect the sizes

of intracranial arteries, anatomy and physiology, composition of the blood and of the arteries that may affect the degree of angiographic vasospasm produced.

3. The cost, time and equipment required and complexity are greater compared to rodents.
4. There are few genetically-modified monkeys. Only a few such animals have been generated. Targeted gene knockout with TALEN and CRISPR/Cas methods may alleviate this in part, although cost, time and ultimately, benefit over existing models are still barriers [34].
5. The early effects of SAH are not modeled. The blood clot is placed surgically without increasing the intracranial pressure. The aneurysm rupture and intracranial pressure changes that occur during rupture and that are hypothesized to contribute to early brain injury do not occur. Theoretically this could allow dissection of the effects of the early brain injury processes from those of the subarachnoid blood. The closed-skull endovascular perforation model of SAH in rats mimics the acute effects of SAH [15, 16] but has important disadvantages in that there is great variability in the volume of SAH, mortality is high, severe angiographic vasospasm is not produced and there frequently is a permanent middle cerebral artery occlusion associated with the initial SAH, making this a model of SAH plus acute ischemia not SAH alone.
6. The arachnoid membrane is opened over the Sylvian fissure but usually not removed from the adjacent brain. Thus, most of the blood clot is actually in the subdural not the subarachnoid space. Pluta, et al., described a modification of the model aimed at alleviating this [21].
7. Young female animals without other diseases are used. Few investigators have used old animals for SAH research or even stroke studies so this is a problem common to the entire field [37].
8. Many old- and new-world monkeys, including the cynomolgus monkey, carry simian B virus (Cercopithecine herpesvirus 1) [38]. They also can be infected with tuberculosis, tropical diseases and even filoviruses such as Ebola and Marburg viruses [39]. Infection of humans with simian B virus is fatal in about 70% of cases and the hemorrhagic fever associated with filoviruses is worse. Most animals used now in medical research are bred in captivity and repeatedly tested and used only if they are B and filovirus-negative.
9. There are ethical issues about the use of animals, especially non-human primates, in medical research [40].

Materials

There are several sets of guidelines for conduct and reporting of experimental studies [33, 41, 42]. Since no treatments proven to be efficacious in humans with stroke have followed these guidelines, their value or validity is open to question. Nevertheless, many of them like randomization and blinding, would best be adhered to, while others like using old animals with medical comorbidities are difficult to implement.

Juvenile female cynomolgus monkeys (*Macaca fasicularis*) weighing 2.5–4.0 kg are used. Regulatory requirements vary but generally over 1 month quarantine is required. Substantial preplanning is required since animals also usually cannot be shipped when the ambient temperature is too low or high. The animals are housed in crush cages and rooms complying with all federal requirements. The procedures are carried out in a veterinary operating room equipped in much the same fashion as a human operating room.

The following general equipment is needed:

1. Small animal ventilator to deliver a tidal volume of approximately 8–10 mL/kg at a rate of 10–20 breaths per minute.
2. Anesthetic machine equipped to deliver isoflurane and oxygen.
3. Thermostatically-controlled heating blanket or warmed-air blanket to maintain body temperature.
4. Angiography machine such as a standard, high-resolution fluoroscope machine capable of performing digital subtraction angiography.
5. Radiolucent animal operating table.
6. Physiological monitoring machines for percutaneous oxygen saturation, end-tidal PCO_2, heart rate, noninvasive blood pressure.
7. An operating microscope, more than a simple dissecting microscope, with good illumination and binocular vision. Extra light bulbs must be available.
8. Monopolar surgical coagulator, grounding pad and bipolar coagulation machines.
9. Suction.
10. Tape measure and magnification standard for angiography.
11. Electronic scale to weigh the blood clot.

For conduct of the angiography and creation of SAH, the following sterile instruments are recommended (Fig. 1):

1. Four small hemostats.
2. One regular size needle driver.
3. One curved, 4 or 5″ micro needle driver.
4. One Mayo scissors.
5. One short Metzenbaum scissors.
6. Two fine vascular forceps.
7. Two Adson toothed forceps.
8. Scalpel handle.
9. Self-retaining brain retractor and retractor blade.
10. Small rongeur for making craniectomy.
11. Trephine and pointed tip to make initial burr hole. An air-powered surgical drill also would work.
12. #5 Frazier suction tube.
13. Piece of angiography guide wire to fit an 18 guage angiography catheter.
14. Two small metal beakers.
15. Penfield microdissector.

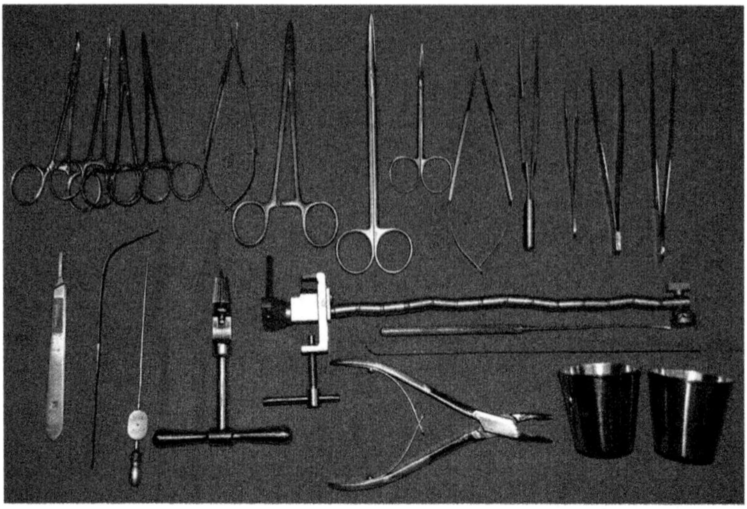

Fig. 1 Photograph of surgical instruments used for this model. From left to right on the top row are four hemostats, micro needle driver, standard needle driver, three types of scissors and four types of forceps. The bottom row from left to right is a scalpel handle, brain retractor, suction, trephine, self-retaining retractor arm, microdissector, angiography guide wire, ronguer and two metal cups. Bipolar and monopolar cautery and a sterile suction tubing are packaged separately

16. Pin head holder.
17. Arachnoid knife.
18. Bipolar forceps with cord.

The disposable sterile supplies that are used include (Fig. 2):

1. Sterile weighing boat to weigh blood clot.
2. Red-top tubes to put arterial blood from the angiography catheter into to clot.
3. 4 × 4 sterile gauze.
4. Cotton patties.
5. Sterile drapes.
6. 18 gauge angiography catheter.
7. 3-way stopcock.
8. One mL syringe with 25 gauge needle for blood gases, needle is bent and used as a dural hook.
9. #10 scalpel blade.
10. Two 10 mL syringes for catheter work and blood for SAH.
11. Non-absorbable monofilament 3-0 suture on a cutting needle for skin. Perfect scalp closure is required. We have not found that stapling devices can provide this.
12. 6-0 monofilament nylon or polypropylene suture on a non-cutting needle for suturing dura.
13. 4-0 silk suture for ligating the brachial artery.

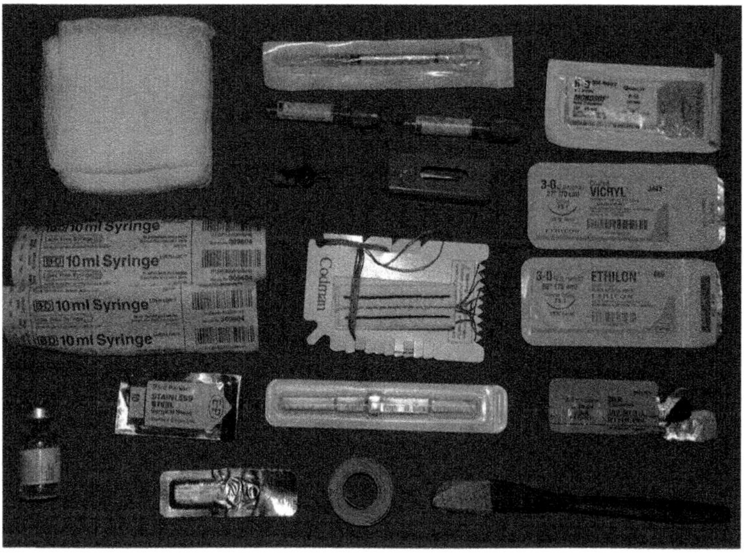

Fig. 2 Disposable supplies for angiography and creation of SAH by clot placement. Starting from the top left, there is guaze, 1 mL syringe with 25 guage needle, red-top tubes for blood, 3-way stopcock, arachnoid knife, 6-0 monofilament nonabsorbable suture, 3-0 absorbable suture, two 10 mL syringes, patties, 3-0 monofilament nonabsorbable suture, heparin solution, scalpel blade, bone wax, 18 gauge 1.8 in. angiocatheter, tape measure, silk ligatures and a coin taped to a stick to be placed behind the monkey's head during angiography as a magnification standard. Not shown are mannitol, contrast, weighing boat, surgical gloves and drapes and sterile surgeon's gown

14. Absorbable 3-0 braided absorbable suture (such as polyglactin or an equivalent) on a noncutting needle for suturing the muscle and fascia.
15. Bone wax.
16. Surgical gloves.
17. Heparin (1:1000 solution, for blood gas syringe only), mannitol (20% for intravenous injection), physiological saline and contrast agent such as 60% Hypaque (diatrizoate meglumine USP, 60%, Amersham Health, Princeton, NJ).

Procedures

Anesthesia

The monkey is sedated in the crush cage with ketamine hydrochloride, 10 mg/kg intramuscularly. It is weighed and an intravenous catheter is placed in a calf vein or other accessible superficial vein. The animal is intubated transorally with an unarmoured number 4 pediatric endotracheal tube. An antibiotic such as enrofloxacin, 2.5 mg/kg, is administered intramuscularly. The head, both groins and the right axilla are shaved with a clipper and the animal then transported to the operating room. Ventilation is with oxygen and 1–3% isoflurane using a small animal respirator. End-tidal PCO_2, body temperature, transcutaneous oxygen saturation, blood pressure, and heart rate are monitored and maintained at physiological levels. For the first stages when angiography is performed it is important to maintain normal arterial PCO_2. Body temperature will drop precipitously if the animal is not kept covered. It is best maintained with a forced-air warming blanket.

Angiography

The animal is placed supine and the head is taped straight up with the neck flexed to permit an anteroposterior angiogram projection. A magnification standard is placed under the neck and the right hand is taped in an abducted and externally rotated posture to the operating Table. A rolled gauze under the triceps will rotate the arm externally and expose the grove between the biceps anteriorly and triceps posteriorly where the neurovascular bundle resides. The right axilla and upper arm are prepared with povidone iodine and a disposable, sterile drape with a 2 cm hole cut in the middle of it placed over the proximal neurovascular bundle. The surgeon puts on a lead apron and sterile gloves. Under the operating microscope, a 1 cm incision is made over the artery which is exposed using microscissors and fine vascular forceps. With experience, this tedious dissection can be accomplished in minutes. The venous plexus around the artery is fragile but easily freed and does not need to be cauterized. Excess manipulation or cauterization will make the brachial artery spastic so it is best not to start any bleeding to begin with. The artery should be exposed proximal to the profunda brachii artery since distal to this it is quite small and difficult to catheterize. The artery is tied distally with a 4-0 silk suture. A 4-0 silk suture is placed around the artery proximally and it is catheterized with a sterile 18 gauge polypropylene catheter connected to a stop cock through which a guidewire is preplaced.

The guidewire is removed and a 10 mL syringe connected to the stopcock. Eight milliliter blood is withdrawn from the catheter, placed in red-top tubes and allowed to clot for unilateral SAH. Do not put heparin in or even near the catheter or stopcock as the minutest wiff of it will prevent the blood from clotting. At this point, a 1 mL syringe with a small amount of heparin in it is used to withdraw 300 μL blood for analysis of blood gases and hematological values.

The operating microscope is removed and the fluoroscope brought in and centered over the head. The isoflurane concentration may be below 1% but should be increased because we have routinely observed that injection of contrast causes the animal to move or vomit and if the animal vomits the procedure has to be aborted otherwise it will die. The distance from the radiographic source to the operating table is measured and kept constant for every angiogram. The closer the head is to the radiographic detector, the clearer the image will be. Maximum magnification is used and exposure factors for the catheter runs should always be the same. We obtain a single, anteroposterior midarterial phase angiogram by manual injection of 5–8 mL 60% diatrizoate meglumine USP under digital subtraction video fluoroscopy.

The catheter is withdrawn from the brachial artery and the artery ligated proximally with 4-0 silk. We have not found it necessary to reconstruct the artery and have cannulated it more proximally for subsequent angiograms that have been done up to three more times as survival procedures. The incision is closed with interrupted, monofilament 3-0 nylon sutures. The catheter can be left in place, heparinized and used to continuously monitor blood pressure for nonsurvival procedures if desired.

Craniectomy and Clot Placement

Mannitol, 0.5 g/kg is administered intravenously and anesthetic depth monitored to make sure pin placement and craniectomy can be safely performed. The head is fixed in a 3-point pin fixation device and the right scalp prepared with povidone-iodine solution and then draped in sterile fashion (Fig. 3). The head position is key to being able to see down the fissure as well as along the anterior cerebral artery. Ventilation is adjusted to lower the endtidal PCO_2 to 25–35 mmHg. Mono- and bipolar cautery and suction are readied as well as all of the equipment including patties cut into small strips 5 mm wide and 2 cm long and 2–3 mm balls of bone wax.

Different incisions are possible but the point to keep in mind is that the temporalis muscle is very thick and the purpose of this operation is to get exposure to retract the brain as much as possible and put as much clot next to the arteries as can be done. Instead of a standard pterional, frontotemporal incision, we tend to use a reverse C incision based posteriorly (Fig. 3). The skin is incised and then monopolar cautery is used to incise the temporalis muscle along the same incision. The muscle is reflected posteriorly. The total blood volume in these animals is about 75 mL/kg so excess blood loss has to be avoided. Pay attention to the deep temporal arteries that traverse up under the zygomatic arch deep to the deep temporal fascia. They need to be coagulated. The incision courses along the top of the zygomatic arch and then superiorly along the orbital rim for about 2 cm before turning posteriorly. A cuff of fascia is left attached to the bone for resuturing but all the muscle has to be removed from behind the orbit. The muscle is covered with moist gauze and held posteriorly. We place it under the self-retaining retractor blade that is mounted on the head holder at this stage. The muscle under the deep temporal fascia usually is excised with the fascia and the fascia can be used as a dural patch.

The operating microscope is brought in. A craniotomy is made in the shape of a kidney, about 2 cm long, centered over the pterion and extending to the floor of the anterior and close to the floor of the middle cranial fossa. Take care not to push the dura away too much as extradural bleeding is difficult to control and perfect hemostasis is best. Also the brain underneath is like the most fragile, mushy neonatal

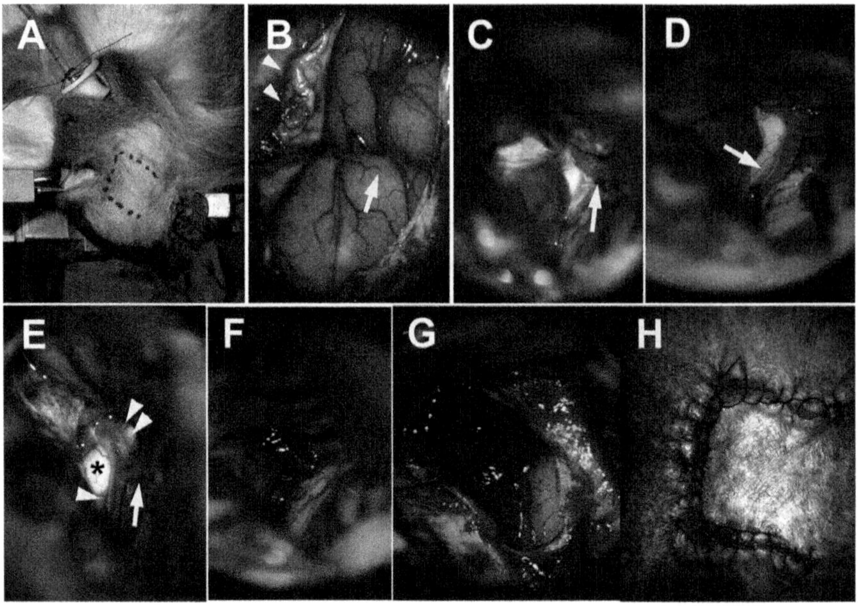

Fig. 3 Step-by-step procedures for creation of SAH. (**a**) The head position in a 3-point pin fixation device viewed from directly above the monkey. The incision is marked. (**b**) The craniectomy has been performed and the dura opened (arrow heads show free edge of dura next to the sphenoid ridge. The arrow shows the Sylvian fissure with the temporal lobe above and the frontal lobe inferior. (**c**) A retractor has been placed to retract the temporal lobe posteriorly. The arrow shows the middle cerebral artery exposed by opening of the arachnoid down to the carotid bifurcation. (**d**) After complete opening of the arachnoid the anterior cerebral artery is well seen (arrow) as the frontal lobe is elevated with a microdissector. (**e**) The intradural segment of the internal carotid (double arrow heads, the upper arrow head is over the posterior communicating artery), the precommunicating segment of the anterior cerebral (single arrow head) and the sphenoidal segment of the middle cerebral artery (arrow) are exposed. The optic nerve (asterisk) is relatively larger than the arteries in this species compared to in humans. (**f**) The clotted autologous blood has begun to be placed against the arteries of the right side of the circle of Willis. (**g**) Clot placement is complete. (**h**) The wound was closed in layers and in this case with a running monofilament 3-0 nylon suture

brain ever and will bruise easily by compression through the dura. The sphenoid ridge is relatively flat compared to the human, but should be ronguered or drilled away for several mm. Bleeding from the diploic space can be marked from the inferior part of the craniectomy where the bone thickens, and is controlled with bone wax.

The dura mater is opened with a bent, 25 gauge needle used a hook and a scalpel blade (Fig. 3). The opening is completed in a C-shape based on the sphenoid ridge and the dura is held anteriorly with a 6-0 suture. In some cases, a large venous sinus will be seen in the temporal dura. This heralds the presence of a draining vein from the Sylvian fissure to the dura of the sphenoid ridge. The temporal lobe should not be retracted or the vein will be avulsed and one will have to struggle to control the bleeding. Bridging veins should be looked for, coagulated and cut.

Next, we place strips of patty 5 mm wide by about 2 cm long on the frontal and temporal lobes. Early deep retraction contuses the brain and causes intracerebral hematomas so we start the dissection laterally in the Sylvian fissure. The arachnoid is first incised at the lateral end of the fissure and craniectomy. We use an arachnoid knife and Jeweler's forceps. The brain is softer than toothpaste and overly aggressive suction will disrupt the pia and brain instantly. Under high magnification, the frontal and temporal operculae are separated and the distal middle cerebral artery exposed. The initial exposure will usually be beyond the bifurcation. Cerebrospinal fluid (CSF) has to be suctioned and small veins bridging across the fissure are coagulated. As the dissection progresses deeper, two fine vascular forceps can be used to pull apart the arachnoid and tease out the middle cerebral artery. At this point, enough CSF will have been suctioned to relax the brain. A 5 × 5 mm patty segment is put distally between the frontal and temporal operculae and the frontal lobe is elevated with the retractor. The arachnoid is opened widely down over the intradural segment of the internal carotid, the precommunicating segment of anterior cerebral and the sphenoidal segment of the middle cerebral artery. It usually is possible to follow the posterior communicating artery back to the membrane of Lilliequist. The retractor is repositioned to retract the temporal lobe posteriorly.

The clotted autologous blood is poured out of the red-top tubes onto a sterile gauze and cut into small pieces. It is weighed and then placed piece-meal along the exposed cerebral arteries of the right side of the circle of Willis. Pieces can be put under the middle cerebral artery where it is free of perforators and can be lifted off the brain. The retractor and patties are removed as more clot is packed in.

Once as much clot as possible is packed in, the dura is closed with interrupted 6-0 sutures. The temporalis fascia is closed with interrupted absorbable 3-0 sutures and the skin with interrupted monofilament sutures (Fig. 3).

Postoperative Treatment and Euthanasia

The isoflurane is turned off and the animal ventilated and then extubated when awake. Buprenorphine 10–30 μg/kg is given intramuscularly at this time and then repeated every 12 h for 48 h and then after that as needed. Enrofloxacin, 2.5 mg/kg, intramuscularly is continued twice daily for 24 h unless there are signs of infection in which case it is continued. The animals are returned to their cages where they are fed their usual diet of biscuits, fruit, and water. They are observed every 12 h and all monitoring and interventions recorded in the animal's records. If they are ill, monitoring may be more frequent as determined by severity of illness.

After 7 days or whatever time is scheduled for the experiment, the monkey is placed under general anesthesia again and cerebral angiography is repeated. Euthanasia is then done according to what endpoints are needed. Perfusion through the left ventricle with phosphate-buffered saline (PBS) followed by paraformaldehyde is used for histologic endpoints, and with ice-cold PBS for molecular endpoints or isometric tension studies.

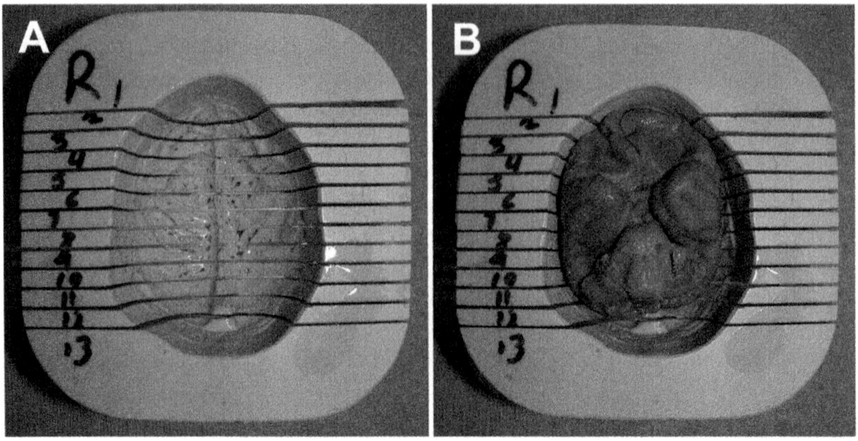

Fig. 4 (**a**) A custom-built brain mold. (**b**) Fixed monkey brain in the mold. The brain is smaller than the mold due to fixation; the fresh brain fills the mold

The brain is removed by reflecting the scalp and removing the cranium with an autopsy saw. The brain is immersed in ice-cold PBS and the cerebral arteries and clot removed. The remaining clot can be weighed. Dissect the cerebral arteries (internal carotid, precommunicating segments of the anterior cerebral, sphenoidal segment of middle cerebral, basilar artery) free under the operating microscope and place in ice-cold Krebs-Henseleit buffer. The brain can be placed in a custom-built brain mold (Fig. 4) and sectioned into 12–13 coronal sections of equal thickness.

Data Analysis

Angiography

To assess arterial diameters from cerebral angiograms (Fig. 5), an optical micrometer can be used to measure diameters at predetermined points along the extradural and intradural internal carotid, anterior and middle cerebral and basilar arteries [29, 43, 44]. When we use digital subtraction angiography, the images are saved electronically and can be measured using edge-detection programs written in Igor and NiDAQ Tools 1.5 (Wavemetrics, Lake Oswego, OR) although the key aspect is determining where in the penumbral area of the dye column on the angiogram that the edge of the artery will be defined, whether it is done electronically or visually. Thus, the reader should familiarize themselves with the literature on this subject [45]. Measurements should be conducted by an investigator without knowledge of the monkey's group and in random order. Vasospasm can be assessed by comparing angiograms taken before and at some time after induction of SAH and treated as a continuous variable or categorized as absent (0–10% reduction in diameter), mild

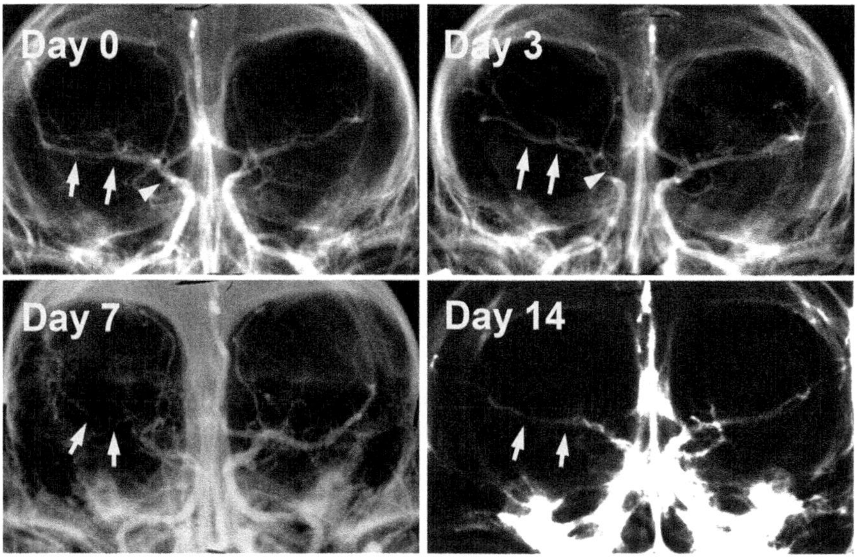

Fig. 5 Representative anteroposterior angiograms from baseline before SAH (**a**) and then 3 (**b**), 7 (**c**) and 14 (**d**) days after SAH showing development and resolution of right-sided vasospasm (arrows point to right middle cerebral artery and arrow head to right internal carotid artery)

(10–25% reduction), moderate (25–50% reduction) or severe (>50% reduction). One also can compare within animals at each time the SAH-side and control-side arteries.

Additional Endpoints

These are limited only by the imagination and needs of the investigator. The authors have used isometric tension studies [46], arterial protein and messenger ribonucleic acid levels [47], second messenger concentrations [48], immunohistochemistry [49], morphometric measurements [50] and studied the brain and blood clot [43]. More detailed studies of the brain itself should be considered, given the importance of DCI, as opposed to just angiographic vasospasm, to outcome after SAH. We have not used behaviour assessment yet.

Applications, Complications and Limitations

The senior author is perhaps the person with the largest experience in the world with this model. This is a technically-demanding procedure where errors are punished by surgical fatality. For craniotomy and SAH, we recently have conducted 129 unilateral SAH, 58 bilateral SAH and 29 clot removals on 187 monkeys. There was unexpected early death in 19 (10%), early illness necessitating planned euthanasia in 8 (4%) and morbidity in 13 (7%). Overall total morbidity and mortality was 41/187 = 22%. Neurological deficits and seizures develop in less than 6% of animals and infection develops in less than 5% of animals.

The model is suitable for a vast array of studies such as intracranial drug delivery by Ommaya reservoir, continuous infusion or biodegradable implants [51, 52], continuous intravenous drug delivery [44] and measurement of intracranial pressure and cerebral blood flow.

Normally no elaborate catheterization techniques are needed but occasionally the anatomy of the aortic arch makes visualization of both carotids impossible by injection of the brachial artery so a transfemoral approach is necessary. This is always an option but we have found the brachial approach simple and fast. The femoral artery can be exposed and catheterized with a 4F angiography catheter that is advanced into the aortic arch or selectively into each common carotid artery. The risks of this procedure are that the iliac artery can become spastic and make removal of the catheter difficult or impossible.

References

1. Fujii M, Yan J, Rolland WB, Soejima Y, Caner B, Zhang JH. Early brain injury, an evolving frontier in subarachnoid hemorrhage research. Transl Stroke Res. 2013;4:432–46.
2. Macdonald RL. Delayed neurological deterioration after subarachnoid haemorrhage. Nat Rev Neurol. 2014;10:44–58.
3. Dreier JP. The role of spreading depression, spreading depolarization and spreading ischemia in neurological disease. Nat Med. 2011;17:439–47.
4. Stein SC, Browne KD, Chen XH, Smith DH, Graham DI. Thromboembolism and delayed cerebral ischemia after subarachnoid hemorrhage: an autopsy study. Neurosurgery. 2006;59:781–7.
5. Budohoski KP, Czosnyka M, Kirkpatrick PJ, Smielewski P, Steiner LA, Pickard JD. Clinical relevance of cerebral autoregulation following subarachnoid haemorrhage. Nat Rev Neurol. 2013;9:152–63.
6. Ostergaard L, Aamand R, Karabegovic S, et al. The role of the microcirculation in delayed cerebral ischemia and chronic degenerative changes after subarachnoid hemorrhage. J Cereb Blood Flow Metab. 2013;33:1825–37.
7. Zoerle T, Ilodigwe DC, Wan H, et al. Pharmacologic reduction of angiographic vasospasm in experimental subarachnoid hemorrhage: systematic review and meta-analysis. J Cereb Blood Flow Metab. 2012;32:1645–58.
8. Yagi K, Lidington D, Wan H, et al. Therapeutically targeting tumor necrosis factor-alpha/sphingosine-1-phosphate signaling corrects myogenic reactivity in subarachnoid hemorrhage. Stroke. 2015;46:2260–70.
9. Motterlini R, Gonzales A, Foresti R, Clark JE, Green CJ, Winslow RM. Heme oxygenase-1-derived carbon monoxide contributes to the suppression of acute hypertensive responses in vivo. Circ Res. 1998;83:568–77.
10. Edvinsson L, Povlsen GK. Late cerebral ischaemia after subarachnoid haemorrhage: is cerebrovascular receptor upregulation the mechanism behind? Acta Physiol (Oxf). 2011;203:209–24.
11. Sakadzic S, Lee J, Boas DA, Ayata C. High-resolution in vivo optical imaging of stroke injury and repair. Brain Res. 2015;1623:174–92.

12. Macdonald RL, Weir B. Cerebral vasospasm, vol. 2001. San Diego: Academic; 2001.
13. Marbacher S, Fandino J, Kitchen ND. Standard intracranial in vivo animal models of delayed cerebral vasospasm. Br J Neurosurg. 2010;24:415–34.
14. Schuller K, Buhler D, Plesnila N. A murine model of subarachnoid hemorrhage. J Vis Exp. 2013;(81):e50845.
15. Veelken JA, Laing RJ, Jakubowski J. The sheffield model of subarachnoid hemorrhage in rats. Stroke. 1995;26:1279–83.
16. Bederson JB, Germano IM, Guarino L. Cortical blood flow and cerebral perfusion pressure in a new noncraniotomy model of subarachnoid hemorrhage in the rat. Stroke. 1995;26:1086–91.
17. Megyesi JF, Vollrath B, Cook DA, Findlay JM. In vivo animal models of cerebral vasospasm: a review. Neurosurgery. 2000;46:448–60.
18. Weir B, Erasmo R, Miller J, McIntyre J, Secord D, Mielke B. Vasospasm in response to repeated subarachnoid hemorrhages in the monkey. J Neurosurg. 1970;33:395–406.
19. Espinosa F, Weir B, Shnitka T, Overton T, Boisvert D. A randomized placebo-controlled double-blind trial of nimodipine after SAH in monkeys. Part 2: pathological findings. J Neurosurg. 1984;60:1176–85.
20. Fathi AR, Bakhtian KD, Marbacher S, Fandino J, Pluta RM. Blood clot placement model of subarachnoid hemorrhage in non-human primates. Acta Neurochir Suppl. 2015;120:343–6.
21. Pluta RM, Bacher J, Skopets B, Hoffmann V. A non-human primate model of aneurismal subarachnoid hemorrhage (SAH). Transl Stroke Res. 2014;5:681–91.
22. Titova E, Ostrowski RP, Zhang JH, Tang J. Experimental models of subarachnoid hemorrhage for studies of cerebral vasospasm. Neurol Res. 2009;31:568–81.
23. Macdonald RL, Zhang J, Sima B, Johns L. Papaverine-sensitive vasospasm and arterial contractility and compliance after subarachnoid hemorrhage in dogs. Neurosurgery. 1995;37:962–7.
24. Rothberg CS, Weir B, Overton TR. Treatment of subarachnoid hemorrhage with sodium nitroprusside and phenylephrine: an experimental study. Neurosurgery. 1979;5:588–95.
25. Espinosa F, Weir B, Overton T, Castor W, Grace M, Boisvert D. A randomized placebo-controlled double-blind trial of nimodipine after SAH in monkeys. Part 1: clinical and radiological findings. J Neurosurg. 1984;60:1167–75.
26. Steinke DE, Weir BK, Findlay JM, Tanabe T, Grace M, Krushelnycky BW. A trial of the 21-aminosteroid u74006f in a primate model of chronic cerebral vasospasm. Neurosurgery. 1989;24:179–86.
27. Findlay JM, Weir BK, Gordon P, Grace M, Baughman R. Safety and efficacy of intrathecal thrombolytic therapy in a primate model of cerebral vasospasm. Neurosurgery. 1989;24:491–8.
28. Fathi AR, Pluta RM, Bakhtian KD, Qi M, Lonser RR. Reversal of cerebral vasospasm via intravenous sodium nitrite after subarachnoid hemorrhage in primates. J Neurosurg. 2011;115:1213–20.
29. Hino A, Weir BK, Macdonald RL, Thisted RA, Kim CJ, Johns LM. Prospective, randomized, double-blind trial of bq-123 and bosentan for prevention of vasospasm following subarachnoid hemorrhage in monkeys. J Neurosurg. 1995;83:503–9.
30. Findlay JM, Kassell NF, Weir BK, et al. A randomized trial of intraoperative, intracisternal tissue plasminogen activator for the prevention of vasospasm. Neurosurgery. 1995;37:168–76.
31. Macdonald RL, Kakarieka A, Mayer SA, et al. Prevention of cerebral vasospasm after aneurysmal subarachnoid hemorrhage with clazosentan, an endothelin receptor antagonist. Neurosurgery. 2006;59:453. (Abstract).
32. Oldfield EH, Loomba JJ, Monteith SJ, et al. Safety and pharmacokinetics of sodium nitrite in patients with subarachnoid hemorrhage: a phase IIa study. J Neurosurg. 2013;119:634–41.
33. Fisher M, Feuerstein G, Howells DW, et al. Update of the stroke therapy academic industry roundtable preclinical recommendations. Stroke. 2009;40:2244–50.
34. Sasaki E. Prospects for genetically modified non-human primate models, including the common marmoset. Neurosci Res. 2015;93:110–5.
35. van der Worp HB, Howells DW, Sena ES, et al. Can animal models of disease reliably inform human studies? PLoS Med. 2010;7:e1000245.

36. Capitanio JP, Emborg ME. Contributions of non-human primates to neuroscience research. Lancet. 2008;371:1126–35.
37. (STAIR) TSTAIRT. Recommendations for standards regarding preclinical neuroprotective and restorative drug development. Stroke. 1999;30:2752–8.
38. Eberle R, Hilliard J. The simian herpesviruses. Infect Agents Dis. 1995;4:55–70.
39. DeMarcus TA, Tipple MA, Ostrowski SR. Us policy for disease control among imported nonhuman primates. J Infect Dis. 1999;179(Suppl 1):S281–S2.
40. Sughrue ME, Mocco J, Mack WJ, et al. Bioethical considerations in translational research: primate stroke. Am J Bioeth. 2009;9:3–12.
41. Lapchak PA, Zhang JH, Noble-Haeusslein LJ. Rigor guidelines: escalating stair and steps for effective translational research. Transl Stroke Res. 2013;4:279–85.
42. Kilkenny C, Browne W, Cuthill IC, Emerson M, Altman DG. Animal research: reporting in vivo experiments—the arrive guidelines. J Cereb Blood Flow Metab. 2011;31:991–3.
43. Zhang X, Hintze TH. Camp signal transduction cascade, a novel pathway for the regulation of endothelial nitric oxide production in coronary blood vessels. Arterioscler Thromb Vasc Biol. 2001;21:797–803.
44. Macdonald RL, Curry DJ, Aihara Y, Zhang ZD, Jahromi BS, Yassari R. Magnesium and experimental vasospasm. J Neurosurg. 2004;100:106–10.
45. Elisevich K, Cunningham IA, Assis L. Size estimation and magnification error in radiographic imaging: implications for classification of arteriovenous malformations. AJNR Am J Neuroradiol. 1995;16:531–8.
46. Macdonald RL, Zhang J, Marton LS, Weir B. Effects of cell-permeant calcium chelators on contractility in monkey basilar artery. J Neurotrauma. 1999;16:37–47.
47. Hino A, Tokuyama Y, Kobayashi M, et al. Increased expression of endothelin B receptor mRNA following subarachnoid hemorrhage in monkeys. J Cereb Blood Flow Metab. 1996;16:688–97.
48. Aihara Y, Jahromi BS, Yassari R, Sayama T, Macdonald RL. Effects of a nitric oxide donor on and correlation of changes in cyclic nucleotide levels with experimental vasospasm. Neurosurgery. 2003;52:661–7.
49. Ono S, Zhang ZD, Marton LS, et al. Heme oxygenase-1 and ferritin are increased in cerebral arteries after subarachnoid hemorrhage in monkeys. J Cereb Blood Flow Metab. 2000;20:1066–76.
50. Macdonald RL, Weir BK, Grace MG, Martin TP, Doi M, Cook DA. Morphometric analysis of monkey cerebral arteries exposed in vivo to whole blood, oxyhemoglobin, methemoglobin, and bilirubin. Blood Vessels. 1991;28:498–510.
51. Cook DJ, Kan S, Ai J, Kasuya H, Macdonald RL. Cisternal sustained release dihydropyridines for subarachnoid hemorrhage. Curr Neurovasc Res. 2012;9:139–48.
52. Macdonald RL, Zhang ZD, Curry D, et al. Intracisternal sodium nitroprusside fails to prevent vasospasm in nonhuman primates. Neurosurgery. 2002;51:761–8.

Blood Injection Intracerebral Hemorrhage Rat Model

Thomas O'Lynnger, Shanshan Mao, Ya Hua, and Guohua Xi

Abstract The autologous blood injection intracerebral hemorrhage (ICH) model has been used successfully in rats due to the fact that they are large enough to accommodate all of the procedures and protocols and that measurement of pathophysiological parameters in the rat brain can be extrapolated to humans. In this chapter, we present a step-by-step procedure to model intracerebral hemorrhage by injecting autologous whole blood. This model can be readily modified to investigate the effects of infusing different compounds other than autologous whole blood.

Keywords Intracerebral hemorrhage · Brain edema · Blood-brain barrier permeability · Brain atrophy

Model Selection

Blood injection is a suitable model for studying intracerebral hemorrhage. Autologous blood obtained from the femoral artery is infused into the right caudate nucleus to simulate hemorrhage (Fig. 1). Advantages of this model include:

1. In vivo observation of the pathological changes of intracerebral hemorrhage and the results of therapeutic manipulations;
2. A relative homogeneity within strains owing to inbreeding;
3. A small brain size well suited to immunohistochemical and biochemical studies;
4. The ability to study a variety of physiologic parameters that are altered by the hematoma, including intracranial pressure, cerebral blood flow, cerebral perfusion pressure, and neurological function;
5. Ease of reproducibility and rarity of complications;

T. O'Lynnger · S. Mao · Y. Hua · G. Xi (✉)
Department of Neurosurgery, University of Michigan, Ann Arbor, MI, USA
e-mail: guohuaxi@umich.edu

J. Chen et al. (eds.), *Animal Models of Acute Neurological Injury*, Springer Series in Translational Stroke Research, https://doi.org/10.1007/978-3-030-16082-1_18

Fig. 1 Rat positioned in a stereotactic frame prepared for injection in the right caudate. The scalp is incised to expose the area to be injected, and a 1 mm cranial burr hole is drilled to create an entry point for the needle

6. The ability to measure blood-brain permeability, brain edema, and brain atrophy and extrapolate those measurements to the human situation;
7. Modification of the procedure to test the neurotoxic and neuroprotective effects of a virtually unlimited number of infused compounds.

Materials

Rats weighing 250–300 g are used in this model of intracerebral hemorrhage. The materials recommended for the surgical procedure include:

1. Kopf stereotactic frame
2. Infusion pump (Harvard)
3. Dremel drill with 1 mm drill bit
4. Moria spring scissors with very sharp points
5. Scalpel handle with scalpel blades
6. Fine iris scissors
7. Curved Mayo scissors
8. Dumont forceps
9. Dumont angled forceps
10. Adson forceps with teeth
11. Heiss retractor
12. Hemostats
13. Needle holder
14. Cotton tipped applicators
15. 1 mL syringes
16. Polyethylene tubing (PE-50)
17. 23-gauge and 26-gauge needles
18. 5-0 black braided silk (Deknatel)
19. 3-0 Dermalon suture

Procedure

1. Inject sodium pentobarbital (50 mg/kg, i.p.) to anesthetize the rat.
2. Apply ophthalmic ointment to both eyes.
3. Use a heating pad with feedback regulator to maintain body temperature at 37 °C.
4. Use a hair clipper to remove hair from the scalp and right groin regions and apply Betadine to the exposed skin.
5. Make a 2 cm incision in the groin at the region of the right femoral artery.
6. Use the curved Mayo scissors to cut away fat to expose the artery, being careful to avoid cutting any major blood vessels.
7. Expose the femoral artery and vein using cotton tipped applicators and fine iris scissors to cut away fascia.
8. Separate 1.5 cm of the femoral artery from the femoral vein using angled forceps and a cotton tipped applicator, being careful not to tear the artery.
9. Ligate the distal end of the artery using 5-0 black braided silk and clip one end of the silk with a hemostat and place to the side to help expose the artery.
10. Place a thread of silk under the proximal end of the artery and clip both ends with a hemostat to temporarily occlude blood flow.
11. Using spring scissors, prick the artery to create an opening for the catheter without severing the blood vessel.
12. Make sure the PE-50 catheter is trimmed to a point and attached to a 23-gauge needle and 1 mL syringe filled with saline.
13. Insert the tubing into the opening with the pointed side down using two pairs of forceps to hold the tubing and the artery in place.
14. Use a loop of silk to cinch the tubing in place.
15. Inject a small amount of saline to ensure that blood is being pumped into the tubing.
16. Exchange the saline filled syringe for an empty syringe and release the tension on the proximal occluding thread to fill the syringe with 0.2 mL of blood.
17. Using another syringe that has been filled with heparin and then emptied completely, withdraw a 0.2 mL blood sample to analyze glucose, hematocrit and blood gases.
18. After sampling the blood and measuring blood pressure if appropriate, put the tension back on the proximal end of the artery, then gently slide the PE-50 out enough so that the loop originally cinching the tubing to the artery slides off, then tighten it to ligate the artery.
19. Remove the silk on the proximal end and cut all the loose ends of silk, then suture the skin using two to three 3-0 Dermalon sutures.
20. Place the rat into the Kopf stereotactic frame and make a 2 cm sagittal incision over the midline of the scalp.
21. Using cotton tipped applicators, clear away the fascia to expose the bregma, then position the Heiss retractor under the skin to keep the area exposed.

22. Measure 4.0 mm laterally to the right and 0.1 mm anterior to the bregma and mark the area for the cranial burr hole, making sure to stay posterior to the coronal suture.
23. Holding the Dremel drill with 1 mm drill bit vertically over the mark, slowly drill through the cranium, stopping when the pressure is no longer felt in order avoid tearing the dura.
24. Place the 0.2 mL sample of non-heparinized blood in the infusion pump with a 26-gauge needle with the beveled edge pointing away from the midline.
25. Position the needle over the burr hole and insert the tip of the needle into the brain just until the beveled edge is no longer visible, then insert the needle 5.5 mm ventrally into the right caudate.
26. Use the infusion pump to infuse 100 μL of autologous whole blood into the right basal ganglia at a rate of 10 μL/min, leaving the needle in place for 2 min once the infusion is complete.
27. Retract the needle and fill the burr hole with bone wax.
28. Suture the skin using two to three 3-0 Dermalon sutures.

Outcome Evaluations

1. Magnetic resonance imaging: It is an intuitive way to evaluate brain injury by measuring brain swelling, T2 lesion and T2* lesion [1, 2].
2. Brain water content: To examine brain edema formation, which is a major determinant of survival in patients with ICH. It is also a simple, reproducible, and inexpensive measure of brain injury [3–5].
3. Blood-brain barrier (BBB) permeability: The commonly used assessors of BBB disruption are radioisotopes and Evans blue [6, 7]. Albumin leakage is used to evaluate BBB disruption as well.
4. Brain genomics: To determine changes of gene expression [8].
5. Histology: To evaluate the size, location, extent of the hemorrhage [9], neuronal death [10, 11], and brain atrophy [11, 12], coronal sections should be prepared (Figs. 2 and 3).
6. Behavioral testing: Forelimb placing test, the forelimb use asymmetry test, and the corner turn test are sensitive measures of neurological deficits induced by intracerebral hemorrhage [9].

Limitations

1. Lack of a ruptured blood vessel as the source of the ICH.
2. Lacks the hematoma enlargement that occurs in ICH patients.
3. Very low mortality rate.

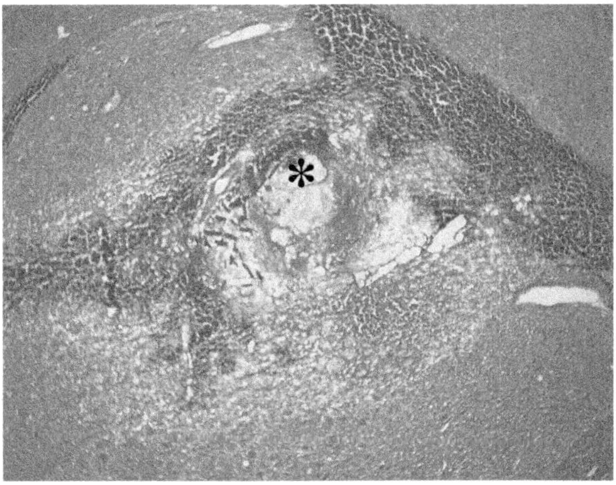

Fig. 2 H&E staining of a coronal slice of the right caudate 3 days after injection of whole blood. Asterisk—hematoma

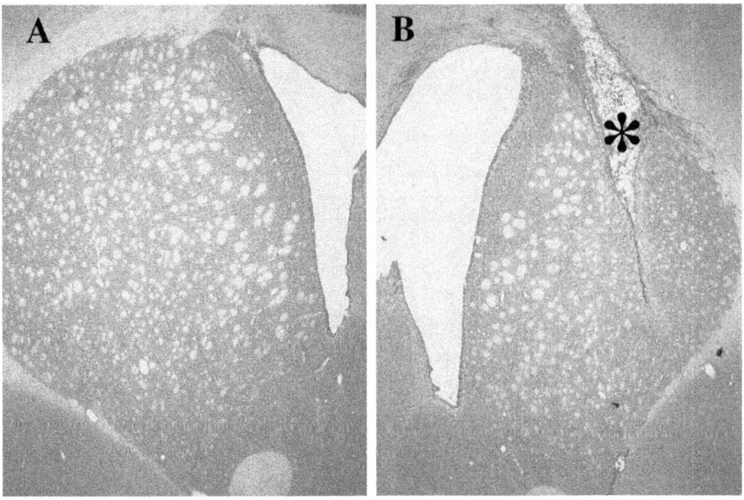

Fig. 3 Brain atrophy after ICH in rats. Coronal brain sections ((**a**) the contralateral caudate; (**b**) the ipsilateral caudate) 2 month after ICH. Note enlargement of ventricle and atrophy of striatum in the ipsilateral side. Asterisk—hematoma cavity

References

1. Wu G, Xi G, Hua Y, Sagher O. T2* magnetic resonance imaging sequences reflect brain tissue iron deposition following intracerebral hemorrhage. Transl Stroke Res. 2010;1:31–4.
2. Dang G, Yang Y, Wu G, Hua Y, Keep RF, Xi G. Early erythrolysis in the hematoma after experimental intracerebral hemorrhage. Transl Stroke Res. 2017;8:174–82.
3. Yang GY, Betz AL, Chenevert TL, Brunberg JA, Hoff JT. Experimental intracerebral hemorrhage: relationship between brain edema, blood flow, and blood-brain barrier permeability in rats. J Neurosurg. 1994;81:93–102.

4. Xi G, Wanger KR, Hua Y, deCourten-Myers GM, Myers RE. Plasma but not packed red blood cells infused in pig cerebral white matter induces rapidly developing, marked edema. Soc Neurosci Abstr. 1996;22:1424.
5. Xi G, Keep RF, Hoff JT. Erythrocytes and delayed brain edema formation following intracerebral hemorrhage in rats. J Neurosurg. 1998;89:991–6.
6. Xi G, Hua Y, Bhasin RR, Ennis SR, Keep RF, Hoff JT. Mechanisms of edema formation after intracerebral hemorrhage: effects of extravasated red blood cells on blood flow and blood-brain barrier integrity. Stroke. 2001;32:2932–8.
7. Belayev L, Saul I, Busto R, Danielyan K, Vigdorchik A, Khoutorova L, et al. Albumin treatment reduces neurological deficit and protects blood-brain barrier integrity after acute intracortical hematoma in the rat. Stroke. 2005;36:326–31.
8. Lu A, Tang Y, Ran R, Ardizzone TD, Wagner KR, Sharp FR. Brain genomics of intracerebral hemorrhage. J Cereb Blood Flow Metab. 2006;26:230–52.
9. Hua Y, Schallert T, Keep RF, Wu J, Hoff JT, Xi G. Behavioral tests after intracerebral hemorrhage in the rat. Stroke. 2002;33:2478–84.
10. Wu J, Hua Y, Keep RF, Schallert T, Hoff JT, Xi G. Oxidative brain injury from extravasated erythrocytes after intracerebral hemorrhage. Brain Res. 2002;953:45–52.
11. Felberg RA, Grotta JC, Shirzadi AL, Strong R, Narayana P, Hill-Felberg SJ, et al. Cell death in experimental intracerebral hemorrhage: the "black hole" model of hemorrhagic damage. Ann Neurol. 2002;51:517–24.
12. Hua Y, Nakamura T, Keep RF, Wu J, Schallert T, Hoff JT, et al. Long-term effects of experimental intracerebral hemorrhage: the role of iron. J Neurosurg. 2006;104:305–12.

Collagenase Model of Intracerebral Hemorrhage in Rats

Anatol Manaenko, Derek Nowrangi, Reis Cesar, Jiping Tang, and John H. Zhang

Abstract Intracerebral hemorrhage is the most devastating stroke subtype with a high mortality rate. Currently there are no clinically approved therapeutic strategies decreasing mortality and improving quality of life for the ICH patients. Preclinical models mimicking human pathophysiology of intracerebral hemorrhage are necessary to evaluate the potential effectiveness of new therapeutic approaches. Rodents are the most commonly used species for modeling stroke, including hemorrhagic stroke. This chapter focuses on describing the "collagenase" model of intracerebral hemorrhage on rats.

Keywords Intracerebral hemorrhage · Stroke · Animal model · Rat · Collagenase

Introduction

Currently, intracerebral hemorrhage (ICH) has the highest mortality rate of all stroke subtypes. With the incidence of 10–30 per 100,000 individuals annually, it accounts for 10–15% of stokes worldwide, and is one of the leading causes of stroke related mortality and morbidity worldwide [1, 2]. Depending on the underlying cause of bleeding, one can differentiate between primary and secondary ICH. While secondary ICH is associated with trauma, vascular abnormalities, tumors or impaired coagulation, it affects a relatively small number of the patients. The most common

Electronic supplementary material: The online version of this chapter (https://doi.org/10.1007/978-3-030-16082-1_19) contains supplementary material, which is available to authorized users.

A. Manaenko · D. Nowrangi · J. Tang · J. H. Zhang (✉)
Department of Physiology and Pharmacology, Loma Linda University School of Medicine, Loma Linda, CA, USA
e-mail: jhzhang@llu.edu

R. Cesar
Departments of Anesthesiology and Neurosurgery, Loma Linda University School of Medicine, Loma Linda, CA, USA

J. Chen et al. (eds.), *Animal Models of Acute Neurological Injury*, Springer Series in Translational Stroke Research, https://doi.org/10.1007/978-3-030-16082-1_19

ICH originates from the spontaneous rupture of small vessels damaged by chronic hypertension or amyloid angiopathy and accounts for 78–88% of ICH cases [3].

Brain injury after ICH can be distinguished as primary or secondary. Despite differences in the origin of bleeding, both primary and secondary intracerebral hemorrhage have similar underlying pathological changes [4]. Primary brain injury occurs as an inevitable, mass effect of the growing hematoma in the affected tissue. Secondary brain injury is caused by toxic blood products released by a hematoma [2] and are more often the target of modern therapeutic approaches.

Typical locations for hypertensive ICH include the basal ganglia, thalamus, cerebellum, pontine tegmentum, and deep lobar white matter [5]. Whereas striatal bleeding is the most common location for humans, in this chapter we describe the animal model mimicking the bleeding into the basal ganglia.

There are two popular animal models of ICH. While the "blood" model is defined by the volume of blood that will be infused into the brain region of interest, the "collagenase" model employs an injection of bacterial collagenase in order to produce a hematoma. Collagenases are proteolytic enzymes present intracellularly in an inactive form. Authors of most ICH studies use collagenase produced by bacteria *Clostridium histolyticum,* which is able to degrade both native and denatured collagen. Collagen is an insoluble and fibrous protein and functions as a major component of extracellular connective tissue. It is the most abundant protein in mammals, making up anywhere from 25 to 35% of the whole-body protein. In addition, collagen is known to be present in the basal lamina of cerebral blood vessels. Injection of bacterial collagenase into the brain breaks down the basal lamina of blood vessels and multifocal damage occurs. Bleeding occurs from in situ vessels, where the blood can first be seen around 10 min following the intracerebral injection and develops slowly into a full hematoma at about 4–24 h later [6]. Re-bleeding and the hematoma expansion following initial bleeding mimic the re-bleeding phenomenon, which is observed in more the 20% of human patients [7].

The collagenase model in rats was originally described by Rosenberg in 1990 [6] and the technique has received only minor modifications since then. Here we will describe the technical details, limitations, and provide video material about the model.

Materials

Major Surgical Equipment (Fig. 1)

Stereotactic Frame (Fig. 1a)

The ICH will be induced by injection of collagenase under stereotactic guidance. Therefore, a stereotactic frame suitable for rat surgery (similar to those mentioned below) will be needed.

(a) http://www.stoeltingco.com/neuroscience/stereotaxic/rat/stoeltings-classic-lab-standard-stereotaxic-instrument-185.html
(b) http://www.bilaney.com/kopf/dkicatalog/900.html

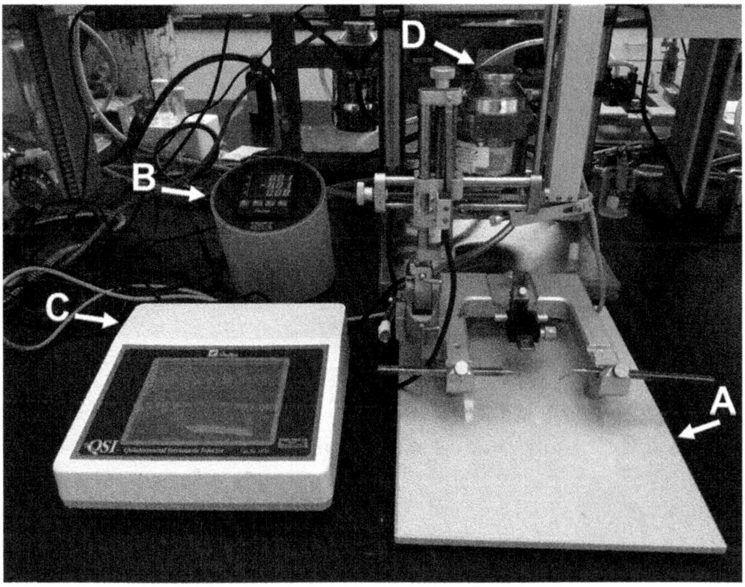

Fig. 1 Major surgical equipment needed to perform ICH surgery

Digital Equipment for Stereotactic Surgery (Fig. 1b)

Using digital stereotactic frame equipment, similar to that mentioned in the hyperlink below, is recommended and makes the surgery significantly easier.

http://www.stoeltingco.com/neuroscience/stereotaxic/rat-mouse/stoeltings-digital-new-standard-stereotaxic-instrument.html

However, there is an alternate and less costly option. The mechanical coordinates system, similar to a Vernier Caliper, is integrated into the stereotactic frame. Vernier Calipers are suitable for fine measurements and precise enough for this particular kind of surgery. The information about how to use a Vernier Caliber can be found here:

https://www.miniphysics.com/how-to-read-a-vernier-caliper.html

Infusion Micro Pump (Fig. 1c)

In order to minimize the mass effect of the injected collagenase, it is necessary to inject a small volume of collagenase (not more the 1 μL) at a slow rate (0.2 μL/min). A programmable infusion pump (syringe pump) similar to that mentioned in the hyperlink below, is an essential part of the equipment.

http://www.harvardapparatus.com/webapp/wcs/stores/servlet/haisku2_10001_11051_67471_-1_HAI_ProductDetail__37313_37314

Induction and Recovery Chambers with Heating Blanket and Feedback Rodent Surgical Probe

Small rodents, both rats and mice, are unable to control their body temperature under anesthesia, and undercooling has significant neuroprotective effects. Therefore, it is strongly recommended to place the animals in a pre-heated induction chamber immediately after initiation of the anesthesia. It is also necessary to use a rodent surgical heating pad with feed-back rectal thermal probe throughout the entire operation.

http://www.stoeltingco.com/neuroscience/surgical-supplies-62/accessories/homeothermic-blanket-system-1747.html

A Micro-Drill with 1 mm Diameter Drill Bits (Fig. 2a)

During the surgery, a small burr hole must be made in the skull so that the syringe can be lowered into the basal ganglia to infuse the collagenase. Using a micro-drill kit such as the one listed below is needed.

http://www.cellpointscientific.com/Products/Ideal-Micro-Drill/67-1200

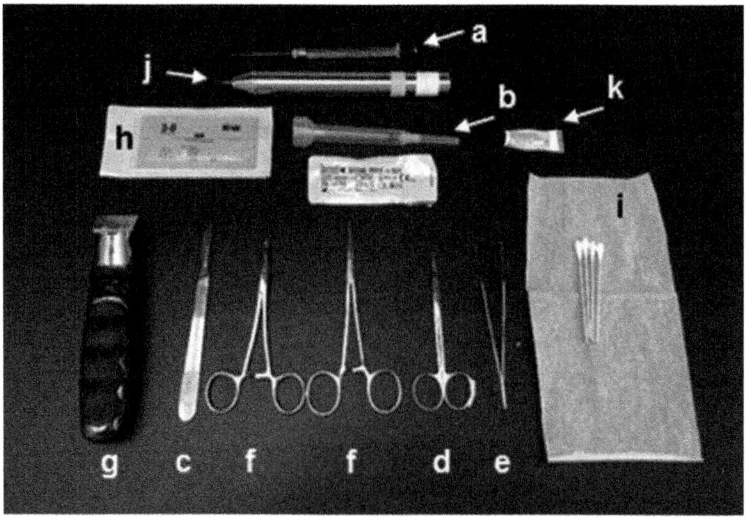

Fig. 2 Minor surgical equipment needed to perform ICH surgery

Equipment for the Inhalation Anesthesia (Optional) (Fig. 1d)

The surgery can be performed under both injection and inhalation of anesthesia. For inhalation anesthesia the equipment for initiation and maintenance of the anesthesia is needed.

Minor Surgical Equipment (Fig. 2)

(a) 10 μL Hamilton syringe with 26 gauge needle
(b) 1 mL 26 gauge syringes
(c) scalpel blade # 10
(d) 4-1/2 inch iris scissors
(e) 6 inch straight forceps
(f) Needle drivers
(g) Electric shaver
(h) 3-0 silk sutures
(i) Cotton-tipped applicators
(j) A micro-drill with 1 mm diameter drill bits
(k) Styptic powder
(l) Restraint cones or container (optional)
(m) Bone wax (optional)

Collagenase

In order to avoid the degradation and inactivation of the collagenase, all preparation steps must be carried out on ice and all solutions should be pre-cooled and used ice-cold. Add sterile ice-cold PBS to generate the concentration of 0.2 U of collagenase in a 1 μL. Aliquot the solution into 0.2 or 0.5 mL micro-centrifuge tubes at 10–15 μL per tube and store in a −20 °C freezer quickly. For the surgery, take one tube from the freezer directly before injection. Use one tube per animal.

Reagents

Atropine

An atropine injection will be given before anesthesia in order to decrease mucus secretions, such as saliva. Use 0.2 mg per kg of the body mass 10 min before induction of anesthesia.

Anesthetic Cocktail of Ketamine/Xylazine

Before starting surgeries, we recommend calculating the total volume of the anesthesia needed for all animals being operated on that day. Ketamine will be given at 100 mg per kg of body mass and Xylazine at 10 mg per kg of body mass. A cocktail mix of Ketamine and Xylazine can be made in a sterile container and aliquoted according to the body weight of the animals.

Betadine

Betadine will be used for sterilization of the operative window and surgical tools.

Ophthalmic Lubricant

An ophthalmic lubricant will be used to avoid the drying the eyes during the surgery.

Saline

Due to decreased activity of animals after surgery there is a significant risk of dehydration. In order to avoid dehydration, 1 mL of pre-warmed saline (37 °C) will be deposit subcutaneously immediately after surgery.

Buprenorphine (Optional)

Pain belongs to the post-stroke clinical picture. However, it is strongly recommended to administer an anti-pain medication (Buprenorphine) immediately after surgery. This should also be given when animals demonstrate pain, as indicated by their behavior.

Methods

1. Before surgery animals will be weighed and 0.2 mg/kg of atropine will be injected subcutaneously (s.c.).
2. Ten minutes after injection, the calculated volume of the Ketamine/Xylazine cocktail (1.5 mL/kg) will be injected intraperitoneally (i.p.). For the injection of the anesthesia cocktail, it is possible to use a restraining cone (see photo). However, since the positioning of animals in the restrainer can be difficult to control, and injecting at the wrong position can result in the damage of inner

organs or in insufficiently deep anesthesia, we recommend restraining the animal by-hand, preferably carried out by two people (see video).

Briefly: avoid abrupt movement or noises, which can place stress on the animal. Place the rat on a hard and flat surface and hold it carefully by the tail. The animal will intuitively try to escape. Do not apply pressure on the tail, and place your index and middle finger on its shoulders and your palm on the back of animal. While avoiding squeezing, carefully wrap your fingers around the neck. Keep holding the tail and flip the animal so that its stomach is revealed. The second person memorizes the position of knees. Next, hold the rear feet and gently stretch the animal, holding the head lower than its body.

The second person draws the imaginary line between the original positions of knees before stretching. The midline of the animal will now be clearly visible. The needle will be inserted along the imaginary line on the animal's right side and close to the midline with an angle of approximately 30° toward the abdomen (Fig. 3). Not following this rule can result in the damage of muscle tissue and, even more, the anesthesia might not get injected into the peritoneal cavity. This, in turn, can result in an insufficiently deep anesthesia or in an unusual long duration before the anesthetic takes effect.

Always aspirate the needle (pull the plunger back) before injection. If blood or other body liquid is visible in the syringe, immediately remove the needle and try again with a new needle and solution. Push the plunger until the drug is fully administered. Place the animal in the pre-warmed chamber, protecting it from any kind of stimulating irritations. After the animal stops moving,

Fig. 3 Correct needle position for intraperitoneal injection

check its muscle tone. If animal reacts to touching and tone is present, return the animal to the chamber. Once the muscles are relaxed, test its reflexes by toe-pinching. Lack of reflexes is a sign of deep anesthesia and that the surgery may continue.

3. During the entire surgical procedure, observe the animal closely for heart and breathing rate. An increase of breathing rate, slight movement of vibrissae or tail, or movement of eyes are clear signals of anesthesia subsiding and an additional injection is needed. The additional anesthesia (50% of the initial amount) will be administrated sc.
4. Insert the rectal temperature probe (approximately 2 cm) and place the rat on the heating pad, programmed to maintain a core body temperature around 37 °C.
5. Apply betadine on the head of animal to disinfect the surgical site and make shaving of head easier.
6. Shave the head to reveal the scalp.
7. Disinfect the skin with alcohol.
8. Use a scalpel to make a vertical midline incision on the scalp. Using cotton-tipped applicators, remove soft tissue by blunt dissection until the bregma and lambda regions on the skull are clearly visible. These two locations will be used as reference points for the surgery. If bleeding occurs, use styptic powder to stop the bleeding completely. Even small amounts of blood can make using the reference points difficult and will cause inconsistencies in the results.
9. Moisten the eyes with ophthalmic lubricant to prevent drying during surgery. Move the animal to the stereotactic frame and position in the prone position.
10. Open the mouth with a forceps and clear the airway by pulling its tongue to the side gently. Put the front teeth into the opening of the stereotactic frame mouthpiece, and then gently pull back the tail to secure the teeth in this position.
11. Insert one ear bar into the ear until you feel resistance from the skull. Level the animal's head using the mouthpiece and gently insert the second ear bar in the ear until the fixation is completed. Apply only a moderate amount of pressure when adjusting the ear bars to avoid injuring the animal. Using the wooden end of a cotton-tipped applicator, gently tap the rat on the skull to ensure that the head is fixed and does not move. Further spread the surgical window and ensure both bregma and lambda are visible.
12. Fit the Hamilton syringe on the injector and lower the needle until it touches bregma (Fig. 4). Memorize the coordinates and repeat the procedure until needle tip touches lambda. If the vertical difference between bregma and lambda is within 0.1 mm of each other, the head is leveled. If not, release the ear bars and re-level the head.
13. Move the needle to the target position (0.2 mm anterior and 2.9 mm lateral to the bregma) and mark the position below the needle tip. Using the micro-drill, thin the skull over the target area by horizontal movement of the spinning drill bit. Apply only slight pressure downward. Once the bone is very thin, stop drilling.

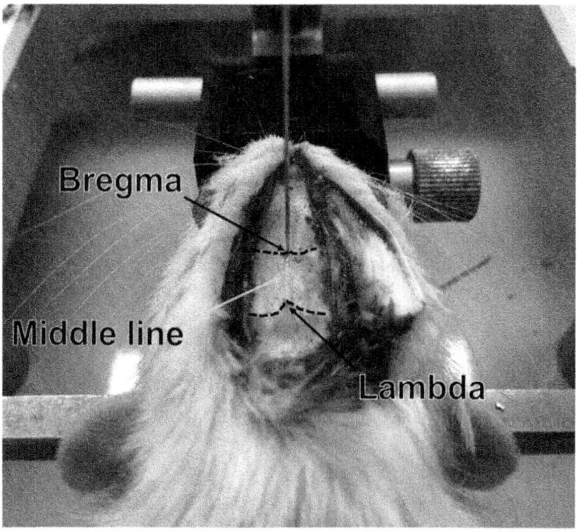

Fig. 4 Position of the needle on the bregma

14. Clean the Hamilton syringe with distilled water and eject the water from the syringe. Thaw the collagenase and flush the syringe with it at least five times. When filling the syringe, ensure a constant column of collagenase without air bubbles in the syringe. Fit the syringe on the injector again and arrange the bevel towards the middle line (if you inject collagenase on the right side, the bevel should be arrange to the left) and using the high-speed mode of the pump, eject the collagenase to make sure there are no air bubbles in the needle as well. Remove the collagenase drop from the top of the needle using the wooden end of a cotton-tipped applicator.
15. Program the infusion pump to deliver 1.0 μL at a rate of 0.2 μL/min. Move the syringe back to the drilled burr hole. Move the needle tip to the vertical reference point, which is the position when the needle bevel is flush with the top of the skull (Fig. 5).
16. Advance the needle to the depth of 5.6 mm. Make sure that the needle moves straight down the hole. If the drilled hole is not straight but at a slant, the flexible needle hits the bone wall and moves along it. Deflection of the needle going down can cause significant changes and variations from the desired coordinates.
17. Inject collagenase and wait 10 min after completion in order to avoid the back flow of collagenase. Remove needle slowly (1 mm/min) and apply bone wax immediately after until hole is completely closed. Loosen the ear bars and release animal from the frame. Close the wound and inject saline (1 mL) into the flank of the hind limb. Because of animal size, injection of 1 mL of cold saline can change the body temperature. It is recommended to pre-warm the saline to 37 °C before injection. Inject buprenorphine to reduce pain of animal.

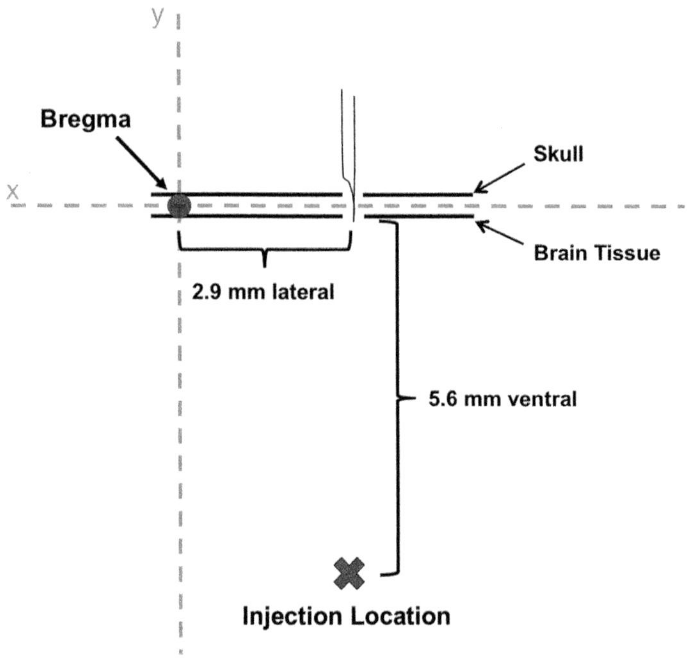

Fig. 5 The bevel arrangement. The bevel is to pose towards the middle line. When the bevel is hidden by skull, this is the reference point "0" for "Z" axis

18. Transfer the animal to the recovery chamber. Once the rat is awake and moving, it can be placed into the original cage. Keep observing the animal and apply an additional dose of buprenorphine as needed.

Sham Operated Animals

In order to exclude influence of surgery related factors such as needle trauma, postoperative pain, effects of anesthesia, etc. on the results, usage of sham-operated animals is strongly recommended. The sham operation will be carried out as described above for the ICH surgery. However, instead of injecting collagenase, 1 μL of sterile saline will be injected. All other methods will be followed the same.

Pitfalls

The coordinates mentioned above can vary between experimenters. We therefore recommend adjusting the coordinates as needed so that the location of the hemorrhage remains consistent. To this end surgical training should be carried out as

Fig. 6 A successful induction of ICH. Photos was taken 24 h after collagenase injection

described above. One to 2 h after injection of the collagenase, animals will be euthanized, and the brain removed and cut through the needle trail, which can be clearly seen at this time point on the brain surface. At this time point, the hematoma is small and its location can be easily seen. Real coordinates of the injection point can then be calculated according to the rat brain atlas, and the injection coordinates can be adjusted according to this information. An example of the correct hematoma location are presented in Fig. 6.

Outcomes

Brain Water Content

The extent of brain edema following the collagenase injection will be evaluated by calculation of brain water content via the wet/dry method. To this end animals will be euthanized. The brain will be harvested and a tissue block of 4 mm (2 mm in front and 2 mm behind of injection point) will be sectioned. The tissue will be divided into right and left hemisphere. For each hemisphere, the basal ganglia and cortex will be separated [8, 9]. Each part of the brain will be weighed immediately with the results representing the wet weight (WW). Brain samples will be dried in the oven at 100 °C for 24 h and weighed again to obtain the dry weight (DW). The brain water content will be calculated and presented as a percentage using

$$Brain\ Water\ Content\left(\%\right)=\left[\frac{WW-DW}{WW}\right]*100$$

No spreading of brain edema into the cerebellum has been observed in this model. Therefore, the cerebellum water content will be used as an internal control [10–12].

It is also possible to use the whole hemispheres without separating the cortex and basal ganglia as some authors did before [13, 14]. However, the most significant elevation of brain water content can be observed in the ipsilateral basal ganglia. Contamination of the basal ganglia with the cortex tissue decreases sensitivity of the assay.

Neurobehavioral Testing

Neurobehavioral testing is a useful tool that provides necessary information for brain injury studies. There are several rules that help us to retrieve and interpret this information correctly. After surgery, animals exhibit neurological deficits shortly after anesthesia subsides. Abnormal behavior shortly after surgery may stem from post-operative stress and pain, long lasting effects of anesthesia, or brain trauma caused by the needle insertion. Even more, sham operated animals sometimes exhibit abnormal behavior shortly after surgery, thereby making correct interpretations of the results at earlier time points challenging. It is advisable to wait 24 h before evaluating the effects of brain injury or performing treatment on neurological functions. Even at 24 h, the tester should observe animals closely before testing. Animals that exhibit signs of pain or distress often do not cooperate. Their behavior is atypical and may not be related to the brain injury.

Noise or bright light can distract animals and affect results of testing. A separate room used only for animal testing is advisable. People not directly involved in the testing are not welcome into the testing area. Phones, alarms etc. must be switched off. If no animal testing room is available, animals should be brought into the testing area for accommodation at least 1 h before testing. As rats use their nose for a great deal of positioning, dirty and smelly equipment can significantly affect the behavior of animals. Make sure the apparatus is cleaned after every animal.

The most extensively used neurobehavioral assessment after ICH include forelimb placing, corner turning, and modified Garcia tests. The effectiveness of these tests in the "collagenase" model of ICH was described previously [15–17]. The tests are also effective in the ICH 'blood' model on rats and in mice models of ICH [18, 19]. Using these tests, neurological dysfunction can be detected starting at 24 h and monitored up to 10 weeks after ICH induction [15, 18, 20].

There are several significant advantages to these neurobehavior tests. For instance, there is no need for any special apparatus. The tests do not require the pre-training of animals. However, non-operated animals can demonstrate a preference for one limb more than the other. We recommend performing a pre-test on the animals before ICH induction and exclude those showing preference for one limb more than the other.

Forelimb Placement Test

The forelimb placement test is used to evaluate the ability of the animal to respond by stretching their forelimbs to ipsilateral vibrissae stimulation. Unilateral brain damage impairs the motor function of the contralateral side. The animals after ICH will demonstrate normal placing of the limb on the ipsilateral side, but the limb contralateral to the injury location will show impairment in stretching and placement of the limb on the surface after sensation.

Method

Hold the animal gently but firmly by the torso. Allow forelimbs to hang free. Give the animal adequate time. Once its muscles are relaxed and the animal does not move or show resistance, move the animal vertically along the table surface, brushing the vibrissae. Repeat ten times on the contralateral and ipsilateral side and record the number of successful placements on each side. We recommend alternating between contralateral and ipsilateral sides. Absence of reaction on both sides can be a sign of wrong handling of animals. In this case we recommend returning animal back to the cage and performing a re-test in 30 min.

Corner Turning

Similar to the forelimb placing, the corner turning test is used to evaluate body asymmetry, which is a consequence of the unilateral brain injury. Animals in the test prefer an ipsilateral turn due to impairment of limbs on the contralateral side.

Method

The test will be performed using a 30° corner. Animals will be placed between the walls and allowed to advance into the corner. Once animal realizes there is no way through, it will try to change direction. In order to do so, it will place its forepaws on the wall and turn around by shifting its weight. Record the direction the animal turns. Trials should be repeated 10–20 times with up to 30 s being allowed between trials. The results will be presented as a percentage of right turns (RT) divided by the total sum of right and left turns (LT).

$$Corner\ Turn\ Score(\%) = \left[\frac{RT}{RT + LT}\right] * 100$$

In order to encourage animals to advance or to turn, it is acceptable to tap animal gently on the tail or to move the corner slowly towards the animal.

Modified Garcia Exam

The Modified Garcia Test consists of six sub-tests. Each sub-test is scored out of 3 points. Animals without neurological deficits get a maximum of 3 points for each sub-test, totaling 18 points. The sub-tests are as follows:

(a) "Spontaneous Activity" test—Animal will be placed in an empty cage and allowed to explore the area. Record the number of attempts the animal makes to raise itself on its hind limbs to explore the walls of the cage. The test duration is 3 min. Animals which approach 3–4 walls and raise themselves on their hind limbs will score 3 points. Animals which approach 1–2 walls only will be scored 2 points. Animals that demonstrate only slow movement without approaching any walls get 1 point while animals unable to move receive 0 points.

Pitfalls: Spontaneous activity could be significantly affected by the environment (noise, light etc.).

(b) "Side stroking" test—Animals will be lightly touched on the side of their body with a cotton-tipped applicator. Reactions of the animal, such as looking towards the direction of the stimuli, escaping the stroke, changes in vibrissae movements, or changes to breathing frequency will be monitored. Animals with a brisk bilateral response will be scored with 3 points. Animals with a weak bilateral or with a brisk ipsilateral or weak contralateral response will be scored with 2 point. Unilateral responses to the stroke will score 1 point and animals without a reflexive response receive 0 points.

Before testing operated animals it is advisable to test healthy rats and evaluate which intensity of stroking is sufficient to cause animal reactions.

(c) "Vibrissae touch" test—This test will be performed similar to the "side stroking" test. Vibrissae will be stimulated with a cotton-tipped applicator. Avoid letting the animals see the cotton-tipped applicator being brought to them to allow a reaction based on touching only. The scoring system is the same as the "Side stroking" test.

Pitfalls: Animals have different sensitivity on the tips and bottoms of the vibrissae. Be consistent and touch tip of the vibrissae in all animal testing.

(d) "Limb symmetry" test—Animals will be carefully held by the tail and lifted directly above a flat surface so that it is immediately out of reach for the rat. The rat will attempt to reach out to the table using its limbs the symmetry of the limbs will be observed. Animals extending both fore-and hindlimbs will be scored with 3 points. Animals with medium flexion of the forelimb and/or the hindlimb will get 2 points. Animals with a full flexion of the contralateral forelimb or hindlimb will receive 1 point. Animals with a full flexion of the contralateral forelimb and hindlimb will receive 0 points.

Pitfalls: The test can be affected by lateral turning of the body. Place the animal close to surface (floor or table). This will stimulate the animal to stretch its body. Watch limb symmetry in the moment of body stretching.

(e) "Lateral turning" test—Animals will be carefully held by their tail and lifted above the surface. The ability of the animal to turn their body will be monitored.

Animals which are able to turn to both side more than 45° get 3 points.

If animals are able to turn to both sides but less than 45° get 2 points. Animals that display an unequal ability to turn to either side are given 1 point. Those that lack the ability to turn either direction get 0 points.

Pitfalls: You might need to touch the animal on its side with a cotton-tipped applicator to stimulate them to turn. Please note, animals with severe injuries will get exhausted very quickly and stop turning. This should not be interpreted as "unequal turning" or "no turning". Stimulate animal on one side and watch its response. Give the rat a couple of minutes to rest. Hold it by the tail again and immediately start stimulation on the opposite side from previously tested.

(f) "Climbing" test—Animals will be placed in the middle of a 60° apparatus and checked for their ability to climb upwards towards the top. The strength of grip will be evaluated by allowing the animal to grip the edge of the apparatus and gently pulling on the tail of the animal. Quality of grip (if it is weak or strong) can be evaluated by comparing the strengths of the animal before and after brain injury. The duration of the test is 1 min. Animals which will climb on the top of the apparatus and which have a strong grip will be scored with 3 points. Animals that climb to the top of the apparatus and have weak grip or do not climb on the top but have a strong grip will receive 2 points. Animals trying to climb but move laterally or downwards will get 1 point. Animals with a weak grip that are unable to climb or fall down get 0 points.

Pitfalls: Animals with a mild brain injury or without injury can be active and move downwards while they explore the area. This can be mistakenly described as "moving laterally or downwards."

The quality of grip being strong or weak is dependent on comparing the strength between an uninjured and injured animal. Therefore, without experience it is difficult to assess the grip quality when only testing one animal.

Advantages and Disadvantages of the Model

Compared to another popular model of ICH, the "blood" model, collagenase model is technically easier to perform. With this model there is little to no risk of the blood back flow during or after injection. There is also no risk of blood coagulation in the syringe before or during the injection. In the blood injection model, blood is placed into the syringe and then injected into the brain. However, the longer the blood stays in the syringe, the greater possibility that the blood will coagulate and the amount of blood injected reduced due to a blockage. These factors significantly decrease the consistency of a hematoma production in the "blood" model of ICH. There is more control of the hematoma volume in the "collagenase" model which depends on the volume of injected collagenase [21]. In addition, the collagenase model is the only model that mimics the spontaneous disruption of blood vessels and enables researchers to investigate systemic effects (such as blood lowering or coagulation dependency) on the development of secondary brain injury after ICH.

The collagenase model has also displayed more severe brain damage. Measured blood volume, detected in the brain of rats after performing the "blood" and "collagenase" model, observed that the "collagenase" model sustained a larger lesion, greater BBB disruption, and consequently resulted in more significant neurological deficits without spontaneous resolution [20]. This makes the model suitable for evaluation of long-term effects of the therapeutic strategies.

A significant disadvantage of the collagenase model is the mechanism of the hematoma formation. While in ICH patients a solid hematoma is a result of bleeding from an arterial source in brain tissue, bleeding in the collagenase model is diffuse and results from rupture of small blood vessels around the site of the collagenase injection. Another critical point of the model, often discussed in the literature, is the neurotoxicity of collagenase resulting in apoptosis and atypical inflammatory reactions caused by the collagenase itself. It is, however, important to note that collagenase in doses comparable to those used in animals did not cause apoptosis in primary neuronal cultures [22]. Moreover, the presence of apoptotic cells was demonstrated in the blood model as well [22–24].

There is also indication that atypical inflammation in blood model on rats is more present then in collagenase model. Atypical post-ICH inflammation is a consequence of hemoglobin crystallization. Unlike the collagenase model, the blood model showed hemoglobin crystals appearing as early as 5 h after ICH induction. 24 h after ICH, four times more crystals are present in blood compared collagenase model. Fourth-eight hours after ICH, eight times more crystals are present in blood model compared to collagenase injection. Consequently, significantly more infiltrated neutrophils and activated microglia are observed in blood injection model, indicating that collagenase model is suitable to study post-ICH inflammatory reactions [25].

References

1. Qureshi AI, Suri MF, Nasar A, et al. Changes in cost and outcome among US patients with stroke hospitalized in 1990 to 1991 and those hospitalized in 2000 to 2001. Stroke. 2007;38(7):2180–4.
2. Qureshi AI, Mendelow AD, Hanley DF. Intracerebral haemorrhage. Lancet. 2009;373(9675):1632–44.
3. Qureshi AI, Tuhrim S, Broderick JP, Batjer HH, Hondo H, Hanley DF. Spontaneous intracerebral hemorrhage. N Engl J Med. 2001;344(19):1450–60.
4. Steiner T, Rosand J, Diringer M. Intracerebral hemorrhage associated with oral anticoagulant therapy: current practices and unresolved questions. Stroke. 2006;37(1):256–62.
5. Elijovich L, Patel PV, Hemphill JC III. Intracerebral hemorrhage. Semin Neurol. 2008;28(5):657–67.
6. Rosenberg GA, Mun-Bryce S, Wesley M, Kornfeld M. Collagenase-induced intracerebral hemorrhage in rats. Stroke. 1990;21(5):801–7.
7. Kazui S, Naritomi H, Yamamoto H, Sawada T, Yamaguchi T. Enlargement of spontaneous intracerebral hemorrhage. Incidence and time course. Stroke. 1996;27(10):1783–7.
8. Sang LY, Liang YX, Li Y, et al. A self-assembling nanomaterial reduces acute brain injury and enhances functional recovery in a rat model of intracerebral hemorrhage. Nanomedicine. 2015;11(3):611–20.

9. Zhao X, Sun G, Zhang J, Ting SM, Gonzales N, Aronowski J. Dimethyl fumarate protects brain from damage produced by intracerebral hemorrhage by mechanism involving Nrf2. Stroke. 2015;46(7):1923–8.
10. Manaenko A, Fathali N, Chen H, et al. Heat shock protein 70 upregulation by geldanamycin reduces brain injury in a mouse model of intracerebral hemorrhage. Neurochem Int. 2010;57(7):844–50.
11. Manaenko A, Fathali N, Khatibi NH, et al. Arginine-vasopressin V1a receptor inhibition improves neurologic outcomes following an intracerebral hemorrhagic brain injury. Neurochem Int. 2011;58(4):542–8.
12. Manaenko A, Lekic T, Ma Q, Zhang JH, Tang J. Hydrogen inhalation ameliorated mast cell-mediated brain injury after intracerebral hemorrhage in mice. Crit Care Med. 2013;41(5):1266–75.
13. Chen M, Li X, Zhang X, et al. The inhibitory effect of mesenchymal stem cell on blood-brain barrier disruption following intracerebral hemorrhage in rats: contribution of TSG-6. J Neuroinflammation. 2015;12:61.
14. Zhang Y, Yi B, Ma J, et al. Quercetin promotes neuronal and behavioral recovery by suppressing inflammatory response and apoptosis in a rat model of intracerebral hemorrhage. Neurochem Res. 2015;40(1):195–203.
15. Hartman RE, Rojas H, Tang J, Zhang J. Long-term behavioral characterization of a rat model of intracerebral hemorrhage. Acta Neurochir Suppl. 2008;105:125–6.
16. Geng X, Ren C, Wang T, et al. Effect of remote ischemic postconditioning on an intracerebral hemorrhage stroke model in rats. Neurol Res. 2012;34(2):143–8.
17. Wu G, Bao X, Xi G, Keep RF, Thompson BG, Hua Y. Brain injury after intracerebral hemorrhage in spontaneously hypertensive rats. J Neurosurg. 2011 Jun;114(6):1805–11.
18. Hua Y, Schallert T, Keep RF, Wu J, Hoff JT, Xi G. Behavioral tests after intracerebral hemorrhage in the rat. Stroke. 2002 Oct;33(10):2478–84.
19. Nakamura T, Xi G, Hua Y, Schallert T, Hoff JT, Keep RF. Intracerebral hemorrhage in mice: model characterization and application for genetically modified mice. J Cereb Blood Flow Metab. 2004 May;24(5):487–94.
20. MacLellan CL, Silasi G, Poon CC, et al. Intracerebral hemorrhage models in rat: comparing collagenase to blood infusion. J Cereb Blood Flow Metab. 2008;28(3):516–25.
21. Lekic T, Tang J, Zhang JH. Rat model of intracerebellar hemorrhage. Acta Neurochir Suppl. 2008;105:131–4.
22. Matsushita K, Meng W, Wang X, et al. Evidence for apoptosis after intercerebral hemorrhage in rat striatum. J Cereb Blood Flow Metab. 2000;20(2):396–404.
23. Karwacki Z, Kowianski P, Dziewatkowski J, et al. Apoptosis in the course of experimental intracerebral haemorrhage in the rat. Folia Morphol (Warsz). 2005;64(4):248–52.
24. Sun H, Li L, Zhou F, et al. The member of high temperature requirement family HtrA2 participates in neuronal apoptosis after intracerebral hemorrhage in adult rats. J Mol Histol. 2013;44(4):369–79.
25. Kleinig TJ, Helps SC, Ghabriel MN, et al. Hemoglobin crystals: a pro-inflammatory potential confounder of rat experimental intracerebral hemorrhage. Brain Res. 2009;1287:164–72.

A Rat Model of Cerebellar Hemorrhage Using Bacterial Collagenase Injection

Devin W. McBride, Tim Lekic, Jiping Tang, and John H. Zhang

Abstract Cerebellar hemorrhage causes unique functional deficits, including ataxic, sensorimotor, and memory deficiencies. Despite the occurrence of CH, most studies have focused on intracerebral hemorrhage. Yet, the unique behavioral deficits of CH necessitates attention for the development of treatments. This chapter details the bacterial collagenase injection model for creating cerebellar hemorrhage. This model is unique in that it leads to reproducible postural deficits caused by a well-defined hemorrhage within the cerebellum. Cerebellar hemorrhage leads to acute increases in brain water content and acute sensorimotor and postural deficits, as well as long-term sensorimotor and memory deficits due to cerebellar tissue atrophy. Herein, the surgical procedures, evaluation of outcomes, caveats, limitations, and alternatives are described.

Keywords Cerebellum · Collagenase · Stroke · Cerebellar hemorrhage · CH · Rat

Electronic supplementary material: The online version of this chapter (https://doi.org/10.1007/978-3-030-16082-1_20) contains supplementary material, which is available to authorized users.

D. W. McBride · T. Lekic · J. Tang
Department of Physiology and Pharmacology, Loma Linda University School of Medicine, Loma Linda, CA, USA

J. H. Zhang (✉)
Department of Physiology and Pharmacology, Loma Linda University School of Medicine, Loma Linda, CA, USA

Department of Neurosurgery, Loma Linda University School of Medicine, Loma Linda, CA, USA

Department of Anesthesiology, Loma Linda University School of Medicine, Loma Linda, CA, USA
e-mail: jhzhang@llu.edu

J. Chen et al. (eds.), *Animal Models of Acute Neurological Injury*, Springer Series in Translational Stroke Research, https://doi.org/10.1007/978-3-030-16082-1_20

Introduction

Intracerebral hemorrhage accounts for approximately 15% of all strokes [1], with between 5 and 10% occurring within the cerebellum [2]. Cerebellar hemorrhage (CH) results in a mortality rate of 20–30% [3], and cerebellar damage causes unique functional deficits and is involved in a number cognitive processes, including motor coordination, balance, and memory [4]. Despite the prevalence of CH, the vast majority of studies have been on understanding the pathophysiology of and therapies for intracerebral hemorrhage [5–12]. A total of five articles have been focused on understanding CH using animal models, two studies using autologous blood injection to create a hematoma [13, 14] and three studies using an injection of bacterial collagenase to create a hematoma [15–17]. The unique behavioral consequences of CH, requires attention for the development of therapies. Herein, a rat model of cerebellar hemorrhage is described, including the caveats, limitations, and alternatives.

Materials

Adult male Sprague-Dawley rats weighing between 290 and 395 g (Harlan, Indianapolis, IN, USA) were used for this model. The following equipment was utilized to complete all surgical procedures:

1. Anesthesia gas blender with dual flow meter tubes and isoflurane vaporizer (Whitemore Enterprises), isoflurane, compressed oxygen (99.7%), compressed medical gas
2. Rodent induction chamber (BrainTree Scientific, Braintree, MA)
3. Rodent stereotactic frame (Stoelting Co., Wood Dale, IL)
4. Standard surgical equipment: scalpel, fine forceps, skin retractors (Fine Science Tools, Foster City, CA)
5. Surgical microscope and fiber optical illuminator (Kent Scientific, Torrington, CT)
6. Hair clippers and skin disinfectant: alcohol prep pads, betadine
7. Cotton tipped applicator or absorbent surgical spears
8. Microdrill with burr (Ideal Micro Drill, CellPoint Scientific, Gaithersburg, MD)
9. Infusion pump (Harvard Apparatus, Holliston, MA, USA) and 1 μL gauge needle (Hamilton Company, Reno NV)
10. Bone wax (Lukens Bone Wax, Medline Industries, Mundelein, IL)
11. 5-0 Suture silk for wound closure
12. Bacterial collagenase (0.2 U/μL saline, VII-S, Sigma, St. Louis, MO)

Procedures

All procedures were approved by the Loma Linda University IACUC. Rats were given access to food and water ad libitum.

Collagenase Preparation

1. On ice, dissolve bacterial collagenase in 3 mL ice cold, sterile PBS (1 M) to a final concentration of 0.2 U/μL on ice
2. Aliquot 20 μL of stock solution into tubes for storage
3. Store the collagenase solutions in −20 °C until ready for use

Surgery

4. Place a rat into an anesthesia induction chamber and expose it to 5% isoflurane delivered in a mixture of oxygen (0.3 L/min) and medical gas (0.7 L/min) for 2–3 min.
5. Upon loss of the paw pinch reflex, place the rat prone onto a standard rodent stereotactic frame under a stereo microscope. Anesthesia is maintained using 2.5% isoflurane administered in a mixture of oxygen (0.3 L/min) and medical gas (0.7 L/min). A thermostat-controlled warming blanket is used to maintain a core temperature of 37.0 ± 0.5 °C.
6. Shave the rat's head and disinfect the skin with an alcohol pad and betadine.
7. A midline incision is made through the skin and connective tissue using a scalpel.
8. The periosteum is separated from the skull to expose the right fissure of the coronal suture and proximally.
9. A burr hole, 1 mm diameter, is created over the right cerebellar hemisphere using a microdrill and burr. Stereotactic coordinates: 11.64 mm posterior and 2.4 mm right lateral to the bregma.
10. A 27 gauge needle is inserted through the burr hole into the brain at a rate of 1 mm/min to a depth of 3.5 mm (Fig. 1).
11. Using a microinfusion pump and Hamilton syringe, 3 μL of 0.2 U/μL bacterial collagenase is infused at a rate of 0.2 μL/min.
12. After complete infusion, the needle is left in place for 10 min to minimize back-flow.
13. The needle is removed at a rate of 1 mm/min.
14. Bone wax is used to seal the burr hole.
15. The incision is sutured and animal allowed to recover.

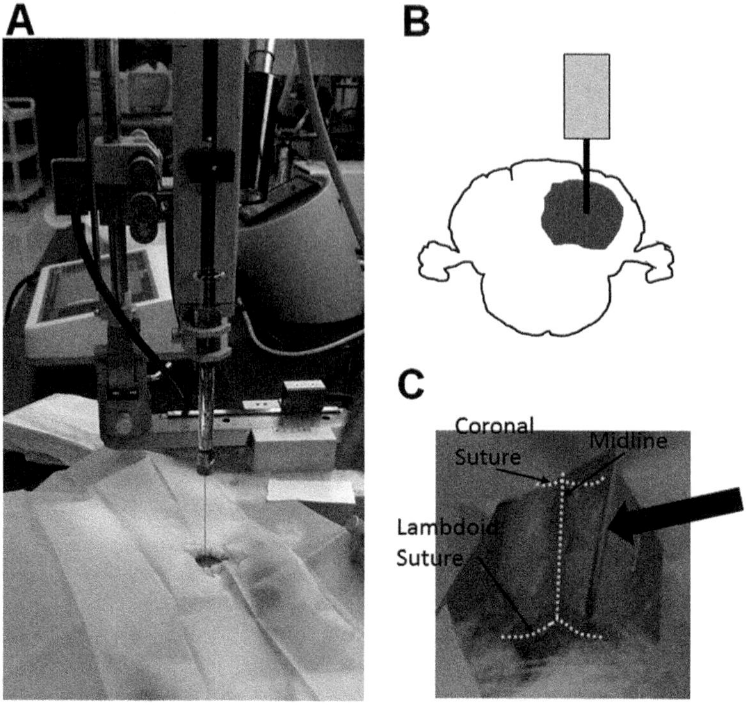

Fig. 1 Cerebellar hemorrhage model. (**a**) A representative image of surgical area. (**b**) A schematic of the injection site for collagenase to induce CH. Stereotactic coordinates: 11.64 mm posterior and 2.4 mm right lateral to the bregma, 3.5 mm deep. (**c**) A representative image of the injection site for CH. Arrow indicates the needle. Sutures are indicated by yellow dashed lines. Blue circle outlines the burr hole

Outcome Evaluations and Anticipated Results

The cerebellar hemorrhage (CH) rat model allows for a number of physiological outcomes to be studied: hematoma volume [15–17], brain edema [16, 17], blood-brain barrier disruption [16, 17], and neuroinflammation. Additionally, while this model presents with the typical sensorimotor unilateral behavioral deficits, animals with CH also display unique neurological deficits [15–17].

Physiological Outcomes

Bacterial collagenase infusion into the right cerebellum results in significant hematoma formation (Fig. 2). Noninvasive imaging by MRI can be used to clearly delineate the hematoma size (Fig. 3). This is especially important, but not required, when animals are to be used for long-term analysis of disease progression and functional

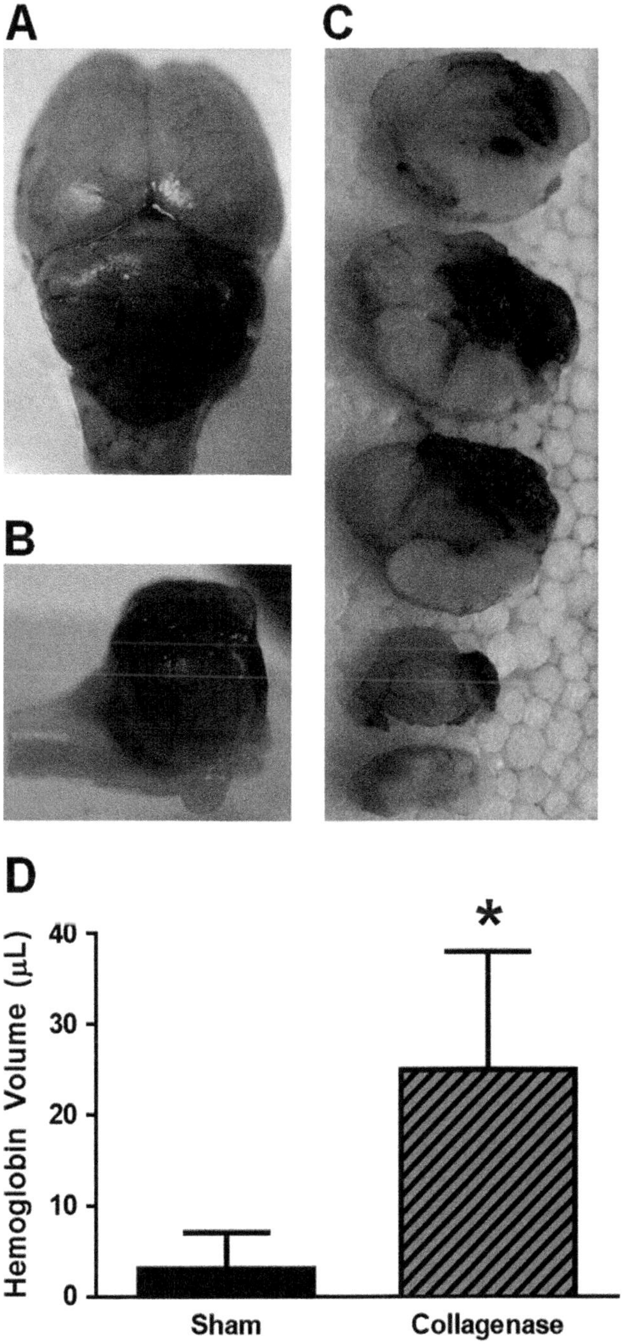

Fig. 2 Cerebellar hemorrhage and hematoma volume. Representative images of the hematoma within the cerebellum of a rat brain 24 h after injury. (**a**) Axial view of the whole brain. (**b**) Sagittal view of the brain. (**c**) Coronal view of brain slices. (**d**) Hematoma volume was assessed using the hemoglobin assay. n = 15/group. Mean ± SD. T-test. $*p < 0.05$ vs. Sham operated animals

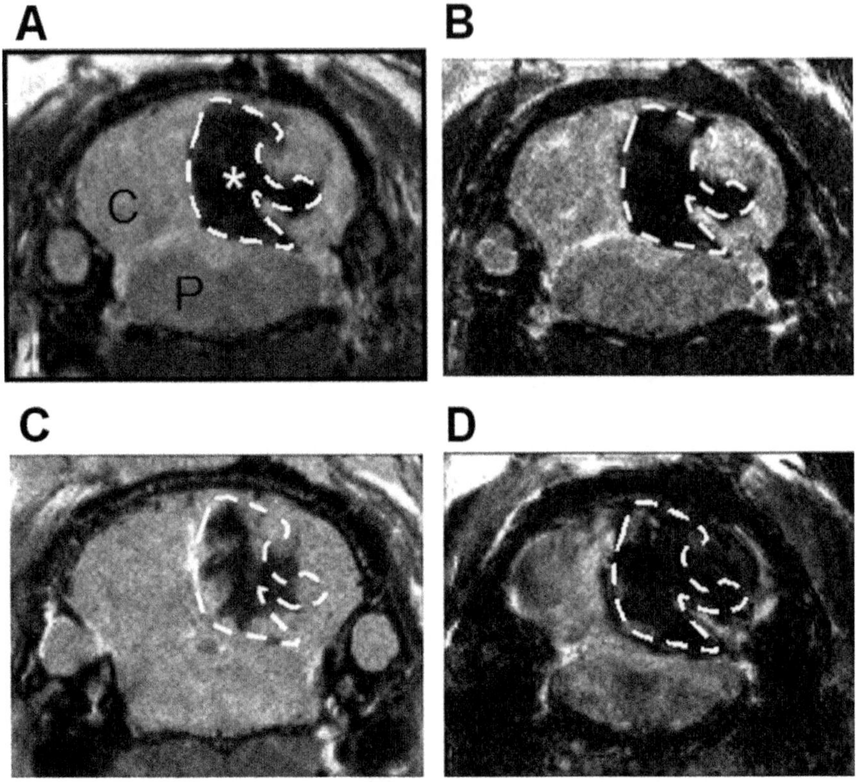

Fig. 3 MRI of cerebellar hematoma. Representative MRI scans of a rat brain 24 h post-ictus. (**a**) T2 MRI image with hematoma (asterisk, white dashed line). *C* cerebellum, *P* pons. (**b**) Diffusion weighted MR image of hematoma (white dashed line corresponding to the T2 MRI hematoma). (**c**) T1 MRI image of hematoma (white dashed line corresponding to the T2 MRI hematoma). (**d**) Susceptibility weighted MR image of hematoma (white dashed line corresponding to the T2 MRI hematoma)

outcomes. Furthermore, MRI can also be utilized to assess hematoma volume (Fig. 4). Short-term analysis of hematoma volume can be analyzed using the hemoglobin assay of the cerebellum, post-mortem [15]. Twenty-four hours after induction of CH in rats, the hematoma volume is significantly higher than that of sham operated animals (needle insertion only) (Fig. 2d).

Additionally, CH results in increased brain water content for the cerebellum as well as blood-brain barrier disruption. Animals subjected to CH have significantly elevated brain water content, measured by the Wet-Dry method [18], of the ipsilateral and contralateral cerebellar lobes 24 h post-ictus compared to sham operated animals (Fig. 5). Cerebellar hemorrhage has no effect on the brain water content of the brain stem. Evans blue assay, a measurement of blood-brain barrier disruption, shows that CH animals have significantly more extravasated Evans blue than sham animals (Fig. 6).

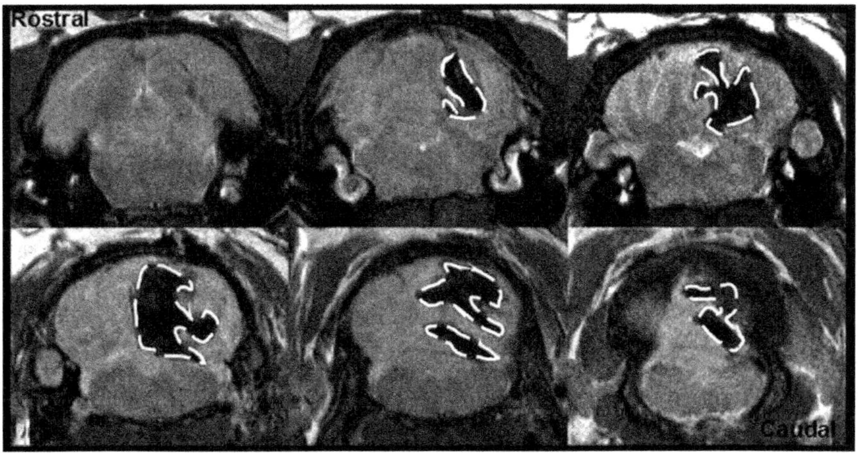

Fig. 4 MRI of cerebellar hematoma volume. A representative T2 MRI scan of rat brain 24 h post-cerebellar hemorrhage. Hematoma is outlined in white

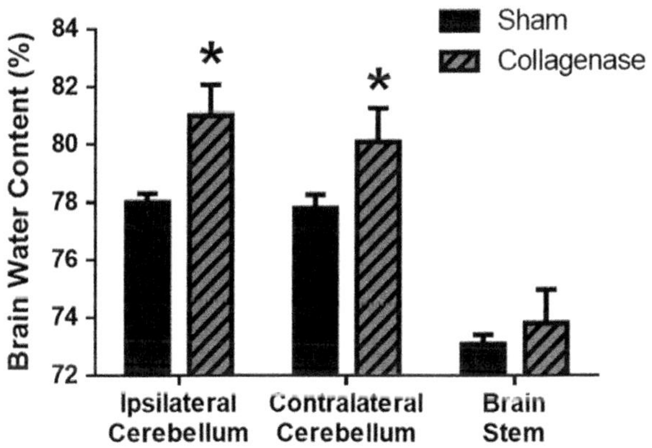

Fig. 5 Cerebellar brain water content 24 h after cerebellar hemorrhage. Brain water content was measured using the Wet-Dry method. n = 10–15/group. Mean ± SD. T-test. $*p < 0.05$ vs. Sham operated animals

Behavioral Outcomes

Neurological deficits are observed after CH and are quantified using a composite neuroscore and a modified Luciani scale [16, 17]. Rats typically present with flexed contralateral forelimbs, decreased body tone, and dyscoordination caused by the unilateral cerebellar injury (Fig. 7a, see video on publisher website). Twenty-four hours post-CH, cerebellar hemorrhage led to significant loss of sensorimotor function, assessed using a composite neuroscore, and ataxic deficits, assessed using the modified Luciani scale (Fig. 7b, c). The composite neuroscore was quantified by averaging the values from the wire suspension, beam balance, and inclined plane

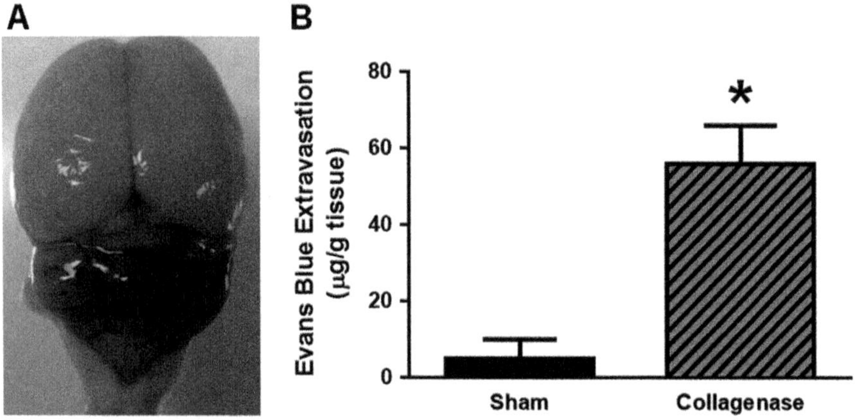

Fig. 6 Evans blue extravasation 24 h after cerebellar hemorrhage. (**a**) A representative rat brain showing Evans blue extravasation after CH. (**b**) Evans blue extravasation 24 h after CH. n = 15/group. Mean ± SD. T-test. ∗p < 0.05 vs. Sham operated animals

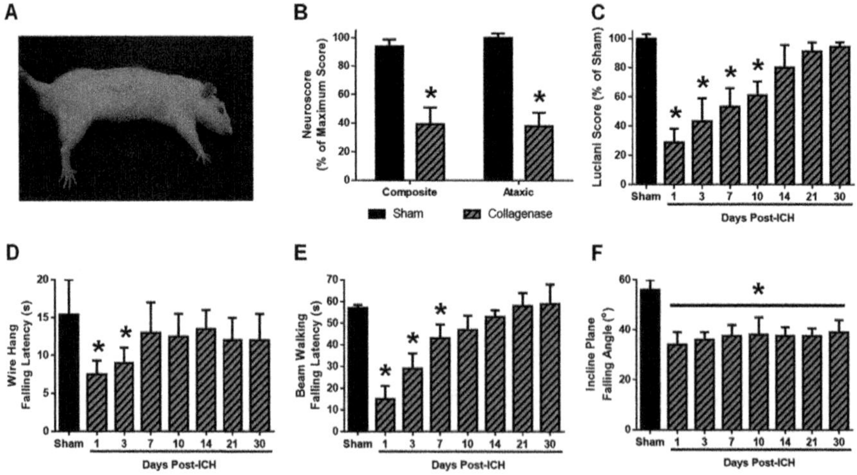

Fig. 7 Neurological impairment after cerebellar hemorrhage. (**a**) A representative image of the postural deficits affecting rat behavior 24 h after CH. (**b**) Composite Garcia neuroscore and Ataxic scoring (via modified Luciani scale) 24 h post-CH. n = 10–15/group. T-test. (**c**) Luciani score assessed up to 30 days post-CH. n = 8/group. One-way ANOVA with Tukey post-hoc test. (**d**) Wire suspension forelimb testing assessed up to 30 days post-CH. n = 8/group. One-way ANOVA with Tukey post-hoc test. (**e**) Beam walking test assessed up to 30 days post-CH. n = 8/group. One-way ANOVA with Tukey post-hoc test. (**f**) Inclined plane test assessed up to 30 days post-CH. n = 8/group. One-way ANOVA with Tukey post-hoc test. Mean ± SD. ∗p < 0.05 vs. Sham operated animals

tests (scored as a percent of the maximum possible score for each test). The ataxic neuroscore is the sum of scores from three sub-tests: body tone, limb extension, and dyscoordination. Each sub-test was scored as either 0 (no deficits), 1 (mild deficits), 2 (moderate deficits), or 3 (severe deficits). In addition to composite neuroscore tests, more sensitive sensorimotor tests have also found deficits in animals subjected to CH [17]. Functional deficits are apparent in the wire suspension forelimb testing, the inclined plane paradigm, and beam walking ability following CH (Fig. 7d–f).

Finally, cerebellar hemorrhage created by collagenase infusion has been shown to cause reproducible long-term functional deficits; sensorimotor deficits have been identified using the rotorod test, decreased working memory has been shown using the T-maze test, and spatial memory deficits were observed using the Morris Water Maze [17]. These deficiencies in neurological function are caused by histopathological analysis. Thirty days after CH, rats subjected to collagenase infusion have greater cerebellar tissue atrophy than sham operated animals (Fig. 8).

Caveats, Limitations, and Alternatives

Bacterial Collagenase Versus Autologous Blood Injection

While herein bacterial collagenase is use to create hemorrhage within the cerebellum. Alternatively, autologous blood may be injected to create an autologous blood injection CH model, described by Cossu et al. [13, 14] A number of papers exist which detail the differences between the use of collagenase and autologous blood for creating intracerebral hemorrhage [19–21]. The differences between the collagenase and autologous blood hemorrhage models likely apply to these two models for cerebellar hemorrhage. It is important to note that the blood injection model creates an instantaneous, exact volume hematoma, whereas in the collagenase model, the hematoma develops over a few hours and may have some variation in blood volume [20, 21]. Although the hematoma volumes are typically similar between the two models, the collagenase model causes greater tissue loss and long-term functional impairment than the blood injection model [20, 21].

Preparation of Bacterial Collagenase

It is critical in preparing the bacterial collagenase to keep everything ice cold. This includes all solutions and tubes. Allowing the collagenase to heat up can quickly decrease the activity and effectiveness of the collagenase, ultimately reducing the severity and reproducibility of the hematoma.

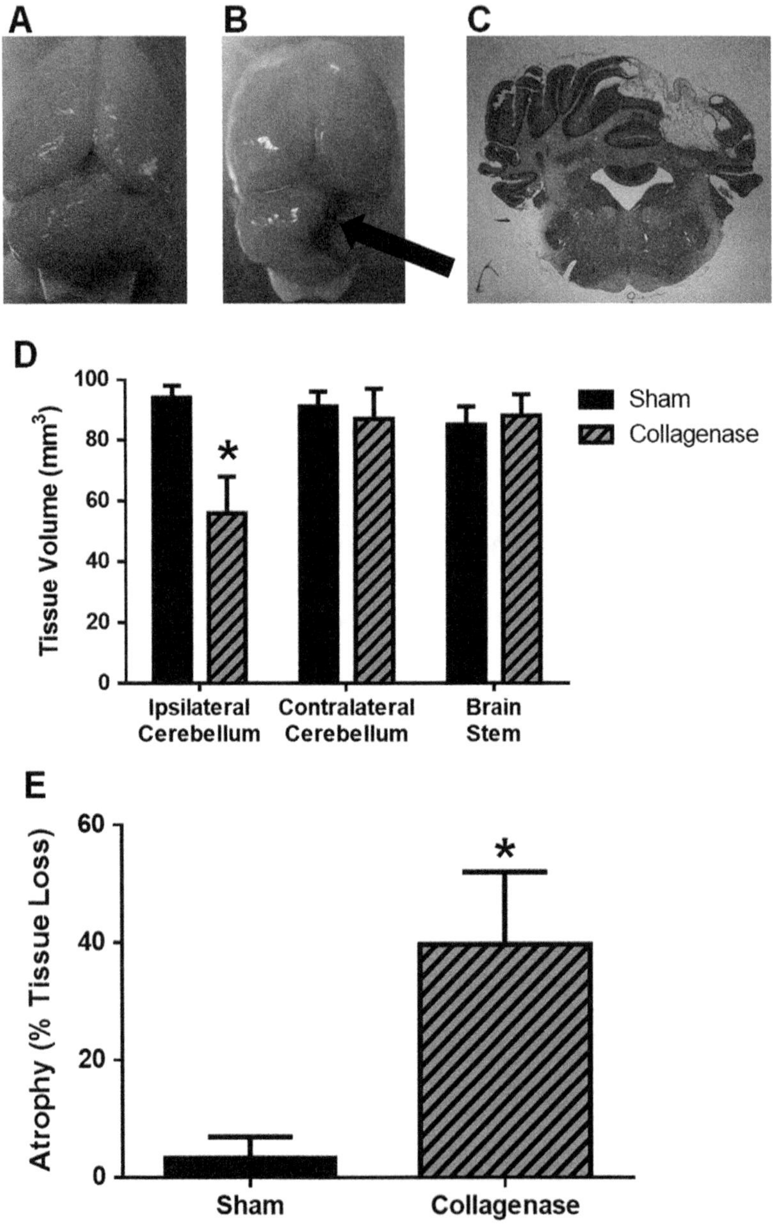

Fig. 8 Brain atrophy 30 days after cerebellar hemorrhage. (**a**) A representative brain of a Sham operated animal. (**b**) A representative brain of an animal subjected to CH showing brain atrophy (arrow). (**c**) A representative Nissl stained post-CH rat brain slice displaying brain atrophy. (**d**) Tissue volume (mm^3) assessed from Nissl stained brain sections. n = 8/group. (**e**) Tissue atrophy (%) assessed from the Nissl stained brain sections. n = 8/group. Mean ± SD. T-test. $*p < 0.05$ vs. Sham operated animals

Surgical Procedures

A number of crucial steps lie in the injection of the collagenase. First, the collagenase should be removed from the freezer and thawed immediately prior to injection. Similar to the preparation of the collagenase solution, allowing the collagenase to warm up reduces its effectiveness. Second, after complete injection of the collagenase, the needle must be allowed to remain in place for 10 min before slow removal to minimize the back-flow of the solution. Evidence of significant back-flow can be observed by hemorrhage along the needle tract. Reducing the amount of back flow yields a more consistent hematoma.

Choice of Species and Age

Herein, young adult male Sprague-Dawley rats were used. However, this model may be adapted for use in other rat species, different age groups, and other species by appropriately adjusting the stereotactic coordinates for collagenase injection.

Summary

The cerebellar hemorrhage model utilizes bacterial collagenase to induce a large and consistent hematoma within the cerebellum of the rat. This model results in acute increases in cerebellum water content, blood-brain barrier disruption, and functional deficits. Of particular interest, and unique to cerebellar hemorrhage, are the postural deficits which are observed 24 h post-ictus. In addition to the loss of posture, acute sensorimotor deficits are observed. Finally, the devastating nature of bacterial collagenase for creating cerebellar hemorrhage also causes long-term histopathological sequelae and neurological impairment (sensorimotor and memory).

References

1. Qureshi AI, Mendelow AD, Hanley DF. Intracerebral haemorrhage. Lancet. 2009;373:1632–44.
2. Sutherland GR, Auer RN. Primary intracerebral hemorrhage. J Clin Neurosci. 2006;13:511–7.
3. Hill MD, Silver FL, Austion PC, Tu JV. Rate of stroke recurrence in patients with primary intracerebral hemorrhage. Stroke. 2000;31:123–7.
4. Baillieux H, De Smet HJ, Paquier PF, De Deyn PP, Marien P. Cerebellar neurocognition: insights into the bottom of the brain. Clin Neurol Neurosurg. 2008;110(8):763–73.
5. Dykstra-Aiello C, Jickling GC, Ander BP, et al. Intracerebral hemorrhage and ischemic stroke of different etiologies have distinct alternatively spliced mRNA profiles in the blood: a pilot RNA-seq study. Transl Stroke Res. 2015;6(4):284–9.

6. Schlunk F, Greenberg SM. The pathophysiology of intracerebral hemorrhage formation and expansion. Transl Stroke Res. 2015;6(4):257–63.
7. Schlunk F, Schulz E, Lauer A, et al. Warfarin pretreatment reduces cell death and MMP-9 activity in experimental intracerebral hemorrhage. Transl Stroke Res. 2015;6(2):133–9.
8. Selim M, Sheth KN. Perihematoma edema: a potential translational target in intracerebral hemorrhage? Transl Stroke Res. 2015;6(2):104–6.
9. Sukumari-Ramesh S, Alleyne CH Jr, Dhandapani KM. The histone deacetylase inhibitor suberoylanilide hydroxamic acid (SAHA) confers acute neuroprotection after intracerebral hemorrhage in mice. Transl Stroke Res. 2016;7(2):141–8.
10. Xiong XY, Yang QW. Rethinking the roles of inflammation in the intracerebral hemorrhage. Transl Stroke Res. 2015;6(5):339–41.
11. Zhao H, Garton T, Keep RF, Hua Y, Xi G. Microglia/macrophage polarization after experimental intracerebral hemorrhage. Transl Stroke Res. 2015;6(6):407–9.
12. Zheng M, Du H, Ni W, et al. Iron-induced necrotic brain cell death in rats with different aerobic capacity. Transl Stroke Res. 2015;6(3):215–23.
13. Cossu M, Dorcaratto A, Pau A, et al. Changes in infratentorial blood flow following experimental cerebellar haemorrhage. A preliminary report. Ital J Neurol Sci. 1991;12(3 Suppl 11):69–73.
14. Cossu M, Pau A, Siccardi D, Viale GL. Infratentorial ischaemia following experimental cerebellar haemorrhage in the rat. Acta Neurochir. 1994;131(1–2):146–50.
15. Lekic T, Tang J, Zhang JH. Rat model of intracerebellar hemorrhage. Acta Neurochir Suppl. 2008;105:131–4.
16. Lekic T, Ostrowski RP, Suzuki H, et al. The postpartum period of pregnancy worsens brain injury and functional outcome after cerebellar hemorrhage in rats. Acta Neurochir Suppl. 2011;111:37–41.
17. Lekic T, Rolland W, Hartman R, et al. Characterization of the brain injury, neurobehavioral profiles, and histopathology in a rat model of cerebellar hemorrhage. Exp Neurol. 2011;227(1):96–103.
18. Tang JP, Liu J, Zhou CM, et al. MMP-9 deficiency enhances collagenase-induced intracerebral hemorrhage and brain injury in mutant mice. J Cereb Blood Flow Metab. 2004;24(10):1133–45.
19. Lei B, Sheng H, Wang H, et al. Intrastriatal injection of autologous blood or clostridial collagenase as murine models of intracerebral hemorrhage. J Vis Exp. 2014;(89). https://doi.org/10.3791/51439.
20. MacLellan CL, Silasi G, Auriat AM, Colbourne F. Rodent models of intracerebral hemorrhage. Stroke. 2010;41(10 Suppl):S95–8.
21. MacLellan CL, Silasi G, Poon CC, et al. Intracerebral hemorrhage models in rat: comparing collagenase to blood infusion. J Cereb Blood Flow Metab. 2008;28(3):516–25.

Blood Injection Intracerebral Hemorrhage Mouse Model

Ludmila Belayev

Abstract A major limitation to intracerebral hemorrhage (ICH) research is the lack of a reproducible animal model. Several animal models of ICH have been developed in mice. In this chapter, we introduce a step-by-step procedure to perform a blood injection method of ICH in the mouse. The present ICH model produces a consistent neurological deficit, hematoma volume and brain swelling. This model closely mimics human hypertensive basal ganglionic ICH and can be useful to evaluate the pathophysiological, biochemical and molecular processes after ICH as well as potential new treatments.

Keywords Stroke · Experimental · Behavior · Histopathology · Gender · Physiological monitoring

Experimental Model Selection

Intracerebral hemorrhage (ICH) accounts for 10–15% of all strokes and is associated with high morbidity and mortality. The most frequent sites of primary spontaneous ICH in humans are the putamen (50%), thalamus (15%), pons (10–15%), and cerebellum (10%) [14]. Several animal models of ICH have been developed in mice, rats, rabbits, cats and primates [18]. The study of intracerebral hemorrhage in the mouse has attracted increasing attention, particularly because the availability of genetically modified (transgenic and knockout) mouse strains provides a unique opportunity to study specific pathways or injury processes [15]. A widely-used ICH model, developed by Rosenberg and colleagues [21] in the rat, uses a local injection of bacterial collagenase into the basal ganglia and was subsequently studied in the mouse [9]. Collagenase dissolves the extracellular matrix and leads to blood vessel rupture and intracerebral bleeding. Although the collagenase method is a simple

L. Belayev (✉)
School of Medicine, Neuroscience Center of Excellence, LSU Health New Orleans, New Orleans, LA, USA
e-mail: lbelay@lsuhsc.edu

J. Chen et al. (eds.), *Animal Models of Acute Neurological Injury*, Springer Series in Translational Stroke Research, https://doi.org/10.1007/978-3-030-16082-1_21

means of producing hemorrhaging and is reproducible, several reports demonstrated that bacterial collagenase causes a significant inflammatory reaction and likely differs from the mechanism that produces ICH in humans [12, 29].

Another commonly used ICH model employs the injection (or infusion) of autologous blood into the brain parenchyma of mice and rats [19]. This model was designed to mimic the natural events that occur with spontaneous ICH in humans. Intraparenchymal injection of blood is an effective technique for producing an intracerebral hematoma. A disadvantage of the blood infusion model is the potential for ventricular rupture as well as backflow of infused blood along the needle track [2]. A double-injection ICH model in rats and mice has been developed, in which a small amount of blood is infused into the striatum at a slow rate to allow blood clotting along the needle track; the remaining blood is then infused to generate the hematoma [11, 17]. Major advantages of this model include reproducible hematoma volume, consistent neurological deficit, hypoperfusion and brain swelling [5].

Materials

Male C57BL/6 mice weighing 25–30 g are the animals of choice.

The following instruments and materials are recommended for the ICH experiment:

1. GAS Anesthesia Instruments (Stoelting): Open isoflurane system (50251), isoflurane/halothane scavenger (50206) and induction chamber (50216).
2. Cranial temperature controller (Omega Engineering): monogram benchtop controller CSC32T (unit set for type T only) and temperature probe: HYP-30-1/2 inch (30-gauge).
3. Rectal temperature controller (CMA/microdialysis AB): CMA/150 temperature controller (8315000), heating pad (8315001), temperature probe (8315002) and insulation pad (8315003).
4. Digi-Med blood pressure analyzer, model 400, Micro-Med Inc.
5. Blood gas system (model ABL 5, Radiometer America Inc).
6. Infusion pump (KDS-100, kd Scientific).
7. Stereotactic instrument (model 900), mouse adaptor (model 926), mouse anesthesia mask (907) from David Kopf Instruments.
8. Micromotor drill (51,453, Stoelting).
9. Stainless steel cannula (30-gauge, 3/4″, W/30° PT, 7460, Popper & Sons Inc) and Luer stub adaptor, 23-gauge.
10. Intramedic polyethylene tubing (Becton Dickinson): PE 10 (427401) and PE 20 (427406).
11. Glass syringe (250 μL, Hamilton company).
12. Gelfoam sponges (9,083,300, Henry Schein Medical).

Procedures

1. Anesthetize a mouse with isoflurane (3% for induction and 1% for maintenance of anesthesia during surgical procedure) in a mixture of 70% nitrous oxide and 30% oxygen delivered by face-mask. Shave the hairs over the head and the right groin area.
2. Insert temperature probes into the rectum and the left temporalis muscle and use separate warming lamps to maintain rectal and temporalis muscle temperature at 36.0–37.5 °C.
3. Insert PE 10 catheter into the right femoral artery for continuous blood pressure monitoring and periodic blood sampling for arterial gases and pH.
4. Place mouse on the stereotactic frame (Fig. 1).
5. Make a midline scalp incision.
6. Make a burr hole in the skull (2 mm lateral to midline, 1 mm anterior to bregma, depth 4 mm below the surface of the skull) using a drill.
7. Take blood from the heart of the donor mouse with a 1 mL syringe, which was flushed with heparin before blood withdrawal.
8. Introduce a 30-gauge stainless steel cannula through a burr hole into the left striatum.
9. Inject 5 μL of whole blood from the donor mouse over 3 min, followed 7 min later by 10 μL injected over 5 min, using a microinfusion pump.
10. Slowly withdraw the injection cannula 10 min after the second injection.
11. Seal the bone hole with bone wax.
12. Suture the wounds and place the animal in a temperature controlled incubator with free access to food and water.

Outcome Evaluations

Neurological Deficit

Animals begin to exhibit neurological signs of an ICH within 60 min. A standardized battery of behavioral tests is recommended to quantify sensorimotor neurological function after ICH. The battery consists of two tests that have been used previously to evaluate various aspects of neurologic function: (1) the postural reflex test, to examine upper body posture while the animal is suspended by the tail; and (2) the forelimb placing test, to examine sensorimotor integration in forelimb placing responses to visual, tactile and proprioceptive stimuli [6]. Behavioral tests should be conducted by an observer blinded to the identity of the individual animals.

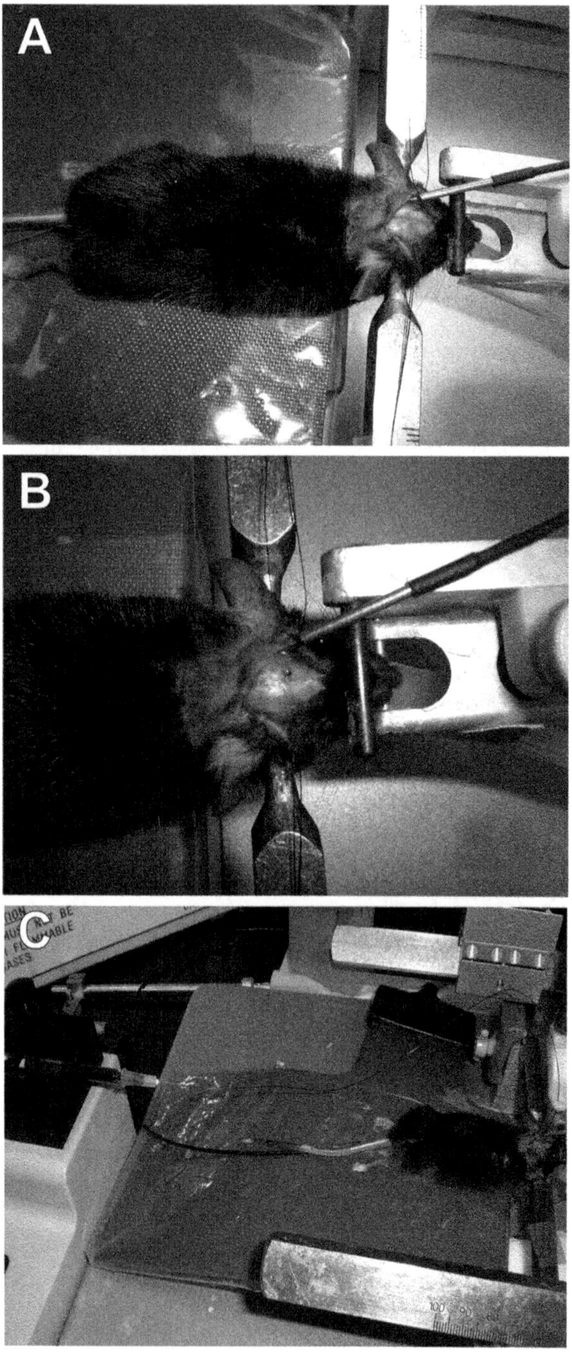

Fig. 1 Step-by-step procedures: Panel (**a**) Place mouse on the stereotactic frame, insert temperature probes into the rectum and the left temporalis muscle and make a midline scalp incision. Panel (**b**) Make a burr hole in the skull using a drill. Panel (**c**) Introduce a stainless steel cannula through a burr hole into the left striatum and inject blood using a microinfusion pump

Postural Reflex Test

The mouse is suspended by the tail 1 m above the floor. The intact mouse extends both forelimbs toward the floor; the animal displaying this behavior is assigned a score of 0. A score of 1 is given if the mouse flexes one or both forelimbs. The animal is then given the lateral push test, which involves placing the mouse on a sheet of plastic-coated paper and applying lateral pressure behind the shoulders in the left and right directions. If the mouse is unable to resist the force equally in both directions, it receives a score of 2 [6].

Elicited Forelimb Placing

For visual placing, the animal is cupped in the experimenter's hands with the forelimb hanging free and is slowly tilted and lowered from each side toward a tabletop. The intact mouse reaches for the table with both forelimbs extended. By moving the animal laterally toward the table edge, sideways visual placing could be assessed as well. Tactile placing is judged by lightly contacting the dorsal and then the lateral surface of the mouse's forepaw to the table edge while slightly elevating its head to obscure its view of the table. The intact mouse immediately places its paw on the tabletop, while the impaired mouse either is slow to place or does not place at all. Proprioceptive placing is also assessed by pressing the mouse's paw against the table edge to stimulate limb muscles. For each of these tests, a score of 0 is given for normal, immediate placing. A score of 1 is given if the placing was delayed or incomplete. A score of 2 indicates placing was absent [6].

Total Neurological Score

Total neurological score is calculated as the sum of scores on forward visual placing (range, 0–2), lateral visual placing (0–2), dorsal (0–2) and lateral (0–2) tactile placing, proprioceptive placing (0–2) and postural reflex (0–2). Thus, the maximal possible score is 12, as we previously described for mice and rats [6].

Histopathological Examination

After completion of the experiment, the animal is perfused with a mixture of 40% formaldehyde, glacial acetic acid, and methanol (FAM, 1:1:8 by volume), as previously described [4]. Then brain blocks are embedded in paraffin, and 10-μm-thick sections are cut in the coronal plane and stained with hematoxylin and eosin. Brain sections are then digitized (MCID core imaging software; InterFocus Imaging Ltd., Cambridge, England) at eight standardized coronal levels (bregma levels: +1.7, +1.1, +0.62, +0.26, +0.46, −1.22, −1.94 and −2.92 mm) [13] using a CCD camera

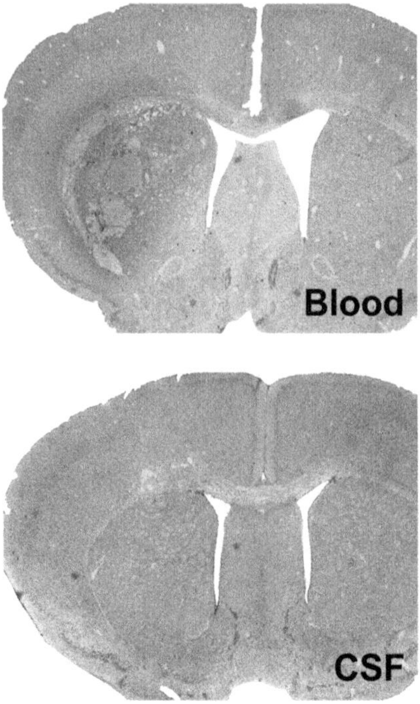

Fig. 2 Paraffin-embedded coronal brain sections, stained with hematoxylin and eosin 2 days after infusion of blood (top) or CSF (bottom) into the mouse brain. A moderate-sized striatal hematoma is evident after blood injection. In contrast, the CSF injected mouse showed only a small, non-hemorrhagic lesion

(QICAM Fast 1394, QIMAGING, British Columbia, Canada). An investigator blinded to the experimental groups then outlines the ipsilateral and contralateral hemispheres and area of hematoma of each section (Fig. 2). Hematoma size is calculated as the integrated product of the cross-sectional area and inter-sectional distance. Brain sections are imaged on a motorized microscope BX61VS (Olympus, Japan) at 20× objective.

Applications, Complications and Limitations

Applications for Animal Model of ICH

The blood-injection method replicates the localization and size of ICH observed clinically but differs in the respect that, in patients with ICH, abnormal microvascular fragility probably plays an important predisposing role. The brain tissue responses to ICH include blood-brain barrier (BBB) disruption, edema development, inflammation, tissue necrosis, clot absorption and scar or cavity formation [31]. Animal models provide a new understanding of the pathophysiology underlying ICH-induced injury, including information on the roles of mass effect, elevated intracranial pressure, alterations in cerebral blood flow and metabolism and impact of specific blood components on brain edema formation [20, 25]. We present a double-injection ICH model that satisfies the need of highly reproducible hematomas in the mouse. Major advantages of this model include: reproducible hematoma volume, consistent neurological deficit together with consistent BBB dysfunction and brain swelling. This model closely mimics human hypertensive basal ganglionic ICH and should be useful for the evaluation of pharmaceutical therapies.

Rapid Injection Versus Slow Infusion

As opposed to the rapid injection of blood used in previous models, this model uses a slow infusion of fresh blood into the brain parenchyma. The presence of a 15-μL blood clot produces a localized striatal hematoma in mice. We think that the slow infusion of blood limits its extravasation into the subarachnoid and ventricular spaces and more closely mimics the natural process. Furthermore, it also avoids undesired non-physiological pressure injury to adjacent tissue.

Autologous Versus Donor Blood

Brain edema is an important clinical endpoint for ICH studies. Hematoma within brain parenchyma triggers a series of events leading to BBB disruption and edema. Brain water content is significantly increased in the ipsilateral basal ganglia and cortex in C57BL/6 mice at 1 and 3 days after injection of autologous blood [19]. Interestingly, donor blood induces significantly more brain edema than autologous blood on day 1 after ICH [19]. This model can be recommended for the further investigations of brain edema and to test agents that reduce edema formation in the ICH model.

Application for Genetically Modified Mice

The development of the mouse ICH model has attracted increasing attention because the availability of gene knockout or transgenic animals may assist in elucidating the mechanisms of brain injury after ICH. Considerable evidence demonstrates that ICH induces alterations in gene expression, enzyme and protein degradation, which contribute to cellular injury after ICH. Studies using knockout and transgenic mice in ICH models include the heme oxygenase-1 (HO-1−/−) [26] and HO-2−/− [8] mice, complement C3-deficient mice [28], matrix metalloproteinase-9 deficient mice [23], tPA-deficient mice [24] and many others [16, 30].

Influence of Gender

Sex differences in incidence, pathophysiology, and outcome of primary ICH have been reported recently—men having higher incidence of ICH compared to women [22]. In ischemic stroke models, behavioral score and brain infarcts are significantly reduced in females as opposed to males [1]. In the ICH mouse model female mice showed a faster recovery for behavioral tests and reduced brain edema compared to male animals [19]. The greater neuroprotection afforded to females in ischemia and ICH is likely because of the effects of circulating estrogens and progestins [22]. Estrogen has also been shown to be neuroprotective by improving blood flow during and after ischemic cerebrovascular events [7].

Importance of Temperature Regulation

It is now well-recognized that even small decreases in brain temperature protect the brain from ischemic injury [27]. Mice after ICH fail to regulate their temperatures, becoming severely hypothermic during the postoperative period, and that these temperature changes may have a profound effect on the outcome of ICH. We recommend close monitoring of cranial and body temperature during surgical procedure and also allowing mice to recover from surgery in a temperature-controlled incubator for 48 h.

Importance of Neurological Evaluation

Observations of neurological deficits are important, not only in clinical stroke patients, but also in animal models of ICH [3]. Two sensorimotor tests (the postural reflex and the forelimb placing tests) applied here appear to be particularly sensitive to this model [6]. They can help to detect a neurological deficit and also to identify animals in which ICH model was not successfully reproduced. In addition, grid-walking, spontaneous-forelimb (cylinder) and single-pellet skilled-reaching tasks can be assessed in mice as described previously [10].

Bleeding Problems

Bleeding may occur during the surgical procedure. To help with this problem, keep the incision of the skin over the skull to the midline and use a haemostatic sponge or wax to stop bleeding.

Acknowledgements I want to thank all of our past and present team members who have contributed to developing the model represented in this chapter.

References

1. Alkayed NJ, Harukuni I, Kimes AS, et al. Gender-linked brain injury in experimental stroke. Stroke. 1998;29:159–66.
2. Andaluz N, Zuccarello M, Wagner KR. Experimental animal models of intracerebral hemorrhage. Neurosurg Clin N Am. 2002;13:385–39.
3. Barratt HE, Lanman TA, Carmichael ST. Mouse intracerebral hemorrhage models produce different degrees of initial and delayed damage, axonal sprouting, and recovery. J Cereb Blood Flow Metab. 2014;34:1463–71.
4. Belayev L, Saul I, Curbelo K, et al. Experimental intracerebral hemorrhage in the mouse: histological, behavioral, and hemodynamic characterization of a double-injection model. Stroke. 2003;34:2221–7.
5. Belayev L, Saul I, Busto R, et al. Albumin treatment reduces neurological deficit and protects blood-brain barrier integrity after acute intracortical hematoma in the rat. Stroke. 2005;36:326–31.
6. Belayev L, Obenaus A, Zhao W, et al. Experimental intracerebral hematoma in the rat: characterization by sequential magnetic resonance imaging, behavior, and histopathology. Effect of albumin therapy. Brain Res. 2007;1157:146–55.
7. Carswell HV, Anderson NH, Morton JJ, et al. Investigation of estrogen status and increased stroke sensitivity on cerebral blood flow after a focal ischemic insult. J Cereb Blood Flow Metab. 2000;20:931–6.
8. Chen-Roetling J, Cai Y, Regan RF. Neuroprotective effect of heme oxygenase-2 knockout in the blood injection model of intracerebral hemorrhage. BMC Res Notes. 2014;7:561.
9. Clark W, Gunion-Rinker L, Lessov N, et al. Citicoline treatment for experimental intracerebral hemorrhage in mice. Stroke. 1998;29:2136–40.
10. Clarkson AN, Overman JJ, Zhong S, et al. AMPA receptor-induced local brain-derived neurotrophic factor signaling mediates motor recovery after stroke. J Neurosci. 2011;31:3766–75.
11. Deinsberger W, Vogel J, Kuschinsky W, et al. Experimental intracerebral hemorrhage: description of a double injection model in rats. Neurol Res. 1996;18:475–7.
12. Del Bigio MR, Yan HJ, Buist R, et al. Experimental intracerebral hemorrhage in rats. Magnetic resonance imaging and histopathological correlates. Stroke. 1996;27:2312–9.
13. Franklin KBJ, Paxinos G. The mouse brain in stereotaxic coordinates. San Diego: Academic; 2008.
14. Hemphill JC III, Greenberg SM, Anderson CS, et al. Guidelines for the management of spontaneous intracerebral hemorrhage: a guideline for healthcare professionals from the American Heart Association/American Stroke Association. Stroke. 2015;46:2032–60.
15. Kathirvelu B, Carmichael ST. Intracerebral hemorrhage in mouse models: therapeutic interventions and functional recovery. Metab Brain Dis. 2015;30:449–59.
16. Leclerc JL, Lampert AS, Diller MA, et al. Prostaglandin E2 EP2 receptor deletion attenuates intracerebral hemorrhage-induced brain injury and improves functional recovery. ASN Neuro. 2015; https://doi.org/10.1177/1759091415578713.
17. Ma B, Zhang J. Nimodipine treatment to assess a modified mouse model of intracerebral hemorrhage. Brain Res. 2006;1078:182–8.
18. Ma Q, Khatibi NH, Chen H, et al. History of preclinical models of intracerebral hemorrhage. Acta Neurochir Suppl. 2011;111:3–8.
19. Nakamura T, Xi G, Hua Y, et al. Intracerebral hemorrhage in mice: model characterization and application for genetically modified mice. J Cereb Blood Flow Metab. 2004;24:487–94.

20. Ni W, Zheng M, Xi G, et al. Role of lipocalin-2 in brain injury after intracerebral hemorrhage. J Cereb Blood Flow Metab. 2015;35:1454–61.
21. Rosenberg GA, Mun-Bryce S, Wesley M, et al. Collagenase-induced intracerebral hemorrhage in rats. Stroke. 1990;21:801–7.
22. Sankalp G, Caplan LR, James ML. Sex differences in incidence, pathophysiology, and outcome of primary intracerebral hemorrhage. Stroke. 2015;46:886–92.
23. Tang J, Liu J, Zhou C, et al. Mmp-9 deficiency enhances collagenase-induced intracerebral hemorrhage and brain injury in mutant mice. J Cereb Blood Flow Metab. 2004;24:1133–45.
24. Thiex R, Mayfrank L, Rohde V, et al. The role of endogenous versus exogenous tPA on edema formation in murine ICH. Exp Neurol. 2004;189:25–32.
25. Wagner KR. Modeling intracerebral hemorrhage: glutamate, nuclear factor-kappa B signaling and cytokines. Stroke. 2007;38:753–8.
26. Wang J, Dore S. Heme oxygenase-1 exacerbates early brain injury after intracerebral haemorrhage. Brain. 2007;130:1643–52.
27. Wei S, Sun J, Li J, et al. Acute and delayed protective effects of pharmacologically induced hypothermia in an intracerebral hemorrhage stroke model of mice. Neuroscience. 2013;252:489–500.
28. Yang S, Nakamura T, Hua Y, et al. Intracerebral hemorrhage in complement C3-deficient mice. Acta Neurochir Suppl. 2006;96:227–31.
29. Zhao X, Sun G, Zhang J, et al. Hematoma resolution as a target for intracerebral hemorrhage treatment: role for peroxisome proliferator-activated receptor gamma in microglia/macrophages. Ann Neurol. 2007;61:352–62.
30. Zhao X, Sun G, Zhang J, et al. Dimethyl fumarate protects brain from damage produced by intracerebral hemorrhage by mechanism involving Nrf2. Stroke. 2015;46:1923–8.
31. Zhou W, Marinescu M, Veltkamp R. Only very early oxygen therapy attenuates posthemorrhagic edema formation and blood-brain barrier disruption in murine intracerebral hemorrhage. Neurocrit Care. 2015;22:121–32.

Blood Injection Intracerebral Hemorrhage Pig Model

Yuxiang Gu, Shenglong Cao, Guohua Xi, and Ya Hua

Abstract The autologous blood injection ICH model has been used successfully in pigs. The larger brain volume and well-developed white matter in the pig brain (in contrast to the rat and the mouse) is good for studies of white matter injury after ICH. Swine provides an excellent model for the study of edema formation, white matter brain damage and hematoma resolution process. Of the non-primate mammals, pigs have a cardiovascular system which is most similar to humans. In this chapter, we present a step-by-step procedure to produce ICH by injecting autologous whole blood.

Keywords Intracerebral hemorrhage · Brain edema · White matter injury · Swine

Model Selection

Experimental models of intracerebral hemorrhage (ICH) have been available since the 1960s and commonly involve the injection of autologous blood into the frontal lobe of dogs, cats, pigs or monkeys [1–5]. Intracerebral blood injection model is a useful model for studying ICH-induced brain injury in pigs (Fig. 1) [6–8]. Autologous blood obtained from the femoral artery is infused into the white matter in the frontal lobe to simulate hemorrhage (Figs. 2 and 3). Below are advantages of this model:

1. Well developed white matter to determine white matter brain damage induced by ICH [4, 9];
2. A nice ICH model well suited to clot aspiration and hematoma resolution studies [10–13];
3. Large brain size suitable to magnetic resonance imaging studies [14];

Y. Gu · S. Cao · G. Xi · Y. Hua (✉)
Department of Neurosurgery, University of Michigan, Ann Arbor, MI, USA
e-mail: guohuaxi@umich.edu; yahua@umich.edu

J. Chen et al. (eds.), *Animal Models of Acute Neurological Injury*, Springer Series in Translational Stroke Research, https://doi.org/10.1007/978-3-030-16082-1_22

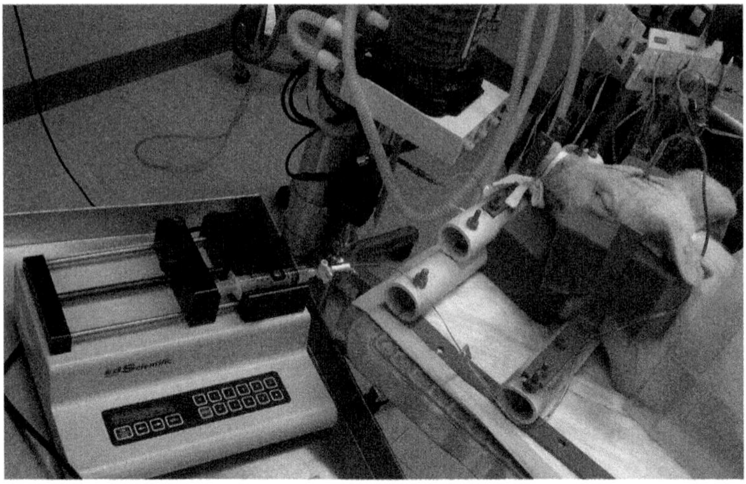

Fig. 1 Autologous whole blood is injected into the white matter of the right frontal lobe in the pig

Fig. 2 A pig brain 1 week after an intracerebral injection of 2.5 mL autologous whole blood

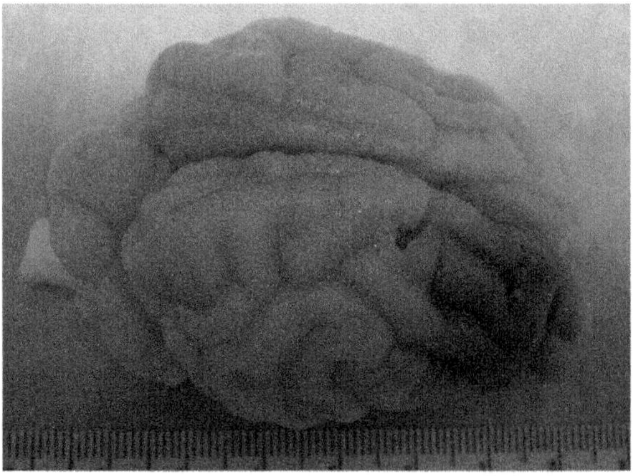

4. The ability to study a variety of physiologic parameters that are altered by the hematoma, including intracranial pressure, cerebral blood flow and cerebral perfusion pressure;
5. Ease of reproducibility and rarity of complications;
6. Modification of the procedure to test the neurotoxic effects of thrombin and red blood cells [15].

Materials

Immature pigs weighing ~10 kg are used in this model of intracerebral hemorrhage. The materials recommended for the surgical procedure include:

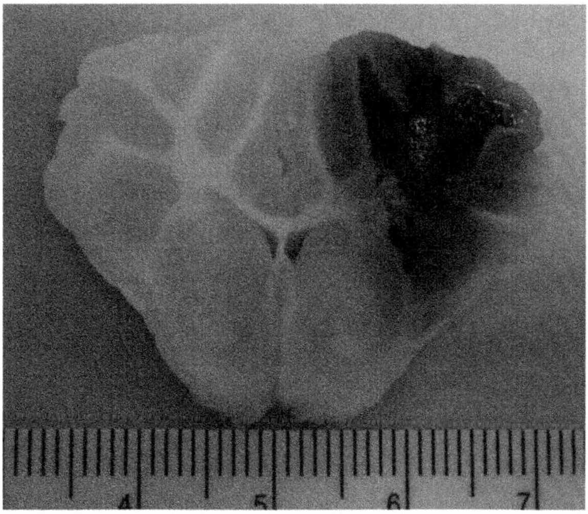

Fig. 3 A coronal section of pig brain 7 days after an intracerebral injection of 2.5 mL autologous whole blood. Note red blood cell lysis and reddish brain tissue around the clot

1. Anesthesia machine
2. Respirator
3. Biophysical recorders with pressure transducers
4. Blood gas analyzer
5. Warm water blanket
6. Glucose analyzer
7. Cuffed endotracheal tube (6.0 mm I.D.)
8. Silicone elastomer tubing (Sil-med)
9. Infusion pump
10. Dremel drill with 1.5 mm drill bit
11. Moria spring scissors with sharp points
12. Scalpel handle with scalpel blades
13. Fine iris scissors
14. Curved Mayo scissors
15. Dumont forceps
16. Dumont angled forceps
17. Adson forceps with teeth
18. Heiss retractor
19. Three-way stopcocks
20. Hemostats
21. Needle holder
22. Cotton tipped applicators
23. 1 and 5 mL syringes
24. Polyethylene tubing (PE-160)
25. 20-gauge needles
26. 20-gauge sterile plastic catheters
27. 4-0 black braided silk (Deknatel)
28. 3-0 Dermalon suture

Procedure

1. Immature swine (~10 kg) will be fasted overnight with water ad libitum.
2. Inject Telazol (6 mg/kg, IM) and Xylazine (2.2 mg/kg, IM) for induction of anesthesia and endotracheal intubation.
3. Intubate the animal orotracheally when a surgical plane of anesthesia is reached.
4. Use isoflurane in oxygen as the gas anesthetic for swine.
5. Use a warm water blanket to maintain body temperature at 37 °C.
6. Use a hair clipper to remove hair from the scalp and right groin regions and apply Betadine to the exposed skin.
7. Make a 4 cm incision in the groin at the region of the right femoral artery.
8. Expose the femoral artery and vein using hemostats.
9. Separate 2 cm of the femoral artery from the femoral vein using angled forceps and a cotton tipped applicator, being careful not to tear the artery.
10. Ligate the distal end of the artery using 4-0 black braided silk and clip one end of the silk with a hemostat and place to the side to help expose the artery.
11. Place a thread of silk under the proximal end of the artery and clip both ends with a hemostat to temporarily occlude blood flow.
12. Using spring scissors, prick the artery to create an opening for the catheter without severing the blood vessel.
13. Make sure the PE-160 catheter is trimmed to a point and attached to a three-way stopcock.
14. Insert the tubing into the opening with the pointed side down using two pairs of forceps to hold the tubing and the artery in place.
15. Use a loop of silk to cinch the tubing in place.
16. Inject a small amount of saline to ensure that blood is being pumped into the tubing.
17. Exchange the saline-filled syringe for an empty syringe and release the tension on the proximal occluding thread to fill the syringe with 5 mL of blood.
18. Using another syringe that has been filled with heparin and then emptied completely, withdraw a 0.2 mL blood sample to analyze glucose, hematocrit and blood gases.
19. After sampling the blood and measuring blood pressure if appropriate, put the tension back on the proximal end of the artery, then gently slide the PE-160 out enough so that the loop originally cinching the tubing to the artery slides off, then tighten it to ligate the artery.
20. Remove the silk on the proximal end and cut all the loose ends of silk, then suture the skin using 3-0 Dermalon sutures.
21. Make a 4 cm sagittal incision over the midline of the scalp.
22. Using cotton tipped applicators, clear away the fascia to expose the coronal suture, then position the Heiss retractor under the skin to keep the area exposed.
23. Measure 10 mm laterally to the right and 10 mm anterior to the coronal suture and mark the area for the cranial burr hole.
24. Holding the Dremel drill with 1.5 mm drill bit vertically over the mark, slowly drill through the cranium, stopping when the pressure is no longer felt in order avoid tearing the dura.

25. A 20-gauge sterile plastic catheter cut to a length of 14 mm will be placed stereotaxically (guided by the pig brain atlas) into the center of the left frontal cerebral white matter (centrum semiovale) at the level of the caudate nucleus and cemented in place.
26. A 30-cm-long silicone elastomer tubing (Sil-med) connected to the arterial catheter will be filled with 10 mL of arterial blood by opening a three-way stopcock.
27. Use the infusion pump to infuse 2.5 mL of autologous whole blood into the brain tissue over 20 min.
28. Retract the plastic catheter and fill the burr hole with bone wax.
29. Suture the skin using 3-0 Dermalon sutures.

Outcome Evaluations

1. Brain edema: Brain edema can be determined by brain water content [4, 12, 16, 17] and magnetic resonance imaging measurements [14, 18, 19].
2. Blood-brain barrier (BBB) permeability: Evans blue is used to determine BBB disruption after ICH in pigs [4, 12].
3. Histology: Brain sections are stained with hematoxylin and eosine (Fig. 4) and luxol fast blue (Fig. 5). Using a computerized image analysis system (NIH image), the areas of hematoma, white matter, ventricle and hemisphere on each section are measured. The total volumes of hematoma, white matter, hemisphere and ventricle are calculated by multiplying area in each section by the distance between sections.
4. Brain metabolic changes: Adenosine triphosphate, phosphocreatine, glycogen, glucose and lactate levels in the perihematomal zone can be determined [20].

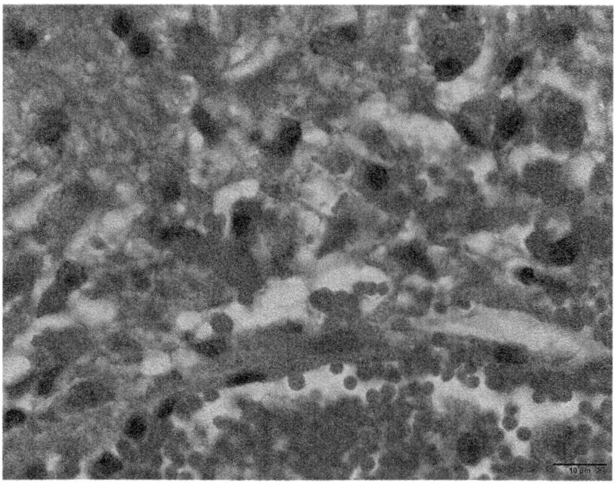

Fig. 4 Hematoxylin and eosine staining showing phagocytosis of erythrocyte by macrophage/microglia and hemosiderin deposition at 7 days after ICH in pigs

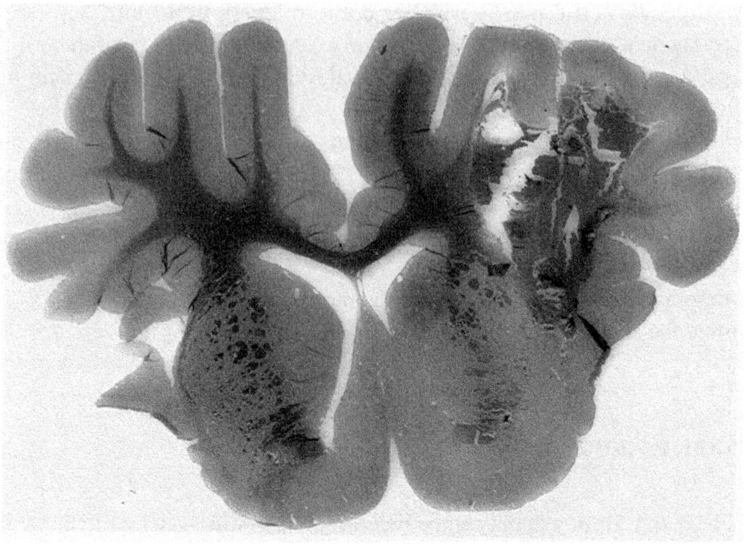

Fig. 5 Luxol fast blue staining showing ICH-induced white matter injury in pigs

Limitations

1. There is a lack of a ruptured blood vessel as the source of the ICH.
2. This model lacks the hematoma enlargement, which occurs in ICH patients.
3. Functional outcome is very important in ICH studies. However, minimal neurological deficits are present in this pig ICH model.
4. The mortality rate is very low.

References

1. Whisnant JP, Sayer GP, Millikan CH. Experimental intracerebral hematoma. Arch Neurol. 1963;9:586–92.
2. Sussman BJ, Barber JB, Goald H. Experimental intracerebral hematoma. Reduction of oxygen tension in brain and cerebrospinal fluid. J Neurosurg. 1974;41:177–86.
3. Takasugi S, Ueda S, Matsumoto K. Chronological changes in spontaneous intracerebral hematoma—an experimental and clinical study. Stroke. 1985;16:651–8.
4. Wagner KR, Xi G, Hua Y, Kleinholz M, de Courten-Myers GM, Myers RE, et al. Lobar intracerebral hemorrhage model in pigs: rapid edema development in perihematomal white matter. Stroke. 1996;27:490–7.
5. Xi G, Keep RF, Hoff JT. Mechanisms of brain injury after intracerebral hemorrhage. Lancet Neurol. 2006;5:53–63.
6. Andaluz N, Zuccarello M, Wagner KR. Experimental animal models of intracerebral hemorrhage. Neurosurg Clin N Am. 2002;13:385–93.

7. Xi G, Wagner KR, Keep RF, Hua Y, de Courten-Myers GM, Broderick JP, et al. The role of blood clot formation on early edema development following experimental intracerebral hemorrhage. Stroke. 1998;29:2580–6.
8. Gu Y, Hua Y, Keep RF, Morgenstern LB, Xi G. Deferoxamine reduces intracerebral hematoma-induced iron accumulation and neuronal death in piglets. Stroke. 2009;40:2241–3.
9. Xie Q, Gu Y, Hua Y, Liu W, Keep RF, Xi G. Deferoxamine attenuates white matter injury in a piglet intracerebral hemorrhage model. Stroke. 2014;45:290–2.
10. Wagner KR, Dwyer BE. Hematoma removal, heme, and heme oxygenase following hemorrhagic stroke. Ann N Y Acad Sci. 2004;1012:237–51.
11. Wagner KR. Modeling intracerebral hemorrhage: glutamate, nuclear factor-kappa b signaling and cytokines. Stroke. 2007;38:753–8.
12. Wagner KR, Xi G, Hua Y, Zuccarello M, de Courten-Myers GM, Broderick JP, et al. Ultra-early clot aspiration after lysis with tissue plasminogen activator in a porcine model of intracerebral hemorrhage: edema reduction and blood-brain barrier protection. J Neurosurg. 1999;90:491–8.
13. Cao S, Zheng M, Hua Y, Chen G, Keep RF, Xi G. Hematoma changes during clot resolution after experimental intracerebral hemorrhage. Stroke. 2016;47:1626–31.
14. Rohde V, Rohde I, Thiex R, Ince A, Jung A, Duckers G, et al. Fibrinolysis therapy achieved with tissue plasminogen activator and aspiration of the liquefied clot after experimental intracerebral hemorrhage: rapid reduction in hematoma volume but intensification of delayed edema formation. J Neurosurg. 2002;97:954–62.
15. Wagner KR, Packard BA, Hall CL, Smulian AG, Linke MJ, De Courten-Myers GM, et al. Protein oxidation and heme oxygenase-1 induction in porcine white matter following intracerebral infusions of whole blood or plasma. Dev Neurosci. 2002;24:154–60.
16. Wagner KR, Dwyer B, Hua Y, Cody M, Broderick JP, Brott TG. Tin-meso-porphyrin reduces intracerebral "mass" in a pig intracerebral hemorrhage. Stroke. 1998;29:332.
17. Wagner KR, Hua Y, de Courten-Myers GM, Broderick JP, Nishimura RN, Lu SY, et al. Tin-mesoporphyrin, a potent heme oxygenase inhibitor, for treatment of intracerebral hemorrhage: in vivo and in vitro studies. Cell Mol Biol. 2000;46:597–608.
18. Thiex R, Kuker W, Muller HD, Rohde I, Schroder JM, Gilsbach JM, et al. The long-term effect of recombinant tissue-plasminogen-activator (rt-pa) on edema formation in a large-animal model of intracerebral hemorrhage. Neurol Res. 2003;25:254–62.
19. Thiex R, Weis J, Krings T, Barreiro S, Yakisikli-Alemi F, Gilsbach JM, et al. Addition of intravenous n-methyl-d-aspartate receptor antagonists to local fibrinolytic therapy for the optimal treatment of experimental intracerebral hemorrhages. J Neurosurg. 2007;106:314–20.
20. Wagner KR, Xi G, Hua Y, Kleinholz M, de Courten-Myers GM, Myers RE. Early metabolic alterations in edematous perihematomal brain regions following experimental intracerebral hemorrhage. J Neurosurg. 1998;88:1058–65.

Collagenase Induced Pontine Hemorrhage in Rats

Loretta O. Iniaghe, Devin W. McBride, Tim Lekic, Jiping Tang, and John H. Zhang

Abstract Pontine hemorrhage is a devastating subtype of stroke with high rates of morbidity and mortality and currently has no treatment. Animal models are useful in elucidating the pathophysiological changes which occur in pontine hemorrhage and to evaluate potential therapeutic interventions. This chapter describes the bacterial collagenase injection in rats which creates a reproducible hematoma in the brain stem of rats. Detailed methods, an overview of the outcomes, and the limitations and alternatives is presented.

Keywords Pontine hemorrhage · Collagenase · Hematoma · Rat model

---

Electronic supplementary material: The online version of this chapter (https://doi.org/10.1007/978-3-030-16082-1_23) contains supplementary material, which is available to authorized users.

L. O. Iniaghe
Department of Physiology and Pharmacology, Loma Linda University School of Medicine, Loma Linda, CA, USA

Department of Pharmacology and Toxicology, University of Benin, Benin, Nigeria

D. W. McBride · T. Lekic · J. Tang
Department of Physiology and Pharmacology, Loma Linda University School of Medicine, Loma Linda, CA, USA

J. H. Zhang (✉)
Department of Physiology and Pharmacology, Loma Linda University School of Medicine, Loma Linda, CA, USA

Department of Anesthesiology, Loma Linda University School of Medicine, Loma Linda, CA, USA

J. Chen et al. (eds.), *Animal Models of Acute Neurological Injury*, Springer Series in Translational Stroke Research, https://doi.org/10.1007/978-3-030-16082-1_23

Introduction

Brain stem or pontine hemorrhage (PH), a form of intracranial hemorrhage which occurs mostly due to poorly controlled chronic hypertension, accounts for 6–10% of all hemorrhagic strokes and has a mortality of about 65% within the first year. PH has been associated with increased likelihood of poor functional recovery [1–3]. Hemorrhagic strokes, and in particular PH, have received less attention and studies than ischemic stroke [4–6]. Current management of PH include strategies aimed at amelioration of secondary injury such as reduction of blood pressure and intracranial pressure [7, 8].

The need for development of appropriate animal models for hemorrhagic stroke cannot be overemphasized. The National Institute of Neurological Disorders and Stroke Workshop (NINDS) recommend three areas of priority for pre-clinical research, one of which is the development of appropriate models of hemorrhage and associated brain tissue injury [9].

Herein we describe the collagenase induced pontine hemorrhage rat model by disrupting the basal lamina of the basilar and pontine arteries, causing leakage of blood into the surrounding brain parenchyma.

Materials

Adult male Sprague Dawley rats weighing 290–350 g
Stereotactic frame for rats (Stoelting, Wood Dale, IL, USA)
Infusion pump (Harvard Apparatus, Holliston, MA, USA)
Anesthesia gas mixer (Whitemore Enterprises, Rancho Cucamonga, CA, USA)
Micro-drill with 1 mm diameter drill bits (CellPoint Scientific, Gaithersburg, MD, USA)
10 μL Hamilton syringe (Hamilton Company, Reno, NV, USA)
Cotton tipped applicators
Induction chamber and recovery chambers
Thermostat controlled heating blanket
Restrainer for rats
Bone wax
Scapel
Forceps
Silk sutures
Iris Scissors
Syringes
Compressed oxygen
Compressed medical gas

Reagents

Atropine
Isoflurane
70% alcohol
Betadine
Ophthalmic lubricant
Type VII-S Collagenase (Sigma Aldrich, St Louis, MO)
Buprenorphine
0.9% Sodium chloride

Procedures

Ethical Approval

Approval was obtained from the Institutional Animal Care and Use Committee of Loma Linda University.

Surgical Procedures

1. Weigh the rat; administer atropine (0.02 mg/kg, subcutaneously) to reduce airway secretions during the surgery. Place the animal in induction chamber, anesthetize with 4% isoflurane delivered in a mixture of oxygen (0.3 L/min) and medical gas (0.7 L/min).
2. When level of anesthesia is sufficiently deep (indicated by the pinch toe reflex), apply eye ointment. Shave the rat's head.
3. Disinfect surgical area with 70% alcohol and betadine.
4. Make a 3 mm midline incision on the scalp, clear away underlying soft tissue with cotton tipped applicators.
5. Place the rat prone on the stereotactic frame. Ensure that the head is flat and firmly secure. Use ear bits to further secure the rat's head on the stereotactic frame (Fig. 1).
6. Maintain anesthesia with 2% isoflurane in a mixture of oxygen (0.3 L/min) and medical gas (0.7 L/min) and core body temperature at 37 ± 0.5 °C using a rectal probe-controlled heating blanket.
7. Drill a 1 mm cranial burr hole at the following coordinates relative to bregma, 10.2 mm caudal, 1.4 mm lateral right.
8. Transfer 0.2 units of bacterial collagenase dissolved in sterile PBS to a 10 μL Hamilton Syringe.

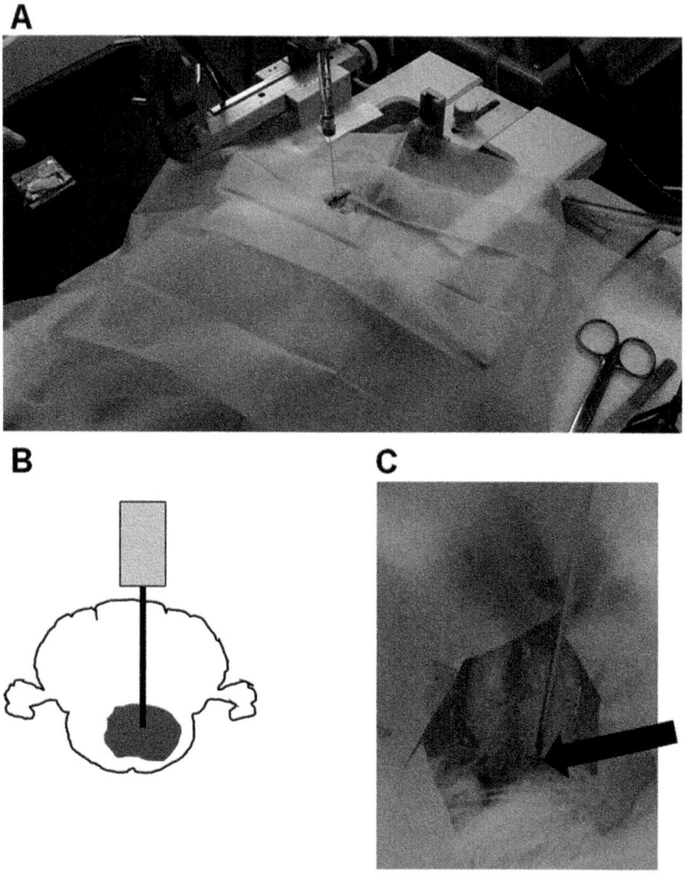

Fig. 1 Pontine hemorrhage model. (**a**) A representative image of the surgical procedure. (**b**) A schematic of the collagenase injection site to induce PH. Stereotactic coordinates: 10.2 mm caudal and 1.4 mm lateral to the bregma, 9.15 mm deep. (**c**) A representative image of the injection site of PH. Arrow points to the needle insertion site for collagenase injection

9. Infuse collagenase (0.75 or 1.5 μL) via the cranial burr hole to a depth of 9.15 mm at a rate of 0.2 μL/min with the aid of an infusion pump.
10. Leave the needle in place for an additional 7 min to prevent back flow along the needle tract then withdraw slowly at the rate of 1 mm/min.
11. Seal the burr hole with bone wax and suture the scalp.
12. Administer normal saline and buprenorphine (0.01 mg/kg, subcutaneously) to reduce post-surgical dehydration and for pain management, respectively.
13. Allow animals to recover under observation.

Sham surgery includes only needle insertion.

Evaluations After Pontine Hemorrhage

Neurological Deficits

Neurological deficits were assessed by individuals blinded to injury and evaluated at 24 and 72 h after induction of PH in rats. Short term neurobehavioral evaluations such as the modified Garcia, composite neuroscore for sensorimotor deficits, and Voetsch scores, evaluation of righting reflexes and circling, can be used to determine the level of injury while long term sensorimotor tests include the use of Morris water maze, for memory testing, and rota-rod, for long-term motor deficits [10–14].

Animals subjected to collagenase induced PH show significant neurological deficits indicated by cranio-caudal turning, minimal responsiveness, loss of righting and corneal reflexes (Fig. 2). Significant functional deficits, shown by lower test scores, are seen in both the modified Voetsch (Fig. 3a, Table 1) and Garcia (Composite) scores (Fig. 3d), in the corner turn test (Fig. 3b) and inclined plane test (Fig. 3c). Infusion of 0.3 U collagenase correlates with worse neurological and functional deficits compared to infusion of 0.15 U collagenase.

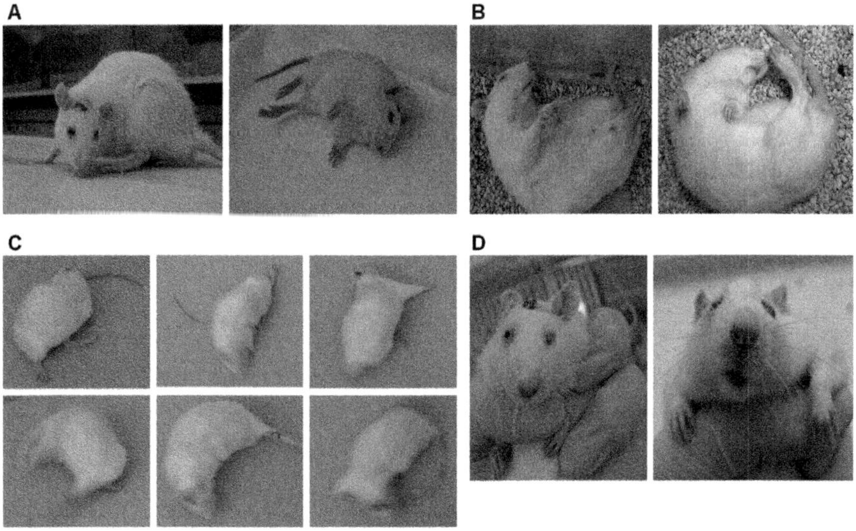

Fig. 2 Neurological impairment after pontine hemorrhage. Representative images of the spectrum of the neurological deficits 24 h after PH. Functional deficits after PH include circling (left) and minimal responsiveness (right) (**a**), loss of righting reflex and abnormal body contortions (**b**), circling (**c**), and ptosis, loss of corneal reflex and facial sensation, and facial weakness (**d**)

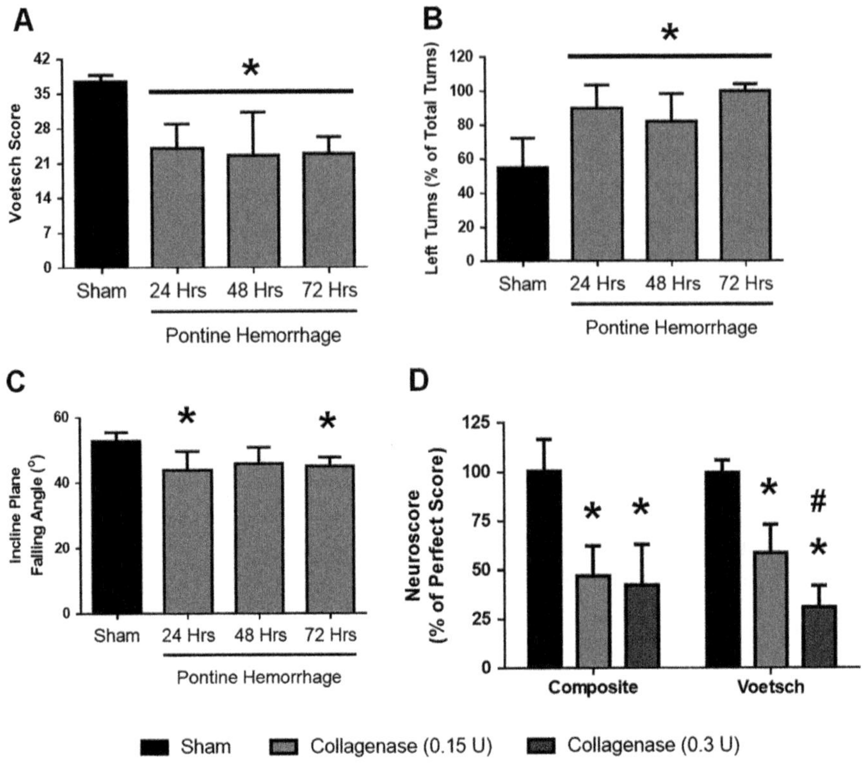

Fig. 3 Functional deficits after pontine hemorrhage. (**a**) Modified Voetsch score 24, 48, and 72 h post-PH. n = 4/group. ANOVA on Ranks with Dunn's post-hoc test. (**b**) Corner turn test. n = 4/group. One-way ANOVA with Tukey post-hoc test. (**c**) Inclined plane test. n = 4/group. One-way ANOVA with Tukey post-hoc test. (**d**) Composite Garcia Neuroscore and modified Voetsch score 24 h post-ictus. One-way ANOVA with Tukey post-hoc test. n = 8/group. Mean ± SD. $*p < 0.05$ vs. Sham animals, $\#p < 0.05$ vs. Collagenase (0.15 U)

Brain Water Content

Brain water content is evaluated via the wet weight/dry weight [15]. Significant increases in the brain stem brain water content of about 2% is observed after PH compared to sham animals for up to 72 h post-ictus (Fig. 4a). Brain stem brain water content is significantly higher when 0.3 U of collagenase are used to create PH compared to when 0.15 U of collagenase are used (Fig. 4b). This increase in brain water content translates to about a 12.5% increase in tissue volume after collagenase PH, which likely leads to neurological deficits [16].

Table 1 Modified Voetsch neuroscore criteria

Test	Score 0	1	2	3
Spontaneous head movement	Flexed to one side	Moves only to one side	Prefers one side	Movement in all directions
Confrontation	No response	Minimal response	Moderately responsive	Full responsiveness
Auditory reflex response	Not startled	Startled by loud clap	Startled by snap of fingers	Startled by finger rub
Ear pinch reflex	No reaction	Slow asymmetrical reaction	Brisk asymmetrical reaction	Brisk symmetrical reaction
Corneal reflex	No reaction	Slow asymmetrical reaction	Brisk asymmetrical reaction	Brisk symmetrical reaction
Vibrissae proprioception	No reaction	Slow asymmetrical reaction	Brisk asymmetrical reaction	Brisk symmetrical reaction
Neck stroking	No reaction	Slow asymmetrical reaction	Brisk asymmetrical reaction	Brisk symmetrical reaction
Exploration (5 min)	No movement	Slow movement	Reaches one wall	Reaches two walls
Circling	Falls to one side	Only turns in one direction	Prefers to turn in one direction	Bilateral turning
Body sensation	No reaction	Slow asymmetrical reaction	Brisk asymmetrical reaction	Brisk symmetrical reaction
4-limb outstretching	Complete paresis	Great hemi-paresis	Slight hemi-paresis	Equal bilateral use
Forelimb outstretching	Complete paresis	Great hemi-paresis	Slight hemi-paresis	Equal bilateral use
Climbing	Falls immediately	Holds on, no movement	Impaired climbing	Climbs to top
Beam balance	Falls immediately	Holds on, no movement	Explores part of the beam	Explores the entire beam

Fourteen sub-tests comprise the modified Voetsch neuroscore test, each receiving a score from 0 to 3 (0 is worst performance, 3 is best performance (i.e. no deficits)). Maximum, perfect score for the modified Voetsch neuroscore is 42 (indicating no neurological deficits)

Hematoma Assay

Injection of collagenase produces a large hematoma within the brain stem (Fig. 5a, b) which can be quantified using the hemoglobin assay [17]. The higher dose of collagenase (0.3 U) results in a significantly greater hematoma volume than the low dose collagenase (0.15 U), yet both doses of collagenase result in hematoma volumes significantly greater than that of sham animals (Fig. 5c).

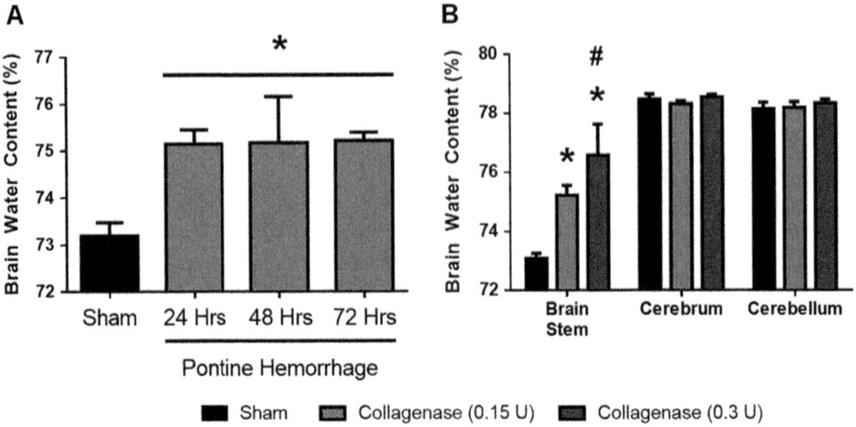

Fig. 4 Brain stem water content 24 h after pontine hemorrhage. Brain water content (%) computed using the Wet-Dry method. (**a**) Brain water content is elevated 24, 48, and 72 h post-ictus. n = 4/group. (**b**) Brain water content 24 h post-ictus. n = 5/group. Mean ± SD. One-way ANOVA with Tukey post-hoc test. ∗p < 0.05 vs. Sham animals, #p < 0.05 vs. Collagenase (0.15 U)

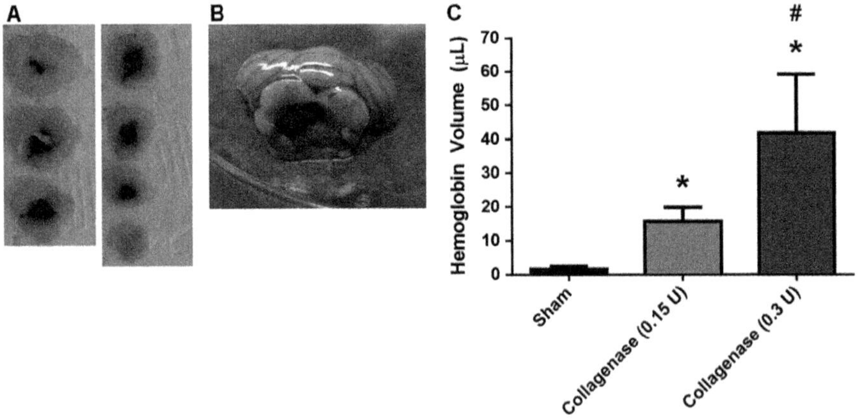

Fig. 5 Pontine hemorrhage and hematoma volume. Representative images of the hematoma within the brain stem of a rat 24 h after injury (**a**) and the whole Brain axial view (**b**). (**c**) Hematoma volume. n = 5–9/group. Mean ± SD. One-way ANOVA with Tukey post-hoc test. ∗p < 0.05 vs. Sham animals, #p < 0.05 vs. Collagenase (0.15 U)

Evans Blue Extravasation Assay

Evaluation of blood brain barrier integrity is typically done using the Evans blue extravasation assay [18, 19]. The amount of extravasated Evans blue dye in animals subjected to collagenase induced PH is significantly different from sham operated animals (Fig. 6). The higher dose of collagenase (0.3 U) results in significantly more extravasation of Evans blue than the low dose collagenase (0.15 U).

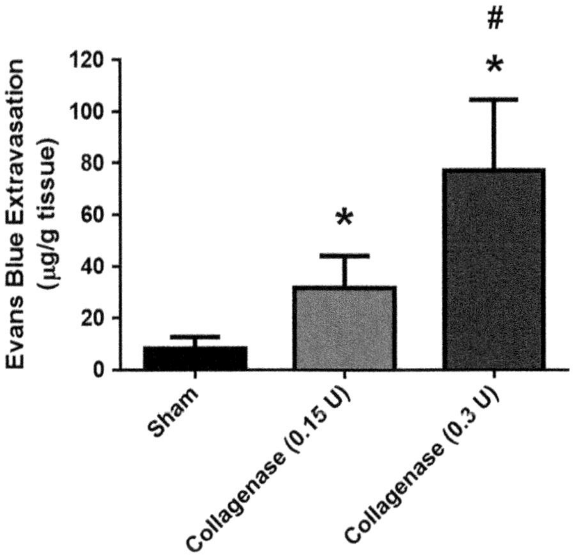

Fig. 6 Evans blue extravasation 24 h after pontine hemorrhage. n = 5/group. Mean ± SD. One-way ANOVA with Tukey post-hoc test. $*p < 0.05$ vs. Sham animals, $\#p < 0.05$ vs. Collagenase (0.15 U)

Magnetic Resonance Imaging (MRI)

Non-invasive neuroimaging techniques provide high resolution views of cerebral injury. MRI can be used to observe location and extent of the hematoma [20, 21]. The hematoma induce by pontine injection of collagenase can be observed with T1, T2, diffusion weighted, and susceptibility weight MRI (Fig. 7).

Notes

- Collagenase should be thawed just before injection. Allowing the collagenase to stand at room temperature leads to inactivation and variation in hematoma volume and other endpoints (i.e. neurological deficits, brain water content).
- When collagenase is transferred into the syringe, it is important to eliminate air bubbles as this can also lead to variations in hematoma volumes. Elimination of air bubbles can be achieved by aspirating the solution several times.
- Anesthesia can also be induced and maintained by co-administration of a Ketamine/Xylazine cocktail (100 and 10 mg/kg, respectively).

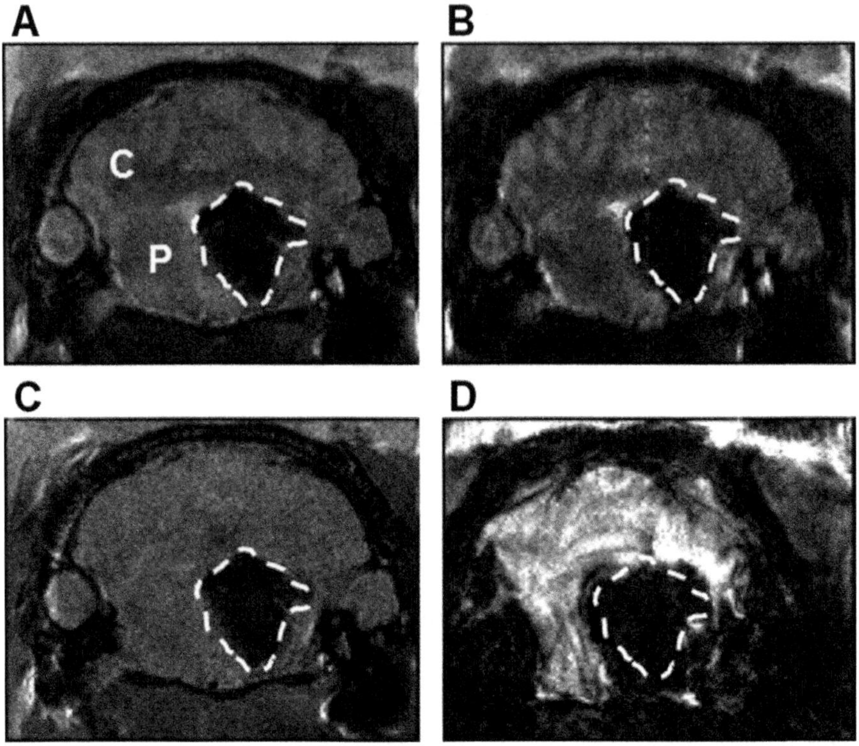

Fig. 7 MRI of pontine hemorrhage. Representative MRI scans of a rat brain 24 h after PH. (**a**) T2 MRI image displaying the hematoma (white dashed line). *C* cerebellum, *P* pons. (**b**) Diffusion weighted MR image displaying the hematoma (white dashed line shows the hematoma area from the T2 MRI scan). (**c**) T1 MRI image displaying the hematoma (white dashed line shows the hematoma area from the T2 MRI scan). (**d**) Susceptibility weighted MR image displaying the hematoma (white dashed line shows the hematoma area from the T2 MRI scan)

- Injection of autologous blood can be used to generate pontine hemorrhage; however, some researchers reported inconsistencies in hematomas produced with the blood model due to the small size of this region. Intra ventricular bleeding has also been documented with the blood-induced PH [4, 5].

References

1. Flaherty ML, Haverbusch M, Sekar P, Kissela B, Kleindorfer D, Moomaw CJ, et al. Long-term mortality after intracerebral hemorrhage. Neurology. 2006;66:1182–6.
2. Flaherty ML, Woo D, Haverbusch M, Sekar P, Khoury J, Sauerbeck L, et al. Racial variations in location and risk of intracerebral hemorrhage. Stroke. 2005;36:934–7.
3. Moncayo J. Pontine infarcts and hemorrhages. Front Neurol Neurosci. 2012;30:162–5.
4. Chung Y, Haines SJ. Experimental brain stem surgery. Neurosurg Clin N Am. 1993;4(3):405–14.

5. Cossu M, Pau A, Siccardi D, Viale GL. Infratentorial ischemia following experimental cerebellar hemorrhage in the rat. Acta Neurochir. 1994;131:146–50.
6. Lekic T, Tang J, Zhang JH. A rat model of pontine hemorrhage. Acta Neurochir Suppl. 2008;000:1–3.
7. Broderick JP, Adams HP Jr, Barsan W, Feinberg W, Feldmann E, Grotta J, et al. Guidelines for the management of spontaneous intracerebral hemorrhage: a statement for healthcare professionals from a special writing group of the Stroke Council, American Heart Association. Stroke. 1999;30:905–15.
8. Broderick JP, Connolly S, Feldmann E, Hanley D, Kase C, Krieger D, et al. Guidelines for the management of spontaneous intracerebral hemorrhage in adults: update: a guideline from the American Heart Association/American Stroke Association Stroke Council, High Blood Pressure Research Council, and the Quality of Care and Outcomes in Research Interdisciplinary Working Group. Stroke. 2007;38:2001–23.
9. NINDS. Priorities for clinical research in intracerebral hemorrhage report from a National Institute of Neurological Disorders and Stroke Workshop. Stroke. 2005;36:e23–41.
10. Fernandez AM, de la Vega AG, Torres-Aleman I. Insulin-like growth factor I restores motor coordination in a rat model of cerebellar ataxia. Proc Natl Acad Sci U S A. 1998;95:1253–8.
11. Garcia JH, Wagner S, Liu KF, Hu XJ. Neurological deficit and extent of neuronal necrosis attributable to middle cerebral artery occlusion in rats. Statistical validation. Stroke. 1995;26:41627–35.
12. Hartman R, Lekic T, Rojas H, Tang J, Zhang JH. Assessing functional outcomes following intracerebral hemorrhage in rats. Brain Res. 2009;1280:148–57.
13. Lekic T, Hartman R, Rojas H, Manaenko A, Chen W, Ayer R, Tang J, Zhang JH. Protective effect of melatonin upon neuropathology, striatal function, and memory ability after intracerebral hemorrhage in rats. J Neurotrauma. 2010;27:627–37.
14. Voetsch B, DeWitt LD, Pessin MS, Caplan LR. Basilar artery occlusive disease in the New England Medical Center Posterior Circulation Registry. Arch Neurol. 2004;61:496–504.
15. Tang J, Liu J, Zhou C, Alexander JS, Nanda A, Granger DN, Zhang JH. MMP-9 deficiency enhances collagenase-induced intracerebral hemorrhage and brain injury in mutant mice. J Cereb Blood Flow Metab. 2004;24:1133–45.
16. Keep RF, Hua Y, Xi G. Brain water content: a misunderstood measurement? Transl Stroke Res. 2012;3:263–5.
17. Choudhri TF, Hoh BL, Solomon RA, Connolly ES Jr, Pinsky DJ. Use of a spectrophotometric hemoglobin assay to objectively quantify intracerebral hemorrhage in mice. Stroke. 1997;28:2296–302.
18. Ma Q, Huang B, Khatibi N, Rolland W II, Suzuki H, Zhang JH, Jiping T. PDGFR-α inhibition preserves blood brain barrier after intracerebral hemorrhage. Ann Neurol. 2011;70:920–31.
19. Saria A, Lundberg JM. Evans blue fluorescence: quantitative and morphological evaluation of vascular permeability in animal tissues. J Neurosci Methods. 1983;8:41–9.
20. Kidwell CS, Wintermark M. Imaging of intracranial hemorrhage. Lancet Neurol. 2008;7:256–67.
21. Wijman CAC, Venkatasubramanian C, Bruins S, Fischbein N, Schwartz N. Utility of early MRI in the diagnosis and management of acute spontaneous intracerebral hemorrhage. Cerebrovasc Dis. 2010;30:456–63.

Germinal Matrix Hemorrhage Neonatal Rat Model Using Bacterial Collagenase Injection

Jerry J. Flores, Damon Klebe, Jiping Tang, and John H. Zhang

Abstract Bacterial collagenase has been used to model neonatal germinal matrix hemorrhage and adult cerebral hemorrhage. This chapter details the bacterial collagenase injection germinal matrix hemorrhage neonatal rat model.

Keywords Child neurology · Neonatal brain hemorrhage · Animal model · Germinal matrix · Experimental

Introduction

Germinal matrix hemorrhage (GMH) is the leading cause of morbidity and mortality in preterm infants in the United States. The microvasculature in preterm infants is underdeveloped, lacking pericyte and astrocyte end-feet coverage, fibronectin, and functional autoregulatory mechanisms, making the vessels prone to rupture from cardiorespiratory and hemodynamic instability associated with prematurity [1]. Those with GMH are often afflicted with long-term neurological deficits such as cerebral palsy, mental retardation, hydrocephalus, and psychiatric disorders [2]. This chapter details the clinically relevant bacterial collagenase injection model for

J. J. Flores · D. Klebe · J. Tang
Department of Physiology and Pharmacology, Loma Linda University School of Medicine, Loma Linda, CA, USA

J. H. Zhang (✉)
Department of Physiology and Pharmacology, Loma Linda University School of Medicine, Loma Linda, CA, USA

Department of Neurosurgery, Loma Linda University School of Medicine, Loma Linda, CA, USA

Department of Anesthesiology, Loma Linda University School of Medicine, Loma Linda, CA, USA
e-mail: jhzhang@llu.edu

J. Chen et al. (eds.), *Animal Models of Acute Neurological Injury*, Springer Series in Translational Stroke Research, https://doi.org/10.1007/978-3-030-16082-1_24

germinal matrix hemorrhage in neonatal rats, which results in acute brain injury, leukomalacia, post-hemorrhagic hydrocephalus, gliosis, and neurological consequences observed in humans [3].

Materials

Day 7 Sprague-Dawley rat pups (P7) (Harlan, Indianapolis, IN, USA) are used for this model. The following equipment was utilized to complete all surgical procedures:

1. Anesthesia gas blender with dual flow meter tubes and isoflurane vaporizer (Whitemore Enterprises), isoflurane, compressed oxygen (99.7%), compressed medical gas.
2. Rodent induction chamber (BrainTree Scientific, Braintree, MA)
3. Digital rodent stereotactic frame with rat pup adaptor (Stoelting Co., Wood Dale, IL)
4. Standard surgical equipment: scalpel, fine forceps, skin retractors, Kelly straight blade forceps, and scissors. (Fine Science Tools, Foster City, CA)
5. Cotton tipped applicator or absorbent surgical spears
6. Microdrill with burr (Ideal Micro Drill, CellPoint Scientific, Gaithersburg, MD)
7. Infusion pump (Harvard Apparatus, Holliston, MA, USA), Quintessential Stereotaxic Injector (Stoelting Co., Wood Dale, IL), and 27 gauge 10 μL syringe (Hamilton Company, Reno NV)
8. Bone wax (Lukens Bone Wax, Medline Industries, Mundelein, IL)
9. 5-0 Suture silk for wound closure
10. Bacterial collagenase (0.3 U/μL saline, VII-S, Sigma, St. Louis, MO)
11. Moist heat pad (Theratherm, Chattanooga Group, Hixson, TN)

Methods

Collagenase Preparation

1. Using ice cold sterile filtered phosphate-buffered saline (1 M), dissolve bacterial collagenase to a final concentration of 0.3 U/μL on ice.
2. Aliquot 20 μL of the bacterial collagenase solution into sterile PCR tubes.
3. Store the collagenase in −20 °C until ready for use.

Surgery

1. The rat pup is placed into an anesthesia induction chamber and exposed to 5% isoflurane. Isoflurane is delivered in a mixture of oxygen (0.3 L/min) and medical gas (0.7 L/min) for 2–3 min until animal is non-responsive.
2. To determine that the animal is non-responsive, a loss of the paw pinch reflex is seen. Gently remove the animal from the induction chamber and place the rat prone onto a stereotactic frame with a mouse/neonate adaptor. Once the animal is firmly mounted on the frame, anesthesia should be maintained at 2.5% isoflurane administered in a mixture of oxygen (0.3 L/min) and medical gas (0.7 L/min). (note: neonatal pups exposed to higher concentrations of isoflurane may be fatal)
3. Disinfect the skin with an alcohol pad and betadine.
4. Make a midline incision and separate the periosteum from the skull to expose lambda and bregma.
5. Prepare Hamilton syringe with 10 μL of bacterial collagenase and attach to the microinfusion pump with the bevel facing the midline.
6. Drill a burr hole (1 mm) at 1.6 mm right lateral and 1.5 mm rostral to bregma, into which a 27 gauge needle will be inserted at a rate of 1 mm/min to a depth of 2.8 mm below the dura.
7. Infuse bacterial collagenase at a rate of 0.333 μL/min for a total of 1 μL.
8. Once the infusion is completed, leave the needle in place for 5 min to minimize the risk of back-flow.
9. Slowly remove the needle at a rate of 0.5 mm/min.
10. Immediately after needle removal, apply bone wax to seal the burr hole.
11. Suture the incision closed using a 5-0 mm suture.
12. Place the neonatal pup in a recovery cage on a thermostat-controlled warming blanket to recover.
13. When the animal has awoken, return it to its mother.

Outcome Evaluations and Anticipated Results

This collagenase-induced GMH rat pup model allows us to study multiple outcomes: hematoma volume, neurocognitive developmental delays, post-hemorrhagic ventricular dilation, brain atrophy, neuroinflammation, and other relevant pathophysiological outcomes [3–5].

Injection of bacterial collagenase in the right hemisphere of the brain results in significant hematoma formation, and bilateral bleeding is common 24 h after induction (Fig. 1A). Short-term hemoglobin assay of the right and left hemisphere of the brain is the primary tool to assess hematoma volume in the brain and to demonstrate that the model has been conducted correctly. Hematoma volume is significantly

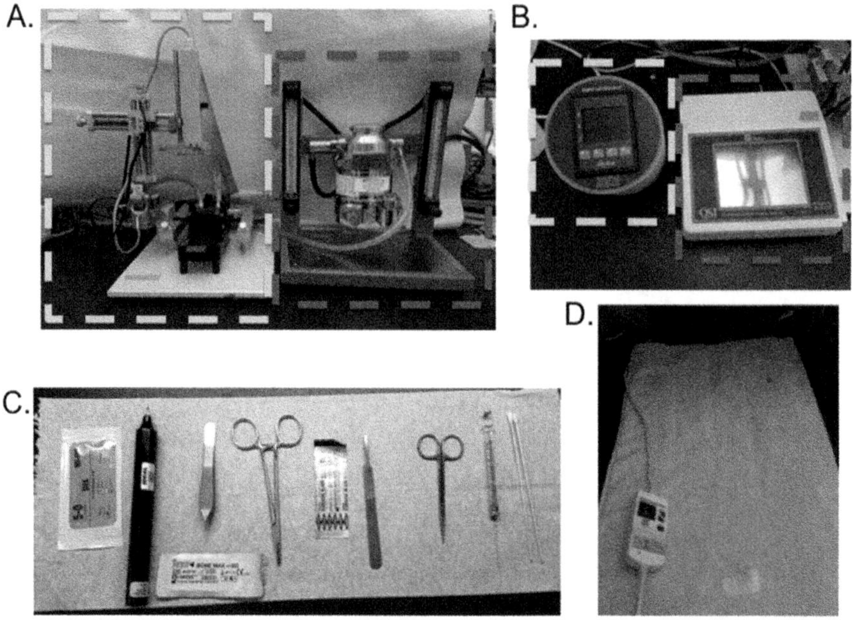

Fig. 1 Tools needed for GMH surgery. (**a**) Stereotactic Frame (left panel) and Anesthesia gas blender with dual flow meter tubes and isoflurane vaporizer (right panel). (**b**) Digital display for the stereotaxic frame (left panel) and Quintessential Stereotaxic Injector (right panel). (**c**) Standard surgical equipment: Suture, Microdril, Fine forceps, Bone Wax, Kelly Straight Blade Forceps, Sterile Scalpel Blade, Scalpel, Scissors, 27 gauge 10 μL syringe, sterile Cotton swabs (**d**) Heating pad for post-operative recovery

increased at 24, 72, and 7 days when compared to animals at 0 h (GMH is not induced) 24 h after GMH induction (Fig. 2B). Non-invasive imaging by MRI can be used to delineate the hematoma volume, yet is not required as hemoglobin assay is a cheaper and faster method to measure volume.

Neurological deficits in the short-term can be assessed up to 10 days after GMH [4]. Such tests include the righting reflex, time needed for neonates to completely roll onto all four limbs when placed on their backs, and negative geotaxis, where the animals is timed for how long it would take it to rotate (180°) after being placed head down on a slope (20° angle).

Long-term neurological deficits after GMH can also be assessed after GMH. Higher order brain function is evaluated usually 3–4 weeks after GMH using T-Maze, to measure short-term (working) memory, and Morris Water maze, which assesses spatial learning and memory on five blocks daily [4]. Sensorimotor deficits can be assessed by Rotarod and Foot-fault evaluations. At 4 weeks after GMH, brain atrophy and post-hemorrhagic ventricular dilation is observed in GMH rats compared to sham operated animals (Fig. 3). Long-term deficits of each animal can be corroborated with histopathological analysis, usually volumetric analysis using Nissl stained coronal brain sections.

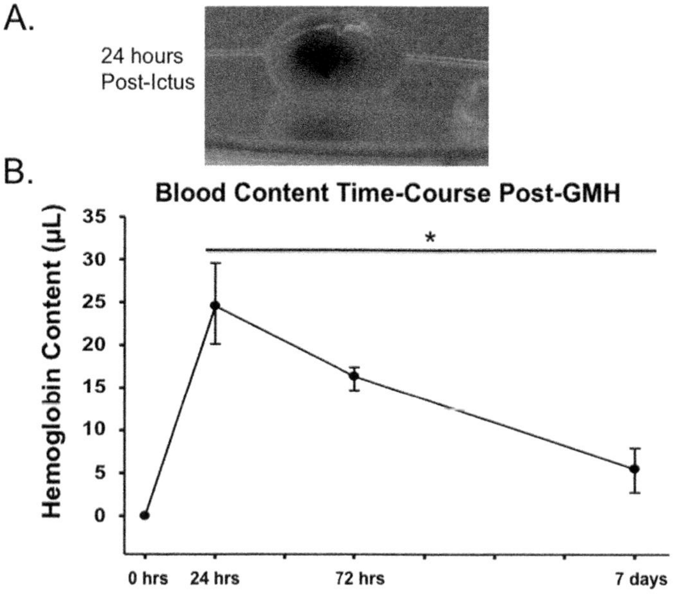

Fig. 2 (**a**) Image of a neonatal brain 24 h after GMH induction. (**b**) Time-course of hematoma volume in neonatal brains at 0, 24, 72 h, and 7 days. Values are expressed as mean ± SD. $^*P < 0.05$ compared with 0 h. N = 4 per time point

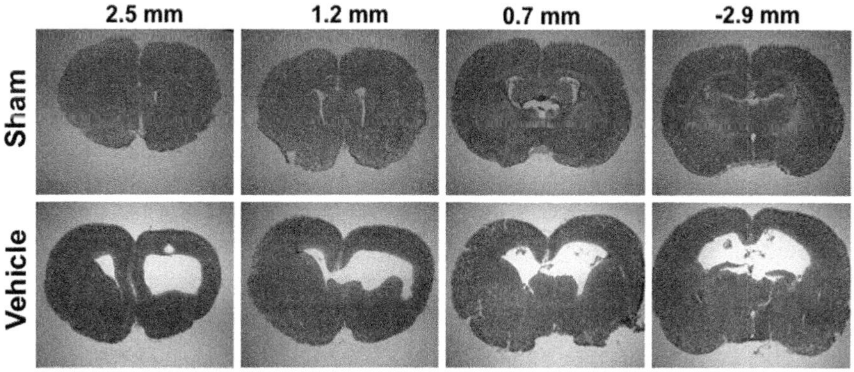

Fig. 3 Twenty-eight days after germinal matrix hemorrhage. Representative micrographs of Nissl-stained brain sections were taken. Note: measurements on the top of the representative pictures indicate the location of the brain section from bregma

Caveats, Limitations, and Alternatives

Preparation of Bacterial Collagenase

It is critical in preparing the bacterial collagenase to keep everything ice cold. This includes all solutions and tubes. Allowing the collagenase to heat up can quickly decrease the activity and effectiveness of the collagenase, ultimately reducing the severity and reproducibility of the hematoma.

Surgical Procedures

A number of crucial steps lie in the injection of the collagenase. First, the collagenase should be removed from the freezer and thawed immediately prior to injection. Similar to the preparation of the collagenase solution, allowing the collagenase to warm up reduces its effectiveness. Second, after complete injection of the collagenase, the needle must be allowed to remain in place for 5 min before slow removal to minimize the back-flow of the solution. Evidence of significant back-flow can be observed by hemorrhage along the needle tract.

Choice of Species and Age

Postnatal day neonatal Sprague-Dawley rats were used for this model because brain development is similar to 28–32 week gestational age humans [3]. However, this model may be adapted for use in other neonatal rodent and other species by appropriately adjusting the stereotactic coordinates and the concentration of bacterial collagenase. Furthermore, this model mimics the post-GMH pathophysiological outcomes, but the pathophysiology causing GMH (proteases degrading blood vessels) is not best presented in this model. Other animal models, such as the rabbit pup glycerol-injection model, beagle pup hypercarbia/hypertensive models, and/or piglet models, deserve consideration as well when designing GMH studies, since they may better present pathophysiological processes causing GMH [6].

Summary

The germinal matrix hemorrhage model utilizes bacterial collagenase to induce a large and consistent hematoma. This model results in large volumes of hemoglobin content and functional deficits. Another advantage of the use of bacterial collagenase is its ability to cause a hemorrhage with long-term histopathological sequelae and neurological deficits (sensorimotor and cognitive). The observed gliosis, leukamalacia, acute brain injury, post-hemorrhagic ventricular dilation, brain atrophy, and neurodevelopmental delays are similar to pathophysiological outcomes in human GMH patients.

References

1. Ballabh P. Pathogenesis and prevention of intraventricular hemorrhage. Clin Perinatol. 2014;41(1):47–67. Epub 2014/02/15.
2. Lekic T, Klebe D, Poblete R, Krafft PR, Rolland WB, Tang J, et al. Neonatal brain hemorrhage (NBH) of prematurity: translational mechanisms of the vascular-neural network. Curr Med Chem. 2015;22(10):1214–38. Epub 2015/01/27.
3. Lekic T, Manaenko A, Rolland W, Krafft PR, Peters R, Hartman RE, et al. Rodent neonatal germinal matrix hemorrhage mimics the human brain injury, neurological consequences, and post-hemorrhagic hydrocephalus. Exp Neurol. 2012;236(1):69–78. Epub 2012/04/25.
4. Lekic T, Manaenko A, Rolland W, Tang J, Zhang JH. A novel preclinical model of germinal matrix hemorrhage using neonatal rats. Acta Neurochir Suppl. 2011;111:55–60. Epub 2011/07/05.
5. Tang J, Chen Q, Guo J, Yang L, Tao Y, Li L, et al. Minocycline attenuates neonatal germinal-matrix-hemorrhage-induced neuroinflammation and brain edema by activating cannabinoid receptor 2. Mol Neurobiol. 2016;53(3):1935–48. Epub 2015/04/03.
6. Balasubramaniam J, Del Bigio MR. Animal models of germinal matrix hemorrhage. J Child Neurol. 2006;21(5):365–71. Epub 2006/08/12.

Part IV
Traumatic CNS Injury Animal Models

Fluid Percussion Injury Model

Rachel K. Rowe, Daniel R. Griffiths, and Jonathan Lifshitz

Abstract Research models of traumatic brain injury (TBI) hold significant validity towards the human condition, with each model replicating a subset of clinical features and symptoms. Hallmarks of traumatic brain injury in man can be faithfully reproduced in the laboratory using fluid percussion injury. Variations in the surgical procedure provide the ability to induce focal diffuse or mixed focal and diffuse brain injury in various laboratory species. Being fully scalable, fluid percussion can induce mild, moderate, or severe brain injury in subjects of either sex, at any age. This chapter outlines the procedures for midline (diffuse) fluid percussion brain injury in adult and post-natal 17 and 35 male Sprague-Dawley rats and lateral (mixed) fluid percussion brain injury in adult male mice. With these procedures, it is possible to generate brain-injured laboratory animals for studies of injury-induced pathophysiology and behavioral deficits, for which rational therapeutic interventions can be implemented.

Keywords Fluid percussion · Rat · Mouse · Juvenile · Adolescent · Brain injury · Concussion · Trauma

R. K. Rowe · J. Lifshitz (✉)
BARROW Neurological Institute at Phoenix Children's Hospital, Phoenix, AZ, USA

Department of Child Health, University of Arizona College of Medicine—Phoenix, Phoenix, AZ, USA

Phoenix Veteran Affairs Healthcare System, Phoenix, AZ, USA
e-mail: jlifshitz@email.arizona.edu

D. R. Griffiths
BARROW Neurological Institute at Phoenix Children's Hospital, Phoenix, AZ, USA

Department of Child Health, University of Arizona College of Medicine—Phoenix, Phoenix, AZ, USA

J. Chen et al. (eds.), *Animal Models of Acute Neurological Injury*, Springer Series in Translational Stroke Research, https://doi.org/10.1007/978-3-030-16082-1_25

Model Selection

Midline fluid percussion permits the study of experimental traumatic brain injury (TBI) in a model that is reproducible, clinically relevant, and scalable between species and injury severities. Brain injury is induced by a rapid (~20 ms) fluid pulse through a craniectomy onto the intact dura that follows the inner curvature of the skull and creates an elastic decompression of the brain [2, 10]. While fluid percussion injury (FPI) necessitates breaching the cranial vault, the skull is sealed to the injury device, recreating a closed system, which approximates a closed head injury with decompressive craniectomy. The mechanical forces disrupt cell membranes, blood vessels, and neuronal processes. By increasing the angle from which the pendulum hammer falls, greater pressures can be generated to travel through the fluid-filled cylinder and impact the brain. At a moderate level of injury 20–25% of animals die as a result of the injury within the acute post-traumatic period (15 min), generally from respiratory failure and pulmonary edema. This is a normal and desired feature of TBI models, as it reflects human TBI.

In laboratories worldwide, subtle variations in surgical and injury procedures reproduce the spectrum of brain injuries found in the human population. Primarily, the location of injury site determines the major features of the injury, where a midline location induces a *diffuse injury* and a lateral location induces a *focal injury with a diffuse component* [4, 5, 12]. Fluid percussion injury reproduces the acute reflex suppression, functional deficits, and histopathology evident after TBI in man [3, 6, 11]. The model continues to be implemented to evaluate pathophysiological mechanisms underlying histological and behavioral deficits, and therapeutic interventions to mitigate degeneration and promote the recovery of function [11].

Materials

Animals

Fluid percussion brain injury has been successfully performed on various species, including cats, rabbits, pigs, rats, and mice. The adaptation of fluid percussion to rats [3, 7, 8] was followed by its implementation in mice [1]. The procedures outlined in this chapter focus on adult male Sprague-Dawley rats (approximately 300–400 g), post-natal day (PND) 17 and PND35 male Sprague-Dawley rats, and 8-week-old adult male C57BL/6 mice (approximately 20–30 g). To maximize the success of brain injury, examine all animals for any signs of ill health (e.g., rough coat, bleeding or dirty eyes, runny or bleeding nose, and scratched around eyes or nose area). Weigh all animals prior to surgery in order to track injury-induced weight loss.

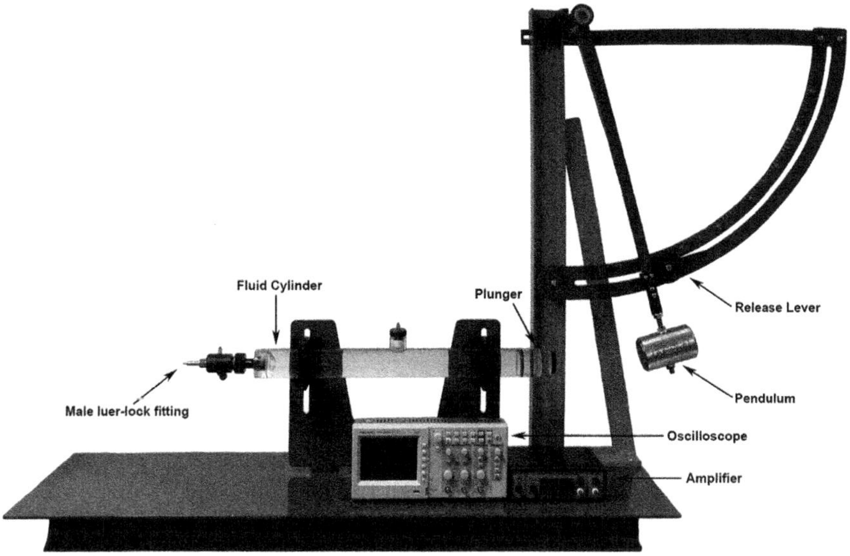

Fig. 1 Fluid percussion injury device. Injury is induced by a 20 ms fluid pulse delivered onto the intact dura via a craniectomy and surgically implanted injury hub. The fluid pulse is generated by the pressure wave produced when the weighted end of the pendulum arm strikes the plunger of a fluid-filled cylinder. The force of the pulse is detected by a transducer and the signal is amplified before being sent to the oscilloscope which outputs the millivolts. The millivolts can then be converted to atmospheres of pressure

Equipment

Injury Device

1. Fluid percussion injury device (Fig. 1)
2. Custom Design and Fabrication
3. Virginia Commonwealth University
4. http://www.radiology.vcu.edu/research/customdesign/fpi.html

Product information including assembly manual, operation manual, and product brochure are provided on the manufacturer's website.

- **Note:** the plunger impact pad on the fluid cylinder should be replaced every 8–12 months. Information and instructions for the setup, cleaning, and maintenance of the FPI device can be found in the FPI Operation Manual: http://www.radiology.vcu.edu/docs/FPIOperationManual.pdf.

- Recording oscilloscope (recommended: Tektronix, Model 1001B)
- Industrial Velcro to secure the device to the bench to prevent movement
- High-vacuum grease (e.g., Fisher Scientific, #14-635-5D)
- Dishwashing solution to clean fluid cylinder
- Jet Dry finishing rinse to minimize air bubbles in the cylinder upon filling

Anesthesia

- Vaporizer for delivery of inhaled anesthesia
- Tubing/petcocks
- Induction chamber
- Isoflurane
- Oxygen and regulator
- Rodent nose cone for inhaled anesthetic that is compatible with the stereotaxic frame.

- **Note:** When using an inhaled anesthetic, it is recommended that all procedures are performed in a well-ventilated area, on a downdraft or similar table, or in a type II biosafety cabinet to minimize anesthesia exposure to the surgeon.

Surgical Supplies

- Gauze sponges
- Cotton tip applicators
- Heating pad (recommended: Deltaphase isothermal heating pad, BrainTree Scientific, #39DP)
- 20 gauge needles (recommended: 1″ length)
- 1-mL syringes
- ≥10-mL syringes, Luer-lock tip
- Small animal trimmer for fur removal (e.g., Wahl, Mini Arco Animal Trimmer)
- Ophthalmic ointment to prevent drying of eyes during surgery
- 4% chlorhexidine solution (or Betadine scrub) for preparation of the incision
- 70% ethanol (or alcohol pads)
- Cyanoacrylate (e.g. Super Glue)
- Perm Reline and Repair resin, liquid and powder (All for Dentist, #H00327)
- Antibiotic Ointment
- Saline-filled syringe, blunted needle bent 90°

Surgical Instruments

- Small animal stereotaxic frame
- Scalpel handle and blade
- Delicate bone scraper (Fine Science Tools, #10075-16)
- Wedelstaedt Chisels ¾ DE (Henry Schein, #600-4972)
- Bull Dog clips (Fine Science Tools, #18050-28, #18051-28)
- Needle holder and scissors

Rat Surgical Instruments

- Dremel tool with engraving cutter #106
- Trephine: 4.7 mm-adult (Miltex, #26-140)

  4.0 mm-PND35; 3.0 mm-PND17 (Fig. 2, custom made; Machine Shop, Arizona State University, Tempe, AZ) contact Rachel Rowe, rkro222@email.arizona.edu

- Fingernail drill with 5/64″ drill bit (Miltex, #33-232)
- Stainless steel skull screws (2-56 × 3/16″) (Small parts Inc., #MX-0256-03B-25)

Mouse Surgical Instruments

- Trephine (3.0 mm) (Fig. 2, Machine Shop, Arizona State University, Tempe, AZ) contact Rachel Rowe, rkro222@email.arizona.edu
- Weed whacker line for cranial disc (1.7 mm diameter)
- Side-grasping forceps (7 × 7) (Henry Schein, #6-124XL)
- 3 M Vetbond tissue adhesive (Henry Schein, #700-3449)

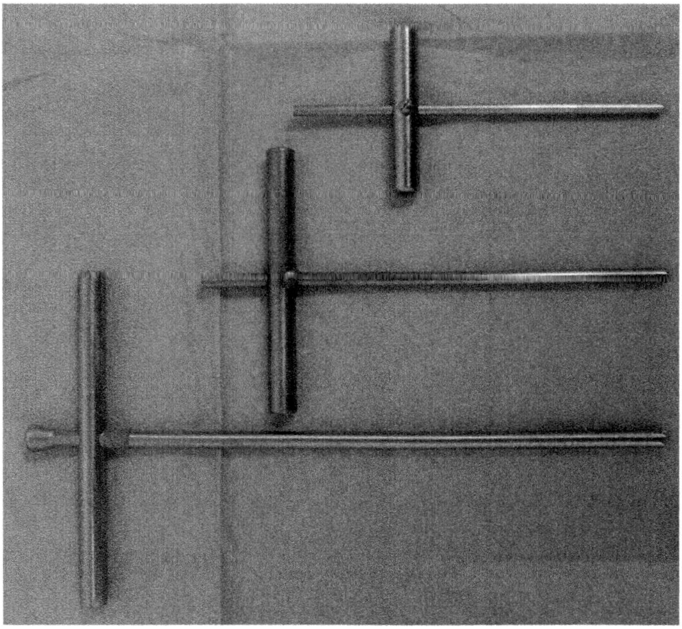

Fig. 2 Custom designed and machined trephines. Trephine of diameter 4.7 mm (bottom) used for the adult rat. 4.0 mm trephine (center) used for PND 35 rats. 3.0 mm trephine (top) used for the mouse PND 17 rat. The trephines displayed are scaled for different aged rodents to accommodate the size difference of their skulls during developmental ages

Injury Hub (Fig. 2)

- 1 ½″ needle (20 gauge) (Becton Dickinson, #305176)
- Syringe (1 cc)
- Razor blades
- Tissue forceps (Henry Schein, #6-114)

Rat Injury Hub

- Cosmetic pencil sharpener

Mouse and PND17 Injury Hub

- Luer-loc extension tubing (Baxter, #2C5643)

Procedure

Administer Anesthesia and Secure in Head Holder

- Follow a standard surgery sheet, recording the progress of the surgery (see Appendix).
- Anesthetize the animal with 5% isoflurane for 5 min in an induction chamber.
- Shave or remove hair from scalp, as appropriate.
- Secure and level the animal in a stereotaxic frame equipped with a nose cone for continuous inhalation of isoflurane (2.0–2.5%). The back of the front incisors should be flush with the bite bar, without tension applied to the teeth. If you observe mouth breathing, check the positioning of the teeth over the bite bar and/or reposition the nose cone to allow for normal respiration.
- Apply ophthalmic ointment to the eyes to keep them moist during the surgery.
- Prepare the surgical area with 70% alcohol and betadine solution (antiseptic).
- Monitor anesthesia by observing muscle relaxation, in addition to assessing the toe pinch reflex. Animals under appropriate anesthesia will have a steady respiration rate.

Cranial Surgery for Hub Placement

- Make a midline sagittal incision extending from between the eyes to the base of the skull, just past the ears. To minimize bleeding, avoid cutting the muscle at the base of the skull (Figs. 3a, 4a, and 5a).

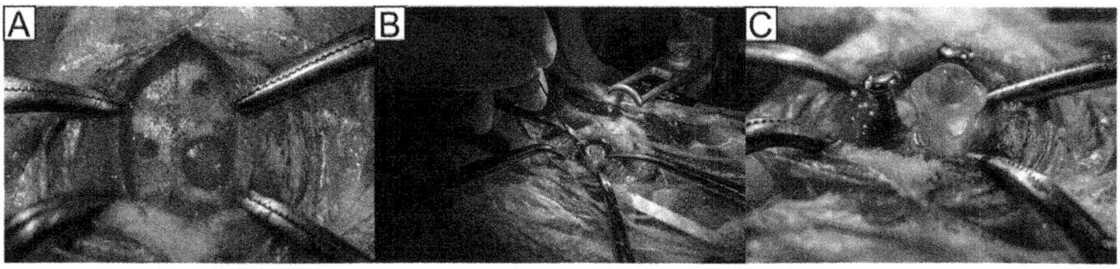

Fig. 3 Cranial surgery for hub placement for lateral FPI in the rat. The rat's head is shaved and the animal is secured in a stereotaxic frame with a continuous flow of isoflourane via a nose cone. A midline incision is made to expose the skull and the overlying fascia is removed. A Dremel tool is used to make three pilot holes. The screw holes are expanded with a finger nail drill and 5/64″ drill bit (**a**). A 4.7 mm diameter trephine is used to create a cranial disc that is removed to expose the underlying dura (**a**). A stainless steel screws are secured into the screw holes (**b**). Small drops of cyanoacrylate gel are placed on the outside of the constructed injury hub, and the hub placed inside the craniectomy (**b**). After the cyanoacrylate gel dries, the injury hub and screw are covered in methyl methacrylate cement and the injury hub is filled with saline (**c**)

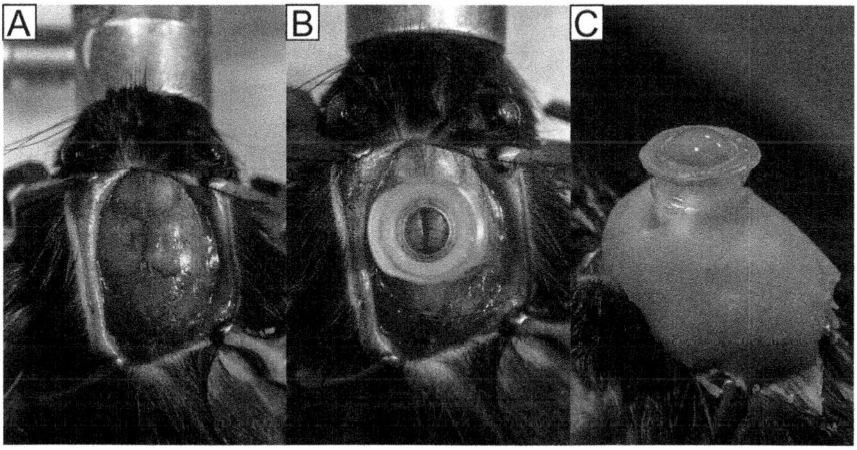

Fig. 4 Cranial surgery for hub placement in the mouse. A midline incision is made to expose the skull and the overlying fascia is removed (**a**). Vetbond tissue adhesive is used to secure a disc shaved from weed whacker line at the location of the craniectomy (**a**). A 3 mm diameter trephine is used to create a cranial disc that is removed to expose the underlying dura (**b**). Small drops of cyanoacrylate gel are placed on the outside of the constructed injury hub, and the hub placed outside the craniectomy (**b**). After the cyanoacrylate gel dries, the injury is covered in methyl methacrylate cement and the injury hub is filled with saline (**c**)

- Expose the skull and scrape the fascia from the skull using a delicate bone scraper, cotton swabs, and gauze. Clear away temporal muscle as necessary. If greater exposure is needed, stretch the skin by applying pressure with the fingers.
- Attach Bull Dog clips to the edges of the incision (two anterior, two posterior) to expose the surgical site. When the Bull Dog clips fall down, the weight will hold the incision open (Figs. 3a, 4a, and 5a).

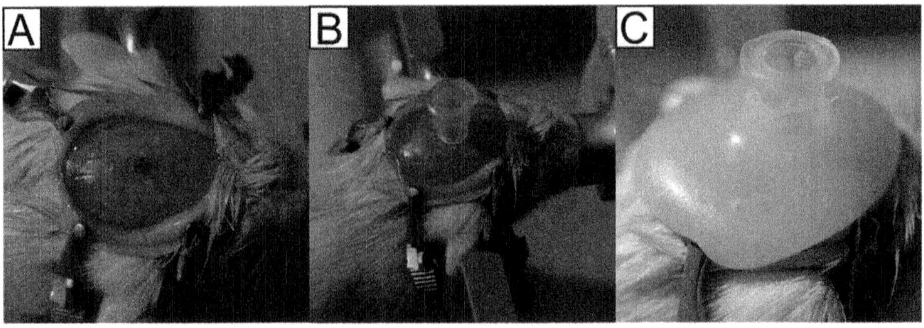

Fig. 5 Cranial surgery for hub placement in PND 17 rat. A midline incision is made to expose the skull and the overlying fascia is removed (**a**). Vetbond tissue adhesive is used to secure a disc shaved from weed whacker line at the location of the craniectomy (**a**). A 3 mm diameter trephine is used to create a cranial disc that is removed to expose the underlying dura. Small drops of cyanoacrylate gel are placed on the outside of the constructed injury hub, and the hub placed outside the craniectomy (**b**). After the cyanoacrylate gel dries, the injury is covered in methyl methacrylate cement and the injury hub is filled with saline (**c**)

Cranial Surgery for Hub Placement-Adult Rat

- Mark the locations on the skull for the screw hole(s) and craniectomy center. The skull screw is used to secure the injury hub in place.
  - For midline FPI, position one screw hole 1 mm lateral to bregma and 1 mm rostral to the coronal suture on the right side. For PND 35 rats, no screw hole is needed.
  - For lateral FPI, position the one screw hole 1 mm lateral to bregma and 1 mm rostral to the coronal suture on the ipsilateral side to the craniectomy; position the second hole midway between bregma and lambda and 1 mm lateral to the central suture contralateral to the craniectomy.
- Drill pilot holes at both markings using the Dremel tool and burr bit.
- Expand the screw hole with a finger nail drill and 5/64″ drill bit (Fig. 3a).
- Place the centering pin inside the 4.7 mm diameter trephine. Anchor the centering pin in the pilot hole at the craniectomy center. For PND 35 rats, place the pin inside the 4.0 mm diameter trephine and anchor in pilot hole.
- Continually turn and spin the trephine to make a craniectomy without disrupting the underlying dura. Keep trephine clean by using a toothbrush to remove bone debris from the trephine teeth. Apply saline to moisten the bone and aid in trephination. As needed, angle the trephine to evenly cut around the craniectomy.
- Frequently check the progress of the craniectomy by applying mild pressure to the center of the craniectomy using forceps or an applicator stick. As the skull thins, the craniectomy will move independently of the skull. The craniectomy is complete when the bone can move freely in all directions.

- Remove the bone piece, working around the circumference using the wedelstaedt and scalpel, or two wedelstaedt instruments without disrupting the dura. When the bone has been removed, gently clear any blood from the craniectomy site using a cotton-tipped applicator (Fig. 3a).
- **Note:** If the surgery site continues to bleed when the skull is removed, lightly remove blood with gauze. Adding saline can create hydrostatic pressure that will reduce bleeding. If the site continues to bleed, control the bleeding with Gelfoam. Excessive wiping or dabbing at the craniectomy site will prevent blood clotting and worsen the bleed.
- Secure a stainless steel screw in the skull screw hole. Hold the screw with tissue forceps and advance the screw with a screwdriver (Fig. 3c).

Cranial Surgery for Hub Placement-Mouse and PND 17 Rat

- Shave weed whacker line with a razor blade as thin as possible to make a circular disc that is an equal thickness on all sides. Disc should be level when placed on the skull (Figs. 4a and 5a).
- Pick up the disc with side-grasping forceps. Dip the cranial disc into a drop of Vetbond tissue adhesive placed on a nonabsorbent surface.
- Place the disc at the location of the craniectomy (midway between bregma and lamda on the sagittal suture). To drop the disc, release the forceps and use a wooden applicator stick to position the disc. Once in position, use a twisted Kimwipe tissue to wick away any excess Vetbond (Figs. 4a and 5a). Allow the Vetbond to fully dry before beginning to trephine.
- Place the 3.0 mm trephine over the disc and perform the craniectomy by continually turning and spinning the trephine without disrupting the underlying dura. Keep trephine clean by using a toothbrush to remove bone debris from trephine teeth. Apply saline to moisten the bone and aid in trephination. As needed, angle the trephine to evenly cut around the craniectomy.
- **Note:** If the disc comes off while trephining, clean excess dried glue from the skull and apply a new disc using Vetbond. However, if the bone can move independently of the skull in an area, use a small dot of superglue to attach a new disc (Vetbond will run and may touch the surface of the dura compromising the surgery).
- Frequently check the progress of the craniectomy by applying mild pressure to the center of the craniectomy with forceps or an applicator stick. As the skull thins, the craniectomy will be able to move independently of the skull. The craniectomy is complete when the bone can move freely in all directions.
- Under magnification, remove the bone piece working around the circumference using the wedelstaedt and scalpel, or two wedelstaedt instruments without disrupting the dura (Fig. 4b). When the bone has been removed, gently clear any blood from the craniectomy site.

- **Note:** If the surgery site continues to bleed when the skull is removed lightly remove blood with gauze. Adding saline can create hydrostatic pressure that will reduce bleeding. If the site continues to bleed, control the bleeding with Gelfoam. Excessive wiping or dabbing at the craniectomy site will prevent blood clotting and worsen the bleed.

Injury Hub

Injury Hub Construction

- Attach a 22 gauge, 1½″ needle to a 1 cc syringe. Place the needle into a laboratory bench pad (Fig. 6a).
- Cut the female Luer-Loc hub from the needle using a razor blade (Fig. 6a). The cut is made parallel to the Luer-loc with an outer diameter of ~4.7 mm for the rat, ~4.0 mm for the PND35 rats, and ~3.0 mm for the mouse and PND17 rat.
- **Note:** To confirm the proper diameter you can place the trephine through the hub and confirm a tight fit.
- Inspect the cut edge of the injury hub and trim to size and level as necessary (Fig. 6b).
- For the rat, bevel the cut edge of the injury hub with a cosmetic pencil sharpener.
- Shave thin burrs around the injury hub starting at the Luer-Loc edge in the direction of the cut edge using a razor blade (Fig. 6c).

Injury Hub Placement

- Hold the hub in tissue forceps (behind the teeth). Apply small drops of cyanoacrylate gel on the outside of the hub, just above the cut end.

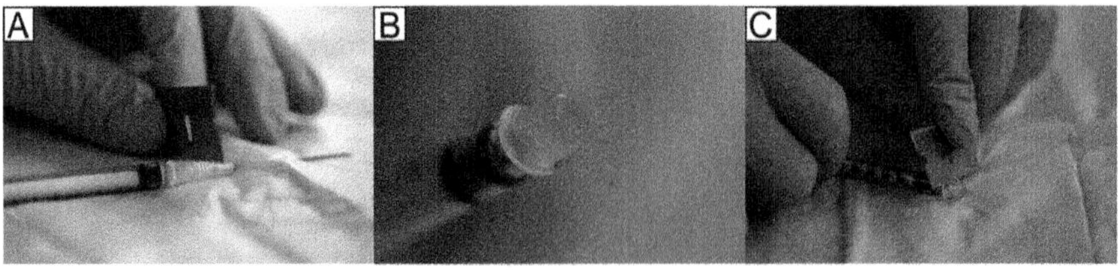

Fig. 6 Injury hub construction. Firmly attach a 20 gauge needle to a 1 cc syringe and insert the needle into a laboratory bench pad to prevent the needle from becoming projected after it is cut (**a**). Use a razor blade to cut off the tip of the needle (**a**). The injury hub is checked and refined for level (**b**). For the rat, the injury hub is beveled using a cosmetic pencil sharpener. Using a razor blade, score the exterior of the hub making burrs at even intervals around the hub (**c**). When finished, the cut end should be flat and even, and parallel to the Luer-Loc plane

- Position the hub over the craniectomy (using magnification for the mouse). For the rat, the injury hub fits inside the craniectomy (Fig. 3b, c). For the mouse and PND 17 rat, the injury hub fits outside the craniectomy (Figs. 4b and 5b).
- Using a wooden applicator stick (cut a sharp angle) gently scrape the cyanoacrylate gel down the injury hub onto the skull. Apply more cyanoacrylate gel if needed to the junction between the injury hub and the skull to firmly adhere the injury hub to the skull in addition to creating a seal.
- After the cyanoacrylate gel dries, cover the injury hub (and screw) in methyl methacrylate cement (Figs. 4c and 5c). Apply the methyl methacrylate cement from a 1 cc syringe when it is thick enough to hold shape.
- When the methacrylate cement has dried, fill the injury hub with saline (Fig. 4c).
- Place a suture at both the anterior and posterior edges of the incision.
- Remove the animal from the stereotaxic frame and anesthesia. Place the animal in a recovery cage on a heating pad until the animal is awake and alert. Monitor animals for outward signs of pain or distress.

Injury

Before using the injury device, check that when the weighted pendulum arm is hanging in a neutral position (at 0°) that it is flush and centered on the foam pad at the end of the plunger. Adjust as needed by injecting or removing fluid from the device. Drop the pendulum hammer several times to prime the device.

- Reanesthetize the animal after an approximately 60 min recovery period from surgery.
- Visually inspect inside the injury hub for debris, blood, or dried dental acrylic. Clean out the injury hub using a small cotton tip applicator or irrigate with saline if necessary.
- Fill the injury hub with sterile saline until a bead of fluid is formed by surface tension. Remove any air bubbles from inside the hub.
- To avoid air between the hub and device, press the plunger so that a drop of fluid is produced at the end of the injury device. Connect the female Luer-Loc injury hub on the animal to the male Luer-Loc fitting on the injury device. Create continuity between the fluid of the cylinder and the fluid in the injury hub.
- **Note:** Rats should be held in your left hand lying on their right side. Attach the rat directly to the device. For mice, attach them to the device using a Luer-Loc extension tube.
- Check the animal for a toe pinch response. Once a normal breathing pattern returns (1–2 breaths per second) and the animal has a positive toe pinch response, release the pendulum to injure the animal. Secure the pendulum after it strikes the plunger once and return it to the catch.
- Immediately after the injury, start a timer to measure the duration of physiological responses.

- Remove the injury hub by pressing on the bridge of the nose for leverage. Visually inspect the hub for obstructions.
- Observe and record the duration and extent of apnea or seizure. Note the condition/appearance of the surgical site and brain tissue beneath the injury site and record brain herniation and hemorrhage. If the dura is breached, the animal should be euthanized and not included in the study.
- Control bleeding if necessary. Leave the craniectomy open. Close the wound (i.e., suture or staple) and apply topical lidocaine and antibiotic ointment.
- Place the animal in a supine position on a heating pad. The time elapsed from the injury until the animal spontaneously rights is recorded as the righting reflex time.
- Once the animal has righted, place it in a designated recovery area equipped with a heating pad.
- When the animal regains normal ambulatory behavior, it can be returned to its home cage.

Postoperative Care

Postoperative Evaluations

- Following injury, animals should be visually monitored for continued recovery every 10 min post-injury (for the first hour). Within 15–20 min after injury, surviving animals should be alert. Within 1 h after injury, animals should be ambulatory. Brain-injured and uninjured control animals typically show no outward effects once they have recovered from anesthesia, and resume normal eating, drinking, and grooming patterns. Typically animals return to sleep, as the injury occurs during their sleep cycle.
- Postoperative evaluations should be done daily (for a minimum of 3 days). Follow the Postoperative Evaluation Sheet to record the external examination, physical examination, suture site, and a pain evaluation. Typically, animals require no special supportive care after surgery. This injury does not produce overt signs of post-operative pain, and does not call for pain monitoring or drugs to manage pain. Caution should be taken in administering such compounds, as they can influence outcome (for review see [9].

Postoperative Weight

- Weigh animals daily. Record weights on the evaluation sheet.
- Animals can lose up to 20% of their body weight after surgery and injury. It is beneficial to prophylactically provide mash (chow + water) and/or place normal chow on the floor of the cage to facilitate weight gain.
- If by the second day post-injury, there is continued weight loss, the animals will likely require fluid injections (0.9% sterile saline) to prevent dehydration. Consult a local veterinarian for advice.

- Weight loss exceeding 20% of pre-injury body weight indicates significant injuries that require intensive post-operative care or euthanasia.

Advantages, Limitations, Complications

Advantages

The clinical relevance of the model allows the study of brain injury phenomena, from behavior to molecular biology. The model achieves face validity by the nature of the mechanical forces applied to the brain, and the ensuing pathophysiological responses. The recovery period after surgery, before injury, enhances the face validity of the brain injury model by returning the animals to a condition that more closely resembles the human condition (in which brain injury occurs in the absence of anesthesia). Also, the open craniectomy reduces the effects of injury-induced elevation of intracranial pressure.

The midline injury results in bilateral diffuse pathology that globally suppresses function, such that therapeutic interventions can be tested against injured control. The lateral injury results in ipsilateral focal injury and contralateral diffuse injury, such that the injury is distinct for every brain region. Only specific lateralized functions may be suppressed after lateral brain injury. However, the ensuing cavity permits targeted therapeutic intervention, especially for cell transplantation and bioengineered matrices.

Limitations

To date, most studies using fluid percussion injury have used a moderate level of injury. Concerns about injury variability are based on the extent of damaged tissue and range of physiological responses among animals. The number of variables that influence the response to brain injury produces a range that can necessitate large group sizes to detect significant effects.

Complications

Air Bubbles

Air bubbles in the FPI device can prevent an accurate measurement of the injury magnitude. When air is present in the device, the oscilloscope reading will have many jagged peaks instead of a smooth curve with one peak. The syringe ports can be used to remove any air that enters the device. One way for air to become trapped in the fluid cylinder is after cleaning of the cylinder. This can be minimized by rinsing with a spot remover solution for the dishwasher (e.g., Jet Dry).

Compromised Surgery

During the cranial surgery, the dura can be compromised by the trephination or removal of the bone. When the injury is induced, pressure from the fluid pulse will cause the dura to tear and the brain will herniate through the craniectomy. If the dura is compromised, the injury becomes inconsistent and should be classified as a technical failure. A dura breach will extend the opening of the blood brain barrier and displace neural tissue. Animals with a dura breach should be excluded from any study.

When placing the injury hub over the craniectomy, cyanoacrylate gel can spread on to the dura. If the cyanoacrylate incompletely forms a seal, then methyl methacrylate can spread under the injury hub and on to the dura. These substances on the dura will change mechanical properties and alter the injury. Visual inspection is necessary to identify cyanoacrylate gel or methyl methacrylate on the dura, as well as any other obstruction over the injury site, such as a blood clot.

References

1. Carbonell WS, Maris DO, McCall T, Grady MS. Adaptation of the fluid percussion injury model to the mouse. J Neurotrauma. 1998;15(3):217–29.
2. Dixon CE, Lighthall JW, Anderson TE. Physiologic, histopathologic, and cineradiographic characterization of a new fluid-percussion model of experimental brain injury in the rat. J Neurotrauma. 1988;5(2):91–104.
3. Dixon CE, Lyeth BG, Povlishock JT, Findling RL, Hamm RJ, Marmarou A, Young HF, Hayes RL. A fluid percussion model of experimental brain injury in the rat. J Neurosurg. 1987;67(1):110–9.
4. Floyd CL, Golden KM, Black RT, Hamm RJ, Lyeth BG. Craniectomy position affects morris water maze performance and hippocampal cell loss after parasagittal fluid percussion. J Neurotrauma. 2002;19(3):303–16.
5. Iwamoto Y, Yamaki T, Murakami N, Umeda M, Tanaka C, Higuchi T, Aoki I, Naruse S, Ueda S. Investigation of morphological change of lateral and midline fluid percussion injury in rats, using magnetic resonance imaging. Neurosurgery. 1997;40(1):163–7.
6. Lifshitz J, Rowe RK, Griffiths DR, Evilsizor MN, Thomas TC, Adelson PD, McIntosh TK. Clinical relevance of midline fluid percussion brain injury: acute deficits, chronic morbidities and the utility of biomarkers. Brain Inj. 2016;30(11):1293–301.
7. McIntosh TK, Noble L, Andrews B, Faden AI. Traumatic brain injury in the rat: characterization of a midline fluid-percussion model. Cent Nerv Syst Trauma. 1987;4(2):119–34.
8. McIntosh TK, Vink R, Noble L, Yamakami I, Fernyak S, Soares H, Faden AL. Traumatic brain injury in the rat: characterization of a lateral fluid-percussion model. Neuroscience. 1989;28(1):233–44.
9. Rowe RK, Harrison JL, Thomas TC, Pauly JR, Adelson PD, Lifshitz J. Using anesthetics and analgesics in experimental traumatic brain injury. Lab Anim (NY). 2013;42(8):286–91.

10. Thibault LE, Meaney DF, Anderson BJ, Marmarou A. Biomechanical aspects of a fluid percussion model of brain injury. J Neurotrauma. 1992;9(4):311–22.
11. Thompson HJ, Lifshitz J, Marklund N, Grady MS, Graham DI, Hovda DA, McIntosh TK. Lateral fluid percussion brain injury: a 15-year review and evaluation. J Neurotrauma. 2005; 22(1):42–75.
12. Vink R, Mullins PG, Temple MD, Bao W, Faden AI. Small shifts in craniotomy position in the lateral fluid percussion injury model are associated with differential lesion development. J Neurotrauma. 2001;18(8):839–47.

The Controlled Cortical Impact Model of Experimental Brain Trauma: Overview, Research Applications, and Protocol

Nicole Osier, Anthony E. Kline, and C. Edward Dixon

Abstract Controlled cortical impact (CCI) is a commonly used and highly-regarded model of brain trauma that uses a pneumatically- or electromagnetically-driven piston to induce reproducible and well-controlled injury. The CCI model was originally used in ferrets and it has since been scaled for use in many other species including mice, rats, swine, and nonhuman primates. This chapter will describe the historical development of the CCI model, compare and contrast the pneumatic and electromagnetic devices, and summarize key short- and long-term consequences of TBI that have been studied. A detailed protocol for producing CCI in rats is provided. Finally, special considerations for using the CCI model will be discussed, including commonly encountered problems and trouble-shooting techniques surrounding craniectomy, tip geometry, and anesthesia induction. In accordance with the recent efforts to promote high-quality evidence though the reporting of common data elements (CDEs), relevant study details that should be reported in CCI studies will be noted.

N. Osier
Safar Center for Resuscitation Research, University of Pittsburgh, Pittsburgh, PA, USA

School of Nursing, University of Texas at Austin, Dell Medical School, TX, USA
e-mail: NicoleOsier@utexas.edu

A. E. Kline
Safar Center for Resuscitation Research, University of Pittsburgh, Pittsburgh, PA, USA

Department of Physical Medicine and Rehabilitation, University of Pittsburgh, Pittsburgh, PA, USA

Department of Critical Care Medicine, University of Pittsburgh, Pittsburgh, PA, USA

Department of Psychology, University of Pittsburgh, Pittsburgh, PA, USA
e-mail: klineae@upmc.edu

C. E. Dixon (✉)
Safar Center for Resuscitation Research, University of Pittsburgh, Pittsburgh, PA, USA

Department of Neurological Surgery, University of Pittsburgh, Pittsburgh, PA, USA

V.A. Pittsburgh Healthcare System, Pittsburgh, PA, USA
e-mail: dixonec@upmc.edu

J. Chen et al. (eds.), *Animal Models of Acute Neurological Injury*, Springer Series in Translational Stroke Research, https://doi.org/10.1007/978-3-030-16082-1_26

Keywords Traumatic brain injury (TBI) · Experimental brain injury · Pre-clinical model · Animal model · Controlled cortical impact (CCI)

Introduction

Animal models have been used to study traumatic brain injury (TBI) for over a century. Use of pre-clinical studies remain widely used today to better understand outcomes of experimental brain trauma, explore the role of genetics in TBI recovery, and test novel therapies [1–4]. Today, pre-clinical TBI researchers have a choice between several models including weight drop injury (WDI), fluid percussion injury (FPI), blast-induced TBI (bTBI) and controlled cortical impact (CCI), the focus of this chapter.

CCI was originally developed to study TBI in ferrets [5], and its desirable properties (e.g. reproducibility; control over injury parameters) led researchers to scale the model and apply it to many other species. The original design uses a pneumatically-driven piston to induce TBI, while a newer alternative added an element of portability by using an electromagnetically-driven piston which is lighter in weight and negates the need for a cylinder filled with compressed N_2 gas.

The purpose of this chapter is to introduce readers to the CCI model so that thoughtful decisions can be made in their own completion of CCI research or consumption of the research literature. In doing so, the following will be discussed: (1) historical development, (2) key features, (3) comparison of the pneumatic and electromagnetic devices, (4) research applications, (5) relevant common data elements, as well as (6) factors that influence outcomes and data quality. A standard protocol will be described for pneumatic CCI in rats, along with a list of all required supplies and equipment.

History of Controlled Cortical Impact and Key Features

Animals have been used to study TBI for over 100 years, with considerable refinement of methodologies within the last three decades [1, 2, 4, 6–14]. In comparison to the many experimental TBI models available today, CCI is relatively new. It was originally developed by J.W. Lighthall and colleagues in the late 1980s and early 1990s with the goal of inducing TBI in ferrets [5, 15]. The desirable properties of CCI including the control over important injury parameters and ability to induce reproducible injury led C.E. Dixon and colleagues to scale the CCI model for use in rats during the early 1990s [16]. Since that time, CCI has been further scaled for use in several other species including mice, pigs, and nonhuman primates (see section "Species Used").

The scalability and other desirable features have resulted in CCI becoming one of the most popular and widely used pre-clinical TBI models. One noteworthy feature is that CCI affords quantitative control over important biomechanical parameters of TBI, in particular, the velocity, depth, and force of the tip are controlled across a wide range of contact velocities; there are also different options for tip size, geometry, and positioning, as discussed later (see section "Tip Geometry"). Taken together, the control and customization of CCI allows researchers to address a multitude of research questions as well as scale the injury as needed to study the histopathological of functional deficit of interest. Reporting the injury parameters is of critical importance when reproducing, interpreting, and comparing published study findings, as described in more detail later in this paper (see section "Common Data Elements").

A detailed protocol for inducing CCI in rats is included in section "Protocol for Inducing CCI in Rats". Protocol for Inducing CCI in Rats, but to set the stage, a brief summary of CCI will be provided first. Traditionally, CCI is an invasive model whereby the exposed cortex is subjected to trauma following a craniectomy; in invasive CCI studies, sham animals are used as controls to ensure the results seen are not due to anesthesia and craniectomy but rather the CCI itself. Following craniectomy, the CCI device is used to transfer mechanical energy onto the intact dura mater, producing a TBI. Traditionally, the tip is pneumatically-driven, though a newer model affords portability by using an electromagnetic device to drive the tip [12]. Both types of CCI allow for control over the tip depth, dwell time, and velocity; additional details about the pneumatic and electromagnetic CCI models are provided in section "CCI Types". CCI Types. It is also worth acknowledging that while invasive CCI remains widely used, CCI has been extended to model closed head injury [17–20]. It is also notable that even with an invasive CCI model, only one surgical procedure is necessary, as opposed to standard FPI which typically requires two surgeries.

CCI Types

Pneumatic CCI

The pneumatic CCI device (Fig. 1) was the first to be developed and it remains the most commonly used today. For these reasons, the pneumatic model will be emphasized in this chapter with brief discussion of how it differs from the electromagnetic alternative. A standard pneumatic CCI device features a small-bore (19.75 mm) reciprocating double-acting pneumatic piston with a 50 mm stroke length. The piston is used to drive a tip of a desired size and geometry into the neural tissue (or in some cases the intact skull) to induce brain trauma. The cylinder is held by a crossbar which can be stereotaxically adjusted for variable mounting positions which allows the tip to be aligned vertically or angled with respect to the brain. The

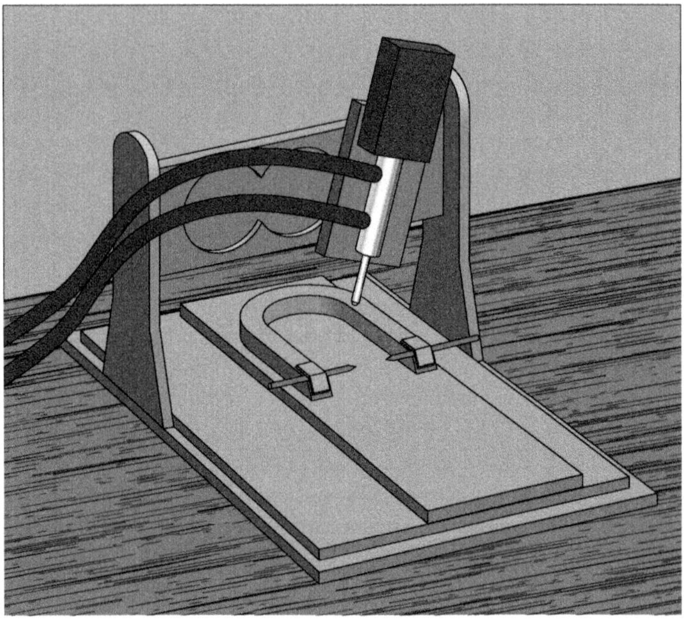

Fig. 1 Diagram depicting a standard pneumatic CCI device (without the associated cylinder of compressed N_2 gas). Depending on the research goals, researchers can choose the ideal tip (e.g., size; geometry) and injury parameters (e.g., depth; dwell time; velocity)

velocity of the piston is monitored by a sensor and can be controlled to promote uniform injury across test animals. Researchers using the CCI model are able to control the depth, duration (i.e. dwell time) and velocity of injury as well as choose what size and shape of tip to use.

Electromagnetic CCI

The electromagnetic CCI device (Fig. 2) is very similar to the pneumatic model and also uses a stereotaxic frame for adjustability. The electromagnetic CCI devices are considered to be portable due to their lighter weight. Another similarity is the availability tips of varying size and geometry (e.g. flat; beveled; round). Generally speaking, tip scaling correlates with animal size; for example 3 mm tips are commonly used for mice, 5–6 mm tips for rats, 10 mm for nonhuman primates, and 15 mm for pigs. Depending on the vendor, the device may come with a variety of tips or there may be additional tips available to purchase separately. Some researchers have also modified tips based on their unique research needs with one group using vulcanized rubber from a lacrosse ball to cover the tip [19]. A list of commercial vendors who sell both electromagnetic and pneumatic CCI devices is included in Table 1.

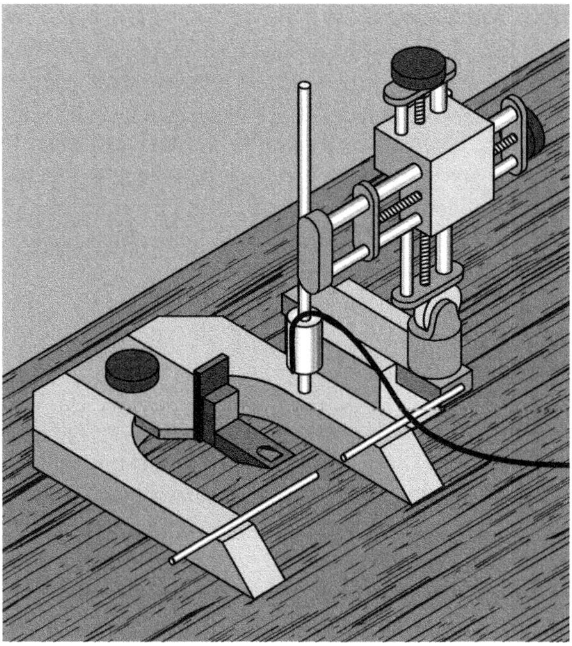

Fig. 2 Diagram depicting a standard electromagnetic CCI device. As with pneumatic CCI, the tip and injury parameters can be adjusted. Unlike with the pneumatic model, the tip is driven by an electromagnetic actuator negating the need for N_2 gas

Table 1 Commercially available pneumatic- and electromagnetic-CCI devices

CCI Type	Company	Model
Pneumatic	Precision Systems & Instrumentation	LLC TBI-0310 Impactor
	Pittsburgh Precision Instruments	Pneumatic Powered Controlled Cortical Impact Device
	AmScien Instruments	Pneumatic (Cortical) Impact Device (Model: AMS 201)
Electromagnetic	Leica	Impact One Stereotaxic Impactor for CCI
	Hatteras Instruments	Pinpoint PCI3000 Precision Cortical Impactor

Suppliers and models are listed for commonly used pneumatic- and electromagnetic-CCI devices

Applications of CCI

Closed Head Injury

Though originally CCI was developed as an invasive TBI model, it has been more recently adapted to study closed head injury (CHI) including repeated concussions. CHI models have become an area of increased research emphasis as the risk to individuals who are in the military or involved in various athletic activities is further appreciated. The aforementioned strengths of CCI make it a popular choice for researchers studying CHI [18–20]. Applications of CCI to study CHI include a study modeling sports-related concussions [19]. In this study, the researchers combined elements of CCI with elements of Marmarou's impact acceleration model in an attempt to enhance control over clinically relevant variables. For example, a foam pad was placed under the rodent to limit rotational acceleration and instead promote linear acceleration. In this study, more deficits were observed after repeated TBI than a single TBI, as assessed using a battery of neurobehavioral tests including measures of cognition, memory, and sleep [19].

Long-Term Outcomes

The high survivability of CCI makes it a good choice for studying the long-term changes associated with TBI. Available evidence suggests that CCI results in chronic and progressive changes. For example, one study which assessed the test animals out to 1 year reported ongoing deficits including progressive tissue loss and ventricular expansion [21]. In addition to the aforementioned histopathological consequences, there are several chronic behavioral deficits that have been reported. For example, after CCI (vs. sham) memory and learning deficits have been found to persist into the long term as assessed using the MWM [21–37]. Notably, persistent MWM deficits are more rarely reported in other models, including lateral FPI [38, 39] and medial FPI [40]. Long-term deficits in motor function have also been reported, as assessed using the foot fault test [33, 35, 41, 42], whereas the authors of this chapter were only able to find one FPI study reporting long-term motor deficits using this measure [43]. That is to say that CCI is a good choice when researchers are interested in exploring long-term motor and memory symptoms after brain injury. Conversely, no CCI studies were identified where motor deficits on the inclined plane task were reported, though such deficits have been reported after lateral FPI [44]. Similarly, no CCI studies were identified where chronic deficits in reversal learning were reported, though this has been reported in several lateral FPI studies [38, 39, 45]. A detailed review of long-term outcomes for the major experimental TBI models is available to interested readers [46].

Protocol for Inducing CCI in Rats

Animals

A strength of the CCI model is that it can be used in many species. In much of our work, adult male Sprague Dawley rats (280–320 g, Charles River Labs, Raleigh, VA, USA, Harlan Sprague Dawley Inc., Indianapolis, IN, USA) are used; thus, this protocol is specific to Sprague Dawley rats. Our animals are routinely housed in a climate-controlled room with a 12 h light/dark cycle and regularly monitored by the Department of Laboratory Animal Research.

Anesthesia

- Induction Dose: 4.0% isoflurane in 2:1 $N_2O:O_2$
- Maintenance Dose: 2% delivered in 2:1 $N_2O:O_2$

Supplies and Equipment

1. Homeothermic heating system (Harvard Apparatus, MA, USA)
2. Stereotaxic frame
3. Isoflurane
4. Cylinders of compressed N_2O and O_2 for isoflurane delivery
5. Cylinder of compressed N_2 to drive pneumatic tip
6. Anesthesia chamber
7. Gas scavenging system
8. Cannula for intubation
9. Laryngoscope to assist with intubation if necessary
10. Animal trimmers
11. Pneumatic drill with drill bits
12. Compressed air for drill
13. Betadine
14. Sterile drape
15. Cotton-tipped applicators
16. Gauze
17. Saline-filled syringe
18. Sterile surgical instruments (scalpel; scissors; periosteal elevator; microdissecting forceps; rongeurs; bulldog clips; etc.)
19. Temperature Probe and associated readout (Harvard Apparatus, MA, USA)
20. MouseOx Plus, blood oxygenation monitoring system and associated readout (Starr Life Sciences Corp., PA, USA)
21. Suture kit
22. Pneumatic CCI device (Pittsburgh Precision Instruments, PA, USA)
23. Tip of desired size and shape

Methods

1. Before starting the surgeries, ensure the CCI device is in good working order. Does the piston fire freely? Is the impact velocity and dwell time consistent with what is set?
2. Place the rat in an anesthesia chamber and induce anesthesia with 4% isoflurane in a 2:1 mixture of $N_2O:O_2$; ensure the animal is sufficiently anesthetized using a toe pinch test.

3. Intubate the rat.
4. Place the anesthetized animal in a stereotaxic frame and secure the incisor and ear bars to keep the animal secure throughout the surgery.
5. Adjust the anesthesia to the maintenance dose of 2% isoflurane. ***(Refer to Note 1)***
6. Assess the animal's level of alertness using the toe pinch test for suppression of pedal response (or another similar test) to ensure sufficient anesthesia is being delivered.
7. Use the hair trimmers to shave the rat's scalp moving both with and against the grain.
8. Use a sterile drape to cover the animal such that the only opening in the drape is directly over the exposed scalp.
9. Use gauze and antiseptic solution (e.g. Betadine) to scrub the scalp and prepare the surgical site
10. Use a scalpel to make a midline incision. ***(Refer to Note 2)***
11. Separate the muscle from the skull using the periosteal elevator and microdissecting forceps.
12. Reflect the skin and fascia to expose the underlying skull and scrub the surface of the skull with a cotton tipped applicator.
13. Use pneumatic drill (hooked up to a compressed air cylinder) to create a craniectomy. Center the craniectomy between the sagittal suture and coronal ridge with the borders near the lambda and bregma for unobstructed tip clearance. ***(Refer to Note 3)***
14. If necessary, use rongeurs to elongate the craniectomy until it is large enough to accommodate the impactor tip; carefully lift away the resulting bone flap, so as to avoid dura breach. ***(Refer to Note 4)***
15. To ensure that the tip is centered over the craniectomy site, manually extend the shaft on the CCI device and gently lower the impactor tip so that it lightly and briefly touches the exposed dura mater.
16. With the piston statically pressurized and in the full stroke position, zero the tip to the cortical surface.
17. Carefully withdraw the tip and adjust the piston assembly to the desired impact depth based on the research goals and study protocol. ***(Refer to Note 5)***
18. Induce injury by actuating the CCI device, discontinue the anesthesia. ***(Refer to Note 6)***
19. Close the surgical site using sutures or another method. Apply topical anesthetic (e.g. lidocaine) to the surgical site to minimize discomfort.
20. Remove the rat from the stereotaxic frame and extubate.
21. Complete any assessments desired in the immediate post-injury period (e.g. righting reflex latency) and post-surgical monitoring.
22. Keep the test animal in a holding cage until it is able to fully recover from anesthesia, as evidenced by return of spontaneous locomotion.
23. Once the animal has fully recovered, return it to the animal housing room and resume normal husbandry.
24. Continue to administer analgesic in accordance with institutional and government guidelines for pain management in laboratory animals.

Species Used

One of the assets of the CCI model is that it can be translated to induce experimental TBI in many species. This is accomplished by scaling the injury parameters so as to maintain the percent of brain volume deformed in relation to total brain volume taking into account desired extent of injury to address the research questions. In order to induce CCI in large animals (e.g. pigs) modifications may be necessary to ensure the piston is high enough. Table 2 provides a summary of the various species that CCI has been applied to including injury parameters commonly used in each species.

Rat

Graded TBI can be easily produced in rats using the CCI model. Indeed, the effects of CCI are well-categorized in published studies using rats, rats where the model has been found to result in diverse histopathological and functional changes consistent with those occurring after clinical TBI, including but not limited to: blood-brain-barrier disruption, derangements in cerebral blood flow and pressure, axonal injury, inflammation, cell death, and edema. It is also worth noting that functional symptoms of TBI (e.g. deficits in learning, memory, and motor function) can be assessed using neurobehavioral testing, which is well-characterized in rats. For example, memory is readily assessed using the Morris water maze (MWM), Barnes maze, or novel object recognition (NOR) task; whereas motor function can be assessed using the beam balance task, beam walking task, rotarod task, wire grip test, as well as the Vermicelli and Capellini handling tests.

Mouse

Shortly after translation of CCI from ferrets to rats, the model was further extended to mice [47–49]. Refinement of CCI in mice has paralleled the increasing application of genetically modified mice to TBI research to explore the role of genes and

Table 2 Commonly used CCI parameters for various species

Species	Injury site	Crani. size (mm)	Tip diameter (mm)	Velocity (m/s)	Dwell time (ms)	Depth (mm)
Mouse	Parietal cortex	4–5	3	4–6	50–250	0.5–2
Rat	Parietal cortex; midline	6–8	5–6	4	50–250	1–3
Primate	Frontal lobe	11–12	10	3.5	150	7
Pig	Frontal lobe	15–18	15	2–4	50–400	12

For each species, injury details commonly used are provided including: injury site, craniectomy size, tip diameter, velocity, dwell time, and depth gene products in brain injury recovery [50]. Generally speaking, to scale CCI down for use in mice means decreasing the injury depth to adjust for the relatively thinner cortex as compared to rats. Another deviation in the protocol is that instead of intubation and mechanical ventilation, anesthetic and other gases are traditionally delivered via a nose cone. Though slightly less well categorized, a plethora of behavioral testing strategies are available for use in mice including the MWM, NOR, and BBT [22, 34, 51, 52].

Pig

To adequately address some research questions, larger, more human-like brains are required, thus necessitating the use of a pig or other large mammal model. The main difference from the rat and mouse models is the considerably larger impactor tip and greater depth to which the neural tissue is deformed (see Table 2). In one study, CCI was scaled to induce TBI in piglets; injury parameters were chosen after adjusting for differences in brain morphology (e.g. size and dimensions) in the study's test animals [53]. As with rodent models of CCI, when pigs are used the histopathological changes mirror what is seen in TBI patients, including, but not limited to, deranged blood flow and changes to the vasculature, ongoing neurodegeneration via a number of mechanisms, and edema. To date, pig models have been used to add to the evidence surrounding TBI biomechanics [54] and as part of an effort to identify clinically relevant biomarkers of underlying brain injury [55]. However, despite these efforts there is a relative dearth of normative data specific to pigs when compared to rodents [56]. Additionally, behavioral testing is more challenging to perform in the pigs due to their larger size and relative intelligence and therefore is less well-characterized.

Non-human Primate

An alternative to the pig model is the nonhuman primate model of CCI, which is typically applied over the frontal cortex [57]. As with the other models, the histopathological changes reported after nonhuman primate CCI, mimic what is seen clinically, including but not limited to edema, macrophage accumulation, and neurodegeneration. Nonhuman primates play a critical role in establishing the safety of novel therapies before translation to humans. Notably, due to the increased ethical considerations, care requirements, and cost, a relatively limited number of research institutions have primate research facilities. Consequently, nonhuman primate studies represent only a small fraction of TBI studies. Indeed, the use of nonhuman primate models is only justified when there are major factors that prohibit the use of a less sentient animal.

Special Considerations, Problems, and Troubleshooting

High-quality CCI research relies on thoughtful study design and careful execution. A few important confounders that have been empirically studied will be addressed below. When appropriate, trouble-shooting strategies will be noted.

Tip Geometry

Commercially available tips come in round or beveled flat shapes of various sizes. When the CCI model was developed in ferrets, round tips were used [5, 15]; though still in use, beveled flat tips have become the norm [47, 50, 58–61] and are especially preferred for mouse CCI. Despite convention, little empirical research has tested the effects of tip geometry on outcomes. In one study, Pleasant et al. [62] compared flat vs. rounded tips in a mouse (C57BL6J) model; they found more extensive cortical hemorrhage and neuronal loss (proportionally) with flat tips. The rate of neocortical loss using a flat tip was faster with a plateau in neurodegeneration 20 h earlier compared to a rounded tip (4 h vs. 24 h). This finding lead to rounded tips being a desirable choice when studying the secondary injury cascades in the subacute period [62].

Anesthesia

Overall, careful choice of whether to use anesthesia and what anesthesia regimen to use is critical. In deciding, researchers should consider the study goals, along with established guidelines for the treatment of research animals. Empirical evidence shows that differences in outcomes of TBI are associated with various anesthesia agents. Notably, isoflurane results in less hippocampal damage than fentanyl as well as fewer behavioral deficits [63]. Another study found pre-injury isoflurane had neuroprotective effects [64]. It is hypothesized that fentanyl contributes to neural suppression, whereas isoflurane reduces excitotoxicity and promotes blood flow [56]. Ketamine also demonstrates neuroprotective properties via antagonism of N-methyl-D-aspartate (NMDA) receptors [65]. Halothane has also been reported to have neuroprotective properties after contusion injury [66]. Use of neuroprotective anesthesia can obscure deficits in performance in all anesthetized groups. For fentanyl, the concern is that the resulting neural suppression would worsen performance and may obscure treatment effects in a drug study.

Despite these concerns, the overwhelming majority of CCI studies use anesthesia, commonly with isoflurane. Volatile gases (e.g., isoflurane; halothane) are often preferred due to their relatively short half-life compared to long-acting options (e.g., pentobarbital), facilitating evaluation of righting reflex shortly after anesthesia discontinuation [16]. Typically a high dose (e.g., 4%) is used to induce anesthesia followed by a reduced maintenance dose (see section "Protocol for Inducing CCI in Rats"). Anesthetized animals should be monitored to ensure consciousness is not regained during surgery. Assessments like the toe pinch can be used to assess sufficiency of anesthesia in accordance with institutional and national policies. One recent study of repetitive closed head injury used un-anaesthetized mice that were instead comfortably restrained to avoid the confounding effects of anesthesia on TBI outcomes and promote clinical relevance [19].

Commonly encountered problems surrounding anesthesia are summarized below. First, if animals are intubated and the animal is fighting the ventilator, changes in tube placement may alleviate the problem. Also, if consciousness is regained the anesthesia induction system should be checked to ensure that the anesthesia is set to the appropriate level and there are no leaks in the tubing. It is also important to consider that if a gaseous anesthesia is used, specialized laboratory equipment is required including appropriate ventilation and scavenging systems to ensure safety of personnel; isoflurane detection systems are available to monitor exposure.

Craniectomy and Sham Procedure

Researchers typically produce craniectomy using a pneumatic or electric drill, though sometimes a handheld trephine is used. Efforts to reduce heat production during the craniectomy can help reduce potential confounders, and sterile saline can be applied to the craniectomy site to reduce the temperature if the procedure is prolonged. Because anesthesia and craniectomy can result in behavioral and functional changes it is important that when invasive CCI is used that control animals be exposed to sham procedures. Empirical evidence shows craniectomy results in inflammation and lesion regardless of whether a trephine or electric drill is used, as compared to naïve rats (anesthesia only). However, the lesions were largest and behavioral deficits most severe in animals that received their craniectomies using a drill [67]. Craniectomy location also affects outcome with midline craniectomy leading to more sagittal bleeding [16], than parasagittal craniectomy [68, 69]. Bilateral craniectomies have been used to enhance lateral movement of tissue and produce subsequent bilateral cortical contusions [70, 71]; producing bilateral craniectomies is also a good way to train individuals on the procedure. Details regarding the control group should be provided in publications [67], including any bleeding, mortality, and inconsistency across animals.

Common Data Elements

The National Institute for Neurological Diseases and Stroke (NINDS) has published a set of common data elements (CDEs) for experimental TBI research including details surrounding the animals used (e.g., species, strain, commercial supplier), demographics (e.g., age; sex), anthropometrics (e.g., weight), animal husbandry, and outcome assessment(s) used (e.g., timing of assessment, measures). Beyond the basic set of CDE's generic to all experimental TBI research, the NINDS recognizes a specific set of CDE's for studies using the CCI model. The CCI-specific CDE's include, but are not limited to: craniectomy size, tip (size, shape, angle, rigidity) and injury parameter settings (depth, dwell time, velocity). Researchers are encouraged to review the current list of CDEs during study planning, grant writing, and dissemination of findings to promote comparison of studies and the conduct of high-quality research [72].

Conclusion

CCI is one of the best-characterized models of experimental TBI and it remains a popular choice for studying the physiologic and functional deficits that occur acutely and chronically following TBI. Traditionally, CCI is an invasive model that is preceded by craniectomy, but recently the model has been applied to study concussion and other types of closed head injury. The original CCI model was pneumatically driven but more recently, an electromagnetic alternative has been introduced which affords increased portability.

Researchers employing the CCI model should give care and attention to study design and selection of injury parameters. Control over important confounding variables (e.g., hypothermia; hyperthermia) is critical to adequately address the research questions. The first step is to thoroughly explore the literature to consider how various injury parameters have panned out with respect to histopathological and functional consequences in the past. Researchers are also encouraged to conduct pilot work in order to tailor the experimental design (e.g., injury parameters, tip size, anesthesia type) to facilitate addressing the research goals. Pilot research also provides a valuable opportunity to ensure that the device is in proper working order and calibrated.

This chapter introduced the CCI model including a brief overview of its development and extension to various species and research applications. A list of required supplies and equipment was provided as well as a detailed protocol for pneumatic CCI in rats. Discussion of confounding factors and trouble-shooting methods were briefly discussed. Lastly, the importance of CDE's was described and exemplars of CDE's specific to CCI were noted. This introductory chapter will enable readers to be thoughtful consumers of publications describing CCI research and have the requisite knowledge needed to design and conduct a CCI study.

Notes

1. Traditionally, the authors of this chapter use a maintenance dose of 2% isoflurane in a 2:1 mixture of $N_2O:O_2$ titrating up the dose if the animal is showing signs of regaining consciousness.

2. The incision made in our labs is approximately 20 mm long for rats (shorter for mice).
3. The authors strive to make consistent craniectomies that are approximately 6 mm in diameter to facilitate clearance of a 5 mm diameter tip.
4. It is common practice to discard the bone flap rather than attempt to reattach it, as this can lead to secondary injury (e.g., increased intracranial pressure).
5. In our labs, we induce moderate TBI using a 5 mm tip to deform the neural tissue of a rat to a depth of 2.8 mm at a velocity of 4 m/s.
6. Depending on the preference of the researchers and the method used to close the surgical site, anesthesia can be discontinued before or after wound site closure.

Acknowledgements Support for this chapter comes from the following government funding sources: Department of Veterans Affairs grant RR&D B1127-I; NIH-NINDS grants R01-NS079061, R01-HD069620, R01-HD069620-S1, and R01NS084967; and NIH-NINR grants 1F31NR014957 and T32NR009759. Additional support for this chapter comes from the following foundations and professional societies: The Pittsburgh Foundation, Sigma Theta Tau International Eta Chapter, the International Society for Nurses in Genetics, and the American Association of Neuroscience Nursing/Neuroscience Nursing Foundation. We would also like to acknowledge Mr. Michael D. Farmer for his time in generating the figures and Mrs. Marilyn K. Farmer for her continued editorial support.

References

1. Denny-Brown D, Russell W. Experimental cerebral concussion. Brain. 1941;64:93–184.
2. Kramer SP. A contribution to the theory of cerebral concussion. Ann Surg. 1896;23:163–73.
3. Lindgren S, Rinder L. Experimental studies in head injury. I. Some factors influencing results of model experiments. Biophysik. 1965;2(5):320–9.
4. Rinder L, Olsson Y. Studies on vascular permeability changes in experimental brain concussion. I. Distribution of circulating fluorescent indicators in brain and cervical cord after sudden mechanical loading of the brain. Acta Neuropathol. 1968;11(3):183–200.
5. Lighthall JW. Controlled cortical impact: a new experimental brain injury model. J Neurotrauma. 1988;5(1):1–15.
6. Cannon WB. Cerebral pressure following trauma. Am J Physiol. 1901;6:91–121.
7. Gennarelli TA, et al. Diffuse axonal injury and traumatic coma in the primate. Ann Neurol. 1982;12(6):564–74.
8. Govons SR, et al. Brain concussion in the rat. Exp Neurol. 1972;34(1):121–8.
9. Nilsson B, Pontén U, Voigt G. Experimental head injury in the rat. Part 1: mechanics, pathophysiology, and morphology in an impact acceleration trauma model. J Neurosurg. 1977;47(2):241–51.
10. Ommaya AK, Geller A, Parsons LC. The effect of experimental head injury on one-trial learning in rats. Int J Neurosci. 1971;1(6):371–8.
11. Ommaya AK, Gennarelli TA. Cerebral concussion and traumatic unconsciousness. Correlation of experimental and clinical observations of blunt head injuries. Brain. 1974;97(4):633–54.
12. Onyszchuk G, et al. A mouse model of sensorimotor controlled cortical impact: characterization using longitudinal magnetic resonance imaging, behavioral assessments and histology. J Neurosci Methods. 2007;160(2):187–96.
13. Parkinson D, West M, Pathiraja T. Concussion: comparison of humans and rats. Neurosurgery. 1978;3(2):176–80.

14. Sullivan HG, et al. Fluid-percussion model of mechanical brain injury in the cat. J Neurosurg. 1976;45(5):521–34.
15. Lighthall JW, Goshgarian HG, Pinderski CR. Characterization of axonal injury produced by controlled cortical impact. J Neurotrauma. 1990;7(2):65–76.
16. Dixon C, et al. A controlled cortical impact model of traumatic brain injury in the rat. J Neurosci Methods. 1991;39(3):253–62.
17. Fidan E, et al. Repetitive mild traumatic brain injury in the developing brain: effects on long-term functional outcome and neuropathology. J Neurotrauma. 2016;33(7):641–51.
18. Klemenhagen KC, O'Brien SP, Brody DL. Repetitive concussive traumatic brain injury interacts with post-injury foot shock stress to worsen social and depression-like behavior in mice. PLoS One. 2013;8(9):e74510.
19. Petraglia AL, et al. The spectrum of neurobehavioral sequelae after repetitive mild traumatic brain injury: a novel mouse model of chronic traumatic encephalopathy. J Neurotrauma. 2014;31(13):1211–24.
20. Shitaka Y, et al. Repetitive closed-skull traumatic brain injury in mice causes persistent multifocal axonal injury and microglial reactivity. J Neuropathol Exp Neurol. 2011;70(7):551–67.
21. Dixon C, et al. One-year study of spatial memory performance, brain morphology, and cholinergic markers after moderate controlled cortical impact in rats. J Neurotrauma. 1999;16(2):109–22.
22. Byrnes KR, et al. Delayed mGluR5 activation limits neuroinflammation and neurodegeneration after traumatic brain injury. J Neuroinflammation. 2012;9:43.
23. Chauhan NB, Gatto R. Restoration of cognitive deficits after statin feeding in TBI. Restor Neurol Neurosci. 2011;29(1):23–34.
24. Chauhan NB, Gatto R. Synergistic benefits of erythropoietin and simvastatin after traumatic brain injury. Brain Res. 2010;1360:177–92.
25. Cheng JP, et al. A relatively brief exposure to environmental enrichment after experimental traumatic brain injury confers long-term cognitive benefits. J Neurotrauma. 2012;29(17):2684–8.
26. Dixon CE, et al. Increased anticholinergic sensitivity following closed skull impact and controlled cortical impact traumatic brain injury in the rat. J Neurotrauma. 1994;11(3):275–87.
27. Fox GB, Faden AI. Traumatic brain injury causes delayed motor and cognitive impairment in a mutant mouse strain known to exhibit delayed Wallerian degeneration. J Neurosci Res. 1998;53(6):718–27.
28. Han R-Z, et al. NMDA receptor antagonist MK-801 reduces neuronal damage and preserves learning and memory in a rat model of traumatic brain injury. Neurosci Bull. 2009;25(6):367–75.
29. Lindner MD, et al. Dissociable long-term cognitive deficits after frontal versus sensorimotor cortical contusions. J Neurotrauma. 1998;15(3):199–216.
30. Longhi L, et al. Ex vivo gene therapy using targeted engraftment of NGF-expressing human NT2N neurons attenuates cognitive deficits following traumatic brain injury in mice. J Neurotrauma. 2004;21(12):1723–36.
31. Longhi L, et al. Long-lasting protection in brain trauma by endotoxin preconditioning. J Cereb Blood Flow Metab. 2011;31(9):1919–29.
32. Marklund N, et al. Functional outcome is impaired following traumatic brain injury in aging Nogo-A/B-deficient mice. Neuroscience. 2009;163(2):540–51.
33. Meng Y, et al. Dose-dependent neurorestorative effects of delayed treatment of traumatic brain injury with recombinant human erythropoietin in rats. J Neurosurg. 2011;115(3):550–60.
34. Tomasevic G, et al. Delayed neuromotor recovery and increased memory acquisition dysfunction following experimental brain trauma in mice lacking the DNA repair gene XPA. J Neurosurg. 2012;116(6):1368–78.
35. Xiong Y, Zhang Y, et al. Erythropoietin mediates neurobehavioral recovery and neurovascular remodeling following traumatic brain injury in rats by increasing expression of vascular endothelial growth factor. Transl Stroke Res. 2011b;2(4):619–32.
36. Xiong Y, et al. Neuroprotective and neurorestorative effects of thymosin β4 treatment initiated 6 hours after traumatic brain injury in rats. J Neurosurg. 2012;116(5):1081–92.

37. Zhang Y, et al. Impact of inhibition of erythropoietin treatment-mediated neurogenesis in the dentate gyrus of the hippocampus on restoration of spatial learning after traumatic brain injury. Exp Neurol. 2012;235(1):336–44.
38. Shultz SR, et al. Repeated mild lateral fluid percussion brain injury in the rat causes cumulative long-term behavioral impairments, neuroinflammation, and cortical loss in an animal model of repeated concussion. J Neurotrauma. 2012;29(2):281–94.
39. Shultz SR, et al. Treatment with an anti-CD11d integrin antibody reduces neuroinflammation and improves outcome in a rat model of repeated concussion. J Neuroinflammation. 2013;10:26.
40. Hamm RJ, et al. The effect of postinjury kindled seizures on cognitive performance of traumatically brain-injured rats. Exp Neurol. 1995;136(2):143–8.
41. Hoane MR. Magnesium therapy and recovery of function in experimental models of brain injury and neurodegenerative disease. Clin Calcium. 2004;14(8):65–70.
42. Xiong Y, Mahmood A, et al. Effects of posttraumatic carbamylated erythropoietin therapy on reducing lesion volume and hippocampal cell loss, enhancing angiogenesis and neurogenesis, and improving functional outcome in rats following traumatic brain injury. J Neurosurg. 2011a;114(2):549–59.
43. Rau TF, et al. Treatment with low-dose methamphetamine improves behavioral and cognitive function after severe traumatic brain injury. J Trauma Acute Care Surg. 2012;73(2 Suppl 1):S165–72.
44. Hallam TM, et al. Comparison of behavioral deficits and acute neuronal degeneration in rat lateral fluid percussion and weight-drop brain injury models. J Neurotrauma. 2004;21(5):521–39.
45. Thompson HJ, et al. Cognitive evaluation of traumatically brain-injured rats using serial testing in the Morris water maze. Restor Neurol Neurosci. 2006;24(2):109–14.
46. Osier ND, et al. Chronic histopathological and behavioral outcomes of experimental traumatic brain injury in adult male animals. J Neurotrauma. 2015;32(23):1861–82.
47. Fox GB, et al. Sustained sensory/motor and cognitive deficits with neuronal apoptosis following controlled cortical impact brain injury in the mouse. J Neurotrauma. 1998;15(8):599–614.
48. Fox GB, LeVasseur RA, Faden AI. Behavioral responses of C57BL/6, FVB/N, and 129/SvEMS mouse strains to traumatic brain injury: implications for gene targeting approaches to neurotrauma. J Neurotrauma. 1999;16(5):377–89.
49. Hannay HJ, et al. Validation of a controlled cortical impact model of head injury in mice. J Neurotrauma. 1999;16(11):1103–14.
50. Smith DH, et al. A model of parasagittal controlled cortical impact in the mouse: cognitive and histopathologic effects. J Neurotrauma. 1995;12(2):169–78.
51. Han X, et al. Imipramine treatment improves cognitive outcome associated with enhanced hippocampal neurogenesis after traumatic brain injury in mice. J Neurotrauma. 2011;28(6):995–1007.
52. Scafidi S, et al. Neuroprotection by acetyl-L-carnitine after traumatic injury to the immature rat brain. Dev Neurosci. 2010;32(5–6):480–7.
53. Duhaime AC, et al. Maturation-dependent response of the piglet brain to scaled cortical impact. J Neurosurg. 2000;93(3):455–62.
54. Manley GT, et al. Controlled cortical impact in swine: pathophysiology and biomechanics. J Neurotrauma. 2006;23(2):128–39.
55. Costine BA, et al. Neuron-specific enolase, but not S100B or myelin basic protein, increases in peripheral blood corresponding to lesion volume after cortical impact in piglets. J Neurotrauma. 2012;29(17):2689–95.
56. Kline AE, Dixon CE. Contemporary in vivo models of brain trauma and a comparison of injury responses. In: Miller LP, Hayes RL, editors. Head trauma: basic, preclinical, and clinical directions. New York: Wiley; 2001. p. 65–84.
57. King C, et al. Brain temperature profiles during epidural cooling with the ChillerPad in a monkey model of traumatic brain injury. J Neurotrauma. 2010;27(10):1895–903.
58. Dennis AM, et al. Hemorrhagic shock after experimental traumatic brain injury in mice: effect on neuronal death. J Neurotrauma. 2009;26(6):889–99.

59. Hemerka JN, et al. Severe brief pressure-controlled hemorrhagic shock after traumatic brain injury exacerbates functional deficits and long-term neuropathological damage in mice. J Neurotrauma. 2012;29(12):2192–208.
60. Monaco CM, et al. Environmental enrichment promotes robust functional and histological benefits in female rats after controlled cortical impact injury. Exp Neurol. 2013;247:410–8.
61. Sandhir R, Berman NEJ. Age-dependent response of CCAAT/enhancer binding proteins following traumatic brain injury in mice. Neurochem Int. 2010;56(1):188–93.
62. Pleasant JM, et al. Rate of neurodegeneration in the mouse controlled cortical impact model is influenced by impactor tip shape: implications for mechanistic and therapeutic studies. J Neurotrauma. 2011;28(11):2245–62.
63. Statler KD, et al. Isoflurane improves long-term neurologic outcome versus fentanyl after traumatic brain injury in rats. J Neurotrauma. 2000;17(12):1179–89.
64. Statler KD, et al. Isoflurane exerts neuroprotective actions at or near the time of severe traumatic brain injury. Brain Res. 2006;1076(1):216–24.
65. McDonald JW, et al. Quantitative assessment of neuroprotection against NMDA-induced brain injury. Exp Neurol. 1989;106(3):289–96.
66. McPherson RW, et al. The neurobiology of central nervous system trauma. New York: Oxford University Press; 1994.
67. Cole JT, et al. Craniotomy: true sham for traumatic brain injury, or a sham of a sham? J Neurotrauma. 2011;28(3):359–69.
68. Shin SS, et al. The effect of environmental enrichment on substantia nigra gene expression after traumatic brain injury in rats. J Neurotrauma. 2013;30(4):259–70.
69. Shin SS, Bray ER, Dixon CE. Effects of nicotine administration on striatal dopamine signaling after traumatic brain injury in rats. J Neurotrauma. 2012;29(5):843–50.
70. He J, et al. Progesterone and allopregnanolone reduce inflammatory cytokines after traumatic brain injury. Exp Neurol. 2004;189(2):404–12.
71. Meaney DF, et al. Modification of the cortical impact model to produce axonal injury in the rat cerebral cortex. J Neurotrauma. 1994;11(5):599–612.
72. Smith DH, et al. Pre-clinical traumatic brain injury common data elements: toward a common language across laboratories. J Neurotrauma. 2015;32(22):1725–35.

Mild Blast-Induced Traumatic Brain Injury Model

Riyi Shi and Nicholas Race

Abstract Blast injuries from explosive events can be simulated in a laboratory setting through a variety of means. Different elements of blast-induced traumatic brain injuries can be recapitulated in the laboratory setting in tandem or in isolation to investigate a variety of research questions. The research question of interest will dictate choices regarding the modality of blast generation, animal positioning, head fixation, and body protection. Due to the relatively young nature of the field and the various blast injury models, consensus opinions have not been reached with regard to model designs, exposure conditions, and outcome evaluations. This chapter outlines one procedure for administering a diffuse, closed-head mild blast-induced TBI (mild bTBI) to rats via open-ended shock tube and includes discussion on the impact of differing choices with regard to model design, exposure conditions, and outcome evaluations.

Keywords Blast injury · Rat · Brain injury · Trauma

Model Selection

Blast-induced traumatic brain injuries are a form of experimental traumatic brain injury that is clinically relevant for individuals exposed to explosions, namely: active and former military personnel, victims of industrial accidents, and persons who live in areas of sectarian violence or are targeted by terrorism-related activities.

R. Shi (✉)
College of Veterinary Medicine, Department of Basic Medical Sciences, Purdue University, Indianapolis, IN, USA

Weldon School of Biomedical Engineering, Purdue University, Indianapolis, IN, USA
e-mail: riyi@purdue.edu

N. Race
Weldon School of Biomedical Engineering, Purdue University, Indianapolis, IN, USA

School of Medicine, Indiana University, Indianapolis, IN, USA

J. Chen et al. (eds.), *Animal Models of Acute Neurological Injury*, Springer Series in Translational Stroke Research, https://doi.org/10.1007/978-3-030-16082-1_27

Traumatic brain injury from blast can result from four major mechanisms: shock wave blast overpressure (primary), penetrating injuries (secondary), head acceleration and blunt force trauma (tertiary), and heat, radiation, and/or chemical exposures (quaternary). In the laboratory setting, primary and tertiary blast injuries are most commonly investigated and will be the focus of this chapter. For high-explosive, non-nuclear events, primary and tertiary blast injuries principally result from two major blast components: the primary shock wave, a near-instantaneous pressure pulse <10 ms duration, and the tertiary blast wind that follows, throwing individuals and objects through surroundings causing accelerations and collisions [1]. Generally, individuals closer in proximity to the explosion suffer more severe injuries.

Clinical injury grading (mild, moderate, or severe) is established with a combination of clinical symptoms on neurological exam and basic neuroimaging. Mild injuries are clinically classified by the following: loss of consciousness (LOC) ≤ 30 min, Glasgow Coma Scale (GCS) 13–15 following LOC, post-traumatic amnesia <24 h, and no evidence of gross intracranial hemorrhage on CT scan [2]. Accompanying symptoms may include headache, fatigue, poor concentration, lethargy, depression, anxiety, and insomnia [2]. Symptom clusters can overlap with concussion or post-traumatic stress disorder [1, 2]. Mild bTBI are often asymptomatic in the acute post-injury period and occur in absence of a direct blow to the head, thus they are difficult to detect [1, 2]. Unfortunately, mild, moderate, and severe injuries currently do not have a defined relationship with exposure conditions such as overpressure magnitude, duration, impulse, or other mechanical features that are commonly used for establishing grading of other types laboratory-simulated neurotrauma. This in large part is due to the inherent difficulty in understanding and computationally modeling the complex, nonlinear mechanical dynamics of shock waves and their interactions with the cranium. Scaling data is available for pulmonary injuries from blast [3], but inter-species injury scaling laws have yet to be established for bTBI. At present, rodent-equivalent behavioral consistency with human clinical presentation, neuroimaging, and histology are the gold standard for assessing injury severity, and conclusions may only be drawn within the context of a particular model's design and exposure conditions until relationships between injury severity and incident injury mechanics are better understood. Procedures outlined here are for a model known to be consistent with the human experience symptomatically in the acute post-injury period.

Materials

Animals

Blast-induced traumatic brain injury studies have been conducted in mice, rats, pigs, and an assortment of animal and human surrogate models. Currently, the most commonly utilized species are rats, including our model. As such they will be the primary focus for this chapter. Rat neurological 'equivalence' to human adults occurs at approximately post-natal day (PND) 20 [4]. Musculoskeletal development completes more slowly; however, particularly concerning maturation and fusion of cranial sutures. To achieve adequate musculoskeletal development and braincase material properties, rats should be at least PND 90 [4]. In our lab, for Sprague Dawley rats, according to the supplier growth curves from Harlan/Envigo Labs, this corresponds to rats ≥350 g [5]. It has been reported that younger animals with incompletely fused cranial sutures experience significant skull flexure across the immature sutures during bTBI, which may result in mechanisms of injury that differ from the human condition [6].

Equipment

Anesthesia

Syringe and 26G needle
Ketamine (100 mg/mL)
Xylazine (100 mg/mL)
Warming pad

Animal Mounting

Stereotaxic bite bar (Kopf Instruments, Model 920 Rat Adaptor)
Stereotaxic ear bars (Kopf Instruments, Model 957 Rat Ear Bars)
Body protector (custom made)

Blast Device

Shock Tube
Compressed gas (Air, Nitrogen, or Helium)
Diaphragm (McMaster-Carr, Moisture-Resistant Polyester Film 0.010″ Thick)
Animal mounting stage (custom made)
Pressure sensors (Omega, DPX101-250)
Remote detonator (custom made)

Procedure

Apparatus Preparation

- Ensure compressed gas tank is sealed and gas delivery lines are purged while preparing apparatus and mounting animals.

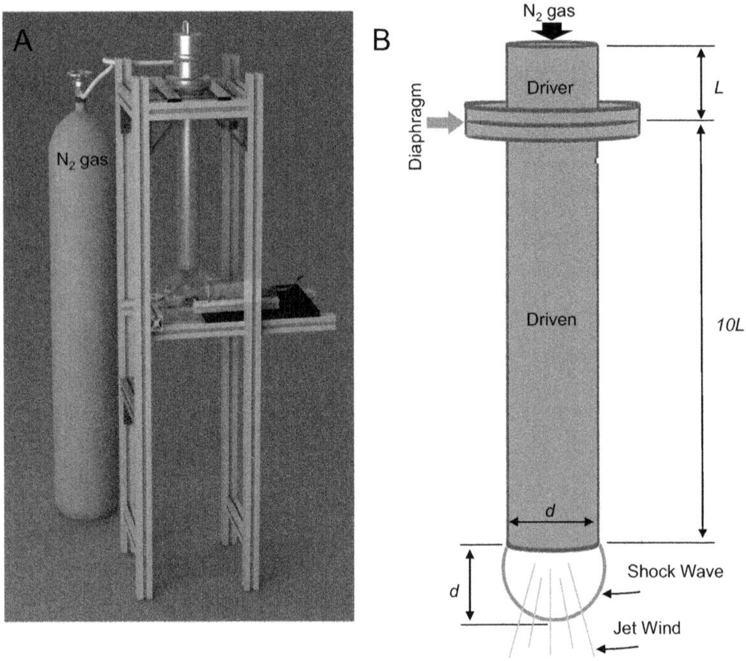

Fig. 1 Open-ended shock tube. (**a**) Overall setup. (**b**) Shock tube components

- Unbolt and remove driver section (closed end where gas enters behind diaphragm).
- Place new PET diaphragm between driver and driven (open end) sections (Fig. 1).
- Replace and re-bolt driver section with diaphragm in place.

Animal Preparation and Positioning

- Administer 80 mg/kg ketamine and 10 mg/kg xylazine cocktail via intraperitoneal injection.
- Check eye-blink and hind limb toe-pinch paw withdrawal reflexes to ensure adequate depth of anesthesia.
- Apply ophthalmic ointment to eyes to keep moist during/after injury.
- Place animal on mounting stage and adjust positioning to desired location either inside or outside of the tube. A few options are illustrated in Fig. 2. If inside the tube, animal should be positioned approximately equidistant from the diaphragm and tube outlet [7]. Doing so will minimize non-ideal reflections and rarefractions from the driver end and outlet. If outside the tube, the animal should be placed within a distance ≤ one tube outlet inner diameter to ensure shock wave presence [8]. If utilizing positions outside the tube and non-ideal outlet effects (particularly jet wind) are a concern for the study of interest, they can be mitigated by positioning the animal off-center from the device outlet or applying head fixation to eliminate head acceleration [7, 9, 10].

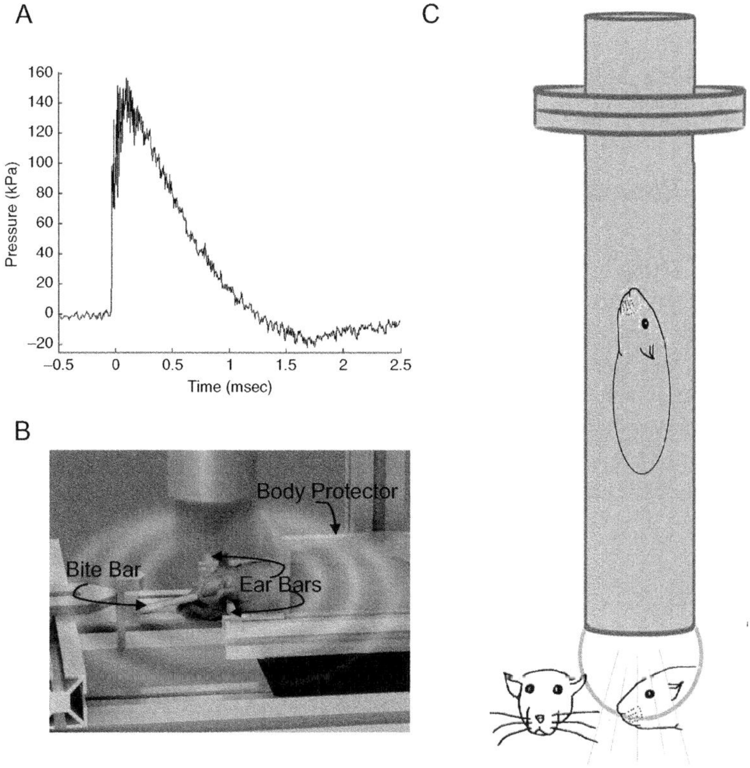

Fig. 2 Open-ended shock tube. (**a**) Representative pressure trace. (**b**) Head fixation and body protector. (**c**) Animal positioning options

- Apply head fixation with stereotaxic frame if desired. Choosing to leave the head free to move or secure it in place depends on study design and desire to study shock wave exposure with or without head acceleration, respectively. Notably, if the head is positioned directly below the shock tube outlet without head fixation, significant injurious head accelerations will occur [9, 11], inciting a composite blast injury as opposed to a primary blast injury [12].
- Apply body protector if desired to simulate military protective equipment. Choosing to protect the body or leave it unprotected depends on study design and desire to study bTBI in isolation or as polytrauma with other organ systems.

Injury

- Scientist should apply appropriate ear protection.
- Turn on compressed gas.
- Exit the room containing the blast device.

- Plug in and activate remote detonator.
- Turn off compressed gas.
- Remove animal from device and rest on warming pad until recovered.

Pressure Recording

- The DPX101-250 is calibrated for 3 mV per kPa (kilopascal). Figure 2 shows a representative pressure trace recorded near the outlet of the driven section of the tube with a peak magnitude of 150 kPa and overpressure duration of 1.5 ms.
- Injury traces should have a sharp rise (near-instantaneous), possible brief plateau, and exponential decay. Depending on sensor positioning, a lower magnitude underpressure may follow the overpressure period.
- Recordings can be viewed/saved using an oscilloscope or data acquisition card/computer with LabView software.

Post-Operative Care

- Visually monitor animals every 5–10 min post-injury. If animals experience post-injury seizures or sustained apnea, the injury was not mild and animals should be euthanized immediately. Thirty to 60 min after injury, mild-injured animals should be awake, alert, and periodically ambulating. Injured and control animals should be undistinguishable after recovery, with neither group showing effects [10]. Eating, drinking, and grooming should resume as per normal behavior.
- Monitor post-operatively for 48 h for signs of pain. Generally, animals do not require supportive care or drugs to manage pain.
- Weigh animals daily. Sustained weight loss in excess of 10% indicates an abnormal response requiring increased attention to supportive care and may suggest injury was more severe than intended.

Outcome Evaluation

Mortality and Acute Complications

In our experience, mortality for the mild exposure conditions in our model occurs only as a result of anesthesia complications. These include but are not limited to excessive dosing if unresponsive to first injection or accidental intravascular injection instead of intraperitoneal injection. In the absence of anesthesia complications, no mortality should be observed if the blast exposure is indeed mild. Five to 10% of animals may experience traumatic tympanic membrane rupture and vestibular injury. This is consistent with the human experience and we have yet to successfully eliminate its incidence completely in our model. Animals suffering these injuries will not perform well on tasks involving hearing or motor coordination and should be excluded from such studies.

Behavioral Evaluation

In the acute post-injury period (first 24 h), per the guidelines set forth by the CDC/DoD, mild bTBI will exhibit little to no symptom presentation in humans as mentioned previously [2]. Mild, nonspecific constitutional symptoms may be present, but no lasting neurocognitive impairment should be observed beyond 24 h [2]. If injury is suspected, neurocognitive testing of basic neural functions, working memory, and motor coordination may be conducted. For mild injuries, the results of these tests are typically negative. Consistent with such observations, the exposure conditions outlined in this chapter result in no observable post-injury alterations in general open box locomotor activity [10], rotarod maximum motor capacity [10], and motor learning (unpublished data), short-term memory via novel object recognition [10], or weight and welfare [10]. Extensive reviews exist that cover many aspects of post-TBI behavioral deficits if more testing is desired including neurological scores, the Morris water maze, and beam balance [13, 14]. Emerging topics of interest include investigating fear and other psychological phenomena post-injury [15]. These deficits generally present at later time points (weeks to months) and have yet to be extensively studied in the context of mild bTBI. Reviews of behavioral tests for such phenomena including fear conditioning/extinction, elevated plus maze, and others are also readily available [16, 17].

Histology and Biochemistry

At acute time points, BBB disruption, oxidative stress, and macrophage/microglia activation have been consistently observed by our group [10] and others [18–21] in the 24 h following mild bTBI exposure. For mild exposure conditions as utilized in this model, neuronal degeneration as assessed by Fluoro-Jade assays (unpublished data) was not readily observable after a single mild bTBI exposure. Additional findings reported include changes indicative of neurodegenerative processes [12, 22, 23], reorganization of cytoskeletal and membrane elements [24, 25], and changes in numerous blood and serum-based biomarker candidates [26, 27] that merit further exploration.

Neuroimaging

As mentioned previously, one of the classifiers dictating mild exposure levels is an absence of obvious abnormalities on classic neuroimaging. From a clinical standpoint, this largely refers to absence of gross intracranial hemorrhage on CT examination. We have confirmed absence of intracranial bleeding via gross brain examination after animal sacrifice and also by CT (unpublished data). Research investigations utilizing MRI for structural investigations in blast-injured animals and humans have largely provided unremarkable findings to date, as might be expected when considering the mild, diffuse nature of the injury. Significant changes in T1 or T2 imaging are not present. Diffusion tensor imaging fractional anisotropy and tractography focal aberrations have been reported [9], but further efforts are needed to generate findings translatable to clinical application or mechanistic scientific discovery. Despite the lack of structural neuroimaging findings, emerging functional MRI (fMRI) [28–30] and ^{1}H MRS [31, 32] techniques have demonstrated recent promise for usage in diagnostic imaging and scientific investigation.

Evoked Potentials

Altered functional connectivity as assessed by evoked potentials has been reported after bTBI. Evoked potentials offer an alternative to fMRI as a noninvasive method for assessing post-injury brain function utilizing subdermal electrodes to record brain activity in response to applied stimuli. Stimuli can include tactile, auditory, visual, and olfactory exposures, among others. Some key advantages provided by this technique are: high temporal resolution, pathway specificity (selected by chosen stimulus), and ease of experimentation, particularly if an MRI machine is not accessible or available. We have conducted work utilizing this technique in response to auditory stimuli which has demonstrated significantly altered functional behavior within central auditory nervous system brain and brainstem regions after mild bTBI [33].

Advantages, Complications, Limitations, Alternatives

Advantages

The presented model parameters result in a post-injury symptom profile consistent with clinically mild bTBI in humans, opening a door for studying pathophysiology from molecular biology to behavior. Both our method and modified versions utilized by other groups incite a diffuse brain injury that does not require any surgical procedures or significant technical expertise, making injury application quite simple, even for novice users. The skull remains intact, not only mimicking clinical situation (mild TBI), but also allowing for post-injury changes in intracranial pressure to run their course uninterrupted by the experimental procedure. Incident injury parameters are easily modifiable by changing diaphragm thickness and/or animal positioning and fixation. Usage of protective equipment can be adjusted to allow or prevent body-wide polytrauma and simulate military (body protection) or civilian (non-protected) blast exposure.

Under the exposure conditions presented here, lack of significant motor or memory deficits [10] allow for in-depth study of brain function in behavioral tasks that would be confounded by the presence of such impairments. It is also a model that opens the door for investigation of imaging, serum, or urine biomarkers. The body of knowledge regarding causes and consequences of bTBI, particularly mild bTBI, are poorly understood and the opportunities for novel research are vast.

Complications

Tympanic membrane, middle + inner ear, and vestibular injury may occur as a result of shock wave exposure. The most obvious indication that this has occurred is observation of bleeding from the ears immediately after injury. If the study of interest involves tests of motor function, coordination, or the auditory system, these animals should be removed from the study. Animals should also be removed from chronic studies to avoid undue suffering for these animals. Acute and subacute studies of brain neurochemistry and biology can safely proceed to completion.

If not properly maintained, o rings, fasteners, and tubing utilized to build the shock tube system can wear over time and fail. Securing all connections firmly and ensuring appropriate lubrication of movable parts is necessary. At minimum, the device should be fully inspected at least monthly.

Limitations

Unfortunately, shock wave exposure conditions for rodents are not well-understood or agreed upon in the neurotrauma community, particularly with respect to mechanisms of injury resulting from primary blast and the differential neuropathological processes incited by primary vs. tertiary blast exposure. While scaling laws exist for pulmonary injuries from blast [3], equivalent relationships for brain injury have yet to be established. Relationships between incident injury parameters and subsequent biological, functional, and behavioral changes also unclear and, as a result, dose-response relationships and injury grading are not straightforward as in other models of neurotrauma.

Alternatives & Variations

One of the major differences between labs are with regards to shock wave parameters (overpressure magnitude, duration, impulse) that result from slight differences in shock tube design. Other variations in the exposure choices exist, as outlined in the 'Procedure' section of this document, particularly with regards to animal positioning, head fixation, and body protection. Different anesthesia regimens may also be used, such as inhaled isoflurane, which primarily results in differences in anesthesia recovery time and cerebrovascular properties.

Other species have been investigated as well, with mice being the most common [11, 34–36]. Pigs [37], goats [38], and even whales [39] have also been used in blast studies. Less supporting literature exists for each of these species, so great care must be taken in experimental design, particularly with regard to shock wave exposure conditions.

References

1. Bass CR, Panzer MB, Rafaels KA, Wood G, Shridharani J, Capehart B. Brain injuries from blast. Ann Biomed Eng. 2012;40(1):185–202. https://doi.org/10.1007/s10439-011-0424-0.
2. Centers for Disease Control and Prevention, National Center for Injury Prevention and Control, Division of Injury Response. Blast injuries: fact sheets for professionals; 2013.
3. Bass CR, Rafaels KA, Salzar RS. Pulmonary injury risk assessment for short-duration blasts. J Trauma Acute Care Surg. 2008;65(3):604–15.
4. Gefen A, Gefen N, Zhu Q, Raghupathi R, Margulies SS. Age-dependent changes in material properties of the brain and braincase of the rat. J Neurotrauma. 2003;20(11):1163–77.
5. Envigo.com. (2019). Sprague Dawley® outbred rat | Envigo. [online] Available at: https://www.envigo.com/products-services/research-models-services/models/research-models/rats/outbred/sprague-dawley-outbred-rat/. https://www.envigo.com/resources/growth-curves/sprague-dawley-gc-sept2015.pdf
6. Bolander R, Mathie B, Bir C, Ritzel D, VandeVord P. Skull flexure as a contributing factor in the mechanism of injury in the rat when exposed to a shock wave. Ann Biomed Eng. 2011;39(10):2550–9. https://doi.org/10.1007/s10439-011-0343-0.
7. Chandra N, Ganpule S, Kleinschmit NN, Feng R, Holmberg AD, Sundaramurthy A, Selvan V, Alai A. Evolution of blast wave profiles in simulated air blasts: experiment and computational modeling. Shock Waves. 2012;22(5):403–15. https://doi.org/10.1007/s00193-012-0399-2.
8. Newman AJ, Mollendorf JC. The peak overpressure field resulting from shocks emerging from circular shock tubes. J Fluids Eng. 2010;132(8):081204. https://doi.org/10.1115/1.4002183.
9. Budde MD, Shah A, McCrea M, Cullinan WE, Pintar FA, Stemper BD. Primary blast traumatic brain injury in the rat: relating diffusion tensor imaging and behavior. Front Neurol. 2013;4:154.
10. Walls MK, Race N, Zheng L, Vega-Alvarez SM, Acosta G, Park J, Shi R. Structural and biochemical abnormalities in the absence of acute deficits in mild primary blast-induced head trauma. J Neurosurg. 2016;124(3):675–86. https://doi.org/10.3171/2015.1.JNS141571.
11. Goldstein LE, Fisher AM, Tagge CA, Zhang XL, Velisek L, Sullivan JA, Upreti C, Kracht JM, Ericsson M, Wojnarowicz MW, Goletiani CJ, Maglakelidze GM, Casey N, Moncaster JA, Minaeva O, Moir RD, Nowinski CJ, Stern RA, Cantu RC, Geiling J, Blusztajn JK, Wolozin BL, Ikezu T, Stein TD, Budson AE, Kowall NW, Chargin D, Sharon A, Saman S, Hall GF,

Moss WC, Cleveland RO, Tanzi RE, Stanton PK, McKee AC. Chronic traumatic encephalopathy in blast-exposed military veterans and a blast neurotrauma mouse model. Sci Transl Med. 2012;4(134):134ra60. https://doi.org/10.1126/scitranslmed.3003716. PubMed PMID: 22593173; PMCID: 3739428.
12. Svetlov SI, Prima V, Glushakova O, Svetlov A, Kirk DR, Gutierrez H, Serebruany VL, Curley KC, Wang KK, Hayes RL. Neuro-glial and systemic mechanisms of pathological responses in rat models of primary blast overpressure compared to "composite" blast. Front Neurol. 2012;3:15. https://doi.org/10.3389/fneur.2012.00015. PubMed PMID: 22403567; PMCID: 3275793.
13. Fujimoto ST, Longhi L, Saatman KE, Conte V, Stocchetti N, McIntosh TK. Motor and cognitive function evaluation following experimental traumatic brain injury. Neurosci Biobehav Rev. 2004;28(4):365–78. https://doi.org/10.1016/j.neubiorev.2004.06.002.
14. Hamm RJ. Neurobehavioral assessment of outcome following traumatic brain injury in rats: an evaluation of selected measures. J Neurotrauma. 2001;18(11):1207–16.
15. Elder GA, Dorr NP, De Gasperi R, Gama Sosa MA, Shaughness MC, Maudlin-Jeronimo E, Hall AA, McCarron RM, Ahlers ST. Blast exposure induces post-traumatic stress disorder-related traits in a rat model of mild traumatic brain injury. J Neurotrauma. 2012;29(16):2564–75.
16. Fernando A, Robbins T. Animal models of neuropsychiatric disorders. Annu Rev Clin Psychol. 2011;7:39–61.
17. Lapiz-Bluhm MDS, Bondi CO, Doyen J, Rodriguez G, Bédard-Arana T, Morilak DA. Behavioural assays to model cognitive and affective dimensions of depression and anxiety in rats. J Neuroendocrinol. 2008;20(10):1115–37.
18. Yeoh S, Bell ED, Monson KL. Distribution of blood-brain barrier disruption in primary blast injury. Ann Biomed Eng. 2013;41(10):2206–14. https://doi.org/10.1007/s10439-013-0805-7.
19. Cho HJ, Sajja VS, Vandevord PJ, Lee YW. Blast induces oxidative stress, inflammation, neuronal loss and subsequent short-term memory impairment in rats. Neuroscience. 2013;253:9–20. https://doi.org/10.1016/j.neuroscience.2013.08.037.
20. Garman RH, Jenkins LW, Switzer RC III, Bauman RA, Tong LC, Swauger PV, Parks SA, Ritzel DV, Dixon CE, Clark RS. Blast exposure in rats with body shielding is characterized primarily by diffuse axonal injury. J Neurotrauma. 2011;28(6):947–59.
21. Readnower RD, Chavko M, Adeeb S, Conroy MD, Pauly JR, McCarron RM, Sullivan PG. Increase in blood-brain barrier permeability, oxidative stress, and activated microglia in a rat model of blast-induced traumatic brain injury. J Neurosci Res. 2010;88(16):3530–9. https://doi.org/10.1002/jnr.22510. PubMed PMID: 20882564; PMCID: 2965798.
22. Säljö A, Bao F, Jingshan S, Hamberger A, Hansson H-A, Haglid KG. Exposure to short-lasting impulse noise causes neuronal c-Jun expression and induction of apoptosis in the adult rat brain. J Neurotrauma. 2002;19(8):985–91.
23. Säljö A, Bao F, Shi J, Hamberger A, Hansson H-A, Haglid KG. Expression of c-Fos and c-Myc and deposition of β-APP in neurons in the adult rat brain as a result of exposure to short-lasting impulse noise. J Neurotrauma. 2002;19(3):379–85.
24. Säljö A, Bao F, Haglid KG, Hansson H-A. Blast exposure causes redistribution of phosphorylated neurofilament subunits in neurons of the adult rat brain. J Neurotrauma. 2000;17(8):719–26.
25. Säljö A, Huang Y-L, Hansson H-A. Impulse noise transiently increased the permeability of nerve and glial cell membranes, an effect accentuated by a recent brain injury. J Neurotrauma. 2003;20(8):787–94.
26. Agoston DV, Elsayed M. Serum-based protein biomarkers in blast-induced traumatic brain injury spectrum disorder. Front Neurol. 2012;3:107.
27. Tate CM, Wang KK, Eonta S, Zhang Y, Carr W, Tortella FC, Hayes RL, Kamimori GH. Serum brain biomarker level, neurocognitive performance, and self-reported symptom changes in soldiers repeatedly exposed to low-level blast: a breacher pilot study. J Neurotrauma. 2013;30(19):1620–30. https://doi.org/10.1089/neu.2012.2683.
28. Graner J, Oakes TR, French LM, Riedy G. Functional MRI in the investigation of blast-related traumatic brain injury. Front Neurol. 2013;4:16. https://doi.org/10.3389/fneur.2013.00016. PubMed PMID: 23460082; PMCID: 3586697.

29. Heffernan ME, Huang W, Sicard KM, Bratane BT, Sikoglu EM, Zhang N, Fisher M, King JA. Multi-modal approach for investigating brain and behavior changes in an animal model of traumatic brain injury. J Neurotrauma. 2013;30(11):1007–12. https://doi.org/10.1089/neu.2012.2366. PubMed PMID: 23294038; PMCID: 3684107.
30. Weber R, Ramos-Cabrer P, Wiedermann D, van Camp N, Hoehn M. A fully noninvasive and robust experimental protocol for longitudinal fMRI studies in the rat. NeuroImage. 2006;29(4):1303–10.
31. Hetherington HP, Hamid H, Kulas J, Ling G, Bandak F, de Lanerolle NC, Pan JW. MRSI of the medial temporal lobe at 7 T in explosive blast mild traumatic brain injury. Magn Reson Med. 2014;71(4):1358–67. https://doi.org/10.1002/mrm.24814.
32. Harris JL, Yeh H-W, Choi I-Y, Lee P, Berman NE, Swerdlow RH, Craciunas SC, Brooks WM. Altered neurochemical profile after traumatic brain injury: 1H-MRS biomarkers of pathological mechanisms. J Cereb Blood Flow Metab. 2012;32(12):2122–34.
33. Race N, Lai J, Shi R, Bartlett EL. Differences in postinjury auditory system pathophysiology after mild blast and nonblast acute acoustic trauma. Journal of Neurophysiology. 2017;118(2):782-99. https://doi.org/10.1152/jn.00710.2016
34. Wang H, Zhang YP, Cai J, Shields LBE, Tuchek CA, Shi R, Li J, Shields CB, Xu X. A compact blast-induced traumatic brain injury model in mice. J Neuropathol Exp Neurol. 2016;75(2):183–96.
35. Cernak I, Merkle AC, Koliatsos VE, Bilik JM, Luong QT, Mahota TM, Xu L, Slack N, Windle D, Ahmed FA. The pathobiology of blast injuries and blast-induced neurotrauma as identified using a new experimental model of injury in mice. Neurobiol Dis. 2011;41(2):538–51. https://doi.org/10.1016/j.nbd.2010.10.025.
36. Tweedie D, Rachmany L, Rubovitch V, Zhang Y, Becker KG, Perez E, Hoffer BJ, Pick CG, Greig NH. Changes in mouse cognition and hippocampal gene expression observed in a mild physical- and blast-traumatic brain injury. Neurobiol Dis. 2013;54:1–11. https://doi.org/10.1016/j.nbd.2013.02.006.
37. Säljö A, Arrhén F, Bolouri H, Mayorga M, Hamberger A. Neuropathology and pressure in the pig brain resulting from low-impulse noise exposure. J Neurotrauma. 2008;25(12):1397–406.
38. Li BC, Li Y, Xu C, Wang J, Chen Z, Li G, Zhang J, Hu S, Wang L, Feng H. Blast-induced traumatic brain injury of goats in confined space. Neurol Res. 2014;36(11):974–82.
39. Knudsen SK, Øen EO. Blast-induced neurotrauma in whales. Neurosci Res. 2003;46(3):377–86.

A Rat Model of Surgical Brain Injury

Prativa Sherchan, Devin W. McBride, Lei Huang, Cesar Reis, Onat Akyol, Yuechun Wang, Cherine Kim, Ishan Solaroglu, Jiping Tang, and John H. Zhang

Abstract The rat model of surgical brain injury (SBI) mimics deleterious sequelae resulting from the unavoidable damage to healthy tissue during many neurosurgical procedures, such as peri-operative hemorrhage, brain edema, and neuroinflammation. The SBI model is ideal for evaluating pre-conditioning, pre-treatment, and peri-operative therapies. The SBI model is characterized by partial frontal lobe resection. First, a 5 × 5 mm cranial window in the right frontal bone is made to expose the underlying brain tissue. Next, partial resection of the frontal lobe is performed along margins of the bone window which is followed by saline irrigation and intraoperative packing to ensure complete hemostasis. The total time for completion of the SBI surgery in a rat is 30–40 min. The partial resection of frontal lobe results in contralateral sensorimotor deficits and anxiety-like behavior in rats. Neurological tests to evaluate anxiety-related behavior in SBI rats have been described in this chapter and are recommended for studies using the SBI rat model.

P. Sherchan · D. W. McBride · C. Reis · O. Akyol · Y. Wang · C. Kim · J. Tang
Department of Physiology and Pharmacology, Loma Linda University School of Medicine, Loma Linda, CA, USA

L. Huang
Department of Anesthesiology, Loma Linda University School of Medicine, Loma Linda, CA, USA

I. Solaroglu
Department of Physiology and Pharmacology, Loma Linda University School of Medicine, Loma Linda, CA, USA

Department of Neurosurgery, Koç University, School of Medicine, Istanbul, Turkey

J. H. Zhang (✉)
Department of Physiology and Pharmacology, Loma Linda University School of Medicine, Loma Linda, CA, USA

Department of Anesthesiology, Loma Linda University School of Medicine, Loma Linda, CA, USA

Department of Neurosurgery, Loma Linda University School of Medicine, Loma Linda, CA, USA
e-mail: jhzhang@llu.edu

J. Chen et al. (eds.), *Animal Models of Acute Neurological Injury*, Springer Series in Translational Stroke Research, https://doi.org/10.1007/978-3-030-16082-1_28

The elevated plus maze test and open field test showed greater sensitivity than the forced swim test to detect anxiety-like behavior in rats after SBI. The rat model of SBI allows for investigation of therapeutics that target SBI-induced complications including intraoperative hemorrhage, post-operative hematoma, brain edema, blood brain barrier disruption, neuroinflammation, cell death and oxidative stress which are also briefly described in this chapter.

Keywords Surgical brain injury · SBI · Rat · Animal model · Neurosurgery · Anxiety tests

Introduction

There are approximately one million neurosurgery procedures performed in the United States annually [1]. Regardless of how meticulous a neurosurgical procedure is performed, it is associated with the risk of inherent post-operative complications due to direct trauma, hemorrhage, and brain edema, which increases post-operative patient morbidity and mortality [2–5]. The inevitable injury to the surrounding healthy tissue during neurosurgical procedures is termed surgical brain injury (SBI). This unavoidable injury may exist in many forms including predetermined cortical incisions to access any deeper pathological tissue, retraction of brain lobes or hemispheres, intra-operative bleeding and thermal injury due to electrocoagulation. Although endoscopic surgeries and stereotaxic guided procedures are designed to minimize the invasiveness of neurosurgical procedures, these procedures also lead to inevitable brain injuries and complications.

Development of the Protocol

We have developed a rodent model of surgical brain injury which mimics the inevitable damage to healthy tissue at the perisurgical site during neurosurgical procedures [5]. This is accomplished by partial resection of the right frontal lobe (Fig. 1) which allows for predictable and reproducible primary and secondary brain injury [5–10], as well significant neurological deficits [11–13]. This model allows for investigation of the molecular mechanisms and signaling pathways involved in surgically-induced brain injury and novel therapeutics aimed at minimizing damage to healthy tissue during neurosurgical procedures. Herein a detailed protocol of the SBI model is described. Although the procedures are described for use in young adult male Sprague-Dawley rats, they can be adapted for use in other rat strains [14–17], other species [18–23], female and aged animals by adjusting the amount of resected tissue.

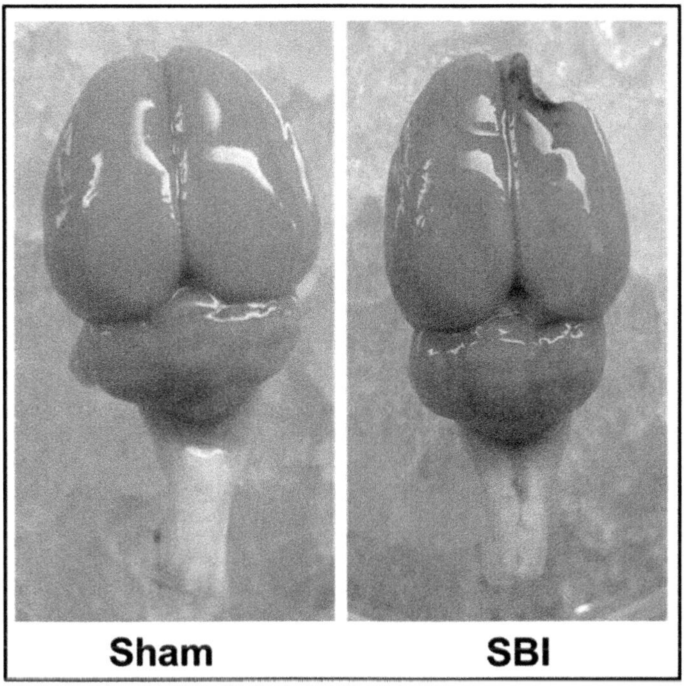

Fig. 1 A representative image of the rat brain after SBI. The picture shows an intact brain sample of a sham rat. In the SBI rat, a portion of the right frontal lobe has been removed, inducing SBI and its sequelae

Materials

The surgical station set up including equipment and surgical tools for making the SBI rat model is shown in Fig. 2.

Animals

Male Sprague-Dawley rats, 8 weeks old, weighing 275–325 g (Harlan Laboratories, Indianapolis IN) are used to make the SBI rat model.

Reagents

Oxygen, Compressed (99.7% O_2, Gilmore Liquid Air Company, South El Monte, CA)
Air, Compressed, UN 1002, Medical gas (Praxair, Danbury, CT)

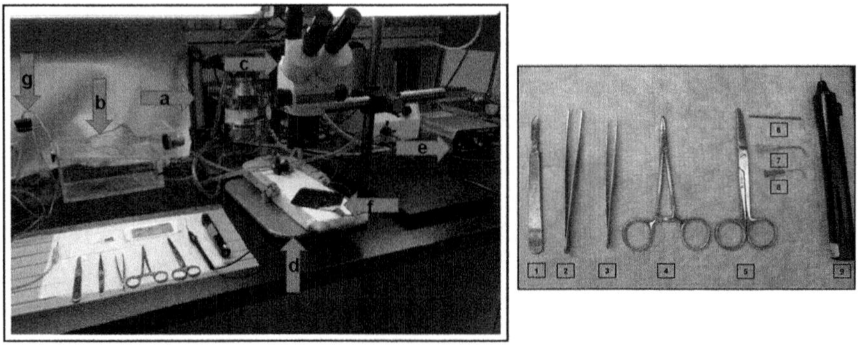

Fig. 2 Surgical set up for making the SBI model. Panel on the left shows the surgical station prepared for making the SBI rat model. (**a**) Isoflurane vaporizer. (**b**) Rodent induction chamber. (**c**) Surgical microscope. (**d**) Rodent stereotactic frame. (**e**) Electrocautery. (**f**) Electrocautery plate. (**g**) Suction vacuum. Panel on right shows surgical tools. From left to right: (**1**) Scalpel. (**2**) Forceps. (**3**) Fine forceps. (**4**) Needle holder. (**5**) Surgical scissors. (**6**) Cautery tip. (**7**) Suction tip made with 18 gauge needle blunt tipped and angled at 90°. (**8**) Resection needle made with 23 gauge needle angled at 90°. (**9**) Microdrill

Isoflurane (cat. no. NDC 66794-013-25, Western Medical Supply, Arcadia, CA)
Sodium chloride 0.9% (cat. no. 0409-7983-03, Hospira, Lake Forest, IL)
Buprenorphine, 0.01–0.03 mg/kg (cat. no. 331802, Buprenex, Reckitt Benckiser Pharmaceuticals, Richmond, VA)

Equipment

Anesthesia gas blender with dual flow meter tubes and isoflurane vaporizer (Whitemore Enterprises, Rancho Cucamonga, CA)
Rodent induction chamber (cat. no. AB 1, BrainTree Scientific, Braintree, MA)
Vacuum (cat. no. 472-001, Erie Medical, Pleasant Prairie, WI)
Surgical microscope (cat. no. KSCXTS-1221, SB2STNDU, Kent Scientific Corporation, Torrington, CT) and fiber optical illuminator (cat. no. KSC5830, Kent Scientific Corporation, Torrington, CT)
Rodent stereotactic frame (cat. no. 51600, Stoelting Co., Wood Dale, IL)
Hair clippers (cat. no. 79250-3401, Wahl Home Products, Sterling, IL)

Surgical Supplies

Betadine (cat. No. 19-815-354, Fisher Scientific Pittsburgh, PA)
Sterile alcohol prep pads (cat. no. 22-363-750, Fisher Scientific, Pittsburgh, PA)
Cotton tipped applicator, 3″ (cat. no. 8884540400, Covidien, Mansfield, MA)

Scalpel (cat. no. RS-9861-36, 65-9843, Roboz Surgical Instrument Co., Gaithersburg, MD)
Microdrill with burr (cat. no. 67-1000 2913, CellPoint Scientific, Gaithersburg, MD)
Micro forceps (cat. no. 11203-23, Fine Science Tools, Foster City, CA)
Needle 18 gauge (cat. no. 305196, BD PrecisionGlide Needle, BD, Franklin Lakes, NJ)
Needle 23 gauge (cat. no. 305193, BD PrecisionGlide Needle, BD, Franklin Lakes, NJ)
Kim wipes (cat. no 06-666-A, Fisher Scientific, Pittsburgh, PA)
Silk suture 3-0 (cat. no. MV-682, Silk Non-absorbable, Oasis, Mettawa, IL)

Procedures

Anesthesia

1. Place the rat in an anesthesia induction chamber and expose it to 4% isoflurane delivered in a mixture of oxygen (0.3 L/min) and medical gas (0.7 L/min) for 5 min.
2. Upon loss of the paw pinch reflex, place the rat prone onto a standard rodent stereotactic frame under a surgical operating microscope. Anesthesia is maintained 2.5% isoflurane administered in a mixture of oxygen (0.3 L/min) and medical gas (0.7 L/min) using anesthesia gas mask during the surgery.

Preparation of the Cranial Window

3. Shave the rat's head and disinfect the skin with betadine and alcohol prep pads.
4. A midline incision is made through the skin and connective tissue using a scalpel.
5. The periosteum is separated from the skull using cotton tipped applicator to expose the bregma and the sagittal and coronal sutures (Fig. 3a).
6. A cranial window (5 × 5 mm edge) is created in the right frontal bone using a microdrill and burr with the lower left corner of the bone window lying towards bregma. The margins of the window are made 2 mm lateral to the sagittal suture and 1 mm proximal to the coronal suture (Fig. 3b). The margins of the bone window are thinned to translucency without penetrating the underlying brain tissue.

   Critical step: To ensure even thinning of the cranial window margin, go around the entire cranial window with the microdrill rather than continuous drilling on one side at a time. This will also prevent puncturing the brain tissue.

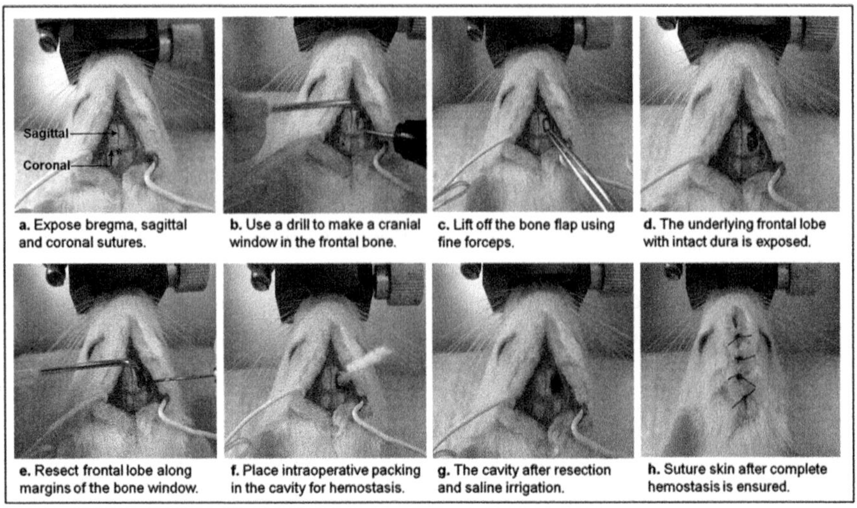

Fig. 3 Steps of the surgical procedure to make the SBI rat model. (**a**) Rat is positioned on the stereotactic frame to expose the bregma (marked with an asterisk), sagittal and coronal sutures. (**b**) Cranial window (5 × 5) mm is drilled in the right frontal bone with the margins of the bone window 2 mm lateral to the sagittal suture and 1 mm anterior to the coronal suture. (**c**) Bone flap is lifted off using fine forceps. (**d**) The underlying frontal lobe with intact dura is exposed. (**e**) The frontal lobe is resected along margins of the bone window, and resected tissue blood is suctioned out. (**f**) Intraoperative packing is kept in the resection cavity for 5 min to ensure complete hemostasis. (**g**) The cavity after the resection is completed and hemostasis is ensured with saline irrigation. (**h**) Skin is sutured after the pack is removed

7. Using fine forceps, the bone flap is lifted off (Fig. 3c), exposing the underlying right frontal lobe with intact dura (Fig. 3d).

   Critical step: When the cranial window has been thinned and several parts of the margins are broken through, use the forceps to grab the bone flap and lift it off. Ensure that the bone margins are thinned out to translucency before lifting off the bone flap. Trying to forcibly pull the bone flap can break the bone flap or lead to dura and brain tissue penetration.

Partial Frontal Lobe Resection

8. The dura is incised using a 23 gauge needle angled at 90° to expose the underlying frontal lobe.
9. Using the 23 gauge needle, incisions is made in the right frontal lobe along the margins of the bone window to perform a partial resection of the frontal lobe. The depth of the incision is extended to the base of the skull.

10. The resected tissue is removed via suction using an 18 gauge needle with the tip made blunt and angled at 90° (Fig. 3e). The needle is hooked up to a vacuum system to remove tissue and blood via suction. Intraoperative blood is collected for measurement and analysis of intraoperative hemorrhage.

    Critical step: When using the vacuum aspirator to suction away intraoperative blood, direct the suction needle tip away from the midline to avoid touching or suctioning away the residual right frontal lobe.

Achieving Hemostasis

11. Saline irrigation and suction is performed to aid in hemostasis. Continue flushing the surgical site with normal saline and use suction to remove the saline with blood.

    Critical step: Avoid forceful or pressured saline irrigation. Re-bleeding can occur if the saline is flushed too vigorously.
12. An intraoperative pack is placed in the resection site for tamponade effect to minimize bleeding. Kim wipes is rolled to make intraoperative packs that can fit in the resection site (Fig. 3f). With intraoperative packing in place, wait approximately 5 min. Gently flush the resection site with normal saline to aid in easy removal of the intraoperative packing. Continue flushing the surgical site with saline until complete hemostasis is achieved (Fig. 3g).

    Critical step: Ensure that the intraoperative packing is kept in the cavity for the same duration for every animal to maintain consistency in the hemostasis time and to ensure adequate time for hemostasis. Care must be taken when removing the intraoperative packing. If not removed gently enough, re-bleeding is likely to occur.
13. If necessary, change the intraoperative packing and continue saline flushing as needed until hemostasis is achieved.
14. When bleeding ceases, remove the intraoperative packing and suction all saline from the resection site and suture close the incision (Fig. 3h). Normal saline (1 mL, subcutaneously) is injected for fluid replacement. Buprenorphine (0.01–0.03 mg/kg, subcutaneously) is injected at the end of surgery for post-operative analgesia. Place the animal in a recovery area and closely observe for post-operative recovery. The rat should wake up within 15–20 min.

Surgery Timing

Steps 3–7. Preparation of the Cranial Window: 10–15 min
Steps 8–10. Partial Frontal Lobectomy: 5 min
Steps 11–14. Achieving Hemostasis: 15–20 min

Critical Steps

Preparation of the cranial window: A square cranial window (5 × 5 mm) is made with the medial margin 2 mm lateral to the sagittal suture and the posterior margin 1 mm proximal to the coronal suture. Since the cranial window will be used for exposure of the frontal lobe, it is critical to make the cranial window the same size in all animals. Thus, it is important to use animals which are weight-matched so that the anatomical landmarks will produce similar cranial windows.

Partial frontal lobe resection: To create a consistent and reproducible surgical brain injury, the amount of brain tissue resected should be identical in all animals. The procedures detailed herein are based on our extensive experience with the SBI model in Sprague-Dawley rats. After exposure of the brain tissue via craniotomy, the exposed frontal lobe tissue is resected along the margins of the cranial window which is 2 mm lateral from the sagittal suture and 1 mm proximal to the coronal suture. The resected tissue is removed via suction through an 18 gauge needle with the tip of the needle made blunt and angled at 90°.

Achieving hemostasis: Due to the resection of a portion of the frontal lobe, bleeding occurs. It is critical to minimize the amount of blood loss. Hemostasis is achieved by irrigation with normal saline maintained at room temperature and by using intraoperative packing with kim wipes for tamponade effect. Following saline irrigation, intraoperative packing is placed at the site of resection for 5 min. The packing is then carefully removed which is followed by saline irrigation and suction removal. All blood lost during resection is collected by suction which can be used as an outcome measurement to assess intraoperative hemorrhagic volumes (Table 1).

Table 1 Troubleshooting surgical steps

Step	Problem	Possible reason	Solution
6–7	Large amount of bleeding	Brain tissue penetration during drilling	View the tissue for visible marks indicating penetration. If present within the tissue to be resected, continue the procedure
8–9	Bleeding obscures the brain tissue	Injury to blood vessels at the skull base	Use intraoperative packing, saline irrigation and suction to stop bleeding
10	Tissue is not being suctioned out of the resection site	Tissue has clogged the suction needle or tubing	Use saline flushing to restore suction
11	Re-bleeding	Saline is being flushed vigorously	Place another intraoperative packing, wait 1–2 min, and gently irrigate with saline
12	Re-bleeding	When the intraoperative packing was removed, a blood clot was displaced	Insert a new intraoperative packing, wait 1–2 min and repeat saline irrigation
13–14	Re-bleeding	See **Steps 11** and **12** above	See **Steps 11** and **12** above

Outcome Evaluation

The SBI rat model allows for several primary outcomes to be studied which mimic the pathophysiology of neurosurgical procedures: intraoperative bleeding, post-operative hematoma [24, 25], brain edema [5, 7, 9, 10, 13, 18–21, 25–37], blood brain barrier disruption [5, 9, 19, 20, 25, 27, 29, 30, 37, 38], and neuroinflammation [7, 14, 16, 18, 22, 23, 25, 27, 29, 37]. Secondary injury which develops post-SBI can induce neuronal apoptosis [5, 9, 15, 17, 19, 32, 35] and oxidative stress [13, 28, 33, 34]. Additionally, neurological tests are performed to evaluate sensorimotor deficits and anxiety related behavior in the SBI rats.

Bleeding Assessment

During SBI, a significant amount of intraoperative bleeding occurs due to direct trauma and resection (Fig. 4a). Intraoperative hemorrhage is analyzed using the hemoglobin assay [39, 40] of all the suctioned blood collected during the procedure

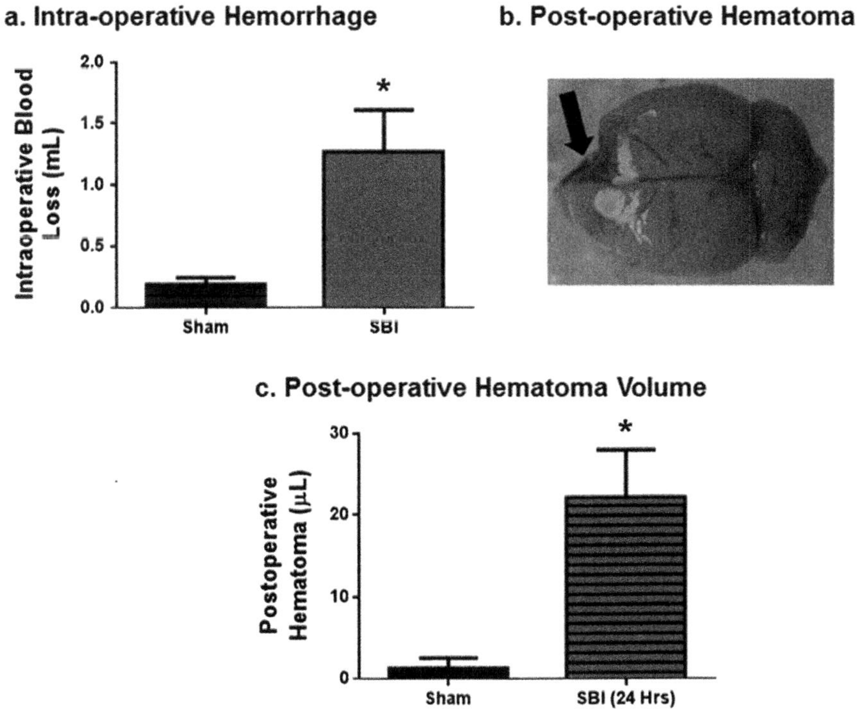

Fig. 4 SBI-induced hemorrhage. (**a**) Intraoperative hemorrhage collected during SBI and measured using the hemoglobin assay. Sham: n = 6; SBI: n = 12. (**b**) Representative image of a brain sample after SBI surgery containing post-operative hematoma (arrow). (**c**) Post-operative hematoma 24 h after SBI, measured using the hemoglobin assay. n = 6/group. Unpaired t-test: *p < 0.05 vs. Sham

as well as the blood collected on the intraoperative packing. Twenty-four hours after SBI, hematoma found along the margins of the perisurgical site brain tissue (Fig. 4b) is quantified using hemoglobin assay [39, 40] to measure postoperative hemorrhagic volumes (Fig. 4c).

Brain Edema Measurement

The SBI animals present with elevated brain water content in the residual right frontal lobe 24 and 72 h post injury (Fig. 5), which can be measured by the Wet-Dry weight method [41]. During sacrifice the brain is quickly removed and dissected into six parts including right frontal, left frontal, right parietal, left parietal, cerebellum and brainstem. The samples are weighed immediately to obtain the wet weight and then weighed again after drying in the oven at 100 °C for 48 h to obtain the dry weight. Brain water content is calculated using the formula [(wet weight − dry weight)/(wet weight)] × 100.

Immunohistochemistry

The SBI rats show increase in inflammatory markers at the perisurgical site 24 h after injury (Fig. 6). Briefly, brain samples are collected after transcardial perfusion of PBS and formalin. After embedding in OCT, the brain samples are sectioned in 10 μm thick slices using a cryostat. The brain sections are incubated overnight with primary antibodies followed by incubation with appropriate secondary antibodies to detect inflammatory markers at the perisurgical site after SBI [42, 43].

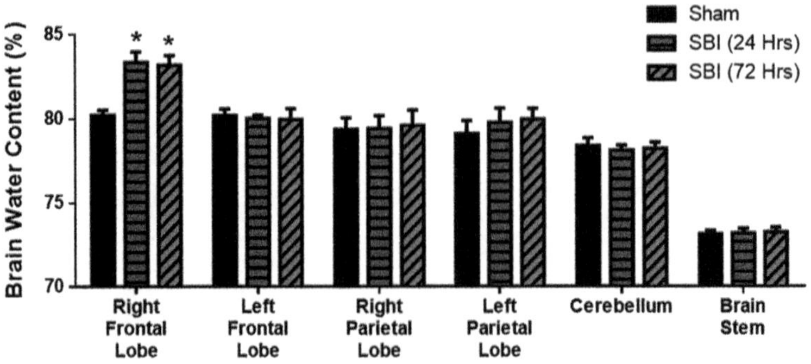

Fig. 5 SBI-induced brain edema. Right frontal lobe brain water content is elevated 24 and 72 h after SBI, but not for the other brain regions. Brain water content was measured for each region of the brain using the Wet-Dry weight method. n = 6/group. One-way ANOVA with Tukey *post-hoc* test: *$p < 0.05$ vs. Sham

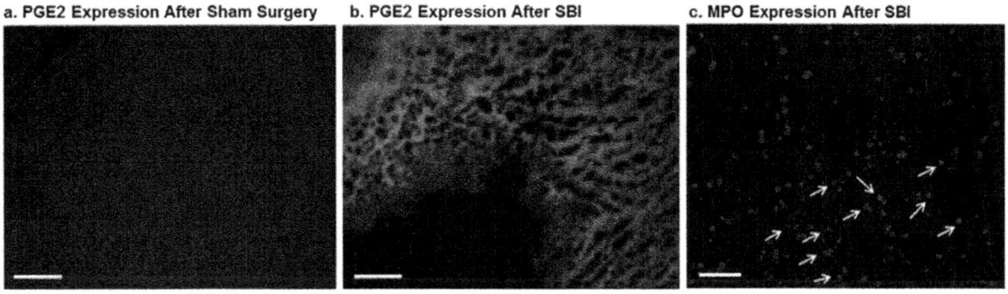

Fig. 6 SBI-induced neuroinflammation. (**a**) Immunohistochemistry picture shows that prostaglandin E2 (PGE2)-FITC/green is not expressed in the right frontal lobe tissue of sham animals. Scale bar: 200 μm. (**b**) The expression of PGE2-FITC/green is significantly elevated in the peri-resection tissue 24 h after SBI. Scale bar: 200 μm. (**c**) Myeloperoxidase (MPO)-Texas Red/red is expressed in the peri-resection tissue 24 h post-SBI. Scale bar: 75 μm

Neurological Tests

Sensorimotor Deficits

Neurological deficits are observed after SBI. Rats typically present with flexed contralateral forelimbs and circling behavior caused by the unilateral injury. Composite neuroscores which can be evaluated using modified Garcia test (Table 2) shows significant deficits at 24 and 72 h after SBI and tends to show recovery by 7 days post-SBI [37]. The corner turn, beam balance, foot fault and forelimb placement tests seem to be the most sensitive tests for evaluating neurological function of rats after SBI (Fig. 7) [11, 13].

Elevated Plus Maze Test

The elevated plus maze test was conducted as previously described [44, 45]. The elevated plus maze consisted of four arms 50 cm long and 10 cm wide. Two of the arms were enclosed by 40 cm high walls (closed arms) and two of the arms were open without any walls (open arms). Each of the arms was supported by a leg that elevated the maze 120 cm off the ground (Fig. 8). Briefly, at 24 h after SBI induction, the rat was placed at the intersection of the four arms facing towards an open arm opposite to where the experimenter is standing. A video tracking system (EthoVision XT 10) with a camera (Euresys PICOLO) mounted overhead on the ceiling was used to track movement of the rats for a duration of 5 min to collect behavioral data. At the end of testing, the rats were returned back to home cages.

Table 2 Modified Garcia test

Parameters	0	1	2	3
Spontaneous activity (in cage for 5 min)	Akinesia	Slow minimal movement	Approaches 1–2 walls	3–4 walls raises on hind limbs to explore over the top
Body Propioception		Unilateral response	Bilateral weak response or brisk on right and weak on left	Bilateral brisk response
Vibrassae Touch		Unilateral response	Bilateral weak response or brisk on right and weak on left	Bilateral brisk response
Limb Symmetry	Left forelimb and hindlimb completely flexed	Left forelimb flexed but hindlimb extended, or vice versa	Left side extended less or more slowly than those on the right	All four limbs extended symmetrically
Lateral Turning	No turning	Unequal turning	Bilateral equal turning <45°	Bilateral equal turning but >45°
Forepaw Outstretching	Left forelimb did not move	Moves in circles	Moves to one side	Briskly walks forward in symmetry
Climbing		Rat failed to climb or circled instead of climbing	Climbs to top with weak grip or good grip but does not climb to top	Climbs to top of cage with strong grip with or without prompting

Data Analysis

The duration that the rat stayed in light during 5 min of observation was recorded as the percent of total time. Additionally, the number of entries into the light, movement between the two closed arms and movement onto the open arms were also recorded.

Results

The SBI rats spent significantly lesser time in the light compared to Sham rats (Fig. 8a). The SBI rats showed a tendency to reduced movement into the light compared to Sham ($p = 0.08$) and crossed significantly fewer number of times between the closed arms compared to Sham (Fig. 8b).

Fig. 7 (continued) from the left front paw increases after SBI, but no difference is observed for the time to remove the adhesive from the right front paw. (**h**) Forelimb placement test. Number of left forelimb placements reduces after SBI. No difference is observed in the animals' ability for right forelimb placement. n = 6/group. (**a**, **b**) ANOVA on Ranks with Dunn's *post-hoc* test. (**c**–**h**) One-way ANOVA with Tukey *post-hoc* test. *$p < 0.05$ vs. Sham

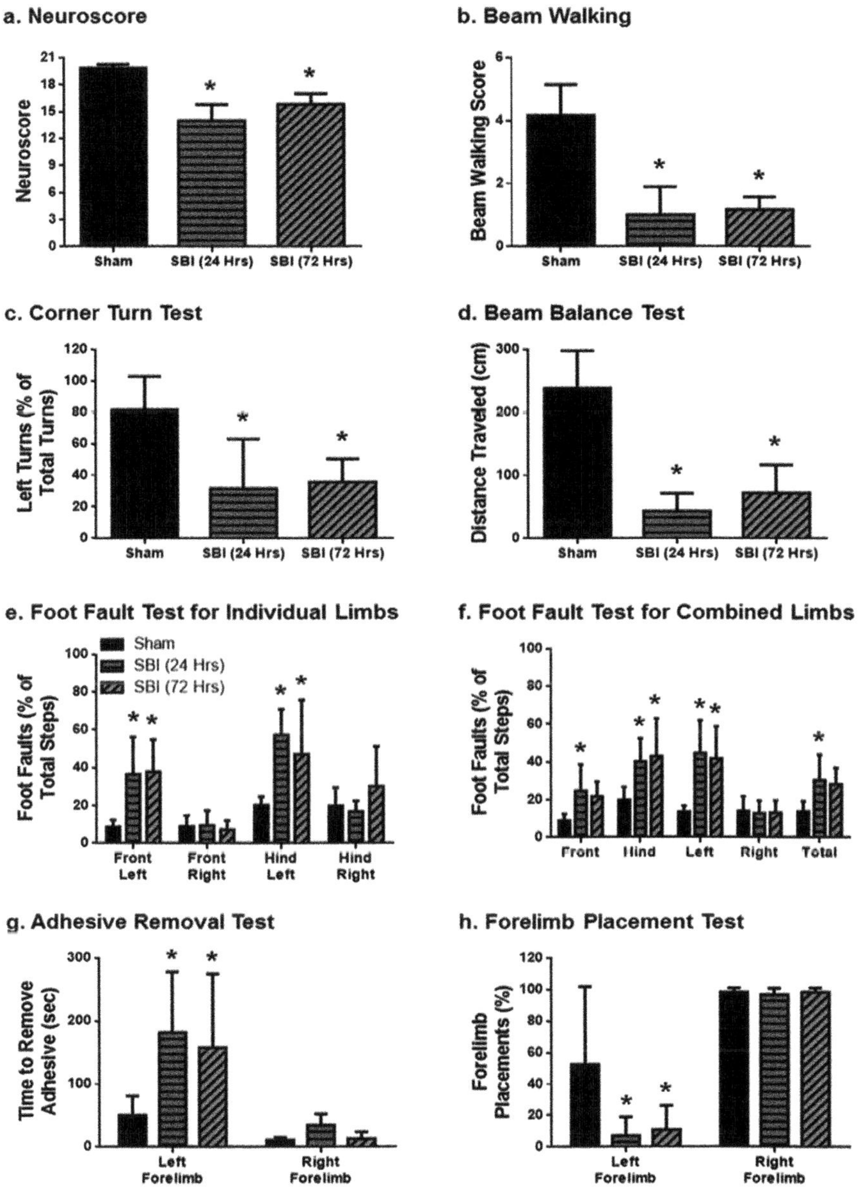

Fig. 7 Neurological deficits after SBI. (**a**) Modified Garcia neuroscore. Composite neuroscore significantly reduces after SBI. (**b**) Beam walking. Ability of animals to traverse a round beam significantly reduces after SBI. (**c**) Corner turn test. Number of left turns significantly reduces after SBI. (**d**) Beam balance test. Ability of animals to traverse a rectangular beam decreases post-ictus. (**e**) Foot fault test for individual limbs. Number of foot faults increase after SBI for the front left and hind left limbs. No inter-group difference is observed for either the front right or the hind right foot faults. (**f**) Foot fault test for combined limbs. Number of foot faults increases, post-ictus, for each combination of limbs except the right limbs. (**g**) Adhesive removal test. Time to remove an adhesive

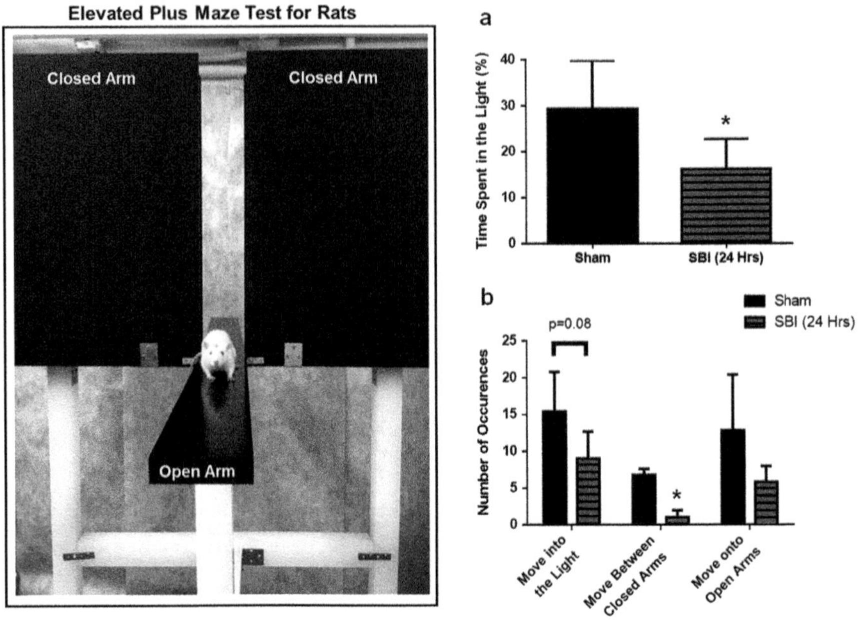

Fig. 8 Elevated plus maze test for rats after SBI. The panel on the left shows the elevated plus maze for rats which consists of four arms, each 50 cm long and 10 cm wide. The two closed arms are enclosed by 30 cm high walls and two of the arms are open without any walls. The maze is elevated by legs 120 cm off the ground. The panel on the right shows the results for the elevated plus maze test after SBI. (**a**) Time spent in light significantly reduces after SBI. (**b**) Movement between closed arms significantly reduces after SBI. n = 5/group, unpaired t-test. *$p < 0.05$ vs. Sham

Open Field Test

The open field test was performed as previously described [46, 47]. The multiple unit open field maze was used which consisted of four square activity chambers each measuring 91 cm (length) × 91 cm (width) × 30 cm (high) (Fig. 9). Briefly, at 24 h after SBI induction, a single rat was placed into the center of each chamber and allowed to freely move and explore for 5 min. The entire duration of testing was recorded with a video camera (Euresys PICOLO) mounted onto the ceiling. A video tracking software (EthoVision XT 10) was used to track the rats during testing for analysis at a later time. At the end of testing, the rat was removed from the maze and returned back to home cage.

Data Analysis

The total distance traveled by the rat during the 5 min observation and the duration of time spent moving was obtained from the video tracking software for analysis.

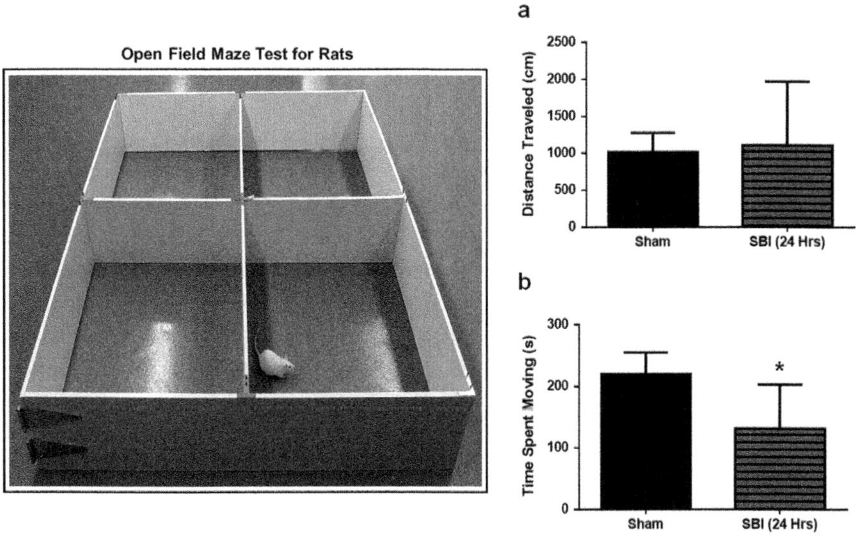

Fig. 9 Open field maze test for rats after SBI. The panel on the left shows the multiple unit open field maze for rats which consists of four square activity chambers, each chamber measuring 91 cm (length) × 91 cm (width) × 30 cm (height). The panel on the right shows the results for open field test after SBI. (**a**) No difference is observed in the distance traveled after SBI compared to sham. (**b**) Time spent moving in the maze significantly reduces after SBI. n = 5/group, unpaired t test. $*p < 0.05$ vs. Sham

Results

The total distance traveled did not significantly differ between Sham and SBI rats (Fig. 9a). However, the SBI rats spent significantly smaller amount of time moving compared to Sham (Fig. 9b).

Modified Forced Swim Test

The modified forced swim test (FST) was performed as previously described [48]. Briefly, rats were subjected to a pretest 24 h before injury and then tested 24 h after injury. The cylinder used for the swim test was 20 cm in diameter, 50 cm deep, and filled with water so the depth of the water was 30 cm.

The rats were subjected to a 15 min swim pretest as previously described [48]. The pretest exposes the rats to FST to familiarize the rats with the testing environment. The rat was gently placed in swim cylinder and allowed to swim for 15 min after which the rat was removed from the water, dried up with towels and then returned back to home cage.

At 24 h after SBI induction, the rat was gently placed in water and the behavior in water was video recorded for 5 min. The behavior was categorized as either escape-directed active behavior (swimming and climbing) or passive behavior (immobility), which was evaluated after the rat was gently dropped in the water. The behaviors were characterized as follows: (1) Swimming-horizontal movement in the water. (2) Climbing-vertical or upward-directed movement of the forepaws. (3) Immobility-minimal movements to keep head above the water. Additionally, the behavior that rats exhibited every time during resurfacing after sinking in the water was also recorded. A sink was characterized by immersion of the nose under water. The rat was removed from water cylinder at the end of 5 min of testing or earlier if the rat sank in water for more than 3 s in which case the rat was immediately removed from the water cylinder and the test was terminated. Video recording was obtained during the testing for subsequent analysis. At the end of testing, the rat was removed from the water, dried up with towels and then returned back to home cages.

Data Analysis

The active (swimming and climbing) and passive (immobile) behavior during the 5 min testing period was analyzed. The percent of time spent being immobile, swimming and climbing during the 5 min observation was scored. Additionally, the number of occurrences of sinks, immobility, swimming and climbing that occurred immediately during resurfacing after a sink was recorded.

Results

The percent of immobility time, swimming time and climbing time during the 5 min observation was not significantly different between Sham and SBI rats (Fig. 10a). During resurfacing after a sink, the SBI rats showed a tendency towards less sinking and reduced immobility compared to Sham (Fig. 10b). There was no significant difference in the occurrences of swimming and climbing between Sham and SBI rats during resurfacing after a sink. A power analysis of test for resurfacing behaviors identified that 8 and 11 animals are required to reach statistical significance between the two groups for the immobile and sinks occurrences, respectively.

Mortality

Approximately 10–15% mortality occurs in the SBI rats. Excessive intraoperative bleeding and early post-operative hemorrhage can contribute to mortality in the early post-operative period. Additionally, increased brain edema at the perisurgical site can lead to mortality up to 24 h after surgery.

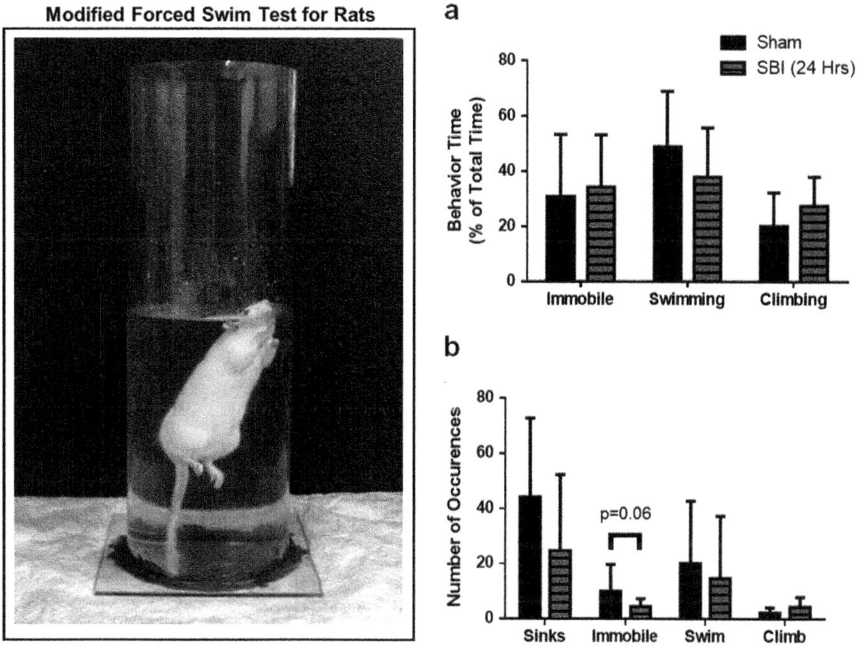

Fig. 10 Modified forced swim test after SBI. Panel on the left shows the swim cylinder 20 cm in diameter, 50 cm deep, and filled with water 30 cm in depth. Panel on the right shows the results for the forced swim test after SBI. (**a**) No significant difference is observed in duration of immobility, swimming or climbing after SBI. (**b**) The number of occurrences of sinks and immobility during resurfacing shows a tendency to decrease after SBI. n = 5–6/group, unpaired t-test. $^*p < 0.05$ vs. Sham

Discussion

Advantages and Limitations of SBI Model

The SBI rat model closely mimics complications encountered after neurosurgical procedures. Since a majority of neurosurgical procedures are elective, the SBI model is ideal for testing, and developing pre-conditioning, pre-treatment, and peroperative therapies [7, 13, 18, 20–22, 26, 29, 33]. Additionally, due to the various pathologies present in rats subjected to SBI, this model provides an ideal platform for testing therapeutics targeting pathologies encountered after brain injury.

The SBI model described utilizes a partial frontal lobectomy to achieve the primary and secondary injuries which accompany brain injury during surgery. Therefore, secondary injury, such as brain edema and inflammation, is localized to the residual right frontal region of the brain. This leads to subtle deficits in composite neuroscore behavior tests acutely [7, 13, 18–22, 26, 27, 29, 31, 33, 35], which may not be observed in long-term neurobehavior examination. Therefore, the use of sensitive sensorimotor tests as well as tests for anxiety behavior is recommended to evaluate neurological deficits in the SBI rat model.

Anxiety Tests for SBI Rat Model

Elevated Plus Maze Test

The elevated plus maze test is a widely used, validated test for anxiety-like behavior in rodents [44, 45]. The test is based on the premise that rats have a tendency to approach dark enclosed spaces and to avoid light and open spaces [49]. The rats when placed in the elevated plus maze are confronted with a conflict between preference for dark enclosed spaces and the motivation to explore a novel environment [44]. A decrease in the number of entries into the open arm or into light, as well as decrease in time spent in open arm or light is indicative of anxious behavior in rodents. Likewise, an increase in either the number of entries into open arm or time spent in the open arm is indicative of anti-anxiety behavior. Our results showed that SBI rats spent significantly less time in the light compared to Sham suggesting that the elevated plus maze test is a sensitive tool to evaluate anxiety-like behavior in the SBI rat model. This test may be useful for understanding the pathology after partial frontal lobe resection. However, a few critical points need to be considered when designing and performing the elevated plus maze test for the SBI rats.

Critical Points

First, although previous studies showed that animals subjected to a battery of other behavioral tests prior to elevated plus maze test did not alter the performance in elevated plus maze [44], it is critical that all neurobehavioral tests be performed in a consistent manner and similar sequence between different animal groups. This is important so that animals during an experiment will be exposed to the same environment and similar experiences, especially if comparing the efficacy of therapeutics. We subjected the rats to the elevated plus maze test, open field test, then the forced swim test in this study.

Second, it has been reported that rodent behavior in the elevated plus maze test is altered when exposed to the test more than once, with an increase in avoidance of the open arms on repeated testing [50, 51]. We therefore, used a single testing session at 24 h after SBI induction. We recommend that a single testing session at 72 h after SBI be performed for animals that are to be evaluated at 72 h after injury.

Open Field Test

The open field test is widely used to evaluate anxiety-like and exploratory behavior in rodents [46, 52]. The open field test showed that SBI rats spent significantly lesser amount of time moving, which is indicative of anxiety-like behavior in rodents. Interestingly, a comparable total distance travelled was observed between SBI rats and Sham rats. Since SBI induces a unilateral injury, mild sensorimotor deficits afflict only the contralateral side after SBI. The ipsilateral side can compensate for the deficits [11] and therefore, normal locomotion is not affected after SBI which may explain the similar distances travelled by Sham and SBI rats. Additionally, it also suggests that the motor component likely did not affect the behavior of rats in open field test and therefore the results observed were attributed to the emotional component of the injury. We did not analyze other parameters such as velocity, thigmotaxis (tendency of the test subject to remain close to the walls), or fecal boli deposits which are indicators of anxiety in rodents [46], and are recommended during testing.

Modified Forced Swim Test

The modified forced swim test (FST) assesses depressive-like behavior in rodents, evaluated by scoring the active and passive behavior when forced to swim in a confined cylinder [48]. The FST is based on the principle that when rats are placed to swim in a cylinder, an initial active attempt to escape-directed behavior comprised of swimming and/or climbing is followed by a passive immobile behavior, either due to failure to persist in the escape-directed behavior during stress or due to loss of mechanisms to cope with stress [48, 53]. There was no significant difference in the percent of immobility time, swimming time and climbing time between Sham and SBI rats in the FST. This suggests that the mild unilateral sensorimotor deficits in SBI rats did not hinder performance in the test. Additionally, the sensorimotor deficits only results in paresis and not complete limb paralysis after SBI and the ipsilateral side may compensate for the deficits [11]. This could explain the lack of significant difference between sham and SBI for the percent time spent in the active behaviors such as swimming and climbing. This finding rules out the possibility of false positive results from motor deficits during the FST. However, we observed that during resurfacing, the SBI rats showed a more escape-directed behavior with a tendency to less sinking and immobility compared to Sham rats. This increase in tendency towards adopting a more active escape-directed behavior by SBI rats suggests that the SBI rats were more persistent in pursuing an escape-directed behavior when placed under stress. We speculate that the development of anxiety in SBI rats following frontal lobe resection makes them continue to persist in trying to escape from the stress inducing environment.

Critical Points

First, motor function can affect performance of rats in the FST. For instance, motor deficits can prolong passive immobile behavior whereas increased locomotor activity can reduce passive behavior during testing. Therefore, it is crucial that the frontal lobe resection be consistent among animal groups when testing for efficacy of therapeutics since the extent of resection can affect motor performance.

Second, compounds and drugs that affect motor activity of the rats can give rise to false positive or false negative outcomes during testing [48]. This factor needs to be taken into consideration when designing experiments and analyzing data. Additionally, other tests for anxiety need to be performed to corroborate the findings.

Conclusions

In conclusion, the rat model of surgical brain injury described here, mimics damage to surrounding healthy tissue which may occur during neurosurgical procedures. We recommend utilizing more sensitive sensorimotor tests and neurological tests for anxiety related behavior to evaluate neurological deficits in the SBI rat model, since the SBI rat model consists of partial resection of right frontal lobe that regulates both sensorimotor function and anxiety-like behavior [54, 55]. Among the neurobehavior tests performed to evaluate anxiety in SBI rats, the elevated plus maze test and open field test were sensitive to detect functional deficits after SBI.

Thus, the SBI model is a clinically relevant model suitable for translational research to test for pre-conditioning or pre-treatment as well as peri-operative therapies for neurosurgical injuries.

Acknowledgments and Disclosure This work was supported by NIH grant R01NS084921 to J.H. Zhang. The authors have no conflicts of interest.

References

1. Dautremont JF, et al. Cost-effectiveness analysis of a postoperative clinical care pathway in head and neck surgery with microvascular reconstruction. J Otolaryngol Head Neck Surg. 2013;42:59.
2. Andrews RJ, Muto RP. Retraction brain ischaemia: cerebral blood flow, evoked potentials, hypotension and hyperventilation in a new animal model. Neurol Res. 1992;14:12–8.
3. Deletis V, Sala F. The role of intraoperative neurophysiology in the protection or documentation of surgically induced injury to the spinal cord. Ann N Y Acad Sci. 2001;939:137–44.
4. Hellwig D, Bertalanffy H, Bauer BL, Tirakotai W. Pontine hemorrhage. J Neurosurg. 2003;99:796; author reply: 796–7.
5. Jadhav V, Solaroglu I, Obenaus A, Zhang JH. Neuroprotection against surgically induced brain injury. Surg Neurol. 2007;67:15–20; discussion 20.

6. Borshchagovskii ML, Dubikaitis IuV. [Clinico-electroencephalographic characteristics of the condition of brain stem systems following surgical and non-surgical brain injury]. Zhurnal nevropatologii i psikhiatrii imeni S.S. Korsakova. 1976;76:337–344.
7. Eckermann JM, et al. Hydrogen is neuroprotective against surgically induced brain injury. Med Gas Res. 2011;1:7.
8. Frontczak-Baniewicz M, Walski M. New vessel formation after surgical brain injury in the rat's cerebral cortex I. Formation of the blood vessels proximally to the surgical injury. Acta Neurobiol Exp. 2003;63:65–75.
9. Jadhav V, Zhang JH. Surgical brain injury: prevention is better than cure. Front Biosci. 2008;13:3793–7.
10. Sherchan P, Kim CH, Zhang JH. Surgical brain injury and edema prevention. Acta Neurochir Suppl. 2013;118:129–33.
11. McBride DW, Wang YC, Sherchan P, Tang JP, Zhang JH. Correlation between subacute sensorimotor deficits and brain water content after surgical brain injury in rats. Behav Brain Res. 2015;290:161–71.
12. Frumberg DB, Fernando MS, Lee DE, Biegon A, Schiffer WK. Metabolic and behavioral deficits following a routine surgical procedure in rats. Brain Res. 2007;1144:209–18.
13. Lee DH, et al. Reproducible and persistent weakness in adult rats after surgical resection of motor cortex: evaluation with limb placement test. Childs Nerv Syst. 2009;25:1547–53.
14. Frontczak-Baniewicz M, et al. Morphological evidence of the beneficial role of immune system cells in a rat model of surgical brain injury. Folia Neuropathol. 2013;51:324–32.
15. Frontczak-Baniewicz M, Walski M, Madejska G, Sulejczak D. MMP2 and MMP9 in immature endothelial cells following surgical injury of rat cerebral cortex—a preliminary study. Folia Neuropathol. 2009;47:338–46.
16. Frontczak-Baniewicz M, Walski M, Sulejczak D. Diversity of immunophenotypes of endothelial cells participating in new vessel formation following surgical rat brain injury. J Physiol Pharmacol. 2007;58(Suppl 5):193–203.
17. Sulejczak D, Grieb P, Walski M, Frontczak-Baniewicz M. Apoptotic death of cortical neurons following surgical brain injury. Folia Neuropathol. 2008;46:213–9.
18. Ayer RE, et al. Preoperative mucosal tolerance to brain antigens and a neuroprotective immune response following surgical brain injury. J Neurosurg. 2012;116:246–53.
19. Bravo TP, et al. Role of histamine in brain protection in surgical brain injury in mice. Brain Res. 2008;1205:100–7.
20. Jadhav V, et al. Hyperbaric oxygen preconditioning reduces postoperative brain edema and improves neurological outcomes after surgical brain injury. Acta Neurochir Suppl. 2010;106:217–20.
21. Jadhav V, et al. Cyclo-oxygenase-2 mediates hyperbaric oxygen preconditioning-induced neuroprotection in the mouse model of surgical brain injury. Stroke. 2009;40:3139–42.
22. Jafarian N, et al. Mucosal tolerance to brain antigens preserves endogenous TGFbeta-1 and improves neurological outcomes following experimental craniotomy. Acta Neurochir Suppl. 2011;111:283–7.
23. Zheng Y, et al. An experimental study on thymus immune tolerance to treat surgical brain injury. Chin Med J. 2014;127:685–90.
24. Frontczak-Baniewicz M, Chrapusta SJ, Sulejczak D. Long-term consequences of surgical brain injury—characteristics of the neurovascular unit and formation and demise of the glial scar in a rat model. Folia Neuropathol. 2011;49:204–18.
25. Xu FF, et al. Effects of progesterone vs. dexamethasone on brain oedema and inflammatory responses following experimental brain resection. Brain Inj. 2014;28:1594–601.
26. Benggon M, Chen H, Applegate R, Martin R, Zhang JH. Effect of dexmedetomidine on brain edema and neurological outcomes in surgical brain injury in rats. Anesth Analg. 2012;115:154–9.
27. Di F, et al. Role of aminoguanidine in brain protection in surgical brain injury in rat. Neurosci Lett. 2008;448:204–7.

28. Hao W, Wu XQ, Xu RT. The molecular mechanism of aminoguanidine-mediated reduction on the brain edema after surgical brain injury in rats. Brain Res. 2009;1282:156–61.
29. Hyong A, et al. Rosiglitazone, a PPAR gamma agonist, attenuates inflammation after surgical brain injury in rodents. Brain Res. 2008;1215:218–24.
30. Jadhav V, Matchett G, Hsu FP, Zhang JH. Inhibition of Src tyrosine kinase and effect on outcomes in a new in vivo model of surgically induced brain injury. J Neurosurg. 2007;106:680–6.
31. Jadhav V, Yamaguchi M, Obenaus A, Zhang JH. Matrix metalloproteinase inhibition attenuates brain edema after surgical brain injury. Acta Neurochir Suppl. 2008;102:357–61.
32. Khatibi NH, et al. Prostaglandin E2 EP1 receptor inhibition fails to provide neuroprotection in surgically induced brain-injured mice. Acta Neurochir Suppl. 2011;111:277–81.
33. Lee S, et al. The antioxidant effects of melatonin in surgical brain injury in rats. Acta Neurochir Suppl. 2008;102:367–71.
34. Lo W, et al. NADPH oxidase inhibition improves neurological outcomes in surgically-induced brain injury. Neurosci Lett. 2007;414:228–32.
35. Manaenko A, et al. PAR-1 antagonist SCH79797 ameliorates apoptosis following surgical brain injury through inhibition of ASK1-JNK in rats. Neurobiol Dis. 2013;50:13–20.
36. Westra D, Chen W, Tsuchiyama R, Colohan A, Zhang JH. Pretreatment with normobaric and hyperbaric oxygenation worsens cerebral edema and neurologic outcomes in a murine model of surgically induced brain injury. Acta Neurochir Suppl. 2011;111:243–51.
37. Yamaguchi M, Jadhav V, Obenaus A, Colohan A, Zhang JH. Matrix metalloproteinase inhibition attenuates brain edema in an in vivo model of surgically-induced brain injury. Neurosurgery. 2007;61:1067–75; discussion 1075–6.
38. Fan D, et al. The protective mechanism for the blood-brain barrier induced by aminoguanidine in surgical brain injury in rats. Cell Mol Neurobiol. 2011;31:1213–9.
39. Asahi M, Asahi K, Wang X, Lo EH. Reduction of tissue plasminogen activator-induced hemorrhage and brain injury by free radical spin trapping after embolic focal cerebral ischemia in rats. J Cereb Blood Flow Metab. 2000;20:452–7.
40. Choudhri TF, Hoh BL, Solomon RA, Connolly ES Jr, Pinsky DJ. Use of a spectrophotometric hemoglobin assay to objectively quantify intracerebral hemorrhage in mice. Stroke. 1997;28:2296–302.
41. Tang JP, et al. MMP-9 deficiency enhances collagenase-induced intracerebral hemorrhage and brain injury in mutant mice. J Cereb Blood Flow Metab. 2004;24:1133–45.
42. Sherchan P, et al. Recombinant Slit2 attenuates neuroinflammation after surgical brain injury by inhibiting peripheral immune cell infiltration via Robo1-srGAP1 pathway in a rat model. Neurobiol Dis. 2016;85:164–73.
43. Huang L, et al. Phosphoinositide 3-kinase gamma contributes to neuroinflammation in a rat model of surgical brain injury. J Neurosci. 2015;35:10390–401.
44. Walf AA, Frye CA. The use of the elevated plus maze as an assay of anxiety-related behavior in rodents. Nat Protoc. 2007;2:322–8.
45. File SE, Lippa AS, Beer B Lippa MT. Animal tests of anxiety. Curr Protoc Neurosci. 2004;8:8.3.
46. Seibenhener ML, Wooten MC. Use of the open field maze to measure locomotor and anxiety-like behavior in mice. J Vis Exp. 2015;(96):e52434.
47. Abdollahnejad F, et al. Investigation of sedative and hypnotic effects of Amygdalus communis L. extract: behavioral assessments and EEG studies on rat. J Nat Med. 2016;70(2):190–7.
48. Slattery DA, Cryan JF. Using the rat forced swim test to assess antidepressant-like activity in rodents. Nat Protoc. 2012;7:1009–14.
49. Ohl F. Testing for anxiety. Clin Neurosci Res. 2003;3:233–8.
50. Bertoglio LJ, Carobrez AP. Previous maze experience required to increase open arms avoidance in rats submitted to the elevated plus-maze model of anxiety. Behav Brain Res. 2000;108:197–203.
51. Bertoglio LJ, Carobrez AP. Anxiolytic effects of ethanol and phenobarbital are abolished in test-experienced rats submitted to the elevated plus maze. Pharmacol Biochem Behav. 2002;73:963–9.

52. Tatem KS, et al. Behavioral and locomotor measurements using an open field activity monitoring system for skeletal muscle diseases. J Vis Exp. 2014;(91):51785.
53. Cryan JF, Valentino RJ, Lucki I. Assessing substrates underlying the behavioral effects of antidepressants using the modified rat forced swimming test. Neurosci Biobehav Rev. 2005;29:547–69.
54. Adhikari A, et al. Basomedial amygdala mediates top-down control of anxiety and fear. Nature. 2015;527:179–85.
55. Uylings HB, Groenewegen HJ, Kolb B. Do rats have a prefrontal cortex? Behav Brain Res. 2003;146:3–17.

The MASCIS Spinal Cord Contusion Model

Wise Young

Abstract The Multicenter Animal Spinal Injury Study (MASCIS) Model is a validated animal model for inducing contusion using the MASCIS Impactor. The model defines procedures for anesthesia, surgery, injury timing, induction, antibiotics, resuscitation methods, fluid replacement, postoperative care, and chronic animal care. This model uses Long-Evans hooded rats obtained at least 2 weeks before surgery. Presurgery preparation, laminectomy, and spinal cord contusion are done under sterile conditions. Rats are suspended with clamps holding T9 and T11 vertebrae, with the chest resting on a clean foam block. The impactor is positioned and is digitally set, and the impact is delivered 60 min after starting anesthesia. After injury, the rat is positioned on the operating table and the skin is closed over the impact site. Impact parameters recorded include impact velocity, compression depth, compression time, and compression rate. Outcome measures include lesion volume, BBB locomotor score, and spared white-matter area.

Keywords Multicenter Animal Spinal Injury Study (MASCIS) model · Spinal cord contusion · Validated animal contusion model

Introduction

The Multicenter Animal Spinal Injury Study (MASCIS) Impactor was first developed in the early 1990s, in response to the demand for a standardized rat spinal cord injury model. In 1993, the National Institutes of Health funded the first multicenter animal spinal cord injury study (MASCIS) to validate the model for multicenter studies of therapies. Eight leading spinal cord injury centers worked together to test approximately 800 rats per year, testing and validating all aspects of the model. The model has been described by several papers [1–4].

W. Young (✉)
W. M. Keck Center for Collaborative Neuroscience, Rutgers University, Piscataway, NJ, USA
e-mail: wisey@pipeline.com

J. Chen et al. (eds.), *Animal Models of Acute Neurological Injury*, Springer Series in Translational Stroke Research, https://doi.org/10.1007/978-3-030-16082-1_29

The MASCIS model is more than a device for contusing the spinal cord. The model included detailed criteria for the animals, including strain, body weight, and pre-operative care. It defines the anesthetic procedure, the surgery, the timing of the injury after anesthesia induction, antibiotics, resuscitation methods, fluid replacement, and post-operative care. In addition, the model defines the care and methods of assessing tissue damage and locomotion of rats after the injury. The model differs from all others in its definition of not only the mechanical but biological parameters of the injury.

Briefly, the model is carried out in 77 ± 1 day old rats (Long-Evan's hooded) anesthetized with pentobarbital (45 mg/kg female, 65 mg/kg male). The injury site is exposed with a T9–10 laminectomy, removing half of T9 and all of T10. The injury is carried out by dropping a 10 g rod 6.25, 12.5, 25.0, or 50.0 mm onto the spinal cord between T9 and T10. The impactor head is 2.1 mm. Animals are monitored for respiratory depression by observing the color of their feet and revived by tracheal suction if cyanosis is present.

Post-operatively, the rats all receive a prophylactic antibiotic, i.e. cephalozolin (daily 25 mg/kg injections subcutaneous) for 7 days. If recurrent urinary tract infections occur, the rats get a 10-day course of Baytril (2.5 mg/kg/day subcutaneous). The rats are inspected twice daily and their bladders are expressed once a day. If the rats show evidence of autophagia (biting of dermatomes below the injury site), they are treated with 67.5 mg/kg of acetaminophen (oral). The animals are weighed and scored weekly for BBB locomotor scores. Depending on the experimental plan, the rats are euthanized either by decapitation or intracardiac perfusion of formaldehyde.

The following sections will summarize the preoperative preparation, the surgery, the injury, post-injury care, and various outcome measures that have been developed for the model, including the BBB locomotor score and spared white matter measurements. Some of the information has been published in various papers and chapters but this is the first article that summarizes, the procedures in detail.

Animals

The MASCIS model stipulates Long-Evans hooded rats that are 77 ± 1 day old at the time of injury. The animals should be obtained at least 2 weeks before surgery and accommodated to the animal holding facility. The rats are not fasted before the injury. The animals are anesthetized with intraperitoneal pentobarbital. Isoflurane is used for other procedures such as transplantation.

Rat strain. Long-Evan's hooded rats are robust, outbred, and widely available from many commercial sources. While other rat strains can be used, investigators should recognize that inflammation, immune response, and responses to injury and treatment differ amongst rat strains. Appropriate controls should be used.

Age. The size of the spinal cord of rats correlate linearly with age of the rat and not to body weight. Male and female rats differ substantially in their body weight as they mature. Male Long-Evans hooded rats reach 320 g by 77 days old while female rats tend to be smaller at 240 g. However, their spinal cords are very similar in size and weight.

Accomodation. Shipping is quite stressful to rats. Most rats do not eat during the shipping period and they tend to respond to injuries differently. Therefore, rats should be ordered and delivered at least 2 weeks before surgery. They must be accommodated to animal facility.

No fasting. The rats should not be fasted before surgery. They should be allowed access to food and water ad libitum. The reason is because rats become hypoglycemic (blood glucose <50 mg/dL) if they do not feed for 12 h. Hypoglycemia increases severity of injuries. Rats do not vomit or aspirate. There is no rationale of preoperative fasting.

Pentobarbital. The rats are anethetized with intraperitoneal pentobarbital (45 mg/kg for female and 65 mg/kg for males). To inject the rats, it is useful to cover the rat's head with the end of a sock and gently immobilized until the rat is relaxed. The injection is then made on the right, two fingers above the pelvis.

Isoflurane. For procedures to inject cells or other manipulations of the spinal cord besides spinal cord injury, isoflurane is recommended. Anesthesia is induced by exposing the animals to 5% isoflurane. When the animal is deeply anesthetized, the anesthesia is maintained with 2% isoflurane by spontaneous breathing.

Testing anesthesia. The rat should become deeply anesthetized within 5 min. The anesthetic depth is tested by manual pinching of the middle of the tail. A well-anesthetized rat show a transient response of chest muscles and breathing response but no limb movement to a firm mid-tail pinch.

Surgery

The surgery should be carried out under sterile conditions. The anesthetized animals are first shaved and prepared for surgery. A standardized sterile instrument pack should contain everything necessary for the surgery. A laminectomy is carried out to expose the spinal cord. After the spinal cord has been hit with impactor, the wound is closed.

Preparation. The rats are shaved in a hood with an electrical hair cutter, exposing the back for the laminectomy and, if necessary, the neck for placement of intravenous lines. Care should be taken to avoid exposure to rat hair to prevent development of allergies. The skin is prepared with a betadine solution. The surgeon is masked, gowned, and gloved.

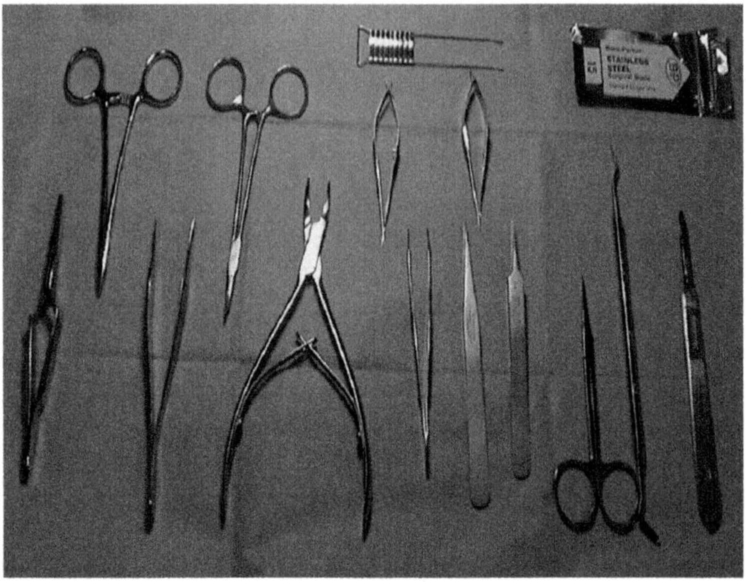

Fig. 1 Surgical instruments used in the operation of the rat. From left to right, they are stainless steel clip applier, mosquito clamp, large toothed tissue forcep, mosquito clamp, mini-rongeur, stainless steel clips (top), microscissors, fine toothed forceps, fine forceps, microscissor, fine forceps, small scissor, scalpel blade, bone elevator, scalpel handle

Surgical Instruments. The following surgical instruments are part of the essential instrument pack, as shown in Fig. 1. These include scalpel, large and small forceps, mini-rongeur for removing bone, mosquito clamps, a dental elevator for detaching muscle from the vertebral processes, microscissors and scissors, and stainless steel clips and applicator. In addition, sterile gauze pads, cotton pads, and drapes should be available, as well as squirt bottles of sterile saline solution and 95% alcohol solutions. All instruments that touch the inside of the wound must be sterile. The surgeon should wear sterile gloves. To use the instruments in sequential surgeries, they are re-sterilized by inserting the tips into a hot beads sterilizer.

Laminectomy. The skin is incised with the scalpel at the midline between T5 and L1 (identified by counting the ribs). Muscles are detached sharply from the dorsal processes while the gauze pads are used to apply pressure to bleeding points. The bone elevator is used to clear the muscle from the lateral processes of T9–T10. Using the forcep with teeth to grasp the T9 dorsal process, the microrongeur is used to nibble at bone between T9 and 10, to remove the half of T9 and all of T10. The laminecotmy opening should be rectangular and 2.5 mm in width. Care is taken to avoid damaging the spinal cord and the T8 and T11 dorsal vertebral processes during the laminectomy. The dura is left intact. The T12 cord is exposed (Fig. 2).

Wound Closure. After hitting the spinal cord with the Impactor (see next section), the rat is unclamped. Using cotton balls moistened with sterile saline, the dural

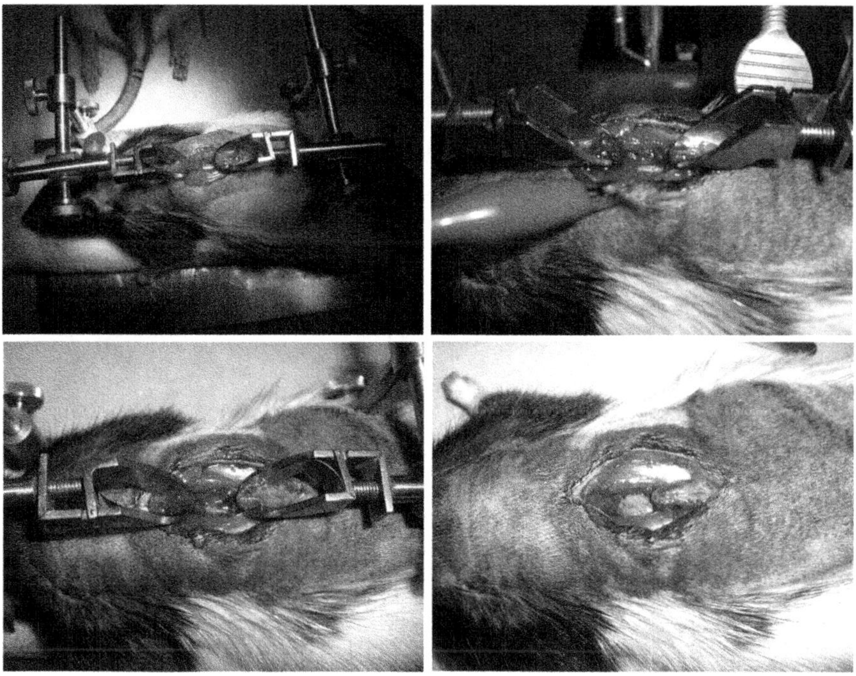

Fig. 2 Laminectomy and impact site: Top left: the T11 spinal cord has been exposed by a T9–10 laminectomy. Top right: closeup picture of the surgery site before placement and centering of the impactor rod. Bottom left: the spinal cord shortly after impact. Bottom right: the spinal cord covered with a piece of fat, just before closure

surface of the spinal cord and the laminectomy site should be cleaned so that the cord can be clearly seen through the dura. Muscle is sutured with silk and the skin is closed with stainless steel clips. The rat is then placed in a small transparent placed on a heating pad and monitored until it wakes up. If the rats show any signs of cyanosis, tracheal suction will be carried out to remove secretions and to stimulate respiration.

Impactor

The rat is suspended with clamps holding T9 and T11. Two types of clamps are available. The first type of clamp is held by right angle screw blocks and is designed to hold the T9 and T11 vertebral processes. If the laminectomy is done at the incorrect levels, these clamps may not fit. The second type is an angled clamp that allows room for instruments on top and can be used to clamp other vertebral processes. A foam block (covered with a clean material) is used to support the chest of the rat (Figs. 3 and 4).

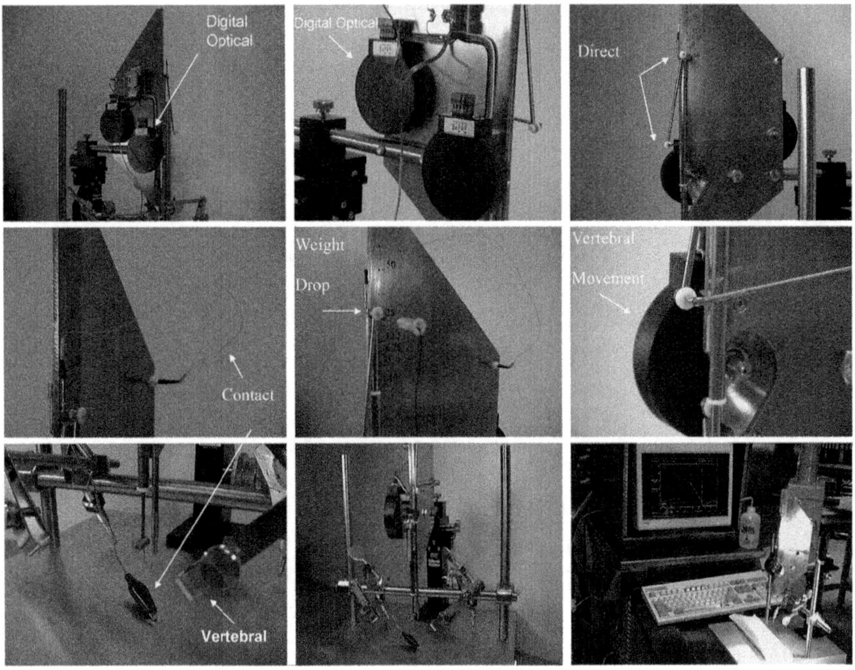

Fig. 3 Impactor. Top: digital optical potentiometers and linkage to the impactor rod. Middle: contact detector wire, weight drop release, vertebral movement detector. Bottom: vertebral clamps, frontal view of the impactor, and the setup in the laboratory

Zero the Impactor. The Impactor is positioned with the Impactor rod placed at zero mm and the Impactor head centered on the spinal cord between T9 and T10. The ground alligator is placed on the clamps (in continuity with the animal) to detecting contact between the impactor rod and spinal cord. Lower the impactor till the head touches the spinal cord, producing a buzzing sound and a light. The impactor rod is raised slightly until the sound and light stops. The impactor is then lowered slowly with micromanipulator until the sound just occurs.

Set the Impactor. The computer program for the Impactor should have been started. Note the computer runs on MS-DOS and should be a 486 computer. The program name is MAS. All commands are single key click commands. Set the height of the weight drop, the name of the file for recording the impact, and set the computer to record the impact. The impactor rod is raised to the intended level for weight drop, i.e. 6.25, 12.5, 25.0, or 50.0 mm weight drop. The pin is inserted into the hole to hold the rod up and the tip is pushed so that it is flush to the Impactor steel plate.

Contusion. The injury should take place at 60 min after induction of anesthesia. An impact done 15 min early can result in 20% greater lesion volume. Ensure that the impactor is centered on the spinal cord. Pull the pin and let the rod drop. The rat should

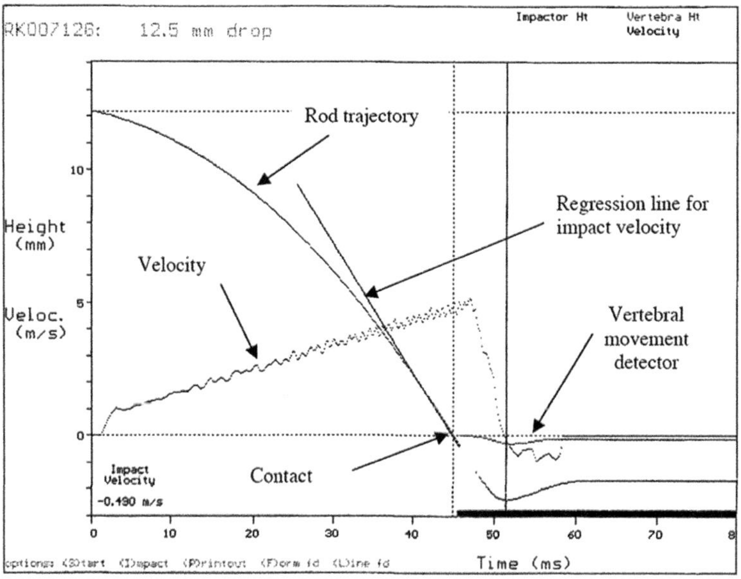

Fig. 4 Impactor trajectory. The graph shows the data from a rod dropped 12.5 mm onto a wet sponge. The Impactor program collects high resolution data (±20 μm and ±1 μs) and generates a graph. The one above has been labelled to show the different components. The y-axis indicates the height of the weight drop, the rod trajectory shows the rod falling until it contacts the spinal cord. The program estimates impact velocity from a linear regression of the rod position during the 2 ms before contact

jump slightly, the legs should go into extension, and the tail should whip back and forth several times. Remove the impactor rod from the cord surface. The computer should display the impactor parameters. Inspect the impact site. There may be subdural hemorrhage evident. Release the clamps. Bring the rat back to the operating table.

Skin Closure. Using saline-soaked cotton balls, wipe the dural surface clean. Pinch a small piece of subcutaneous fat and place it on the dural surface to inhibit adhesions and formation of a collagenous scar over the laminectomy site. If long term inhibition of adhesions are desired, stretch place a piece of Parafilm® over the fat. Using silk and a cutting needle, suture the muscles at the midline at two points above and below the impact site while avoiding the injury site. Use three stainless steel clamps to close the skin, one at the middle and the other two between the middle and the ends of the wound.

Impact Parameters

The Impactor provides a detailed record of the falling rod and the displacement of the spinal cord. The measurements are accurate to ±10 μm and ±10 μs. In addition, the impact records the time and position of contact, the deepest displacement of

RK00725H PARAMETER CALCULATIONS Impactor v7.0

Basic Impact Parameters (relative to baseline / t=0 ms)

Variable	Sample	MM	Time (ms)
Baseline	7457	-0.002	71.979
Drop start	1	24.823	0.010
Impact	7451	0.038	71.921
Cord_Rev	7926	-2.565	76.506
Vert_Rev	8058	-0.468	77.780
Crd-Surf_Rev	7919	-2.270	76.438
Compression		2.308	

Impact Height and Velocity Calculations

Variable	Height (mm)	Velocity (m/sec)	Time (msec)
Expected :	24.903	-0.700	71.392
Actual :	24.785	-0.682	71.911
Error :	0.118	0.018	0.519
Error (%):	0.5	2.6	0.727

Fig. 5 Impact parameters. The picture shows the graphic display of the impact parameters. The top set of numbers have three columns named Sample, MM, and Time (ms). Sample is the number of pulses received by the computer. MM is the conversion of those pulses to mm. Time is in milliseconds. Cd is estimated from MM (Impact-Crd-Surf Rev). Ct is estimated from Time (Impact-Crd-Surf). The lower table lists the expected and actual height, velocity and time of the impacts. The absolute and percentage errors are listed

spinal cord. It does linear regression to find the slope of the line during the 2 ms just before contact of the spinal cord, giving the impact velocity (Vi). The Impactor Program provides three other parameters: the compression depth (Cd), compression time (Ct), and compression rate (Cr) (Fig. 5).

Impact Velocity (Vi). Vi is the what the impactor delivers. Under ideal circumstances, the impact velocity can be predicted by Galileo's formula where $H = (g \cdot T^2)/2$ where H is the height of the weight drop and T is the time required for the rod to fall that height. If so, $T = \text{sqrt}(2*H/g)$, $g = 9800$ m/ms^2, and $V = g*T$. From these equations, we can readily calculate that the idealized impact velocities should be 0.3571, 0.4945, 0.6993, and 0.9889 for the rod dropped from 6.25, 12.5, 25.0, 50.0 mm heights. If the impact velocity is close to the ideal velocities, this means that the impactor is working.

Compression Depth (Cd). Cd is how much the dropping rod compressed the spinal cord. It is typically in the range of 2–3 mm and increases with greater height of the weight drop. If Cd is much greater than 3 mm, this usually means that the spinal cord was not well clamped. If it is much less than 2 mm, it usually means that the impactor rod has hit bone along the way.

Compression Time (Ct). Ct is the time required for the dropping rod to compress the spinal cord to the deepest point. Usually expressed in milliseconds (ms), Ct is typically on the order of several milliseconds. The time of compression tends to increase with greater weight drop heights. The greatest utility of Ct, however, is its use for calculating the cord compression rate.

Compression Rate (Cr). Cr is the average rate at which the spinal cord is being compressed by the falling weight, calculated from Cd/Ct. Cr correlates linearly with Vi, as well as lesion volume, spared white matter, and BBB locomotor scores. In general, it is 70–80% of the value of Vi. The difference in Vi and Cr is proportional to the energy absorbed by the spinal cord.

The Impactor program is designed to give a warning the impact parameters deviate from expected amounts by more than 5%. This is an arbitrary criteria. The best measure of impactor consistency is the relationship between Vi and Cr. During their first 15–30 impacts, most investigators have correlation coefficients on the order of 0.60 but they often reach 0.80 after 30 impacts. Very experienced surgeons have correlation coefficients of 0.95 or better.

Postoperative Care

The rats should be kept on a warm pad and placed in a prominent place in the operating room where it can be seen by many people. During the immediate post-injury, the rat is most susceptible to respiratory depression and therefore must be frequently inspected.

Respiratory Depression. While we use to use arterial blood gases (from a tail artery) to determine respiratory status of the animal, we discovered that the color of the feet is just as good or better determinant of the oxygenation status of the rats. If the rat is well oxygenated, the feet are bright pink. If they are dusky or even blue, this means the rats must get respiratory stimulation.

Tracheal Suction. We use a thin soft catheter on a 5 mL plastic syringe. After inserting the catheter into the trachea, we pull back on the 5 mL syringe, producing a vacuum. As the catheter is slowly withdraw, it sucks up mucus and also stimulates the rat to breathe. This method is by far the most effective of any that we have tried. If the rat is not responsive to the tracheal suction, it may be necessary to use the fingers to provide manual ventilation and then do the tracheal suction again.

Fluid Replacement. The need for fluid replacement depends on the rat. To test the hydration status of the rats, pinch the skin on the back and let go. Normally, the skin goes flat within a second. If the skin remains elevated and takes several minutes to fall, this means that the rat is dehydrated. We usually inject 5 mL of sterile saline subcutaneously for females and 10 mL for males.

Prophylactic Antibiotic. To prevent wound and urinary tract infections, all rats receive a daily dose of cephalosporins (25 mg/kg/day) for 7 days. This is a broad spectrum antibiotic that eliminates most common bacteria. When recurrent infections occur, they are usually of the gram-negative variety, particularly hemophilus strains that are sensitive to fluroquinolones.

Daily Bladder Expression. The rats should receive daily bladder expression. This is done by holding the rats around the thorax with one hand and the thumb and first two fingers of the other hand are used to gently express urine from the bladder. The urine is collected in a test-tube to measure the volume of the urine expressed every morning.

Hemorrhagic Cystitis. This is a major cause of death in the postoperative period. It manifests with bloody urine during the first day after injury. This condition results from stress and may be associated with excessive hemorrhage during surgery and bladder infection. The animal is already on antibiotics at the time. When there is blood, some urine should be left in the bladder to avoid clot formation.

Chronic Animal Care

The most common complications that cause morbidity and mortality in rats with chronic spinal cord injury are recurrent urinary tract infections and autophagia.

Recurrent Urinary Tract Infections. Recurrent urinary tract infections occur in about 25% of the animals and usually at 1–2 weeks after injury. We treat the animals with a 10-day course of Baytril (enrofloxacin 2.5 mg/kg/day). About 10% of the animals get recurrent infections after the Baytril treatments. These animals usually have a bladder or renal stone. This drug is also effective against mycoplasma and is used primarily for veterinary purposes.

Flaccid bladders. A third or more rats may develop a flaccid bladder after a 25 mm weight drop, perhaps due to spread of the damage from the T9–10 impact site to the sacral sympathetic and parasympathetic nuclei. In any case, the first sign of a flaccid bladder is difficulty expressing them. The bladders often become quite big and thin-walled and must very carefully expressed or else the bladder will burst.

Autophagia. One third of the rats will develop a behavior called autophagia, usually starting 2–3 weeks after injury. They bite themselves on dermatomes below the injury site. The first sign is reddening of the skin and hair loss (Stage 1). This is soon followed by biting that penetrates skin exposing subcutaneous layers (Stage 2), exposing muscle (Stage 3), and sometimes viscera (Stage 4).

Autotomy. About 5% of the rats will bite themselves in foot and the toes. The first manifestation of this is often bloody toes and may rapidly progress to the point that the rat will eat its foot. Acetaminophen may stop this behavior but we usually euthanize the rat if there is significant damage to the foot because this would interfere with evaluation of its locomotor function.

Autophagia treatment. An effective treatment of autophagia is 65 mg/kg of acetaminophen (oral, children's Tylenol) per day until the lesion is completely healed. The rats like the cherry-flavored fluid and will readily eat it from a dropper. Daily treatment, if started as soon as level 1 autophagia is seen and continued until the lesion is completely healed, is effective in stopping autophagic behavior. If the treatment is stopped prematurely, the rats may engage in a rebound autophagia that rapidly progressed to stage 4.

Failure to Thrive. Animals may fail to thrive for a variety of reasons. The condition can be documented by weekly weighing. From ages 77 to 100 days old, rats are gaining weight from 220 to 350 g for females and 320 to 600 g for males. However, despite the huge difference body weight between males and females, spinal cord weights remain the same.

Outcome Measures

We developed several outcome measures to assess spinal cord injuries in rats.

Lesion volume. To measure lesion volume, we cut the rat spinal cord into five 5-mm long pieces. The first piece is centered on the impact site, called IMP. Adjacent proximal and distal 5-mm pieces are P1 and D1. The pieces proximal and distal to P1 and D1 are P2 and D2. We homogenize the samples with ultrasound in 10 mL of deionized water and analyze 10 μL aliquots of the tissue solution with atomic absorption spectroscopy for Na and K [1, 5, 6]. Differences of tissue sodium ($[Na]_t$) and potassium ($[K]_t$) concentration linearly predict the ratio of intracellular and total tissue volume (Vi/Vt) or cell volume fraction. If intracellular and extracellular fluids are isotonic and Na and K account for >90% of tissue fluid osmolarity, the sums of tissue, extracellular, and intracellular Na and K concentrations should be similar, i.e. [Na]t + [K] t = [Na]e + [K]e = [Na]i + [K]i and [Na]e − [Na]i = [K]i − [K]e = G. If so, [Na]t − [K] t is linearly related to Vi/Vt with a slope of 2G and a y-intercept of [Na]e − [K]e, i.e. [Na]t − [K]t = [Na]e − [K]e + 2*G*(Vi/Vt). We estimate change in cell volume by subtracting normal Vi/Vt from Vi/Vt of the injured cord, i.e. ΔVi/Vt. Since 1 mg of wet tissue is about 1 μL, multiplying ΔVi/Vt by tissue wet weight (W in mg) gives loss of cell volume in microliters. The lesion volume can also be estimated from $[K]_t$, i.e. $V_{ii}/V_t = ([K]_t - 4)/120$ and is expressed in microliters of cells lost from the spinal cord within 1.25 cm from the contusion center.

BBB Locomotor Score. The BBB score was developed to evaluate locomotor recovery in rats after spinal cord contusion [7]. Based on observations of 10 behaviors and the rate at which they recovered after a contusion injury, the score ranks behaviors by how late they occurred, generating a progressive recovery curve. The score is reliable [3] and correlates with spared white matter [8]. Team scores were significantly more reliable and closer to those given experienced scorers [3]. Thus, the method recommends a pair of observers situated opposite of each other, observing the rat for 4 min, and coming up with a consensus score. In case of disagreement, the lower score is used. The components of the BBB scores are recorded as well as the final consensus score. Although some of the behaviors are based on one side, the final BBB score is a single score that reflects both left and right side.

Spared White Matter. In general, gray matter is lost at the contusion site and a thin rim of spared white matter typically remains. We measure spared white matter by staining coronal sections of the injury site with Luxol Fast Blue. We usually make the measurements from the coronal section representing the contusion epicenter with the greatest white matter loss. Spinal cord stained with luxol fast blue reflect the presence of myelinated axons. The spared white matter is expressed in square mm.

References

1. Constantini S, Young W. The effects of methylprednisolone and the ganglioside GM1 on acute spinal cord injury in rats. J Neurosurg. 1994;80(1):97–111.
2. Beattie MS, Bresnahan JC, Komon J, et al. Endogenous repair after spinal cord contusion injuries in the rat. Exp Neurol. 1997;148(2):453–63.
3. Basso DM, Beattie MS, Bresnahan JC, et al. MASCIS evaluation of open field locomotor scores: effects of experience and teamwork on reliability. Multicenter Animal Spinal Cord Injury Study. J Neurotrauma. 1996;13(7):343–59.
4. Young W. Spinal cord contusion models. Prog Brain Res. 2002;137:231–55.
5. Young W, Kume-Kick J, Constantini S. Glucocorticoid therapy of spinal cord injury. Ann N Y Acad Sci. 1994;743:241–63; discussion 63–5
6. Young W, Constantini S. Chapter 11. Ionic and water shifts in injured central nervous tissues. In: Salzman S, Faden AI, editors. Neurobiology of central nervous system trauma. Oxford: Oxford University Press; 1993. p. 221–33.
7. Basso DM, Beattie MS, Bresnahan JC. A sensitive and reliable locomotor rating scale for open field testing in rats. J Neurotrauma. 1995;12(1):1–21.
8. Basso DM, Beattie MS, Bresnahan JC. Graded histological and locomotor outcomes after spinal cord contusion using the NYU weight-drop device versus transection. Exp Neurol. 1996;139(2):244–56.

The Ohio State University ESCID Spinal Cord Contusion Model

Lyn B. Jakeman, Dana M. McTigue, Patricia Walters, and Bradford T. Stokes

Abstract Approximately half of the clinical cases of spinal cord injury are caused by a rapid contusive impact to the soft tissue of the spinal cord. Reproducible models of contusion injury are produced in rodents using a variety of approaches following the exposure of the intact dura with a dorsal laminectomy. In the contusion model developed at The Ohio State University (OSU), an impact to the spinal cord is mediated by the rapid and calibrated displacement of a vertical shaft, using a sensitive electromagnetic shaker controlled by interactive software. Key elements of the technology are the initiation of the impact from a fixed and measurable starting force or "touch" feature, investigator control of the slope, duration and amplitude of peak displacement, rapid retraction of the impactor after a single impact to prevent bounce effects, and the direct and real time measurement of displacement and force of impact upon the spinal compartment. To date, there has not been a comprehensive description of the development of this device and the methods used to obtain reproducible injuries to the spinal cord with this device. In this chapter, we will briefly describe the history of the design and the procedures that are used to calibrate, perform, and evaluate controlled contusion injuries using the OSU electromagnetic spinal cord injury device (ESCID). Final comments emphasize the advantages and limitations of the device and provide suggestions for future uses in advancing spinal cord injury research.

L. B. Jakeman (✉) · P. Walters
Department of Physiology and Cell Biology, The Ohio State University, College of Medicine, Columbus, OH, USA

Department of Neuroscience, The Ohio State University, College of Medicine, Columbus, OH, USA
e-mail: jakeman.1@osu.edu

D. M. McTigue
Department of Neuroscience, The Ohio State University, College of Medicine, Columbus, OH, USA

B. T. Stokes
Department of Physiology and Cell Biology, The Ohio State University, College of Medicine, Columbus, OH, USA

J. Chen et al. (eds.), *Animal Models of Acute Neurological Injury*, Springer Series in Translational Stroke Research, https://doi.org/10.1007/978-3-030-16082-1_30

Keywords Contusion · Mouse · Rat · Vertebral column · Electromagnet · Laminectomy

History and Evolution of the OSU Controlled Impactor

The Use of the Electromagnetic Impounder

Several models of spinal cord contusion injury have been developed in small animals following the initial work of Allen [1]. Using a simple design with a vertical tube and free-falling mass, a rapid impact to the canine spinal cord was produced, using gravity as the primary force of impact to the surface of the exposed spinal cord following a dorsal laminectomy of a single vertebra. The initial studies demonstrated that mild and moderate contusion injury pathology that closely resembles the clinical condition could be produced in animal models.

In subsequent work done several decades later, the concept of a gravity-driven impact was adapted to rat models, where contusion injury was produced by dropping a calibrated mass from a defined distance above the exposed spinal cord, using a variety of vertical tube designs [2, 3]. In these models, a weight is rapidly imparted either directly onto the surface of the exposed dura, or onto a platform that rests on the dura, providing a fairly standardized impact force to the spinal compartment. Important modifications of the experimental procedures further improved reliability by standardizing the laminectomy size, stabilizing the spinal cord vertebrae to reduce respiratory artifacts, and standardizing of the impact device components across laboratories [4–6]. The most widely distributed of these designs was initially developed at New York University (NYU) and was adopted for the Multi Animal Spinal Cord Injury Studies (MASCIS) program, which expanded spinal cord injury research through a decade of collaborative research [4, 6, 7]. The NYU device incorporated a transducer to record the effect of the impact on the vertebral column, providing feedback to the user regarding the biomechanical characteristics of the impact. Weight-drop injury devices have been shown to produce fairly reliable and graded injuries in rats and mice over the last two decades. Many basic mechanistic and pre-clinical studies have been pursued with these devices, which reproduce the basic hemorrhagic pathology observed in human contusion injuries. However, the desire to control the parameters of impact beyond height and mass has supported interest in alternative approaches.

At Ohio State University (OSU), a unique device was designed in the late 1980s with a goal to better control some of the mechanical variables of contusion impact [8–12]. The "OSU Injury Device" incorporated an electromagnetic driver capable of rapid and forceful vibration (Ling Shaker) and a pattern generator and amplifier to allow compression of the dorsal surface of the spinal cord with a single impact and retraction (Fig. 1). User control of the compression pattern was incorporated to reduce variability in the mechanics of displacement. In addition, the elimination of

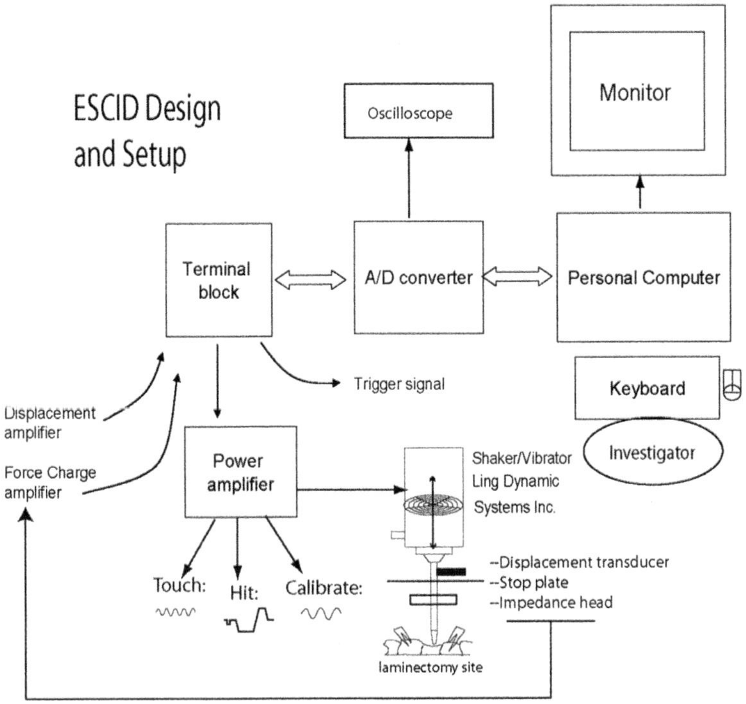

Fig. 1 The OSU injury device consists of a personal computer (PC), data acquisition board, terminal block, power amplifier and the shaker/vibrator (Ling Dynamic Systems). The personal computer runs a control program, using software that was created using LabView™ programming language. The control program generates one of three different signals, sends them through the terminal block via an interface cable, to a power amplifier. The resulting signal from the power amplifier causes the coil and armature to move axially in the field of the magnet. The control program also initiates a trigger pulse that is used in recording and measuring output data from the transducers for displacement, force and acceleration of the moving probe. Three transducers are located in line with the vertical shaft of the impactor. The displacement is tightly controlled and protected from electronic overshoot by a mechanical Stop plate in line with the impactor

the long vertical tube that is required for a weight-drop design helped to decrease concerns with maintenance and consistency over time associated with friction and angle of descent. Displacement and force/acceleration transducers (impedance head) were incorporated into the vertical shaft of the OSU device to provide direct and instant information about the precise dynamics of the impact on the spinal cord. Initially, a closed loop feedback-based design was developed to permit dynamic control of the shaker displacement during the course of the impact [8, 13], so that impact force parameters could be incorporated into the signal generation. However, the feedback features proved difficult to control in the short time frame necessary for a contusion mimicking the clinical condition, and so the amplitude of displacement was chosen as the independent variable for producing injuries of graded severity, and the design modified to a open-feedback system [14]. The OSU device was also modified to incorporate a mechanical *stop plate* that effectively prevented electronic overshoot and better control of the slope and duration of the displacement [10, 11, 15]. The resulting device was thoroughly validated and shown to produce reliable graded injuries in rats with reproducible biomechanical, behavioral, and histological outcomes predicted by the choice of displacement distance [16, 17]. The OSU Injury Device was thus used exclusively at Ohio State for several years [18–26].

Revision and Updating: The ESCID 2000

The advantages of a user-controlled and single impact design were evident to many investigators from the initial descriptions, but reproduction of the OSU device was complicated by the unique design of the pattern generator, which had been created by using carefully selected electronic components and a unique engineered design [13]. In the late 1990s, three major developments came together that inspired further revision of the Ling Shaker-based OSU Injury Device. The first was the evolution and affordability of interface boards and user-friendly programming software for personal computers, such as those made available commercially from National Instruments, Inc. This permitted the production of a controlled analog output signal with unlimited flexibility and reproduction, and simultaneous data collection from transducers to a standard ASCII data file. The second development was the increased need for reproducible mouse models of contusion injury, and a need to easily alter user defined parameters for adaptation to this small species. Finally, the financial backing for revision of the impactor was provided by a non-profit organization, the International Spinal Research Trust. The objective of the project was to establish a device that could be reproduced by academic machine shops and programming staffs anywhere in the world, using commercially available components and software. The resulting design was renamed the Electromagnetic Spinal Cord Injury Device 2000 (ESCID 2000) to reflect its design, standardization, and era of development [27]. The ESCID successfully reproduced the essential features of the OSU Impactor while providing the flexibility for establishment of the model in a number of other institutions, including several major spinal cord injury research centers [28–30].

OSU/ESCID Features

The key design features and components that are used for the ESCID impactor are illustrated in Fig. 1. The user interface is comprised of a PC based computer with an analog/digital processor card and LabView™ programming software. The resulting analog signal is forwarded to a terminal block and amplifier needed to drive the electromagnetic impact pattern.

The principle characteristic of the OSU injury device design is the use of three distinct electronic signal patterns to ensure biomechanical reproducibility of the impact (Fig. 1). The first of these is a "Touch" signal, comprised of a small (0.4 V) sinusoidal displacement (±15 μm) at 60 Hz. This is used in conjunction with the force transducer to obtain real-time measures of the dynamic force and acceleration from the impact probe as changes in static impedance. As the impact probe is slowly lowered onto the exposed dura, the recorded force is used to indicate when a precise amount of initial resistance is encountered, consistently indicating the location of the cord surface. Thus, the "Touch" signal is used to establish the displacement start point (or 0 distance) from which to initiate the injury sequence. The ESCID is the only current model that incorporates this starting point feature. By starting the injury sequence with a small and reproducible dimpling of the cord surface, the initial force is also standardized. The effect of this small force on the shape of the sinusoidal pattern measured from the force transducer provides immediate feedback regarding the initial mechanical properties of the specimen. This slight pressure on the surface of the spinal cord causes no detectable behavioral deficits or histological damage in rats or mice, although a very mild microglial hypertrophy can be observed in the dorsal columns at 2–3 days post-surgery with either laminectomy alone or laminectomy plus the touch pattern.

The second computer generated signal sequence produced by the ESCID is a "Hit" pattern, with the slope and dwell time dictated by the investigator through a software interface. Control of this sequence provides the necessary voltage ramp settings for the impact and retraction phases of impact in order to cause a single, rapid contusion injury. Reduction in the slope of impact or increasing the duration of dwell can be used to modify the total energy or impulse imparted upon the spinal cord with injury. Using these parameters, the extent of hemorrhage and ischemic insult may be independently manipulated.

The dependence of the user interface on the conversion of force and displacement measures to voltages requires a means to exclude daily variation from the intended injury characteristics. To account for this required precision, a third signal sequence, composed of a 0.64 V sinusoidal "Calibration" pattern, is provided to permit calibration of the voltage output from the force/acceleration transducer and ensure consistent measures over time.

The ability to easily generate these standardized signal patterns and transmit them to the injury probe on the shaft of the electromagnet provides exquisite and precise control of the independent displacement variable and flexibility necessary to incorporate variations in the sequence of injury.

Recording the Biomechanical Variables at Impact

The use of sensitive monitoring devices to permit evaluation of biomechanical variables during the injury process has proven to be important for developing and standardizing injuries across animals. With the compression rate and amplitude of displacement tightly controlled, inter-animal variation is principally a function of

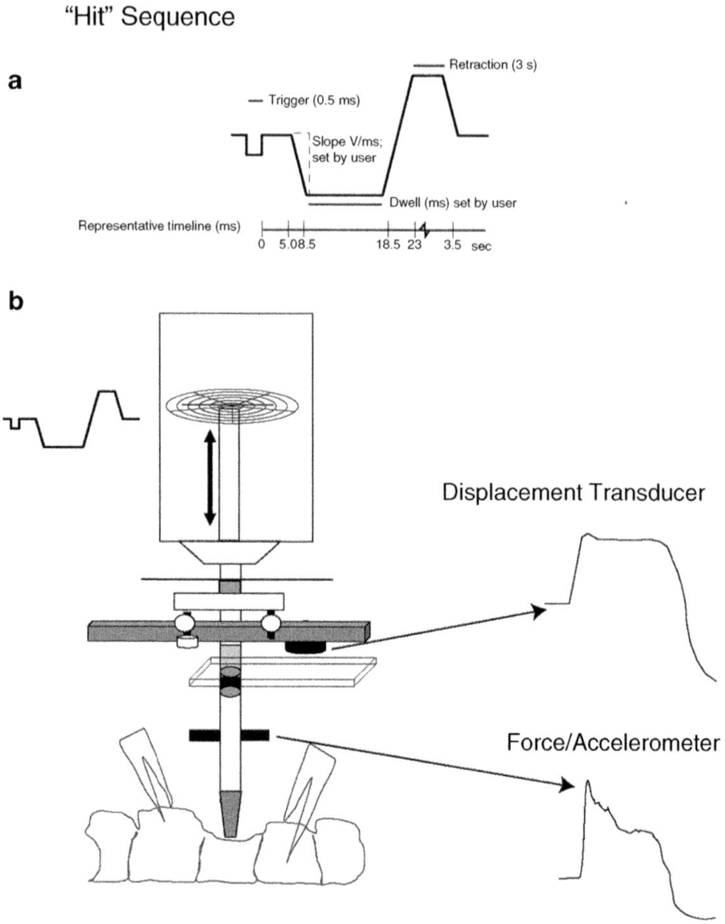

Fig. 2 (**a**) The "Hit" sequence is defined by the user in the control program menu. The sequence consists of a trigger pulse followed by downward displacement with user-defined slope and dwell at maximal displacement. After a single impact, the probe is retracted beyond the start point and held for 3 full seconds, which allows the investigator to remove the subject from the injury table. (**b**) Illustration of the "Hit" sequence as defined by the user and the impactor components. The voltage signal is converted to an electromagnet, causing vertical displacement of the central shaft. In its current configuration, the displacement transducer is located below the stop plate with the reference plate just below it. The impedance head (force and acceleration transducer) is mounted directly on the shaft. Traces on the right illustrate typical readings from the displacement and force transducers, respectively. Scales for the recordings are included in Fig. 3

differences within the spinal cord compartment [31]. The OSU Injury Device allows the investigator to record the actual displacement and force imparted to the spinal cord during the injury (Fig. 2). Displacement of the vertical probe shaft, by definition, reflects the independent input by the investigator, and shows little to no variation under experimental conditions. In contrast, variations between animals are reflected primarily in the peak force measurement and in the shape of the recorded force curve.

In the OSU device design, this parameter is a precise force measurement that is corrected for acceleration of the mass below the probe [13]. Evaluation of the changes in force over time gives very sensitive feedback regarding biological variability that can be used to assess subtle characteristics of the injury event [27]. Additional biomechanical parameters are derived from displacement and force parameters, including impulse, velocity, power and energy (Fig. 3). The calculation and display of these biomechanical measurements provide the investigator with immediate feedback that can be used to identify and exclude subjects with injury conditions that are predicted to lie outside of a predetermined range of variation (see below).

Device Setup and Validation

As designed, the ESCID can be assembled by a qualified electronics lab from commercially available components according to specifications available from the authors. The device is controlled using a standard personal computer with data acquisition board (National Instruments, Inc., Austin, TX), terminal block, and Labview application builder programming software installed (National Instruments, Inc.). A Labview control program and documentation, developed by Dr. Rodger Dwonczyk and updated by Dr. John Buford in collaboration with the authors, is also available on request (©1997, 2006 The Ohio State University). The impactor unit assembly consists of the Shaker unit and lateral holder (V203/S, T-200; Ling Dynamics Systems, Los Angeles, CA), power supply and power amplifier, a steel vertical threaded rod, reference plate and transducer mount for the displacement transducer (KDM-7200 4SB, Kaman Instrumentation Corp., Colorado Springs, CO), impedance head mount and force/acceleration transducer (Wilcoxin Research, Bethesda, MD), impedance charge amplifier (Model 104, Endevco, Pasadena, CA), displacement transducer pre-amp (Endevco), aluminum displacement manual adjustment disk assembly, steel mechanical stop plate, and a lucite impactor tip sized for rat or mouse injuries (Fig. 4g).

The investigators must devise and assemble appropriate equipment to support the instrument components, an oscilloscope to record the transducer output in real time, a lab jack or adjustable table to raise and lower the experimental animal in a prone position, and a frame and clamping mechanism to stabilize the vertebral column of the experimental animal before, during and immediately after the impact. A dial displacement indicator is required for routine calibration of the displacement/voltage conversion (Model 2046-11; Mitutoyo, City of Industry, CA). With few modifications, the support system and accessories will permit reproducible injuries in rats and mice, and potentially translate to larger animals as well.

Once the components have been assembled and the power amplifier calibrated and adjusted, several steps must be completed to establish the reliability and reproducibility of the injury device. The investigators should first familiarize themselves with the dialog screens and graphics manipulations of the control program. Using the documentation provided with the control program software and prepared sham

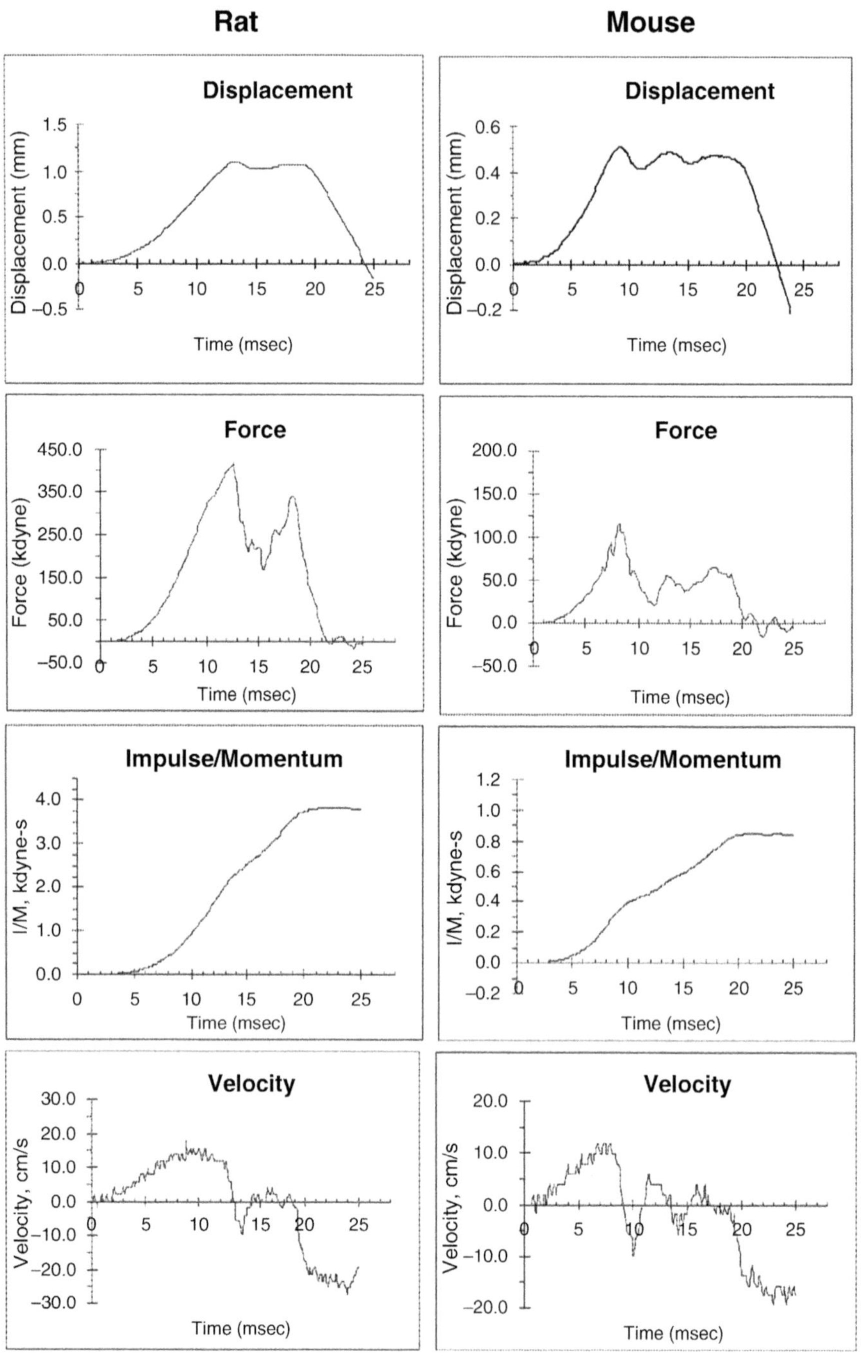

Fig. 3 Direct measures of displacement (top row) and force (second row) were obtained from representative moderate injuries performed in rat (left; 1.0 mm displacement) and mouse (right; 0.5 mm displacement) subjects. Impulse/momentum represents the calculated area under the force curve (third row), and velocity amplitude and direction were calculated as displacement per unit time (bottom row). Note the initial acceleration of the impact head to a maximum of 10–20 cm/s at the time of impact

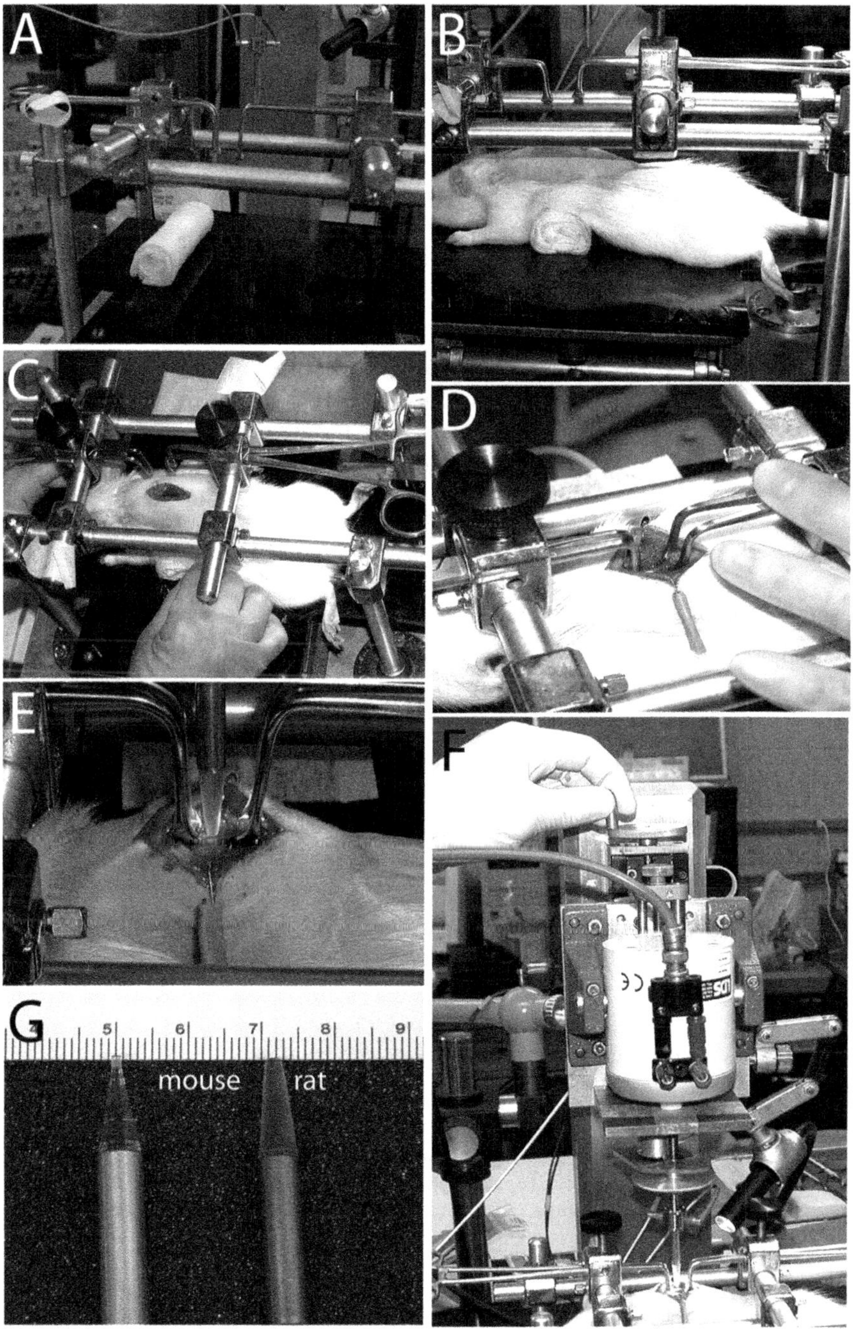

Fig. 4 Photographs of the procedures used to stabilize and impart injury to the rodent spinal cord. (**a**) The labjack is set below the impactor with a guaze pillow positioned to support the subject during positioning. Allis clamps with 90° bends are used to stabilize the rostral and caudal **dorsal**

tubing materials prepared and provided by the author's laboratory, sample touch, air hit, and injury sequences should be initiated and the resulting output data recorded for comparison with examples in the documentation. It is important to ensure that the output signals are converted to achieve the anticipated magnitude and direction of impactor displacement.

Displacement Voltage Calibration

The independent variable of impact displacement is achieved with precision by applying a defined voltage ramp to the impactor head. Likewise, the precise measurement of actual displacement in time is recorded as a voltage and converted to mm by the Labview software. Variations in voltage readings are known to occur over time, and the direct conversion from mm to voltage is not precisely linear. Therefore, it is important to calibrate displacement on a regular basis. The following steps are used to measure and establish the conversion factor for the displacement transducer. These steps are performed each time the device and transducers are powered on, and whenever the desired displacement setting is established for a group of subjects. The voltage calibration is also checked over the course of the day and adjusted if needed. The voltage calibration is performed as follows:

1. Place the displacement dial indicator directly below the impactor probe tip and lower the impactor to approximately 1 mm above the top of the indicator.
2. Using the Displacement Calibration screen of the Labview interface, initiate the "touch" signal profile to establish a recording from the impedance head/force transducer.
3. Slowly lower the impactor device using the adjustment handle until the force transducer reading on an oscilloscope reveals spikes that indicate the 0 or starting point of the indicator plate. Select the "Stop Touch" button on the Labview screen to stop the vibration. Then, select the "Calibrate" button to record the calibration procedure.
4. Manually depress the impactor ring until the indicator reaches the desired displacement setting for the study that will be performed (0.30–1.300 mm). The

Fig. 4 (continued) vertebral processes in the rat. (**b**) A rat is positioned directly under the clamps and impactor. (**c**) Dorsal view of the positioning of the rat. (**d**) The rostral and caudal vertebral processes are secured with modified Allis Clamps, while the skin is retracted laterally. (**e**) The impactor probe is lowered down to within 2 mm of the spinal cord surface at the laminectomy site. Fiber optic lights are used to help guide the investigator during this fine positioning step. (**f**) With the control program running a sinusoidal "touch" pattern, the entire impactor device is slowly lowered with the top wheel assembly while viewing the output from the force transducer on an oscilloscope. At a fixed starting force that just barely dimples the surface of the spinal cord (3000 dynes for rat, 1500 dynes for mouse), the "touch" pattern is stopped and the "Hit" pattern initiated. (**g**) Tips of the mouse and rat ESCID impactor probes shown to scale. The rat impactor is ~1.85 mm in diameter; the mouse probe is ~1.3 mm in diameter central shaft must descend precisely vertically. This is accomplished by using the thumbs of both hands, each thumb placed on opposing sides of the impactor ring. Hold the ring at the desired displacement for 3–5 s and release.

5. Using the cursors on the Displacement Calibration screen, determine the voltage difference that corresponds to the manual displacement profile. Adjust the transducer output until the voltage difference corresponds to the manual displacement (1 mm = 1 mV). Repeat steps 1–5 until five consecutive voltage readings are within 0.005 mV. Select "Return" to the main control program screen.
6. Make final adjustments to the mechanical stop plate. Confirm that there is nothing in the path of the impactor to prepare for air hits. On the Labview main control screen, select "Hit", and record actual displacement. Using the fine adjustment, set the position of the stop plate until actual displacement corresponds to the desired value. The ESCID is now ready for use.

Suspension and Impact Procedure

Once the injury device is operational and displacement has been calibrated, the procedures for obtaining reproducible injuries are straightforward. These steps assume that the investigator has anesthetized the animal appropriately and prepared a laminectomy at the desired spinal level. The laminectomy site must be sufficiently large to allow the impact probe tip to freely contact the spinal surface. Residual perisoteal membrane on the surface of the dura should be removed prior to injury, and the dura must be kept moist with sterile saline to prevent changes in elasticity and turgor. A suitable suspension apparatus will be chosen, and the surgical site prepared to secure the vertebrae by the dorsal or lateral processes as chosen by the investigator. The steps used to proceed with the injury procedure is as follows:

1. Prior to positioning the animal at the site of the injury device, check the anesthesia level and manually void the animal's bladder. The Lab-jack, chest pillow, retractor hooks, and vertebral clamps should be in position (Fig. 4a).
2. Place the animal on the Lab-jack or adjustable table with a chest pillow if needed (Fig. 4b, c). Be sure the laminectomy incision is parallel to long axis of the animal and the table. Use retractors or hooks to pull overlying muscles laterally and open the laminectomy incision for maximal visability. Confirm hemostasis and hydration, and ensure that no muscle, bone chips or other objects will obstruct a clear displacement trajectory for the probe.
3. Using the desired suspension and stabilization apparatus, clamp the vertebrae rostral and caudal to the laminectomy site (Fig. 4d). Again, maintain alignment, hemostasis, and hydration. Do not stretch or compress the surrounding vertebrae in the steps used to suspend and stabilize the vertebrae.
4. When the vertebrae are secure and stable, lower the support table and remove any pillows. This is important for reducing variations due to respiratory artifacts and to keep the airway clear.

5. Position the laminectomy site under the probe. Once the animal is in position, lower the probe until it is 2–3 mm above the exposed spinal cord surface and well centered beneath the probe (Fig. 4f, g).
6. Initiate the "touch" signal to the impactor probe and manually lower the probe toward the cord. By observing the sinusoidal output of the force transducer, determine when the probe touches the surface of the spinal cord. This will be reflected by a brief horizontal reading of the force transducer, followed by a continued steady sinusoidal wave. If the sine wave appears jagged or skewed to one side, the probe is not evenly apposed to spinal cord tissue or fluid. Raise the impactor head and reposition the vertebrae if needed to center the probe, or remove any obstructing tissues. In some cases, the animal will need to be removed to the surgical table and the laminectomy site enlarged.
7. Once a clear path is obtained, continue descending slowly, to just cause a small dimpling of the dural surface. When the amplitude of touch wave increases to 3 kilodynes (for rat) or 1.5 kilodynes (for mice), stop the "touch" signal and initiate the "hit" sequence without delay. The sequence will end with the retraction of the probe to a position approximately 1.5 mm above the start, or 0 displacement position.
8. Immediately following the hit, use the impactor crank arm to raise the probe approximately 1 cm to easily clear the top of the incision site.
9. Raise the Lab-Jack or table and release the vertebral clamps. The investigator must carefully hold the animal's hindquarters and tail while the animal is released. Frequently, reflex kicking will occur as an immediate response to the injury.
10. When the subject is clear of all obstructions and any reflex movements have ceased, support the body and return the animal to the surgery table.
11. Observe the injury site under a surgical microscope; record the appearance of the dorsal cord surface. Bruising is typically rated as mild (slight discoloration or surface vessel damage), moderate (discoloration or hemorrhage visible and bilateral) or severe (extensive subdural hemorrhage, occasionally accompanied by swelling).
12. Close the incision site and perform post-operative procedures as dictated by an institutionally approved survival surgery and care protocol.

Discussion

The OSU/ESCID spinal cord injury device produces a reliable and reproducible contusion injury in rats and mice. The choice of displacement as the independent variable for the design is supported by early data indicating that displacement can be fixed with low variability, and shows a high correlation with anatomical sparing and functional recovery [16, 18, 22, 32, 33]. The initiation of the injury sequence from a standardized "Touch" force recorded directly from the impact probe ensures the reliability of the displacement/force relationship across animals.

The user controlled impact or "Hit" sequence serves to strike a balance between the goals of producing highly reproducible injury pathophysiology of injuries of graded severity while maintaining biological conditions that closely mimic the biomechanical consequences of rapid contusive injury in humans. Using the software interface, the user can dictate the slope of the displacement and the dwell time of compression. Recent work with a force-driven spinal cord injury device (IH Impactor) has demonstrated that different pathological and functional events can be obtained by increasing the dwell time of a single impact. While the mechanisms for this are not clear, the longer dwell time may increase the ischemic component of the injury, but this has not yet been studied in detail.

Immediate feedback is available to the user, regarding the biomechanical characteristics of the impact event. The greatest advantage of this feedback is the ability to analyze the impact parameters in comparison with historical and experimental controls. In most studies, the output data can be quickly evaluated to determine if abnormal physical events may contribute to non-standard biological outcome [27]. For example, errors of surgical preparation such as bone or muscle impact effects, dural adhesion, vertebral slipping from the suspension apparatus, or a very dehydrated tissue compartment would be evident as abnormalities in the shape of the recorded waveforms during the "Touch" or "Hit" sequences. Furthermore, in those studies where very low variability is essential for testing a hypothesis regarding injury severity, analysis of the biomechanical conditions within the injury compartment can be used to carefully select a narrow range of variation [32]. A recent study discusses the advantages of setting narrow limits for peak force or displacement parameters to ensure consistent injury outcomes and reduce structural and functional variations with injury [34].

The ESCID has been successfully used for producing reliable injuries in rats and mice with the simple replacement of the impact probe. While most of these contusion models have been developed for thoracic spinal cord levels, this device is also useful for preparing contusive injuries in the cervical spinal cord [30]. To date, the contusion injury studies using this device have employed a standard single impact with a displacement ranging from 0.3 to 1.5 mm, a slope of 0.3–0.5 V/ms, and dwell times of 8–10 ms duration. However, the control program can be adjusted to accommodate more prolonged slope or longer dwell times, allowing the development of more complex models of contusion/compression injury.

There are many advantages to the use of the OSU/ESCID model, including the use of a fixed starting point through the "touch" feature, the ability to adjust impact parameters with a user interface, the precision of the hit sequence that is possible with the mechanical stop, and the immediate feedback available from the displacement and force transducer readings. However, any validated and reproducible contusion injury device will create a lesion with secondary pathology that closely resembles that of contusion injuries in the clinical theatre. Despite its advantages, the one limitation of the OSU/ESCID for modern laboratories to date has been the requirement for extensive assembly and testing of the components and device and the limited technical assistance available. Core and individual laboratories with a long substantive commitment to spinal cord injury research have been very successful in establishing this model. One day in the future, it may be feasible to obtain commercial investment in sales, assembly, and technical support for the injury device. This type of commitment would greatly facilitate the growth of reproducible spinal cord injury models.

Acknowledgements The early discussions, documentation, and first prototype ESCID was developed in conjunction with Drs. Philip Horner and Fred Gage at the Salk Institute. Development of the ESCID was possible through contributions by Roger Dzwonczyk and Ernie Gresh. User methods and portions of Fig. 4 were adapted with permission from the Spinal Cord Injury Training Manual (unpublished), NINDS, Facilities of Research Excellence in Spinal Cord Injury, © The Ohio State University, 2004–2008 with technical help from Zhen Guan and Todd Lash. Collaborating laboratories and consultations regarding ESCID device assembly have included Dr. Joseph Springer (University of Kentucky) and Dr. John Bethea (University of Miami). Funding provided by the International Spinal Research Trust, N01-NS-3-2350, NS045758, and NIH NS043246.

References

1. Allen AR. Surgery of experimental lesion of spinal cord equivalent to crush injury or fracture dislocation of spinal column. A preliminary report. JAMA. 1911;57:870–80.
2. Ford RW. A reproducible spinal cord injury model in the cat. J Neurosurg. 1983;59:268–75.
3. Noble LJ, Wrathall JR. An inexpensive apparatus for producing graded spinal cord contusive injury in the rat. Exp Neurol. 1987;95:530–3.
4. Basso DM, Beattie MS, Bresnahan JC, Anderson DK, Faden AI, Gruner JA, Holford TR, Hsu CY, Noble LJ, Nockels R, Perot PL, Salzman SK, Young W. MASCIS evaluation of open field locomotor scores: effects of experience and teamwork on reliability. J Neurotrauma. 1996;13:343–59.
5. Gruner JA. A monitored contusion model of spinal cord injury in the rat. J Neurotrauma. 1992;9:123–6; discussion 126–8
6. Young W, Decrescito V, Flamm ES, Blight AR, Gruner JA. Pharmacological therapy of acute spinal cord injury: studies of high dose methylprednisolone and naloxone. Clin Neurosurg. 1988;34:675–97.
7. Beattie MS, Bresnahan JC, Komon J, Tovar CA, Van Meter M, Anderson DK, Faden AI, Hsu CY, Noble LJ, Salzman SK, Young W. Endogenous repair after spinal cord contusion injuries in the rat. Exp Neurol. 1997;148:453–63.
8. Noyes DH. Correlation between parameters of spinal cord impact and resultant injury. Exp Neurol. 1987a;95:535–47.
9. Somerson SK, Stokes BT. Functional analysis of an electromechanical spinal cord injury device. Exp Neurol. 1987;96:82–96.
10. Stokes BT. Experimental spinal cord injury: a dynamic and verifiable injury device. J Neurotrauma. 1992;9(2):129–34.
11. Stokes BT, Noyes DH, Behrmann DL. An electromechanical spinal injury device with dynamic sensitivity. J Neurotrauma. 1992;9:187–95.
12. Stokes BT, Somerson SK. Spinal cord microenvironment: can the changes resulting from trauma be graded? Neurochem Pathol. 1987;7:47–55.
13. Noyes DH. Electromechanical impactor for producing experimental spinal cord injury in animals. Med Biol Eng Comput. 1987b;25:335–40.
14. Stokes BT, Horner PJ. Spinal cord injury modeling and functional outcome assessment. In: Narayan RK, Wilberger JE, Povlishock JT, editors. Neurotrauma. New York: McGraw Hill; 1996. p. 1395–403.

15. Anderson TE, Stokes BT. Experimental models for spinal cord injury research: physical and physiological considerations. J Neurotrauma. 1992;9(Suppl 1):S135–42.
16. Behrmann DL, Bresnahan JC, Beattie MS, Shah BR. Spinal cord injury produced by consistent mechanical displacement of the cord in rats: behavioral and histologic analysis. J Neurotrauma. 1992;9:197–217.
17. Bresnahan JC, Beattie MS, Todd FD III, Noyes DH. A behavioral and anatomical analysis of spinal cord injury produced by a feedback-controlled impaction device. Exp Neurol. 1987;95:548–70.
18. Basso DM, Beattie MS, Bresnahan JC. A sensitive and reliable locomotor rating scale for open field testing in rats. J Neurotrauma. 1995;12:1–21.
19. Behrmann DL, Bresnahan JC, Beattie MS. A comparison of YM-14673, U-50488H, and nalmefene after spinal cord injury in the rat. Exp Neurol. 1993;119:258–67.
20. Chen XY, Wolpaw JR, Jakeman LB, Stokes BT. Operant conditioning of H-reflex in spinal cord-injured rats. J Neurotrauma. 1996;13:755–66.
21. Jakeman LB, Wei P, Guan Z, Stokes BT. Brain-derived neurotrophic factor stimulates hindlimb stepping and sprouting of cholinergic fibers after spinal cord injury. Exp Neurol. 1998;154:170–84.
22. Ma M, Basso DM, Walters P, Stokes BT, Jakeman LB. Behavioral and histological outcome following graded contusion injury in C57Bl/6 mice. Exp Neurol. 2001;169:239–54.
23. McTigue DM, Popovich PG, Morgan TE, Stokes BT. Localization of transforming growth factor-b1 and receptor mRNA after experimental spinal cord injury. Exp Neurol. 2000;163:220–30.
24. McTigue DM, Tani M, Krivacic K, Chernosky A, Kelner GS, Maciejewski D, Maki R, Ransohoff RM, Stokes BT. Selective chemokine mRNA accumulation in the rat spinal cord after contusion injury. J Neurosci Res. 1998;53:368–76.
25. Popovich PG, Wei P, Stokes BT. The cellular inflammatory response after spinal cord injury in Sprague-Dawley and Lewis rats. J Comp Neurol. 1997;377:443–64.
26. Stokes BT, Jakeman LB. Experimental modelling of human spinal cord injury: a model that crosses the species barrier and mimics the spectrum of human cytopathology. Spinal Cord. 2002;40:101–9.
27. Jakeman LB, Guan Z, Wei P, Ponnappan R, Dzwonczyk R, Popovich PG, Stokes BT. Traumatic spinal cord injury produced by controlled contusion in mouse. J Neurotrauma. 2000;17:299 319.
28. Brambilla R, Bracchi-Ricard V, Hu WH, Frydel B, Bramwell A, Karmally S, Green EJ, Bethea JR. Inhibition of astroglial nuclear factor kappaB reduces inflammation and improves functional recovery after spinal cord injury. J Exp Med. 2005;202:145–56.
29. McEwen ML, Springer JE. Quantification of locomotor recovery following spinal cord contusion in adult rats. J Neurotrauma. 2006;23:1632–53.
30. Pearse DD, Sanchez AR, Pereira FC, Andrade CM, Puzis R, Pressman Y, Golden K, Kitay BM, Blits B, Wood PM, Bunge MB. Transplantation of Schwann cells and/or olfactory ensheathing glia into the contused spinal cord: survival, migration, axon association, and functional recovery. Glia. 2007;55:976–1000.
31. Blight A. Mechanical factors in experimental spinal cord injury. J Am Paraplegia Soc. 1988;11:26–34.
32. Kloos AD, Fisher LC, Detloff MR, Hassenzahl DL, Basso DM. Stepwise motor and all-or-none sensory recovery is associated with nonlinear sparing after incremental spinal cord injury in rats. Exp Neurol. 2005;191:251–65.
33. Ma M, Wei P, Wei T, Ransohoff RM, Jakeman LB. Enhanced axonal growth into a spinal cord contusion injury site in a strain of mouse (129X1/SvJ) with a diminished inflammatory response. J Comp Neurol. 2004;474:469–86.
34. Ghasemlou N, Kerr BJ, David S. Tissue displacement and impact force are important contributors to outcome after spinal cord contusion injury. Exp Neurol. 2005;196:9–17.

Cervical Hemicontusion Spinal Cord Injury Model

Philippa M. Warren, Basem I. Awad, Davina V. Gutierrez, Kevin C. Hoy, Michael P. Steinmetz, Warren J. Alilain, and John C. Gensel

Abstract This chapter describes a unilateral cervical spinal cord contusion model that causes ipsilateral respiratory and/or forelimb motor deficits. Additional techniques are presented to assess forelimb function via grooming and paw placement tasks, as well as respiratory activity using additional lesion techniques that remove descending compensatory respiratory motor control. Cervical injury is the

P. M. Warren
Department of Neurosciences, MetroHealth Medical Center, Case Western Reserve University School of Medicine, Cleveland, OH, USA

Department of Neurosciences, Case Western Reserve University School of Medicine, Cleveland, OH, USA
e-mail: philippa.warren@case.edu

B. I. Awad
Department of Neurosciences, MetroHealth Medical Center, Case Western Reserve University School of Medicine, Cleveland, OH, USA

Department of Neurological Surgery, Mansoura University School of Medicine, Mansoura, Egypt
e-mail: basemawad@mans.edu.eg

D. V. Gutierrez · K. C. Hoy · M. P. Steinmetz
Department of Neurosciences, MetroHealth Medical Center, Case Western Reserve University School of Medicine, Cleveland, OH, USA
e-mail: dvg1@case.edu; msteinmetz@metrohealth.org

W. J. Alilain (✉)
Department of Neurosciences, MetroHealth Medical Center, Case Western Reserve University School of Medicine, Cleveland, OH, USA

Spinal Cord and Brain Injury Research Center, University of Kentucky, Lexington, KY, USA

Department of Anatomy and Neurobiology, University of Kentucky, Lexington, KY, USA
e-mail: warren.alilain@uky.edu

J. C. Gensel (✉)
Spinal Cord and Brain Injury Research Center, University of Kentucky, Lexington, KY, USA

Department of Physiology, University of Kentucky, Lexington, KY, USA
e-mail: Gensel.1@uky.edu

J. Chen et al. (eds.), *Animal Models of Acute Neurological Injury*, Springer Series in Translational Stroke Research, https://doi.org/10.1007/978-3-030-16082-1_31

most common type of human spinal cord injury. Modeling functions of highest priority for this spinal cord injured population (i.e. respiratory and arm/hand control) provides a translational approach for the evaluation of potentially therapeutic interventions.

Keywords Breathing · Diaphragm · Rat · Phrenic · Hemisection

Model Selection

The majority of human spinal cord injuries (SCIs) are caused by motor vehicle accidents or falls resulting in contusive or compressive damage to spinal tissue [1]. Rapid flexion-extension of the vertebral column during trauma results in destructive vertebral bone impact with the spinal cord. Experimental contusive SCI allows for reproducible injuries with controllable levels of anatomical and functional severity. In 1911, Allen [2] first described a contusive canine SCI model by dropping a calibrated weight through a vented tube onto the exposed spinal cord. A number of weight-drop, force-driven, or displacement operated impactors have since been developed for use in rodents, pigs, cats, monkeys, and other smaller vertebrates. The trauma produced by the contusion causes immediate mechanical damage and triggers a host of secondary injury events common to both rodents and humans [3].

Cervical trauma is the most common type of human SCI [4]. Modeling this type of injury, however, is challenging. Depending upon the impact severity and spinal level, animals may experience severe respiratory and autonomic dysfunction or be unable to locomote, thereby limiting access to food and water. Collectively these hurdles confound our ability as researchers to study experimental cervical SCI. To overcome these challenges, unilateral models of cervical contusion injury have been developed. These models produce an injury isolated to one hemisphere of the spinal cord (or hemi-cord). Depending upon the contusion parameters, a range of injury severities can be produced on one side of the spinal cord which causes substantial ipsilateral functional deficits as a result of damage to both the gray and white matter. However, contralateral tissue and motor function are spared thereby reducing the need for extensive post-SCI animal care and allowing for within subject controls [5, 6]. This unilateral model of cervical contusion can therefore provide investigators with an alternative and precise means to evaluate potentially promising therapeutic interventions aimed at repairing injured upper and lower motor neuron pathways in the damaged spinal cord.

This chapter focuses on the procedures necessary to perform unilateral cervical contusion injury in the rat. Generalized methods are presented so the injury can adapted to evaluated both forelimb motor (C5 injury and below) and respiratory motor performance (C3–C4 injury) after trauma. The methods and additional experimental paradigms for evaluating these outcome measures are also presented in detail. Adapting these procedures to produce injury in mice is possible with the aid of previously published characterizations [7].

Protocol

The protocol and animal usage follows the animal care guidelines of the governing institution (Table 1).

Animal Preparation and Surgical Approach

1. Autoclave all instruments required for the procedure to ensure sterility. Cover any equipment to be handled during the procedure with autoclaved aluminum foil.
   (a) Perform all surgery with aseptic procedures. The investigator should wear a mask, sterile gloves, and sterile gown. Disinfect the surgical area with 70% ethyl alcohol or glutaraldehyde wipes and keep equipment uncluttered to promote asepsis.
2. A Sprague Dawley rat, (male, ~300 g) should be weighed and anesthetized with an intraperitoneal (i.p.) injection of a ketamine (70 mg/kg) and xylazine (7 mg/kg) solution.

   Note: Most researchers use the Sprague Dawley strain, however, this protocol may be applied to adult rats (200+ g) of any strain or sex. The original C5 characterization was performed in female, Long Evans rats [5]. Further, prior to application of anesthetics the animal should be assessed for breathing or heart problems, discoloration to the mucous membrane, discharge from the eyes or nose, and hydration to ensure they are in sufficient health to undergo surgery. A combination of ketamine and xylazine is recommended for this surgery as

Table 1 Materials

Supplies/equipment/reagent	Company	Catalogue #
Infinite Horizon SCI impactor	Precision System & Instrumentation	IH-0400 Impactor
Power 1401	Cambridge Electronic Design	Power1401-3
Spike2	Cambridge Electronic Design	Spike2 v.7
Amplifier	Grass Instruments	QP511
Dumont #5 fine forceps	Fine Scientific Tools	11254-20
Angled spring micro-scissors	Fine Scientific Tools	15006-09
Shea scissor	Fine Scientific Tools	14105-12
Goldstein retractor	Fine Scientific Tools	17003-03
Fridman-Pearson Rongeur	Fine Scientific Tools	16221-14
Adson forceps	Fine Scientific Tools	11006-12
Circulating warm water-heating pad	Kent Scientific	T/Pump, 120v
Paraformaldehyde	Sigma-Aldrich	P6148
Perfusion pump	Ismatec	Ecoline VC-380
Offset bone nippers	Fine Scientific Tools	16101-10

animals maintain a short-term (30–40 min) anesthesia with analgesia, muscle relaxation, agonism α2 adrenergic receptors, antagonism of NMDA receptors, and minimal effects on respiratory function.

(a) Determine depth of anesthesia through the animal's loss of consciousness, suppression of reflexes, muscle relaxation, and suppression of pain perception (including loss of pedal reflex). Perform regular toe pinching throughout the procedure to assess depth of anesthesia. Time to reach a surgical plane of anesthesia will vary depending on the weight of the animal. However, it should take no longer than 15–20 min.

3. Once the animal is anesthetized, shave the dorsal neck area.
4. Scrub the skin three times with alternating povidone-iodine and ethyl-alcohol wipes.
5. Apply ophthalmic lubricant on each eye to prevent corneal dehydration and damage while anesthetized. Reapply lubricant after 45 min if necessary.
6. Carefully transfer the rat to the surgical field avoiding contamination. Apply appropriate heat from a circulating warm water-heating pad to the surgical field to conserve the optimal body temperature for the animal. Body temperature should be monitored at regular intervals throughout the procedure using a temperature gauge and kept within 37 ± 1 °C. If available, a pulse oxymeter should be used throughout the experiment to monitor oxygenation of the animal.
7. Inject 0.4 mL of bupivacaine hydrochloride intramuscularly (0.25%) about the midline of the shaved area. (*The need of this step may vary by institution*).
8. With a #15 scalpel, make a midline skin incision of approximately 4–5 cm, originating at cervical level (C) 1 and ending at the caudal base of the neck.
9. Using Shea scissors, and with as few incisions as possible, cut through the midline of the acromiotrapezius, semispinalis capitus, and rectua capitus posterior muscles followed by blunt dissection using sterile cotton tipped applicators to expose the spinal column (Fig. 1).
10. Insert a retractor deeply between the muscles to spread them apart (Fig. 1).
11. Using the same Shea scissors, detach the muscles from the articular processes bilaterally from the lamina below, at, and above the desired level of injury. Secure the impactor clamps under these bony protrusions (Fig. 1). If hemostasis is needed—use mild pressure with a sterile cotton tipped applicator.
12. Completely expose all bony, anatomical landmarks to ensure secure placement of the impactor clamps (Fig. 2).
13. Locate the spinal vertebra level of desired injury (i.e. C2 or C5). Lift the spinal column with straight-toothed forceps to allow safe insertion of fine tipped Rongeur under the lamina without touching the spinal cord. Carefully perform a dorsal laminectomy by slowly cutting the bone on one side then the other. Based upon the occurrence of the spinal dorsal roots, if not enough of the desired spinal segment has been exposed to perform the contusion repeat the laminectomy to remove the anterior portion of the adjacent vertebra (Figs. 1 and 2).

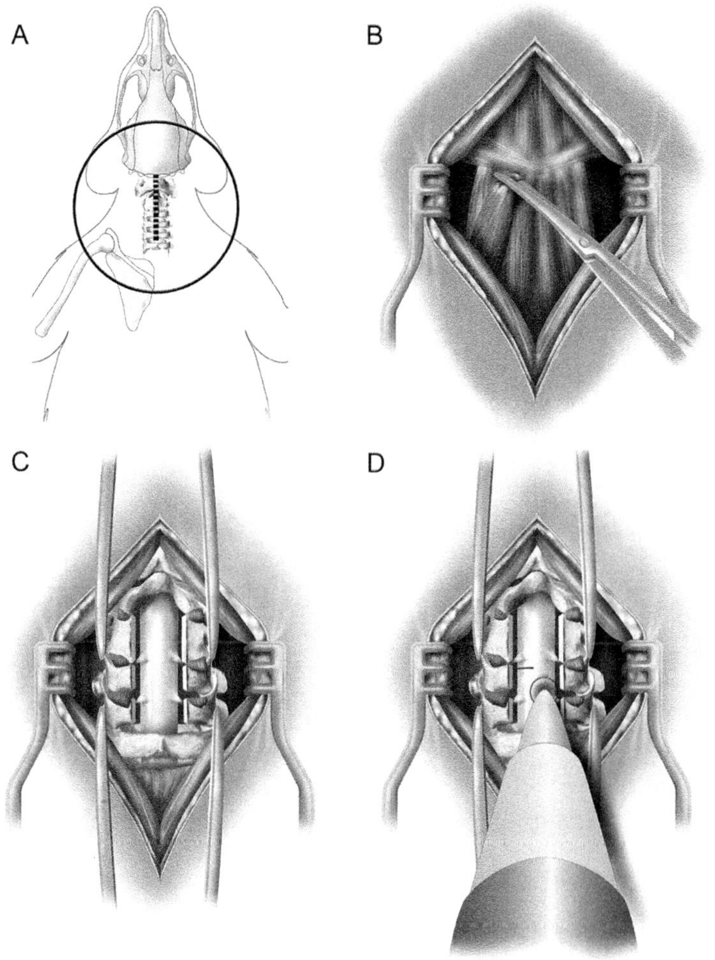

Fig. 1 Illustrations of the laminectomy and impactor placement for C3/C4 unilateral contusion and C2 hemisection. (**a**) Schematic illustrating position of the cervical spinal cord. (**b**) To expose the cervical spinal column a midline incision through theacromiotrapezius, semispinalis capitus, and rectua capitus posterior muscles should be made. (**c**) Wide C2 and C3 laminectomy should be performed to expose the dorsal roots and identify the precise positioning of the impact. The C2 and C3 spinal vertebra should be tightly clamped and held in position to ensure a reliable and consistent contusion. (**d**) Schematic shows the approximate position of impact to achieve a right lateral C3 contusion. The left C2 hemisection has been illustrated (line on the diagram) to show relative position of the second injury to occur at the terminal end-point of the experiment. Illustrations by Tom Dolan, University of Kentucky Analytics and Technologies

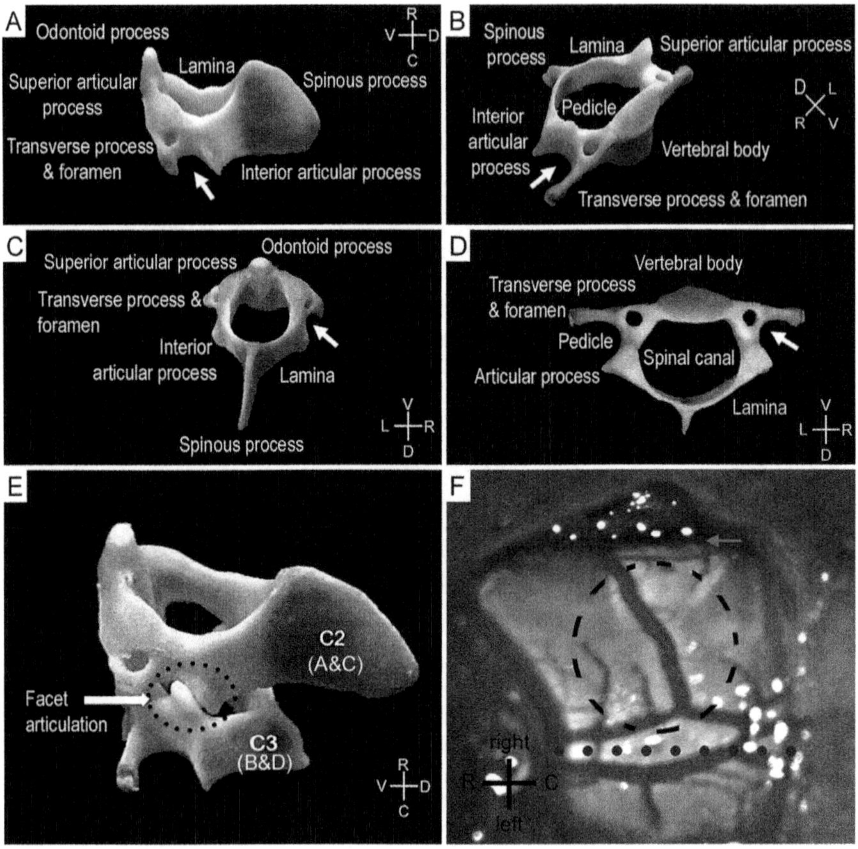

Fig. 2 (**a**, **c**) Illustrated Axis (C2) vertebral anatomy of a Sprague-Dawley rat shown from two orientations (**a**) lateral view and (**c**) axial view. (**b**, **d**) C3 vertebral anatomy of a Sprague-Dawley rat shown from two orientations (**b**) anterolateral view, (**d**) axial view. Arrows in (**a**–**d**) indicate the groove between the articular and transverse processes. (**e**) C2/C3 vertebral anatomy of a Sprague-Dawley rat (lateral view) shows the C2/3 articular joint. Avoid violating this joint to prevent spinal instability during the contusion. (**f**) Representative C5 lateralized laminectomy. Typically 3/4 of the spinal cord is exposed for C5 contusion and a full laminectomy performed for C3/4 contusion. The dotted line demarcates the midline of spinal cord. In this case two dorsal blood vessels are evident instead of the usual one vessel and midline is determined from anatomical and vertebral landmarks. The laminectomy was extended slightly to the right to illustrate the curve of the cord and the lateral completeness of the laminectomy (arrow). It should also be noted that a dorsal root is present in the laminectomy site (white area below the arrow). If possible, avoid contusing the root entry zone. The impactor (dashed circle) is centered over one hemicord with the edge of the impactor just lateral to the midline of the spinal cord. *Note*: impactor probe area not drawn to scale. *D* dorsal, *V* ventral, *R* Rostral, *C* Caudal, *L* left, *R* right

Note: Avoid a downward pressure on the thecal sac, unintentional injury to the spinal cord with the Rongeur tips or violation to the articular processes during the laminectomy (Fig. 2).

14. Use blunt dissection to free the dura from any ligamentous attachment with the surrounding lamina. There should be no leaks of cerebral spinal fluid during any part of the procedure as integrity of the dura should not be compromised.
15. Achieve good hemostasis using mild pressure applied through a cotton tipped applicator or an electrocautery before proceeding with the next step. See Fig. 2 for an example of a C5 laminectomy.

Unilateral C3/4 Contusion Injury (Parenthesis for C5 Injury)

Note: Contusion is described using the commercially available Infinite Horizons Impactor (Precision Systems and Instrumentation) [8], however, a variety of impactors have been utilized to produce unilateral cervical injuries and we have compared injury parameters across different injury devices [9].

1. Move the animal to the contusion device stabilization platform.
2. Use the stabilization clamps to immobilize C2 (C4) and C3 (C6) vertebrae. Bilaterally secure the clamp tips at C2 (C4) and C3 (C6) into the bony groove between the articular and transverse processes (Fig. 2). Ensure that the animal is still deeply anesthetized before attaching the clamps.
3. Tilt both clamps horizontally away from the exposed dura so there is space for the free movement of the impactor probe. In order to avoid unwanted respiratory stress, ensure that the clamps are secured on to the spinal column so that the spinal column remains on the same horizontal plane as compared with the impactor platform i.e. do not force the column up or down from its original position. Clamp animals in a slightly suspended position which allows thoracic expansion and contraction during respiration.
4. After securing the spinal column, use the surgical microscope to ensure a flat horizontal spinal cord (make sure the microscope light is equivalently reflected over the dura).
5. Tighten all the screws on the platform handles and assure that all arms are secure and hold fast under pressure.
6. Position the center of the 1.5 mm diameter impactor probe over the unilateral C3/4 (C5/C6) dura using the vertical and horizontal impactor knobs.
7. Verify that the medial edge of the impactor tip is in line with the midline posterior spinal vessels (Fig. 2).
8. Slowly lower the tip so that it lightly touches the surface of the cord and then retract to a position 5 mm over the dura.
9. Confirm that the exposed spinal cord and dura are dry and free from blood using a clean, sterile cotton tipped applicator.

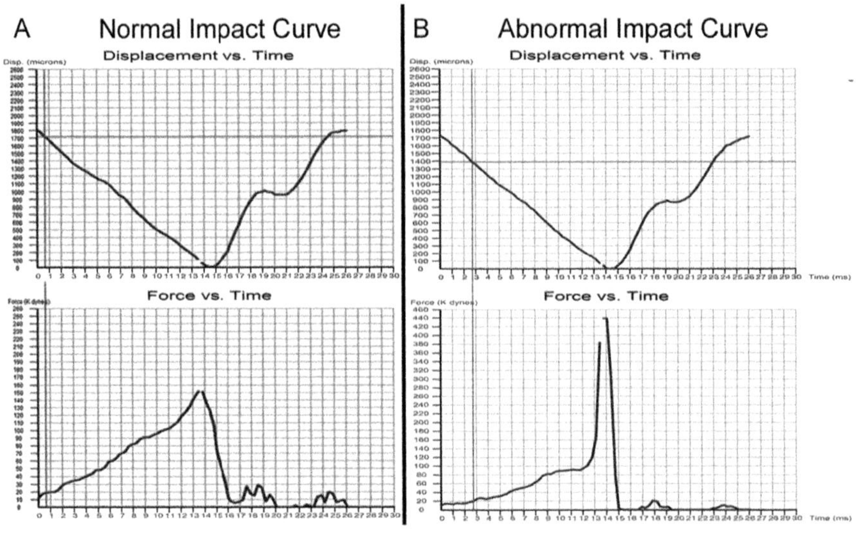

Fig. 3 Representative force and displacement graphs for the Infinite Horizon impactor. The upper graphs represent displacement vs. time, while the lower graphs represent the force vs. time. (**a**) These graphs show that the actual force achieved was 152 kdyn, and the impactor displacement into the cord was measured to be 1746 μm. The slope of the force vs. time curve is indicative of surface tension. In this graph, the steady slope demonstrates that the force was delivered in a consistent manner and the actual force is close to the desired force (152 vs. 150 kydn). (**b**) These graphs are indicative of cord displacement during the delivery of the contusion. The surface tension changes dramatically 12 ms into the impact, likely due to the impactor probe sliding laterally into the bone as the spinal cord twisted with impact. Abnormal curves are indicative of bone hits or cord movement during impact. The displacement values are unreliable when an abnormal curve is recorded and accordingly, an animal receiving this type of injury should be replaced

10. Program the impactor device to deliver the desired force of SCI-usually ranging from 150to 200 kdyn, then initiate the machine to cause a contusion (Fig. 3).
11. Remove the animal from the clamps and stabilization platform.
12. Provide respiratory support and close monitoring directly after injury as the animal will typically be suffering from severe respiratory distress and will be fighting to breathe. To provide respiratory support place the animal on its back and seal a tube over its nose. Pulse air through the tube at a rate of 1 breath per second. Check that the chest is moving and air is not passing into the stomach. Continue this for 5–15 min or until the animal is able to breath strongly with autonomy. If available, the pulse oximeter should be used to monitor blood oxygenation during this time period.

    Note: C5 SCI causes less respiratory distress and support may not be needed.
13. Suture together the opposing muscle layers then close the wound by sealing the skin with wound clips.

Post-operative Care

1. Move the animal onto a circulating warm water-heating pad and allow for recovery until the animal awakens from anesthesia. Following surgery, administer 3 mL of 9% saline (subcutaneously) daily and analgesics (buprenorphine, 0.03 mg/kg s.c.) every 12 h for 5 days post-surgery or as necessary. Yohimbine (1.0 mg/kg s.c.) should be injected after surgery to reverse the effects of the xylazine ensuring faster recovery form anesthesia and elevate the effects of respiratory depression (note: this may not be necessary for C5 contusion). Do not leave an animal unattended until it has regained sufficient consciousness to maintain sternal recumbency. Following the C2/C3 lateral contusion, animals will show slightly diminished respiratory response and lose fine motor control in the forepaw ipsilateral to the injury. Animals will be able to ambulate, eat and drink with autonomy.
2. Once the animal has recovered from the anesthesia, place the animal in a cage over a circulating warm water-heating pad overnight with access to food and water *ad libitum*. Do not house animals in cages with other animals until they have fully recovered from the effects of anesthesia.
3. The following day, return the animal to the standardized housing of the investigators institution. Monitor the animal daily for signs of distress, weight loss, pain, discomfort, and dehydration.

   Note: Treatment application may occur prior to, immediately after or in the days following the contusion injury depending on the investigators experimental design.

Utilizing Lateral C2 Hemisection to Evaluate Respiratory Function

Bulbospinal motor inputs to the phrenic motor neurons (PMNs), are located at the mid-cervical spinal cord segments [10, 11]. The diaphragm is the major muscle used for inspiration and is innervated by the PMNs. Subsequently, a common consequence of high cervical SCI is impaired or paralyzed diaphragm activity, resulting in the inability to breathe independently [12, 13].

Current models of high cervical contusion injury show extensive PMN loss with limited endogenous recovery [14–21]. However, the evaluation of treatment strategies to recover respiratory motor function following cervical contusion has been minimal. Here, we provide technical details on a dual injury model of cervical SCI: unilateral C3/4 contusion followed by a delayed contralateral C2 hemisection (Figs. 1 and 4).

This method of assessing the respiratory motor deficit [21] and conceivable functional recovery following contusion is simple and efficient. Briefly, a unilateral C3/4

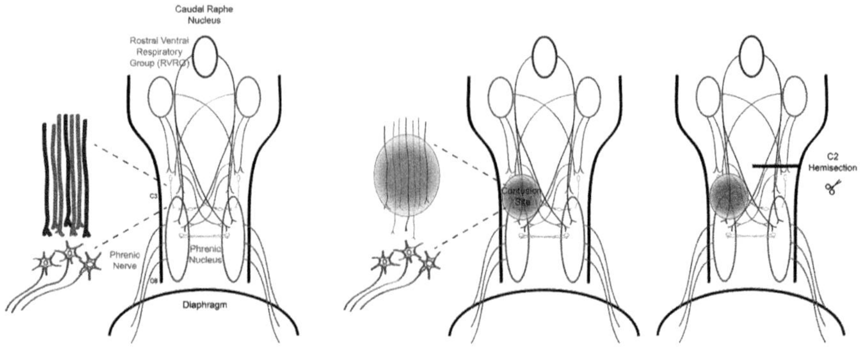

Fig. 4 Schematic diagram for the dual injury model. Compared to the uninjured spinal cord (left panel), a unilateral C3/4 contusion leaves inspiratory bulbospinal pathways at a weakened and fragile state (middle panel). A contralateral C2 hemisection removes modulatory decussating inputs and reveals severely compromised hemidiaphragmatic motor activity through these injured pathways (right panel). Figure adapted from [6]

contusion (as described above) is performed. One week later (or as long as investigators require for experimental design or treatment efficacy), the spinal cord is re-exposed and a contralateral C2 hemisection performed (Fig. 4). The length of time between the contusion and the hemisection can be varied based on the specific experimental design of the investigator. The authors have performed the hemisection at multiple time-points post contusion ranging from 1 day to several weeks. The respiratory deficits described below have been uniformly exhibited at all of these time-points. Completion of the lateral C2 hemisection (LC2H) removes the intact crossed phrenic pathway (CPP) and ensures that the injured animal becomes solely reliant upon the damaged pathways running through the site of the contusion to maintain inspiratory activity of the hemidiaphragm (Fig. 4). Once completely dependent on these injured pathways, the contused animals cannot activate PMNs and the entire diaphragm is paralyzed. However, the time lapse between the initial unilateral contusion and the contralateral hemisection provides a therapeutic "window" to strengthen or enhance these spared and weakened pathways after contusion. If successful in repairing or strengthening inputs, unlike untreated animals, breathing ipsilateral to the hemicontusion should still persist.

Contralateral C2 Hemisection

1. The animal should be anesthetized using urethane (1.2 g/kg i.p.; please see "Animal Preparation and Surgical Approach" for description as to the assessment of anesthesia depth). Similar to that described in section "Model Selection", the animal must have their lower abdomen and dorsal surface of the neck shaved, sterilized, and ophthalmic lubricant applied to each eye before being moved to the surgical field.

Note: Urethane is suggested for the final recording to produce a longer-lasting steady state of anesthesia with minimal effects on cardiovasculary and respiratory function while maintaining spinal reflexes. This will enable investigators to record diaphragmatic electromyograms (diaEMGs) over a considerable period of time (1–2 h) and allow for investigations or drug application should any be required (as fits the nature of the investigators study). However, the procedure described is not dependent upon the use of these specific anesthetics. The use of alternative injectable or inhalant anesthetics would be feasible for this protocol and, indeed, may reduce any neuroprotective effects after the initial contusion [22].

2. Perform a laparotomy by making a 5 cm transverse incision through the abdomen 1 cm from the base of the rib cage cutting through the skin and muscle tissue (rectus abdominis, external oblique and internal oblique muscles) to expose the abdominal surface of the diaphragm. Palpate the area before the incision to ensure the rib cage is not cut.
3. Insert platinum bipolar electrodes bilaterally, connected to an amplifier and data acquisition system, into the crural part of the diaphragm. One electrode should be placed in each hemidiaphragm within the costal origin and ventral to the anterolateral branch of the inferior phrenic artery. From these electrodes obtain diaEMGs by digitally recording the waveform activity of the diaphragm during breathing.
   (a) Recording equipment should be started after a ground electrode is placed on the animals foot. It is recommended to calibrate the equipment before every use using an electronic stimulation of known voltage.
   (b) With the electrodes positioned in the diaphragm, confirm acquisition of a clear diaEMG signal. Electrodes should be repositioned if the signal-to-noise ratio does not allow for a clear recording.
   (c) DiaEMG recordings should be amplified (gain ×5000), band-pass filtered between 30 and 3000 Hz, and digitized at a sampling rate of 2 kHz using an analogue to digital converter recording data points every 10 μs. Save data automatically to disk. Please see 'Representative Results' for a discussion of data analysis.
4. Close the anterior abdominal wall by suturing together the two sides of the rectus abdominis, external oblique and internal oblique muscles. Ensure that the electrodes are secured in place through the wound by tying them to the abdomen musculature with sutures.
5. Carefully move the rat into sternal recumbency.

   Note: Repositioning the animal usually results in dramatic intensification of diaphragmatic activity, so ensure that stable and consistent bilateral baseline diaEMGs are recorded prior to beginning the hemisection.
6. Remove wound clips from previous surgery. Using Shea scissors, open the previous neck skin incision, acromiotrapezius, semispinalis capitus and rectua capitus posterior muscle layers. From the lower edge of the C1 lamina dissect the scar tissue attached to the spinal cord and dura moving caudally to the upper edge of the C3 lamina.

7. After exposing the spinal cord between C1 and C3 lamina, grasp the dura with fine forceps. Perform a midline linear durotomy using angled spring micro-scissors to expose the C2 spinal cord segment.
8. Identify the C2 dorsal root and start the C2 hemisection just caudal to it and contralateral to the contusion.
9. While recording diaEMGs bilaterally, gradually perform the C2 hemisection by cutting the spinal cord with a microblade, knife faced medially, from the lateral edge to the midline.

   Note: If treatment has been unsuccessful in repairing the contused respiratory motor pathways, animals will require ventilation following C2 hemisection if investigators want to perform further investigation. The C2 hemisection has been used as the end point of *in vivo* experimentation for the authors. At this stage animals have received a quadriplegic injury, and are likely to require respiratory and nutritional support as well as hydration, analgesia and bladder expression. Investigators should be aware of this extra care requirement should they wish to maintain the animal for a substantial period of time following the C2 hemisection.

Evaluating Motor Outcome

The major lower motor neuron projections from the C5/C6 spinal cord innervate muscles responsible for gross forelimb movements including the biceps, deltoid, and spinodeltoid [23]. The motor pools for fine digit control are at the C7/C8 spinal level [23]. Damage to the cervical spinal cord disrupts lower motor neuron function at the level of injury, as well as, upper motor neuron projections to caudal spinal cord segments. Therefore, gross forelimb function, digit control, and hindlimb function can be disrupted after unilateral cervical contusion SCI depending upon the spinal level of injury and the injury severity. Researchers have developed a number of functional outcome measures to evaluate motor recovery after unilateral cervical contusion SCI. These have been reviewed previously and include grasping and grip strength tasks for assessing fine digit function; automated, kinematic, or observation-based approaches for assessing hindlimb/locomotor function; sensory-motor tasks including sticker removal tasks and grid or narrow beam walking; and forelimb specific tasks such as the paw preference and grooming tests [24].

We have conducted high order statistical analyses to determine the relationship among functional outcome tests, anatomical measures, and unilateral cervical SCI severity [9, 25]. From these analyses, two functional outcome measures show a high degree of sensitivity to injury severity, lesion size, and tissue sparing [26]. Specifically, grooming function strongly associates with lesion size while paw preference (measured using the cylinder task) is closely associated with white matter sparing. Measures of fore- and hind-limb locomotor function are associated to a lesser degree with injury parameters [26]. For these reasons, we will focus on the procedures for the grooming and paw preference or cylinder tasks as we originally described for unilateral C5 contusion SCI [5].

Grooming Test

The grooming test utilizes a 6-point scoring system adapted from the testing protocol described by Bertelli and Mira to evaluate brachial plexus reconstruction in rats [27]. A single grooming score is given for each forelimb based upon the maximal movement achieved during a normal grooming sequence. A stereotypical grooming sequence includes: (1) licking of the forepaws and face washing; (2) forelimb grooming of the face typically starting with the nose/whiskers then working up the face to include the back of the neck and ears; (3) repetitive licking of the body; and (4) hindpaw scratching [28]. By recording stereotypical grooming sequences, each forelimb can be scored using slow motion video playback. The scoring, originally described in Gensel et al. [5] is as follows:

0-NO FORELIMB CONTACT TO THE HEAD OR FACE
1-FOREPAW CONTACTS THE UNDERSIDE OF THE CHIN OR MOUTH
2-FOREPAW CONTACTS THE TOP OF THE FACE UP TO, BUT NOT INCLUDING, THE EYES
3-FOREPAW CONTACTS THE AREA INCLUDING THE EYES AND UP TO THE FRONT OF THE EARS
4-FOREPAW CONTACTS THE FRONT BUT NOT BACK OF THE EARS
5-FOREPAW CONTACTS THE HEAD BEHIND THE EARS

If the view is unobstructed, animals can be filmed in their home cages. Typically a Plexiglas or plastic unit and mirrors are needed to capture grooming behavior from multiple angles simultaneously. Water can be applied to the head with wet gauze to initiate the grooming behavior. We have found that maximal grooming behavior remains consistent over multiple grooming sequences with a given testing day. Therefore, filming 2–3 stereotypical grooming sequences for each time point is sufficient for scoring. Each limb receives one independent grooming score per time point.

Paw Preference Test

The paw preference test, also known as the cylinder test or forelimb asymmetry test, utilizes normal exploratory activity to evaluate forelimb use. The assessment is sensitive to forelimb asymmetries caused by SCI as well as other neurological disorders [5, 29]. Animals are placed in a clear Plexiglas cylinder (~20 cm in diameter and high enough so the animal cannot reach the top edge, ~30–45 cm height) and spontaneous movement is recorded for 3–10 min, dependent upon activity level. A mirror should be placed behind the cylinder to ensure the paw placements on the cylinder can be recorded when the animal is facing away from the camera. Scoring is done using slow-motion playback to score each paw placement on the cylinder as a left, right, or simultaneous (both) arrangement. Scoring can also be done for floor placements as described by Schallert et al. [29]; however, in our experience, floor placements are harder to discern and are not as predictive of injury severity as wall/cylinder placements.

Scoring is done using slow-motion video playback as described by Liu et al. [30]. A "left" or "right" score is given for every weight-supported contact on the wall of the cylinder made without the other forepaw contacting the cylinder for at least 0.5 s. These individual paw placements correspond to the sides ipsi- or contralateral to the SCI. If both paws are used within the 0.5 s interval, then a score of "both" is recorded. For every two-step "walking" sequence of lateral exploration, while rearing, a score of "both" is tallied with the criteria that both paws changed position (came off the cylinder and re-established contact). A "left" or "right" score is given only after both paws are off the cylinder and a single paw is used for a weight supported movement with at least 0.5 s passing before placement of the second paw.

The total score is generated for each time point by dividing the number of "right", "left" or "both" placements by the total number of placements recorded. Depending upon the severity of the SCI, it is common to record few, if any, ipsilateral paw placements. However, more impaired animals rely on the contralateral limb, therefore, the ratio of contralateral:total placements increases with impairment. Less impaired animals will record a higher rate of "both" placements even if independent ipsilateral forelimb use is rarely observed [5].

Histology

1. Prepare animal for tissue collection. Euthanize the animal's through exsanguination as a by-product of transcardial perfusion with 4% paraformaldehyde.
2. Briefly, open the thoracic cavity of the animal by severing the sutures holding the abdomen together and removing the electrodes (if necessary) then detach the connective tissue at the bottom of the diaphragm to give access to the rib cage. Next, cut through the ribs 3 cm left of midline up to the sternum. Clamp the thoracic cavity open to expose the heart.
3. While holding the still beating heart with forceps, insert and secure a needle in the left ventricle of the animal (extending the needle about 5 mm into the tissue and into the aorta if feasible) and make a cut in the right atrium.
4. Through the needle pump 300 mL of 1×PBS at a rate of 20 mL/min. Re-position the needle if solution is coming from the nostrils mouth or nose or the solution is not flowing at a consistent rate.
5. When 300 mL of solution has been pumped through the animal and the blood has been cleared from the body change the solution to pump 300–500 mL of ice cold 4% paraformaldehyde through the needle. Perfusion is complete when the liver has reached a light-brown colour and spontaneous muscle movement has ceased.
6. Following perfusion, carefully dissect the brain and spinal cord of the animal. From a dorsal approach, remove the brain and spinal column by cutting through the skin, musculature and ribs using an offset bone nipper. Remove the spinal

cord and the brain from the spinal column using the same bone nippers buy cutting through each vertebrae and the dorsal and ventral roots in turn from a ventral approach. Post-fix as appropriate prior to cryoprotection or other tissue processing.

Representative Results, Limitations, and Challenges

Contusion and Histology

This report describes how to induce a consistent unilateral cervical contusion injury with a severe behavioral deficit to the respiratory or other motor systems. The authors recommend a unilateral C3/4 contusion of 150 kdyn force and a force of 150–200 kdyn for C5 contusions for optimal results. However, investigators are able to vary this force depending on experimental design. Nonetheless, it is recommended to histologically confirm laterality of the contusion (Fig. 5), and completeness of the C2 hemisection. Figure 5 demonstrates that this model of cervical contusion causes unilateral trauma to white and gray matter. Grey matter is more fragile to contusive injury than white matter [31]. The observation that this model displays damage to both types of tissue suggests that the impact generated is reasonably severe, illustrating the clinical relevance of this contusion model.

Histological/anatomical data may be analyzed in a number of ways including staining with Luxol fast blue for myelin and counterstained with Cresyl violet for Nissl substance (as in Fig. 5). With this, lesion size and tissue sparing can be assessed through camera lucida drawing and imaging. As such, the injury can be reconstructed, and the degree of myelin and motor neuron loss quantified [5]. As we

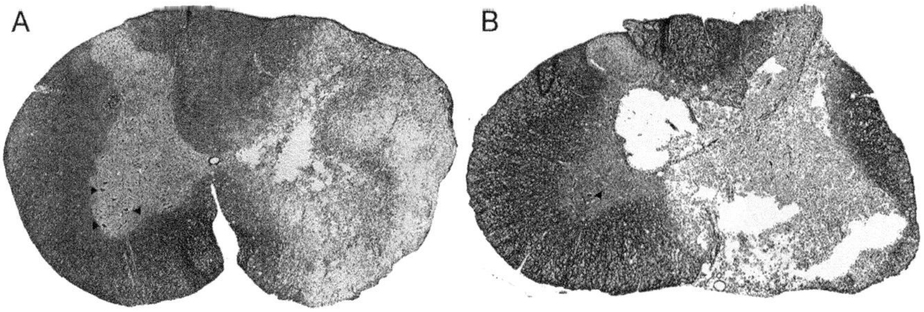

Fig. 5 C5 spinal cord cross sections following unilateral 175 kydn C5 contusion SCI stained with Luxol fast blue for myelin and counterstained with Cresyl violet for Nissl substance. (**a**) When positioned properly, damage is isolated to the ipsilateral hemicord with sparing of neurons (arrowheads) contralateral to the injury. (**b**) More medial or angled impactor probe placement results in loss of gray and white matter contralateral to the injury. Note that only one neuron is spared in (**b**) (arrowhead) vs. the normal appearing gray matter with an abundance of neurons (arrowheads) in (**a**)

have illustrated previously, the contralateral/uninjured side can often be used as a within subject control [5]. Further, sections of the tissue can be stained for any protein of the investigators choice.

Respiratory Outcome Measures

Concerning respiratory outcome measures, 1 week following the initial C3/4 contusion, ipsilateral diaEMG shows a dramatic decrease in activity compared to the non-injured side (Fig. 6). These data show that the functional activity of the complete respiratory motor system following contusion injury can be assessed. This includes compensatory mechanisms from spared or latent pathways and enables the data to be compared with data obtained from alternative C3/4 contusion models.

The contralateral C2 hemisection abolished ipsilateral hemidiaphragm function. However, at the same time, it also induced cessation of respiratory activity in the hemidiaphragm on the same side as the initial contusion (Fig. 6). These data demonstrate how this model clearly defines the functional deficit in respiratory motor activity caused to the contused pathways. Without treatment, these injured and weakened pathways are unable to support respiration alone. This is not the case if the animal had not received the contusion. The ability of treatment strategies applied between the contusion and contralateral hemisection to restore function to the respiratory motor system as a whole, and specifically the contused pathways, will be established through the use of this model.

These data may be analyzed in many ways including: the effect of hemisection on the respiratory motor activity of the diaphragm ipsilateral to the contusion,

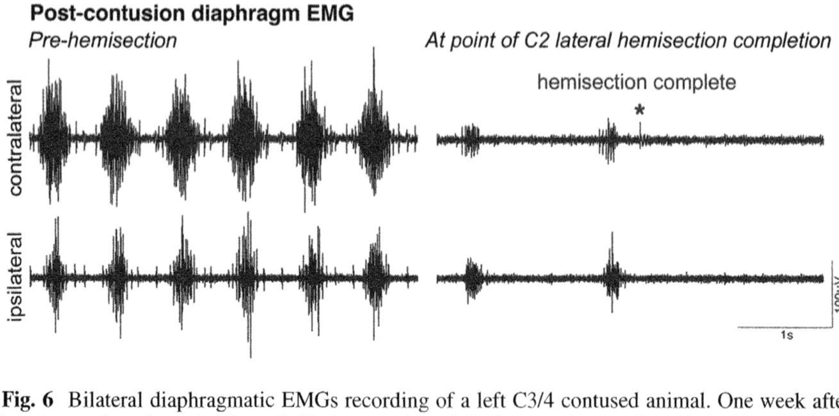

Fig. 6 Bilateral diaphragmatic EMGs recording of a left C3/4 contused animal. One week after hemicontusion, there is bilateral diaphragmatic activity but weaker on the side ipsilateral to the injury (*left panel*). Upon completion of hemisection, and removal of the modulatory decussating inputs, the activity abolishes on both sides of the diaphragm (*right panel*). Traces are from the same animal and all bottom traces in each panel are from the contused left side

respiration rate, burst duration, cycle time, amplitude of breaths, and within breath variability.

Limitations of Utilizing a Unilateral Injury Model

While there are incidences of traumatic unilateral SCI in humans (e.g. Brown-Sequard syndrome) [32], the majority of clinical causes involve damage to both sides of the spinal cord. Presumably the biological response to therapeutic intervention in a unilateral model will have high predictive and translational value for application in bilateral SCI conditions. Nonetheless, it is important to recognize this caveat when using a unilateral SCI model. Indeed, a recent evaluation of a therapeutic intervention showed that efficacy varied as a function of the mechanics of cervical injury with greater protective effects with unilateral vs. bilateral contusion [33, 34]. Unilateral models are also sensitive to small changes in the placement of the impactor tip on the hemicord prior to injury. More medial or angled placements have the potential to cause damage to the contralateral hemi-cord thus increasing the area of primary and secondary damage (see Fig. 5 for example). When unaccounted for, these subtle variations in impactor placement have the potential to increase variability in unilateral models. Despite this, there are a number of clinical trials underway based upon results gathered in unilateral SCI models [35].

Challenges and Limitations in Evaluating Respiratory Function in Experimental SCI

Multiple groups using several trauma models have assessed respiratory motor function following SCI. Researchers have utilized the C2 hemisection model in isolation to gain substantial information regarding respiratory anatomy, receptors mediating repair, and treatment strategies that produce functional respiratory motor recovery [21, 36–39]. However, while providing proof-of-principle data, these studies lack clinical relevance as cervical laceration injures are rare in the human population [40]. Further, the C2 hemisection and recent lateral C2 hemi-contusion [14] models cause trauma largely to white matter and spare caudal areas of the phrenic motor neurons. The more recently described models of respiratory motor dysfunction following C3 contusion [14–19] have made significant progress in assessing the effects of SCI in a more clinically relevant model. However, they fail to demonstrate a significant respiratory motor deficit. The combination contusion/hemisection model described in this chapter overcomes these problems by using a clinically relevant injury and specifically assessing the function of the injured pathways.

Optimizing the location as which to clamp the spinal column prior to contusion can be challenging. Unilateral C3/4 contusions require a C2 laminectomy and removal of the C2 spinous process that is typically used to stabilize the cervical spine during contusion [19, 41]. Characteristically, the impactor requires two points of fixation, above and below the targeted spinal cord segment. The unique anatomical attribute of the Atlas (C1) vertebra makes it practically inaccessible to clamp or utilize as a fixation point. The optimal location described in this manuscript is the bilateral, bony groove between the articular and transverse processes of the C2 and C3 lamina (Fig. 2). These bony landmarks allow for lateral fixation of the impactor clamps and stabilization of the C2–C3 vertebrae without slippage during impact. Violating the C2/3 articular joints during clamping can result in considerable instability during impaction that may end with spikes in the force vs. time curves, causing inconsistency (Fig. 3b). Securing the stabilization clamps correctly is crucial to immobilizing the spine prior to injury and ensuring consistency of this contusion model.

Following the initial contusive impact the animal can require respiratory support for a transient length of time and if provided, greater than 95% of animals successfully survive the injury. However, if not adequately provided, the animal may die. The contusion injury is severe enough that, without treatment or respiratory support, the animal typically dies following the C2 hemisection due to respiratory failure. This clearly shows the functional deficit caused by the initial injury, but may limit any further assessment investigators wish to perform on the animals. To perform further investigations following the hemisection investigators should ventilate the animal prior to, or immediately after, lacerating the spinal cord.

Concluding Remarks

The overall purpose of this report is to provide a detailed protocol and technique that is easily replicated. This clinically relevant unilateral cervical contusion model provides a sustained, reproducible, and robust respiratory deficit to the ipsilateral diaphragm (when delivered at C3/4) and results in predictable fore and hind-limb motor deficits (when delivered at C5). Utilization of a subsequent contralateral C2 hemisection provides an alternative and precise method to administer and assess potential respiratory treatments. This model is currently employed to evaluate various therapeutic interventions aimed at repairing or protecting motor pathways. The use of this technique has the potential to generate meaningful therapeutic data for translation to clinical application.

Acknowledgements This study was supported by grants to WJA from the International Spinal Research Trust and the Craig H. Neilsen Foundation. JCG was supported, in part, by the Paralysis Project of America. Additional support comes from MetroHealth Medical Center in Cleveland, Ohio and the Spinal Cord and Brain Injury Research Center at the University of Kentucky. BIA was supported by the Egyptian Governmental Scholarship and PMW by the International Spinal Research Trust and Wings for Life.

Disclosures: The authors do not have any competing financial interests to disclose.

References

1. DeVivo MJ, Chen Y. Trends in new injuries, prevalent cases, and aging with spinal cord injury. Arch Phys Med Rehabil. 2011;92:332–8. https://doi.org/10.1016/j.apmr.2010.08.031.
2. Allen AR. Surgery of experimental lesion of spinal cord equivalent to crush injury of fracture dislocation of spinal column: a preliminary report. J Am Med Assoc. 1911;LVII(11):878–80.
3. Trivedi A, Olivas AD, Noble-Haeusslein LJ. Inflammation and spinal cord injury: infiltrating leukocytes as determinants of injury and repair processes. Clin Neurosci Res. 2006;6:283–92. https://doi.org/10.1016/j.cnr.2006.09.007.
4. DeVivo MJ. Epidemiology of traumatic spinal cord injury: trends and future implications. Spinal Cord. 2012;50:365–72. https://doi.org/10.1038/sc.2011.178.
5. Gensel JC, Tovar CA, Hamers FPT, et al. Behavioral and histological characterization of unilateral cervical spinal cord contusion injury in rats. J Neurotrauma. 2006;23:36–54. https://doi.org/10.1089/neu.2006.23.36.
6. Awad BI, Warren PM, Steinmetz MP, Alilain WJ. The role of the crossed phrenic pathway after cervical contusion injury and a new model to evaluate therapeutic interventions. Exp Neurol. 2013;248:398–405. https://doi.org/10.1016/j.expneurol.2013.07.009.
7. Streijger F, Beernink TMJ, Lee JHT, et al. Characterization of a cervical spinal cord hemicontusion injury in mice using the infinite horizon impactor. J Neurotrauma. 2013;30:869–83. https://doi.org/10.1089/neu.2012.2405.
8. Scheff SW, Rabchevsky AG, Fugaccia I, et al. Experimental modeling of spinal cord injury: characterization of a force-defined injury device. J Neurotrauma. 2003;20:179–93. https://doi.org/10.1089/08977150360547099.
9. Ferguson AR, Irvine K-A, Gensel JC, et al. Derivation of multivariate syndromic outcome metrics for consistent testing across multiple models of cervical spinal cord injury in rats. PLoS One. 2013;8:e59712. https://doi.org/10.1371/journal.pone.0059712.
10. Goshgarian HG, Ellenberger HH, Feldman JL. Decussation of bulbospinal respiratory axons at the level of the phrenic nuclei in adult rats: a possible substrate for the crossed phrenic phenomenon. Exp Neurol. 1991;111:135–9.
11. Goshgarian HG, Rafols JA. The ultrastructure and synaptic architecture of phrenic motor neurons in the spinal cord of the adult rat. J Neurocytol. 1984;13:85–109.
12. Inskip JA, Ramer LM, Ramer MS, Krassioukov AV. Autonomic assessment of animals with spinal cord injury: tools, techniques and translation. Spinal Cord. 2009;47:2–35. https://doi.org/10.1038/sc.2008.61.
13. Krassioukov A. Autonomic function following cervical spinal cord injury. Respir Physiol Neurobiol. 2009;169:157–64. https://doi.org/10.1016/j.resp.2009.08.003.
14. Baussart B, Stamegna JC, Polentes J, et al. A new model of upper cervical spinal contusion inducing a persistent unilateral diaphragmatic deficit in the adult rat. Neurobiol Dis. 2006;22:562–74. https://doi.org/10.1016/j.nbd.2005.12.019.
15. Choi H, Liao W-L, Newton KM, et al. Respiratory abnormalities resulting from midcervical spinal cord injury and their reversal by serotonin 1A agonists in conscious rats. J Neurosci. 2005;25:4550–9. https://doi.org/10.1523/JNEUROSCI.5135-04.2005.
16. el-Bohy AA, Schrimsher GW, Reier PJ, Goshgarian HG. Quantitative assessment of respiratory function following contusion injury of the cervical spinal cord. Exp Neurol. 1998;150:143–52. https://doi.org/10.1006/exnr.1997.6757.
17. Golder FJ, Fuller DD, Lovett-Barr MR, et al. Breathing patterns after mid-cervical spinal contusion in rats. Exp Neurol. 2011;231(1):97–103. https://doi.org/10.1016/j.expneurol.2011.05.020.
18. Lane MA, Lee K-Z, Salazar K, et al. Respiratory function following bilateral mid-cervical contusion injury in the adult rat. Exp Neurol. 2012;235(1):197–210. https://doi.org/10.1016/j.expneurol.2011.09.024.
19. Nicaise C, Hala TJ, Frank DM, et al. Phrenic motor neuron degeneration compromises phrenic axonal circuitry and diaphragm activity in a unilateral cervical contusion model of spinal cord injury. Exp Neurol. 2012;235:539–52. https://doi.org/10.1016/j.expneurol.2012.03.007.

20. Stamegna JC, Felix MS, Roux-Peyronnet J, et al. Nasal OEC transplantation promotes respiratory recovery in a subchronic rat model of cervical spinal cord contusion. Exp Neurol. 2011;229:120–31. https://doi.org/10.1016/j.expneurol.2010.07.002.
21. Vinit S, Stamegna J-C, Boulenguez P, et al. Restorative respiratory pathways after partial cervical spinal cord injury: role of ipsilateral phrenic afferents. Eur J Neurosci. 2007;25:3551–60. https://doi.org/10.1111/j.1460-9568.2007.05619.x.
22. Teasell RW, Mehta S, Aubut J-AL, et al. A systematic review of pharmacologic treatments of pain after spinal cord injury. Arch Phys Med Rehabil. 2010;91:816–31. https://doi.org/10.1016/j.apmr.2010.01.022.
23. McKenna JE, Prusky GT, Whishaw IQ. Cervical motoneuron topography reflects the proximodistal organization of muscles and movements of the rat forelimb: a retrograde carbocyanine dye analysis. J Comp Neurol. 2000;419:286–96.
24. Šedý J, Urdzíková L, Jendelová P, Syková E. Methods for behavioral testing of spinal cord injured rats. Neurosci Biobehav Rev. 2008;32:550–80. https://doi.org/10.1016/j.neubiorev.2007.10.001.
25. Nielson JL, Guandique CF, Liu AW, et al. Development of a database for translational spinal cord injury research. J Neurotrauma. 2014;31:1789–99. https://doi.org/10.1089/neu.2014.3399.
26. Nielson JL, Paquette J, Liu AW, et al. Topological data analysis for discovery in preclinical spinal cord injury and traumatic brain injury. Nat Commun. 2015;6:8581. https://doi.org/10.1038/ncomms9581.
27. Bertelli JA, Mira JC. Behavioral evaluating methods in the objective clinical assessment of motor function after experimental brachial plexus reconstruction in the rat. J Neurosci Methods. 1993;46:203–8.
28. Berntson GG, Jang JF, Ronca AE. Brainstem systems and grooming behaviors. Ann N Y Acad Sci. 1988;525:350–62.
29. Schallert T, Fleming SM, Leasure JL, et al. CNS plasticity and assessment of forelimb sensorimotor outcome in unilateral rat models of stroke, cortical ablation, parkinsonism and spinal cord injury. Neuropharmacology. 2000;39:777–87.
30. Liu Y, Kim D, Himes BT, et al. Transplants of fibroblasts genetically modified to express BDNF promote regeneration of adult rat rubrospinal axons and recovery of forelimb function. J Neurosci. 1999;19:4370–87.
31. Ichihara K, Taguchi T, Sakuramoto I, et al. Mechanism of the spinal cord injury and the cervical spondylotic myelopathy: new approach based on the mechanical features of the spinal cord white and gray matter. J Neurosurg. 2003;99:278–85.
32. Amendola L, Corghi A, Cappuccio M, De Iure F. Two cases of Brown-Séquard syndrome in penetrating spinal cord injuries. Eur Rev Med Pharmacol Sci. 2014;18:2–7.
33. Popovich PG, Lemeshow S, Gensel JC, Tovar CA. Independent evaluation of the effects of glibenclamide on reducing progressive hemorrhagic necrosis after cervical spinal cord injury. Exp Neurol. 2012;233:615–22. https://doi.org/10.1016/j.expneurol.2010.11.016.
34. Simard JM, Popovich PG, Tsymbalyuk O, Gerzanich V. Spinal cord injury with unilateral versus bilateral primary hemorrhage—effects of glibenclamide. Exp Neurol. 2012;233:829–35. https://doi.org/10.1016/j.expneurol.2011.11.048.
35. Kwon BK, Okon E, Hillyer J, et al. A systematic review of non-invasive pharmacologic neuroprotective treatments for acute spinal cord injury. J Neurotrauma. 2011;28:1545–88. https://doi.org/10.1089/neu.2009.1149.
36. Goshgarian HG. The crossed phrenic phenomenon: a model for plasticity in the respiratory pathways following spinal cord injury. J Appl Physiol. 2003;94:795–810. https://doi.org/10.1152/japplphysiol.00847.2002.
37. Porter WT. The path of the respiratory impulse from the bulb to the phrenic nuclei. J Physiol Lond. 1895;17:455–85. https://doi.org/10.1113/jphysiol.1895.sp000553.
38. Alilain WJ, Silver J. Shedding light on restoring respiratory function after spinal cord injury. Front Mol Neurosci. 2009;2:18. https://doi.org/10.3389/neuro.02.018.2009.

39. Vinit S, Darlot F, Stamegna J-C, et al. Long-term reorganization of respiratory pathways after partial cervical spinal cord injury. Eur J Neurosci. 2008;27:897–908. https://doi.org/10.1111/j.1460-9568.2008.06072.x.
40. Vinit S, Kastner A. Descending bulbospinal pathways and recovery of respiratory motor function following spinal cord injury. Respir Physiol Neurobiol. 2009;169:115–22. https://doi.org/10.1016/j.resp.2009.08.004.
41. Nicaise C, Frank DM, Hala TJ, et al. Early phrenic motor neuron loss and transient respiratory abnormalities after unilateral cervical spinal cord contusion. J Neurotrauma. 2013;30:1092–9. https://doi.org/10.1089/neu.2012.2728.

Applications of the Infinity Horizon Spinal Cord Contusion Injury Model

Samir P. Patel and Alexander G. Rabchevsky

Abstract As described in the inaugural chapter, the Infinite Horizon (IH) impactor was developed and first characterized (Scheff et al., Journal of Neurotrauma 20:179–193, 2003) to produce graded behavioral and correlative morphological changes in adult female Sprague-Dawley rats and their spinal cords following varying severities of contusion spinal cord injuries (SCI) at thoracic spinal level 10 (T10) in adult rats. Since that time, the IH device has become a widely used contusion SCI model, and more importantly, its tractability has allowed its adaptation to elicit different injury severities (impact forces) at different spinal levels, in both rats and mice; and the accompanying distinct behavioral recovery outcome measures.

Keywords Cervical · Thoracic · Lumbar · Hemi-contusion · Tissue sparing · Locomotor recovery

NYU/MASCIS and PSI Infinite Horizon (IH) Impactors

Shortly after standardizing the PSI IH Impactor [21], the authors [18] tested the effects of creatine diet supplementation on behavioral recovery after moderate T10 contusion SCI in adult female Sprague-Dawley rats using both the IH device (150 kdyn) and the New York University (NYU)/MASCIS weight drop device (12.5 mm). Comparatively, both instruments caused similar amounts of gray matter damage in control-fed rats while the NYU device rendered a greater loss of white matter, reflected in more severe hindlimb deficits acutely, followed by a lower plateau of recovery than the IH cohort. A more recent comparative study demonstrated that the loss and recovery of hindlimb locomotor function between the 25 mm NYU/MASCIS group and the 200 kdyn IH group as well as between the 50 mm and

S. P. Patel · A. G. Rabchevsky (✉)
Spinal Cord and Brain Injury Research Center, Department of Physiology, College of Medicine, University of Kentucky, Lexington, KY, USA
e-mail: agrab@uky.edu

J. Chen et al. (eds.), *Animal Models of Acute Neurological Injury*, Springer Series in Translational Stroke Research, https://doi.org/10.1007/978-3-030-16082-1_32

250 kdyn injury groups were very similar between two devices [13]. However, compared to the 25 mm group, rats with 200 kdyn injuries trended to score higher when assessed over time for the ladder rung test, demonstrating reduced functional deficits. Magnetic resonance imaging (MRI) further revealed predominantly dorsal injuries in the 25 mm NYU group whereas in the 200 kdyn IH group the injury cavities were located deeper. Moreover, BDA labeling of dorsal corticospinal tracts in spinal cord sections caudal to the injury was evident in 200 kdyn group but not after 25 mm injuries, likely because the IH impactor was reported to spare more ascending and descending white matter tracts. Collectively, the studies above illustrate that while they produce similar SCI severities, these two devices appear to elicit differential neuroanatomical pathologies.

IH Mid-Thoracic SCI in Rats

It is reported that the phase of the estrus cycle at the time of a moderate (100–190 kdyn) T10 SCI or during the post-injury recovery period does not affect the outcome, indicating that female Sprague-Dawley rats may be used without consideration of the estrus cycle [3]. Others report that more severe (~225 kdyn) T8 contusion in adult female Wistar rats results in delayed estrous cycle, and the amount of delay is not correlated with impact force or total damage at the lesion epicenter [5] (Fig. 1).

After severe (200 kdyn) T10 SCI in adult female Sprague-Dawley rats, mitochondrial respiration is significantly compromised as early as 6 h. post-injury with a progressive significant increase in mitochondrial oxidative stress markers over 24 h. post-injury [24]. Others have shown that 200 kdyn T10 SCI results in the generation of reactive oxygen species and oxidative damage [26], and that scavenging peroxynitrite with the antioxidant, tempol, attenuates both compromised mitochondria function and increased oxidative stress [25]. Using this model it was further demonstrated that targeting mitochondrial dysfunction using alternative electron transport chain biofuels after injury results in protection of mitochondrial integrity and tissue sparing [16].

A recent study characterized how altering piston impact and dwell time of the IH device affects functional recovery [4]. Adult male Sprague-Dawley rats were subjected to T10 SCI using different forces (kdyn) and dwell times (sec) in the following combinations: 150-4, 150-3, 150-2, 150-1, 150-0, 200-0, 90-2. Rats in 150-0 injury group showed the highest recovery of hindlimb locomotion compared to other injury groups while the recovery of spontaneous bladder function was most prolonged in the 150-4 injury group; conversely the 90-2 group had the fastest recovery times [4]. Another recent study showed that 100 or 150 kdyn injuries at T8/T9 spinal segments of female Sprague-Dawley rat elicits expected differences in initial loss and long-term recovery of hindlimb locomotion, and that measures of bladder dysfunction are correlated with intensity of contusion parameters [11].

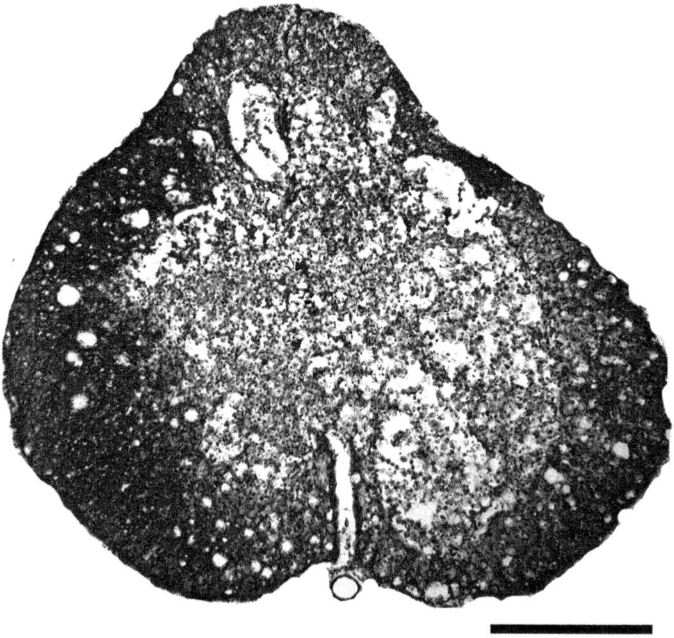

Fig. 1 Photomicrograph represents rat spinal cord section for injury epicenter at 5 weeks following T10 contusion SCI at 200 kdyn force using IH Impactor. Spinal cord sections (20 μm) were stained with eriochrome cyanine/cresyl violet (EC/CV) to identify white matter, grey matter and injury lesion. Scale bar = 500 μm

IH Cervical SCI in Rats: Bilateral and Unilateral

While most SCI studies using the IH Impactor have focuseds on hindlimb recovery, some have evaluated forelimb functions after cervical (C) contusions since spinal trauma at this level is most common in humans. In a study where female Sprague-Dawley rats received moderate (200 kdyn) or severe (250 kdyn) bilateral (center aligned) SCI at C5, C6, or C7/8, both the forelimb and hindlimb functions were severely impaired immediately post-injury, but all rats regained the ability to use their hindlimbs for locomotion [1]. Gripping ability was abolished immediately after injury but recovered partially, depending upon the spinal level and severity of the injury; forepaw gripping function recovered after mild injuries even when the descending dorsal columns were completely interrupted.

Severe (200 kdyn) unilateral C4 SCI in adult female Sprague-Dawley rats results in unilateral forelimb-hindlimb functional deficits immediately after injury with spontaneous recovery over time using BBB locomotor scale and grid walk assessments, and forced exercise transiently improves forelimb and hindlimb function during weeks 2–4 following injury; but not at 8 weeks [19, 20]. Using similar sized impact tips and rotating adult male Sprague-Dawley rats at 22.5° before applying C5 contusion SCI (150 kdyn), highly reproducible functional deficits of forelimbs were established using the horizontal ladder test, cylinder rearing test, grooming test, and modified Montoya staircase test [8]. This latter study also presents a comprehensive discussion on the various contusion impactors employed for cervical SCI in rats and the comparative neuropathology with accompanying deficits.

Experimental cervical contusion at C3–C5 in adult female Sprague-Dawley rats is used to study SCI-induced respiratory dysfunction by targeting the pool of phrenic motor neurons that innervate the diaphragm. Moderate (150 kdyn) and severe (250 kdyn) bilateral C3/C4 contusion SCI results in significant chronically impaired diaphragm activity assessed by bilateral diaphragm EMG recordings [7]. Nevertheless, despite extensive white and gray matter pathology, ventilation is unaffected during both quiet breathing and respiratory challenge (hypercapnia), indicating anatomical reorganization and spinal plasticity following cervical contusion. Alternatively, 150 kdyn C3/C4 hemicontusion in adult male Sprague Dawley rats demonstrates spared pathways sufficient to partially activate ipsilateral phrenic motor neurons, but that contralateral hemisection resulted in a cessation of all activity ipsilateral to the contusion [2].

IH Upper-Lumbar SCI in Rats

The interneuron populations located in the rostral lumbar (L1/L2 spinal levels) intermediate gray matter, also termed the central pattern generator (CPG), are critical to rhythmic hindlimb locomotion in the adult rats [9, 10]. In order to target the CPG and test neuroprotective therapies, adult female Sprague-Dawley rats received L1/L2 SCI with 250 kdyn force using a modified 2 mm diameter tip, with impact made at the caudle part of T12 laminectomy [14]. This resulted in significant hindlimb functional deficits assessed by BBB locomotor scale over 4–6 weeks post injury and terminal gait analyses, and such deficits were significantly attenuated using neuroprotective agents that target mitochondrial dysfunction [15, 17] (Fig. 2).

IH High-Thoracic SCI in Rats

Abnormal cardiovascular control is prominent in individuals with SCI. Recently, a high-thoracic (T3) contusion SCI model at 200 or 400 kdyn was developed to study both hindlimb locomotor impairment and accompanying cardiovascular dysregulation (autonomic dysreflexia) over 1 month post-injury [22]. Using telemetric blood pressure monitoring, T3 SCI produces a severity-dependent decline in both motor and autonomic functions in rats, citing novel relationships between injury site morphometrics and the evolution/extent of autonomic dysreflexia.

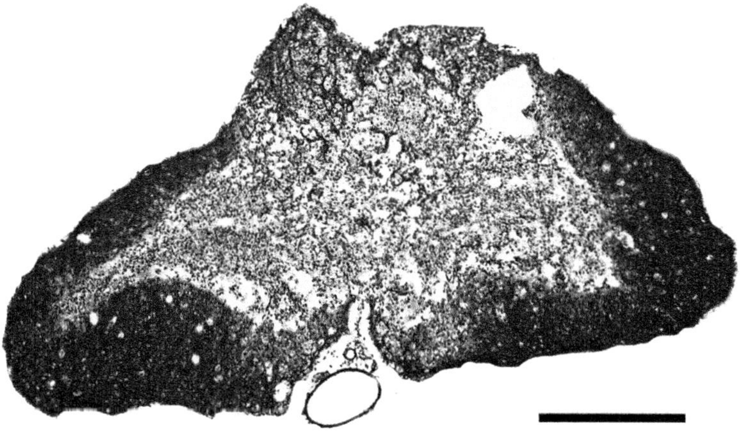

Fig. 2 Photomicrograph represents rat spinal cord section for injury epicenter at 6 weeks following L1/L2 contusion SCI at 250 kdyn force using IH Impactor. Spinal cord sections (20 μm) were stained with eriochrome cyanine/cresyl violet (EC/CV) to identify white matter, grey matter and injury lesion. Scale bar = 500 μm

IH Mid-Thoracic SCI in Mice

Because mice are increasingly used because of the availability of genetically engineered lines, studies are also been carried out to characterize use of IH Impactor to produce contusion SCI in mice. Nishi et al. assessed lesion volume using magnetic resonance imaging (MRI) following a mild (30 kdyn), moderate (50 kdyn), or severe (70 kdyn) contusion injury at the T9 vertebral level in female C57Bl/6 mice [12]. The Basso Mouse Scale (BMS) was used to assess hindlimb locomotor function in injured mice. The BMS, histology and MRI results demonstrated that the IH Impactor produced precise, graded contusion SCI in mice. Lesion volumes were positively correlated with force of impact, and negatively correlated with spared white matter and functional recovery.

Comparative studies in adult female C57Bl/6 mice using three leading rodent SCI impacting devices-0.1 m/s (Infinite Horizon), 0.2 m/s (Ohio State University), and 0.4 m/s (New York University) to assess if different impact speed with fixed impact displacement (0.8 mm) correlated with injury severity [6]. In vivo diffusion tensor imaging (DTI) and post-mortem histology as well as BMS assessments revealed that identical injury severity can be reproduced using speeds from 0.1 to 0.4 m/s, irrespective of device, given a fixed impact displacement (Fig. 3).

IH Cervical SCI in mice: Unilateral

To assess the breathing impairment in mice, cervical (C3–C5) hemi-contusion mouse models are developed using IH impactor. Researchers applied different force with and without dwell time to produce various injury severities. A 75 kdyn C5

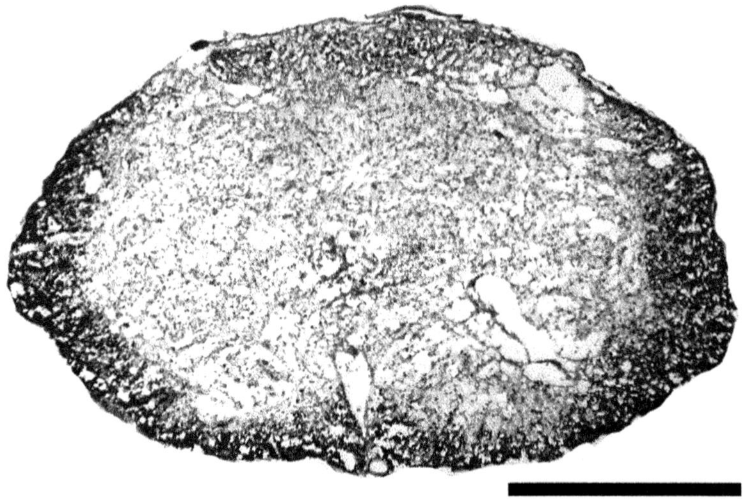

Fig. 3 Photomicrograph represents mouse spinal cord section for injury epicenter at 4 weeks following T9 contusion SCI at 75 kdyn force using IH Impactor. A modified Eriochrome Cyanine (EC) staining was used to stain spinal cord sections (10 μm) to identify white matter, grey matter and injury lesion. Scale bar = 500 μm

hemicontusion injury with or without dwell time in adult male C57BL/6 mice showed sustained deficits of the affected forepaw, reduction in grooming scores, retrieval of significantly fewer pellets on the Montoya staircase test and reduced paw-print size and swing speed assessed by CatWalk analysis. Addition of dwell time of 15 or 30 s worsened these behavioral outcomes as well as reduced residual white and gray matter at the epicenter and rostral-caudal to the injury [23].

Conclusions

In summary, IH Impactor device has been shown to produce reliable spinal cord contusions in both rats and mice. Depending on the study design and therapeutic targets, contusions are able to be produced at different spinal cord levels and at different impact severities, with or without dwell times. Such tractability further lends to the ability to assess various behavioral and physiological outcome measures of sensory, locomotor and/or autonomic function.

References

1. Anderson KD, Sharp KG, Steward O. Bilateral cervical contusion spinal cord injury in rats. Exp Neurol. 2009;220:9–22.
2. Awad BI, Warren PM, Steinmetz MP, Alilain WJ. The role of the crossed phrenic pathway after cervical contusion injury and a new model to evaluate therapeutic interventions. Exp Neurol. 2013;248:398–405.

3. Baker KA, Hagg T. An adult rat spinal cord contusion model of sensory axon degeneration: the estrus cycle or a preconditioning lesion do not affect outcome. J Neurotrauma. 2005;22:415–28.
4. Carter MW, Johnson KM, Lee JY, Hulsebosch CE, Gwak YS. Comparison of mechanical allodynia and recovery of locomotion and bladder function by different parameters of low thoracic spinal contusion injury in rats. Korean J Pain. 2016;29:86–95.
5. Hubscher CH, Armstrong JE, Johnson JR. Effects of spinal cord injury on the rat estrous cycle. Brain Res. 2006;1100:118–24.
6. Kim JH, Tu TW, Bayly PV, Song SK. Impact speed does not determine severity of spinal cord injury in mice with fixed impact displacement. J Neurotrauma. 2009;26:1395–404.
7. Lane MA, Lee KZ, Salazar K, O'Steen BE, Bloom DC, Fuller DD, Reier PJ. Respiratory function following bilateral mid-cervical contusion injury in the adult rat. Exp Neurol. 2012;235:197–210.
8. Lee JH, Streijger F, Tigchelaar S, Maloon M, Liu J, Tetzlaff W, Kwon BK. A contusive model of unilateral cervical spinal cord injury using the infinite horizon impactor. J Vis Exp. 2012;24;(65). pii: 3313. https://doi.org/10.3791/3313.
9. Magnuson DS, Green DM, Sengoku T. Lumbar spinoreticular neurons in the rat: part of the central pattern generator for locomotion? Ann N Y Acad Sci. 1998;860:436–40.
10. Magnuson DS, Lovett R, Coffee C, Gray R, Han Y, Zhang YP, Burke DA. Functional consequences of lumbar spinal cord contusion injuries in the adult rat. J Neurotrauma. 2005;22:529–43.
11. Munoz A. Neurogenic bladder dysfunction does not correlate with astrocyte and microglia activation produced by graded force in a contusion-induced spinal cord injury. Brain Res Bull. 2017;131:18–24.
12. Nishi RA, Liu H, Chu Y, Hamamura M, Su MY, Nalcioglu O, Anderson AJ. Behavioral, histological, and ex vivo magnetic resonance imaging assessment of graded contusion spinal cord injury in mice. J Neurotrauma. 2007;24:674–89.
13. Park JH, Kim JH, Oh SK, Baek SR, Min J, Kim YW, Kim ST, Woo CW, Jeon SR. Analysis of equivalent parameters of two spinal cord injury devices: the New York University impactor versus the Infinite Horizon impactor. Spine J. 2016;16:1392–403.
14. Patel SP, Smith TD, VanRooyen JL, Powell D, Cox DH, Sullivan PG, Rabchevsky AG. Serial diffusion tensor imaging in vivo predicts long-term functional recovery and histopathology in rats following different severities of spinal cord injury. J Neurotrauma. 2016;33:917–28.
15. Patel SP, Sullivan PG, Lyttle TS, Magnuson DS, Rabchevsky AG. Acetyl-L-carnitine treatment following spinal cord injury improves mitochondrial function correlated with remarkable tissue sparing and functional recovery. Neuroscience. 2012;210:296–307.
16. Patel SP, Sullivan PG, Lyttle TS, Rabchevsky AG. Acetyl-L-carnitine ameliorates mitochondrial dysfunction following contusion spinal cord injury. J Neurochem. 2010;114:291–301.
17. Patel SP, Sullivan PG, Pandya JD, Goldstein GA, VanRooyen JL, Yonutas HM, Eldahan KC, Morehouse J, Magnuson DS, Rabchevsky AG. N-acetylcysteine amide preserves mitochondrial bioenergetics and improves functional recovery following spinal trauma. Exp Neurol. 2014;257:95–105.
18. Rabchevsky AG, Sullivan PG, Fugaccia I, Scheff SW. Creatine diet supplement for spinal cord injury: influences on functional recovery and tissue sparing in rats. J Neurotrauma. 2003;20:659–69.
19. Sandrow-Feinberg HR, Izzi J, Shumsky JS, Zhukareva V, Houle JD. Forced exercise as a rehabilitation strategy after unilateral cervical spinal cord contusion injury. J Neurotrauma. 2009;26:721–31.
20. Sandrow-Feinberg HR, Zhukareva V, Santi L, Miller K, Shumsky JS, Baker DP, Houle JD. PEGylated interferon-beta modulates the acute inflammatory response and recovery when combined with forced exercise following cervical spinal contusion injury. Exp Neurol. 2010;223:439–51.
21. Scheff SW, Rabchevsky AG, Fugaccia I, Main JA, Lumpp JE Jr. Experimental modeling of spinal cord injury: characterization of a force-defined injury device. J Neurotrauma. 2003;20:179–93.

22. Squair JW, West CR, Popok D, Assinck P, Liu J, Tetzlaff W, Krassioukov AV. High thoracic contusion model for the investigation of cardiovascular function after spinal cord injury. J Neurotrauma. 2017;34:671–84.
23. Streijger F, Beernink TM, Lee JH, Bhatnagar T, Park S, Kwon BK, Tetzlaff W. Characterization of a cervical spinal cord hemicontusion injury in mice using the infinite horizon impactor. J Neurotrauma. 2013;30:869–83.
24. Sullivan PG, Krishnamurthy S, Patel SP, Pandya JD, Rabchevsky AG. Temporal characterization of mitochondrial bioenergetics after spinal cord injury. J Neurotrauma. 2007;24:991–9.
25. Xiong Y, Hall ED. Pharmacological evidence for a role of peroxynitrite in the pathophysiology of spinal cord injury. Exp Neurol. 2009;216:105–14.
26. Xiong Y, Rabchevsky AG, Hall ED. Role of peroxynitrite in secondary oxidative damage after spinal cord injury. J Neurochem. 2007;100:639–49.

Clip Impact-Compression Model

Charles H. Tator, Peter Poon, and Andrea J. Mothe

Abstract In human spinal cord injury, the most common mechanism is the combination of acute impact and continuing compression. To simulate combined impact-compression, we developed in the 1970s the acute clip impact-compression model, one of the first non-transection models in the rodent. Subsequently, we characterized the relationships between clip strength, duration of compression and neurological recovery, and established dose-response relationships between the forces of clip compression injury, axon evoked potentials, spinal cord blood flow, neurological function, axon counts and retrograde labeling of supraspinal neurons with axonal tracers. In this review, we discuss the defining features of the acute clip impact-compression model of spinal cord injury and outline its advantages and disadvantages. We also briefly discuss other impact-compression/contusion models and non-impact models of spinal cord injury. The method for performing acute impact-compression injury in rats is included. Clip injury is useful for *in vitro* and *in vivo* spinal cord injury studies in rats and mice for cervical, thoracic and lumbar injuries, and is consistent, reliable and relatively inexpensive.

Keywords Spinal cord injury · Injury mechanisms · Rodent models · Impact-compression · Contusion

C. H. Tator (✉)
Krembil Brain Institute, Toronto Western Hospital, University Health Network, Toronto, ON, Canada

Department of Surgery, Division of Neurosurgery, University of Toronto, Toronto, ON, Canada
e-mail: charles.tator@uhn.ca

P. Poon · A. J. Mothe
Krembil Brain Institute, Toronto Western Hospital, University Health Network, Toronto, ON, Canada
e-mail: ppoon@uhnres.utoronto.ca; amothe@uhnres.utoronto.ca

J. Chen et al. (eds.), *Animal Models of Acute Neurological Injury*, Springer Series in Translational Stroke Research, https://doi.org/10.1007/978-3-030-16082-1_33

Introduction

There are many mechanisms involved in producing spinal cord injury (SCI) in humans, although the most common are impact-compression, contusion, laceration, transection and traction of the spinal cord [1]. No single experimental model can simulate all these diverse mechanisms, and therefore, there is a need for more than one model of SCI, especially for screening potential therapies for use in human SCI [2, 3]. There have been several informative reviews of experimental models for SCI [4–6], and in our view, there are several reliable and consistent animal models some of which replicate human injuries reasonably well [6, 7].

The most common mechanism of SCI in humans is a combination of acute impact followed by persisting compression [1, 8], and this combination occurs in fracture-dislocations and burst fractures of the spine, the two most common mechanisms in human SCI. In anterior fracture-dislocation, the most common type, the anterior surface of the spinal cord is compressed between the posterior aspect of the rostral vertebra while the posterior surface of the spinal cord is compressed by the anterior aspect of the lamina of the caudal vertebra. In burst fractures the anterior surface of the spinal cord is compressed by the posterior aspect of the vertebral body while the posterior surface of the spinal cord is compressed by the anterior aspect of the lamina of the same vertebra. To simulate these mechanisms of combined impact-compression, the senior author developed in the 1970s the acute clip impact-compression model, one of the first to simulate the sequence of acute impact followed by persisting compression [9]. Sudden impact is produced by rapid release of the clip. The clip closes rapidly producing bilateral, dorsal and ventral impact followed by persisting compression injury with severity related to the closing force of the clip and the duration of compression. This model was one of the first non-transection models in the rodent [7]. Subsequently, in my laboratory, we characterized the relationships between clip strength, duration of compression and neurological recovery [10, 11]. We also established dose response-curves for the relationships between the force of clip compression injury, axon evoked potentials, and spinal cord blood flow [12], and for the relationships between the force of clip compression injury, neurological function, axon counts and retrograde labeling of supraspinal neurons with retrograde axonal tracers [13]. Since then, most of the studies in the Tator and Fehlings laboratories at the University of Toronto have utilized this model in *in vitro* and *in vivo* SCI studies in rats and mice in cervical, thoracic and lumbar injuries.

In the initial phase of the development of this model of experimental SCI, we found that a 1-min duration of clip compression was the minimum time required to produce a consistent pathological and clinical effect in rats with clips varying in force from 2 to 178 g. With clips of this range of forces, we varied the duration of compression from 3 s to as long as 4 h, and found a complex correlation between both the force and duration of compression in rats at C7-T1 or T7-8 and several histopathological and clinical outcome measures. We also showed similar findings in injuries at T2-3 and L1-2 in the rat [14]. Recently, we characterized acute clip impact-compression injury at the lumbar spinal cord level L1-2 showing neurological and histological correlation with clip strength [15]. The range of clip forces can be varied to represent the range of injury severities associated with incomplete injuries in humans graded by the system of the American Spinal Injury Association (ASIA) as ASIA grades B to D [16]. More severe injuries such as a 50 g injury at T4 in the rat produces the clinical and histological features of a complete SCI [17], graded as ASIA A in the human. An incomplete bilateral clip-compression injury to the cervical cord in the rat can be achieved with an 18 g clip at C6 [18]. Of major importance is that the clip impact-compression SCI can be used to model injuries at all levels of the spinal cord, at a range of severities, and also acute, subacute and chronic SCI [14, 15, 19–22].

In the Fehlings laboratory, the clip was adapted for producing SCI in the mouse, and the mouse clip was found to produce consistent SCI of graded severity similar to what had been previously established in the rat [23, 24]. However, the pathological changes in the mouse spinal cord differ significantly from what has been established in rats and humans. In both rats and humans, major SCI causes ischemia, necrosis and cavitation whereas in the mouse the lesions are generally without cavitation.

It should be noted that many other investigators have used clip compression SCI, some with the same modified Kerr-Lougheed aneurysm clip that we have used [17, 25]. However, many other types of clips have been used in experimental SCI [26]. For example, von Euler et al. used bulldog clips of varying forces, and placed them on the cord vertically, so that the compression force was from side-to-side or lateral rather than in the antero-posterior direction [27]. Similarly, a vascular clip was placed vertically to achieve lateral compression of the mouse spinal cord at thoracic level T9 [28]. In many reports, the authors did not state whether the clips were slowly released to produce compression or rapidly released to produce impact-compression.

Advantages and Disadvantages of the Clip Impact-Compression Model

The main advantages and disadvantages of the clip impact-compression model are as follows:

Advantages

1. Inexpensive.
2. Durable. The same clip can be used for an infinite number of times, although the springs will weaken after multiple openings and closings.

3. Useful for injuries at all levels of the spinal cord-cervical, thoracic and lumbar.
4. Useful for modeling acute, subacute and chronic injuries.
5. Useful in both rats and mice.
6. Does not require opening of the dura.
7. Produces anterior-posterior compression of the spinal cord.
8. Rapid release of the clip produces sudden impact of the spinal cord.
9. Control of time and force of application.
10. Consistent graded injuries can be produced. Both the force and duration of compression can be altered to model injuries of low to high severity, and reliable dose-response curves can be generated for both force and duration of compression.
11. Stabilization and precise positioning of the spine during injury is not required.
12. The same clip can be mounted with different rings to produce injuries of varying severity.
13. Very low mortality and morbidity rates.

Disadvantages

1. After hundreds of openings and closings the springs upon which the force of the clips depends will weaken. Thus, it is wise to calibrate the force of compression before and after each experiment. Calibration is not very complicated and the method of calibration, including a description of the calibration device has been published [29]. Alternatively, the clips can be sent to the Tator or Fehlings laboratories for calibration.
2. Practice and some surgical expertise are required for optimal and consistent placement of the clip around the cord with minimal cord trauma.
3. Laminectomy is required for clip placement.

Other Impact-Compression or Contusion Models

Several other impact-compression models were developed after the acute clip impact-compression model. Although the New York University (NYU) weight-drop impactor [30] is an improvement over the original Allen [31] weight-drop model, none of the weight-drop models produce persisting compression, and thus, they do not simulate the impact-compression of the most common human SCIs. Also, there is inherent variation and wide standard deviations in the results probably because of the randomness with which the weight contacts the walls of the cylinder and the impounder placed on the surface of the cord. The Ohio State University electromagnetic spinal cord injury device [32] is based on cord displacement which can be accurately measured. The same is true of the Infinite Horizons impactor [33, 34], but it is expensive, impacts the spinal cord from the dorsal surface only, has limited dwell-times (i.e. persisting compression for only up to 60 s), and requires both stabilization by forceps and horizontal positioning of the spinal column at the time of SCI. However, the Infinite Horizons device has the advantages of being able to produce SCI of varying displacement and force in rats [34] and mice [35]. The clip compression model is easily adaptable for use at the cervical level, which is not true for several of the contusion models where the adjacent and overlying spinal musculature can interfere with lesion making [36]. Recently, the Ohio State electromagnetic SCI device was modified to produce a consistent and graded hemi-contusion injury at cervical level C4 [37]. In this model, the electromagnetic probe is lowered to the surface of the cord with the same starting force and then a high velocity spinal contusion is produced [37]. Establishing a common starting point for the injury probe reduces variability in impact force.

The temporal effect of impact-compression injury is not achieved in most other contusion or compression models, including the Infinite Horizons model which has a limited dwell time. In our previous studies with the clips, we examined these temporal effects after cervical cord SCI in the rat [10, 11]. The ability to study varying compression times is an advantage of the clip technique as it can be used to examine the duration of cord compression and the value of early decompression which are important issues in human SCI [2]. The clip can remain in place compressing the spinal cord for clinically relevant durations of compression.

Non-impact and Non-contusion Models of SCI

There are many non-impact and non-contusion models, and only a fraction of them will be discussed. A large number of complete and incomplete transection models have been used to great advantage to examine finite pathophysiological events, but laceration and transection are rare in human SCI. Photochemical injury delivered by irradiation of the spinal cord after intravenous administration of the photosensitive rose bengal dye [38–40] produces ischemia and graded infarction of the spinal cord and is useful to examine post-traumatic ischemia and infarction which are common secondary mechanisms after SCI [1, 41, 42]. Also, this model is less invasive because irradiation can reach the spinal cord without laminectomy. Some compression models involve "placement" of a static weight on the cord, and thus do not cause sudden impact [43], and therefore, these have been considered non-impact.

Balloon compression models were first introduced by Tarlov [44] in the 1950s and there have been many subsequent versions. Most are slowly compressing, and non-impact, although the more recent balloon materials can be rapidly inflated without bursting. In general, balloon compression models suffer from a lack of reproducibility due to the uncertainty of the balloon-cord interface.

Forceps compression of specific tracts can be achieved with forceps blades that are inserted into the cord and slowly closed. Most common is forceps compression of the dorsal columns resulting in damage to the ascending sensory axons and descending corticospinal tract [45]. If the forceps blades are rapidly closed, impact-compression can be achieved, although closing force is not consistent or calibrated. These partial lesions are also difficult to produce reliably and consistently, and are dependent on proper placement of the forceps blades.

Methods for Clip Impact-Compression Injury

The following description applies specifically to rats.

Anesthesia

Rats can be anesthetized with a variety of techniques. We have found that the combination of 1.5–2% isofluorane and oxygen produces rapid induction, good analgesia, and lack of movement during SCI. Rats are placed on a heating pad at 36 °C during surgery.

Microscopy

After the skin incision is made the entire dissection, clip placement and removal are performed with a surgical operating microscope.

Laminectomy

A laminectomy is performed at the desired cervical, thoracic or lumbar level. To facilitate the laminectomy a self-retaining retractor is essential for holding laterally the paraspinal muscles (Fig. 1a). Failure to retract the muscles sufficiently far laterally impairs accurate and atraumatic clip placement. We usually remove two laminae to facilitate dissection of the extradural plane. The laminae are removed as far laterally as possible (Fig. 1b). Usually, this involves removal of the pedicles as far anteriorly as possible so that the stumps of the pedicles will not obstruct the passage of the clip (Fig. 1b). Bone removal is performed with a small bone-cutting instrument or rongeur with slender tapering ends (Fig. 2). A nail cutting instrument is ideal. Rongeurs with larger blades work better for laminectomy in the cervical region.

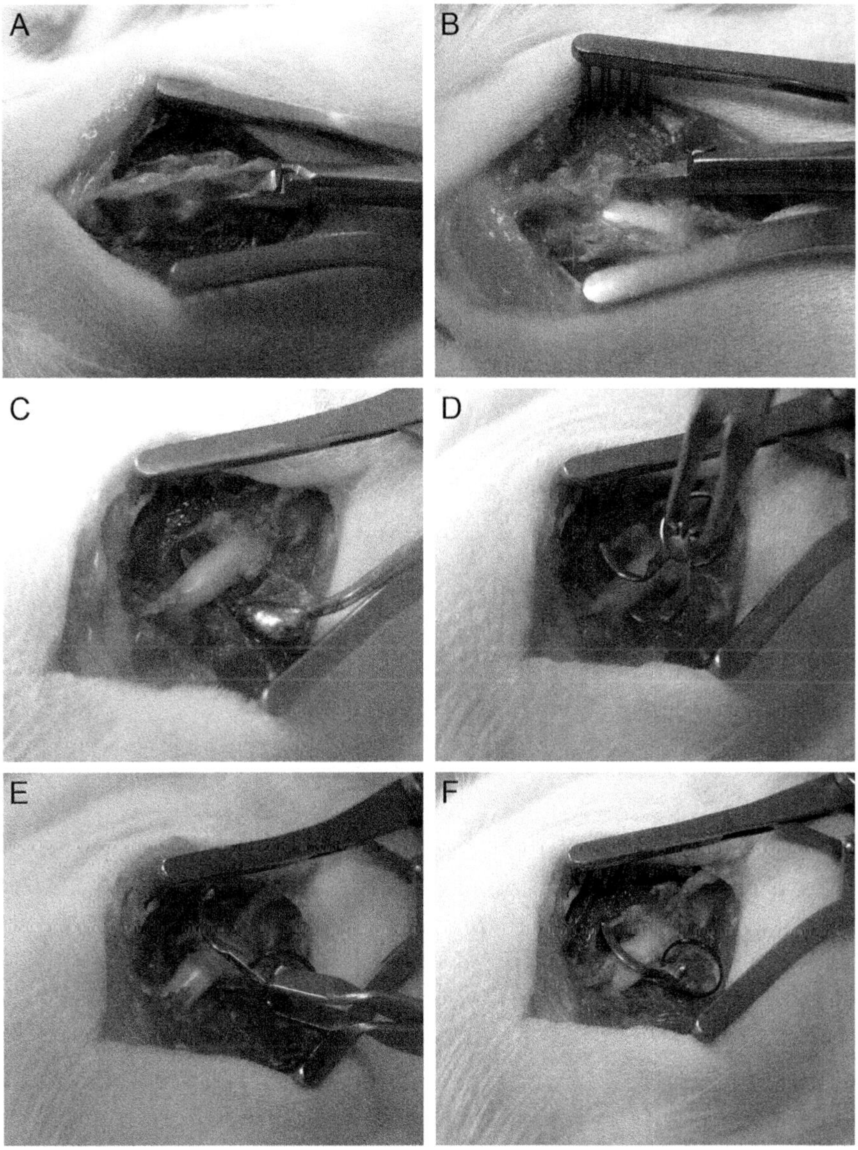

Fig. 1 The operative procedure. The steps are shown in a rat that has been perfused. (**a**) shows the self-retaining retractor in place holding back the paraspinal muscles. (**b**) shows the rongeur removing portions of the pedicle far laterally and anteriorly. (**c**) shows the cord exposed after the two-level laminectomy, and with the passage of the probe anteriorly and extradurally. (**d**) shows the clip held in the clip applicator, with the clip in the fully opened position, just prior to passage of the blade of the clip extradurally. (**e**) shows the blades of the clip anterior and posterior to the cord in the extradural space just prior to release of the clip from the applicator. The clip is still in the fully opened position. (**f**) the clip has been released from the applicator and is compressing the cord

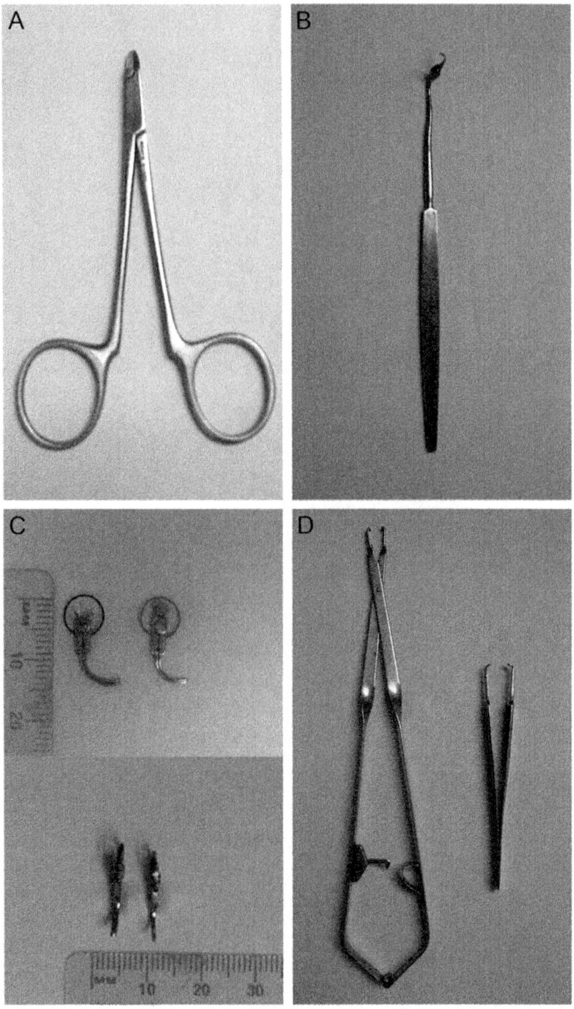

Fig. 2 The instruments for performing clip compression and the clips. (**a**) The bone cutting instrument (rongeur) for removing the laminae and pedicles. (**b**) The probe for exploring the extradural plane is a dental hook bent to conform to the shape of a clip blade. (**c**) The upper panel is a side view of the clips and the lower panel is a top view. In both the upper and lower panels, a rat clip is on the left and a mouse clip is on the right. The length of the blades of the mouse clip is smaller, and the blades and spring of the mouse clip are thinner. (**d**) Two styles of clip applicators. The applicators have ledges at the ends that fit into the grooves at the end of the clip blades to permit firm grasping of the clips

Exploration of the Extradural Plane

A dissecting hook, (Figs. 1c and 2) with similar curvature and thickness as the clip, is used to dissect the extradural plane between the dura and the adjacent vertebrae. The clips, dissecting hook, and clip applicators are manufactured to specification (Fig. 2). The hook is especially important for locating the nerve roots, and for clearing away any extradural adhesions at the optimal site for clip placement which is at the mid-point between adjacent roots. It is important to pass the hook extradurally from both the right and left sides to determine the optimal trajectory for clip placement. Bleeding will often occur at this stage from the epidural veins and can be controlled with application of small cotton pledgets and with gelfoam sponge. Bleeding must be controlled in order to have optimal visualization of the cord during clip placement.

Placement of the Clip

With the clip held in the applicator in its fully opened position (Fig. 1e), the anterior blade of the clip is passed extradurally anterior to the cord with avoidance of damage to the adjacent nerve roots, and with as little cord displacement as possible (Fig. 1e). Some displacement is necessary, but it should be done very slowly to avoid any traction or impact of the cord. The clip held in the applicator can be used to clear away any remaining adhesions that are encountered, but only minimal force can be used, otherwise the clip will be unintentionally and prematurely released from the applicator.

Rapid Release of the Clip to Deliver the Impact-Compression Injury

The clip is then rapidly released from the applicator to produce the acute impact-compression injury (Fig. 1f). This is done by allowing the blades of the applicator to spring open at maximum velocity. The clip is then left compressing the spinal cord for the desired duration of compression.

Removal of the Clip with the Applicator

After the desired interval, the applicator is re-introduced and used to grasp and then open the clip maximally without any further manipulation of the cord. The clip is then removed with as little cord displacement as possible.

Suturing of the Wound

After injury, the muscles are sutured in one layer using 3-0 sutures and the skin closed with either sutures or Michel clips (Fine Science Tool, B.C., Canada). Buprenorphine (0.03 mg/kg, Temgesic, Schering-Plough, U.K.) is given subcutaneously before the animals awake, and then every 12 h for 48 h.

Post-operative Care

The rats are housed singly in a temperature-controlled room at 26 °C for the duration of the experiment with a 12 h light/dark cycle. Bladders are expressed three times daily until spontaneous voiding occurs, and Clavamox is added to the drinking water to prevent urinary tract infection. Water and food are provided *ad libitum*.

Ordering Information

The clips and applicators for rats and mice can be ordered directly from the Tator or Fehlings laboratories. The rat clips are approximately $2000 each and the mouse clips are approximately $1300 each. They can be supplied in a variety of forces of clip closing strength.

The clips for mice known as Fejota (Fehlings-Joshi-Tator) clips and can be ordered from the Fehlings laboratory, and a signed Research Agreement is required. The University Health Network, the group of hospitals in which the Tator and Fehlings laboratories are located is the holder of the patent and trademark on the mouse clip.

Calibration of Clips

Investigators can perform their own force of clip closure calibrations or can request this service from the Tator or Fehlings laboratories.

Outcome Evaluations

A full range of histological (Fig. 3), imaging and functional outcome measures has been used successfully after clip impact-compression injury.

Summary

Our group has used the clip impact-compression injury model to injure the spinal cord at the cervical [46–48], thoracic [19, 20–22, 49, 50], and lumbar [15, 51] levels since the 1970s, and this model has given reliable and consistent results for studying acute, subacute and chronic SCI in rodents. Other groups have also successfully

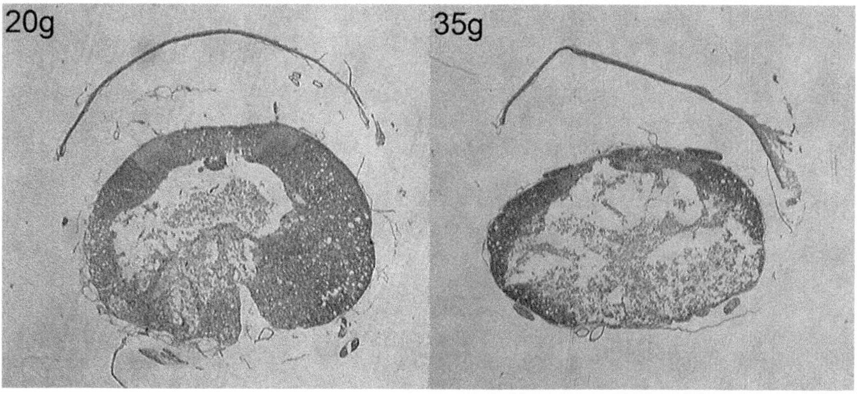

Fig. 3 Histological sections of the rat thoracic cord several weeks after clip compression injury. Hematoxylin and eosin and luxol fast blue staining. The central necrosis and cavitation are much greater after 35 g injury than after 20 g injury

used our clips for experimental SCI in rodents [17, 25, 52]. Several other clip designs were subsequently published and appear to give reliable results [53–56]. The clips have been used for examining a variety of injury mechanisms including inflammatory cytokines [57], secondary injury antagonists [26], and ischemia [12]. The clips have also been used for studying axonal physiology and morphology [58], and neuroprotective and regenerative measures with various agents including localized drug delivery [59–62], biomaterials and scaffolds [20, 63], and cell transplantation [19, 22, 49, 54, 64, 65].

Acknowledgments We thank the following sources for funding. Conquer Paralysis Now and the Sam Schmidt Foundation, Ontario Ministry of Research and Innovation and Medicine by Design, Ontario-China Research and Innovation Fund, Krembil Foundation, Toronto General and Western Hospital Foundation, and Spinal Cord Injury Ontario.

References

1. De Girolami U, Frosch MP, Tator CH. Regional neuropathology: diseases of the spinal cord and vertebral column. In: Greenfield's neuropathology. 7th edn. London: Arnold; 2002.
2. Tator CH. Review of treatment trials in human spinal cord injury: issues, difficulties, and recommendations. Neurosurgery. 2006;59:957–82.
3. Anderson TE, Stokes BT. Experimental models for spinal cord injury research: physical and physiological considerations. J Neurotrauma. 1992;9(Suppl 1):S135–42.
4. Kwon BK, Oxland TR, Tetzlaff W. Animal models used in spinal cord regeneration research. Spine. 2002;27(14):1504–10.
5. Grill RJ. User-defined variables that affect outcome in spinal cord contusion/compression models. Exp Neurol. 2005;196:1–5.
6. Fehlings MG, Tator CH. A review of experimental models of acute spinal cord injury. In: Spinal cord dysfunction: assessment. Oxford: Oxford University; 1988.

7. Wrathall JR. Spinal cord injury models. J Neurotrauma. 1992;9(Suppl 1):S129–34.
8. Tator CH. Spine-spinal cord relationships in spinal cord trauma. Clin Neurosurg. 1983;30: 479–94.
9. Rivlin AS, Tator CH. Effect of duration of acute spinal cord compression in a new acute cord injury model in the rat. Surg Neurol. 1978;10:38–43.
10. Dolan EJ, Tator CH, Endrenyi L. The value of decompression for acute experimental spinal cord compression injury. J Neurosurg. 1980;53:749–55.
11. Guha A, Tator CH, Endrenyi L, Piper I. Decompression of the spinal cord improves recovery after acute experimental spinal cord compression injury. Paraplegia. 1987;25:324–39.
12. Fehlings MG, Tator CH, Linden RD. The relationships among the severity of spinal cord injury, motor and somatosensory evoked potentials and spinal cord blood flow. Electroencephalogr Clin Neurophysiol. 1989;74:241–59.
13. Fehlings MG, Tator CH. The relationships among the severity of spinal cord injury, residual neurological function, axon counts, and counts of retrogradely labeled neurons after experimental spinal cord injury. Exp Neurol. 1995;132:220–8.
14. Poon PC, Gupta D, Shoichet MS, Tator CH. Clip compression model is useful for thoracic spinal cord injuries: histologic and functional correlates. Spine (Phila Pa 1976). 2007;32:2853–9. https://doi.org/10.1097/BRS.0b013e31815b7e6b.
15. Moonen G, Satkunendrarajah K, Wilcox JT, Badner A, Mothe A, Foltz W, Fehlings MG, Tator CH. A new acute impact-compression lumbar spinal cord injury model in the rodent. J Neurotrauma. 2016;33(3):278–89. https://doi.org/10.1089/neu.2015.3937.
16. Geisler FH, Coleman WP, Grieco G, Poonian D. Measurements and recovery patterns in a multicenter study of acute spinal cord injury. Spine. 2001;26(24 Suppl):S68–86.
17. Weaver LC, Verghese P, Bruce JC, Fehlings MG, Krenz NR, Marsh DR. Autonomic dysreflexia and primary afferent sprouting after clip-compression injury of the rat spinal cord. J Neurotrauma. 2001;18:1107–19.
18. Forgione N, Karadimas SK, Foltz WD, Satkunendrarajah K, Lip A, Fehlings MG. Bilateral contusion-compression model of incomplete traumatic cervical spinal cord injury. J Neurotrauma. 2014;31:1776–88. https://doi.org/10.1089/neu.2014.3388.
19. Mothe AJ, Bozkurt G, Catapano J, Zabojova J, Wang X, Keating A, Tator CH. Intrathecal transplantation of stem cells by lumbar puncture for thoracic spinal cord injury in the rat. Spinal Cord. 2011;49:967–73. https://doi.org/10.1038/sc.2011.46.
20. Mothe AJ, Tam RY, Zahir T, Tator CH, Shoichet MS. Repair of the injured spinal cord by transplantation of neural stem cells in a hyaluronan-based hydrogel. Biomaterials. 2013;34:3775–83. https://doi.org/10.1016/j.biomaterials.2013.02.002.
21. Parr AM, Kulbatski I, Tator CH. Transplantation of adult rat spinal cord stem/progenitor cells for spinal cord injury. J Neurotrauma. 2007;24:835–45.
22. Parr AM, Kulbatski I, Zahir T, Wang X, Yue C, Keating A, Tator CH. Transplanted adult spinal cord-derived neural stem/progenitor cells promote early functional recovery after rat spinal cord injury. Neuroscience. 2008;155:760–70. https://doi.org/10.1016/j.neuroscience.2008.05.042.
23. Joshi M, Fehlings MG. Development and characterization of a novel, graded model of clip compressive spinal cord injury in the mouse: part 2. Quantitative neuroanatomical assessment and analysis of the relationships between axonal tracts, residual tissue, and locomotor recovery. J Neurotrauma. 2002;19:191–203.
24. Joshi M, Fehlings MG. Development and characterization of a novel, graded model of clip compressive spinal cord injury in the mouse: part 1. Clip design, behavioral outcomes, and histopathology. J Neurotrauma. 2002;19:175–90.
25. Khan M, Griebel R, Rozdilsky B, Politis M. Hemorrhagic changes in experimental spinal cord injury models. Can J Neurol Sci. 1985;12:259–62.
26. Gorio A, Gokmen N, Erbayraktar S, Yilmaz O, Madaschi L, Cichetti C, Di Giulio AM, Vardar E, Cerami A, Brines M. Recombinant human erythropoietin counteracts secondary injury and markedly enhances neurological recovery from experimental spinal cord trauma. Proc Natl Acad Sci U S A. 2002;99:9450–5.

27. von Euler M, Seiger A, Sundstrom E. Clip compression injury in the spinal cord: a correlative study of neurological and morphological alterations. Exp Neurol. 1997;145:502–10.
28. Marques SA, Garcez VF, Del Bel EA, Martinez AM. A simple, inexpensive and easily reproducible model of spinal cord injury in mice: morphological and functional assessment. J Neurosci Methods. 2009;177:183–93. https://doi.org/10.1016/j.jneumeth.2008.10.015.
29. Dolan EJ, Tator CH. A new method for testing the force of clips for aneurysms or experimental spinal cord compression. J Neurosurg. 1979;51:229–33.
30. Gruner JA. A monitored contusion model of spinal cord injury in the rat. J Neurotrauma. 1992;9:123–6.
31. Allen AR. Surgery of experimental lesions of the spinal cord equivalent to crush injury of fracture dislocation of the spinal column. A preliminary report. JAMA. 1911;57:878–80.
32. Behrmann DL, Bresnahan JC, Beattie MS, Shah BR. Spinal cord injury produced by consistent mechanical displacement of the cord in rats: behavioral and histologic analysis. J Neurotrauma. 1992;9:197–217.
33. Cao Q, Zhang YP, Iannotti C, DeVries WH, Xu XM, Shields CB, Whittemore SR. Functional and electrophysiological changes after graded traumatic spinal cord injury in adult rat. Exp Neurol. 2005;191(Suppl 1):S3–S16.
34. Scheff SW, Rabchevsky AG, Fugaccia I, Main JA, Lumpp JE Jr. Experimental modeling of spinal cord injury: characterization of a force-defined injury device. J Neurotrauma. 2003;20:179–93.
35. Ghasemlou N, Kerr BJ, David S. Tissue displacement and impact force are important contributors to outcome after spinal cord contusion injury. Exp Neurol. 2005;196:9–17.
36. Pearse DD, Lo TP Jr, Cho KS, Lynch MP, Garg MS, Marcillo AE, Sanchez AR, Cruz Y, Dietrich WD. Histopathological and behavioral characterization of a novel cervical spinal cord displacement contusion injury in the rat. J Neurotrauma. 2005;22:680–702.
37. Mondello SE, Sunshine MD, Fischedick AE, Moritz CT, Horner PJ. A cervical hemi-contusion spinal cord injury model for the investigation of novel therapeutics targeting proximal and distal forelimb functional recovery. J Neurotrauma. 2015;32:1994–2007. https://doi.org/10.1089/neu.2014.3792.
38. Watson BD, Prado R, Dietrich WD, Ginsberg MD, Green BA. Photochemically induced spinal cord injury in the rat. Brain Res. 1986;367:296–300.
39. Bunge MB, Holets VR, Bates ML, Clarke TS, Watson BD. Characterization of photochemically induced spinal cord injury in the rat by light and electron microscopy. Exp Neurol. 1994;127:76–93.
40. Verdu E, Garcia-Alias G, Fores J, Vela JM, Cuadras J, Lopez-Vales R, Navarro X. Morphological characterization of photochemical graded spinal cord injury in the rat. J Neurotrauma. 2003;20:483–99.
41. Tator CH. Strategies for recovery and regeneration after brain and spinal cord injury. Inj Prev. 2002;8(Suppl 4):IV33–6.
42. Tator CH. Update on the pathophysiology and pathology of acute spinal cord injury. Brain Pathol. 1995;5:407–13.
43. Koda M, Nishio Y, Hashimoto M, Kamada T, Koshizuka S, Yoshinaga K, Onodera S, Nishihira J, Moriya H, Yamazaki M. Up-regulation of macrophage migration-inhibitory factor expression after compression-induced spinal cord injury in rats. Acta Neuropathol (Berl). 2004;108:31–6.
44. Tarlov IM. Spinal cord compression: mechanisms of paralysis and treatment. Springfield: Thomas; 1957.
45. Gruner JA, Yee AK, Blight AR. Histological and functional evaluation of experimental spinal cord injury: evidence of a stepwise response to graded compression. Brain Res. 1996;729:90–101.
46. Casha S, Yu WR, Fehlings MG. Oligodendroglial apoptosis occurs along degenerating axons and is associated with FAS and p 75 expression following spinal cord injury in the rat. Neuroscience. 2001;103:203–18.

47. Midha R, Fehlings MG, Tator CH, Saint-Cyr JA, Guha A. Assessment of spinal cord injury by counting corticospinal and rubrospinal neurons. Brain Res. 1987;410:299–308.
48. Park YK, Tator CH. Failure of topical DMSO to improve blood flow or evoked potentials in rat spinal cord injury. J Korean Med Sci. 1998;13:638–44.
49. Fan XY, Mothe AJ, Tator CH. Ephrin-B3 decreases the survival of adult rat spinal cord-derived neural stem/progenitor cells in vitro and after transplantation into the injured rat spinal cord. Stem Cells Dev. 2013;22:359–73. https://doi.org/10.1089/scd.2012.0131.
50. Namiki J, Kojima A, Tator CH. Effect of brain-derived neurotrophic factor, nerve growth factor, and neurotrophin-3 on functional recovery and regeneration after spinal cord injury in adult rats. J Neurotrauma. 2000;17:1219–31.
51. Park YK, Tator CH. Prevention of arachnoiditis and postoperative tethering of the spinal cord with Gore-Tex surgical membrane: an experimental study with rats. Neurosurgery. 1998;42:813–23.
52. Schultke E, Kendall E, Kamencic H, Ghong Z, Griebel RW, Juurlink BH. Quercetin promotes functional recovery following acute spinal cord injury. J Neurotrauma. 2003;20:583–91.
53. Baffour R, Achanta K, Kaufman J, Berman J, Garb JL, Rhee S, Friedmann P. Synergistic effect of basic fibroblast growth factor and methylprednisolone on neurological function after experimental spinal cord injury. J Neurosurg. 1995;83:105–10.
54. Boyd JG, Lee J, Skihar V, Doucette R, Kawaja MD. LacZ-expressing olfactory ensheathing cells do not associate with myelinated axons after implantation into the compressed spinal cord. Proc Natl Acad Sci U S A. 2004;101:2162–6.
55. Kureshi IU, Ho SY, Onyiuke HC, Wakefield AE, D'Arrigo JS, Simon RH. The affinity of lipid-coated microbubbles to maturing spinal cord injury sites. Neurosurgery. 1999;44:1047–53.
56. Saporta S, Kim JJ, Willing AE, Fu ES, Davis CD, Sanberg PR. Human umbilical cord blood stem cells infusion in spinal cord injury: engraftment and beneficial influence on behavior. J Hematother Stem Cell Res. 2003;12:271–8.
57. Fu ES, Saporta S. Methylprednisolone inhibits production of interleukin-1beta and interleukin-6 in the spinal cord following compression injury in rats. J Neurosurg Anesthesiol. 2005;17:82–5.
58. Nashmi R, Fehlings MG. Changes in axonal physiology and morphology after chronic compressive injury of the rat thoracic spinal cord. Neuroscience. 2001;104:235–51.
59. Elliott Donaghue I, Tator CH, Shoichet MS. Sustained delivery of bioactive neurotrophin-3 to the injured spinal cord. Biomater Sci. 2015;3:65–72. https://doi.org/10.1039/c4bm00311j.
60. Gupta D, Tator CH, Shoichet MS. Fast-gelling injectable blend of hyaluronan and methylcellulose for intrathecal, localized delivery to the injured spinal cord. Biomaterials. 2006;27:2370–9.
61. Jimenez Hamann MC, Tator CH, Shoichet MS. Injectable intrathecal delivery system for localized administration of EGF and FGF-2 to the injured rat spinal cord. Exp Neurol. 2005;194:106–19.
62. Kang CE, Baumann MD, Tator CH, Shoichet MS. Localized and sustained delivery of fibroblast growth factor-2 from a nanoparticle-hydrogel composite for treatment of spinal cord injury. Cells Tissues Organs. 2013;197:55–63. https://doi.org/10.1159/000339589.
63. Bozkurt G, Mothe AJ, Zahir T, Kim H, Shoichet MS, Tator CH. Chitosan channels containing spinal cord-derived stem/progenitor cells for repair of subacute spinal cord injury in the rat. Neurosurgery. 2010;67:1733–44. https://doi.org/10.1227/NEU.0b013e3181f9af35.
64. Karimi-Abdolrezaee S, Eftekharpour E, Wang J, Morshead CM, Fehlings MG. Delayed transplantation of adult neural precursor cells promotes remyelination and functional neurological recovery after spinal cord injury. J Neurosci. 2006;26:3377–89.
65. Roussos I, Rodriguez M, Villan D, Ariza A, Rodriguez L, Garcia J. Development of a rat model of spinal cord injury and cellular transplantation. Transplant Proc. 2005;37:4127–30.

Method of Spinal Cord Contusion Injury Created by Tissue Displacement

Yi Ping Zhang, Lisa B. E. Shields, Xiao-Ming Xu, and Christopher B. Shields

Abstract Experimental contusion spinal cord injury (SCI) is the most frequently used *in vivo* model to provide insight into the mechanisms of secondary injury. Many criteria have been utilized to control the severity of SCI, including the height of the weight drop, the magnitude of the force striking the spinal cord, and the degree of spinal cord tissue displacement. The tissue displacement method is the most reliable technique of producing experimental SCI after the target spine has been stabilized. We have described the creation of a contusion SCI model using the LISA that is versatile and may be adapted for use in mice, rats, and monkeys with minor adjustments in the mechanical settings. Reproducible and reliable contusion SCI may be produced using this method of tissue displacement that may be characterized by gradable behavioral and morphological assessment.

Keywords Spinal cord injury · Contusion SCI model · Louisville Injury System Apparatus (LISA) · Spinal cord tissue displacement injury

---

Y. P. Zhang · L. B. E. Shields (✉)
Norton Neuroscience Institute, Norton Healthcare, Louisville, KY, USA
e-mail: LBES@earthlink.net

X.-M. Xu
Department of Neurological Surgery and Goodman and Campbell Brain and Spine, Spinal Cord and Brain Injury Research Group, Stark Neurosciences Research Institute, Indiana University School of Medicine, Indianapolis, IN, USA
e-mail: xu26@iupui.edu

C. B. Shields (✉)
Norton Neuroscience Institute, Norton Healthcare, Louisville, KY, USA

Department of Anatomical Sciences and Neurobiology, University of Louisville School of Medicine, Louisville, KY, USA

J. Chen et al. (eds.), *Animal Models of Acute Neurological Injury*, Springer Series in Translational Stroke Research, https://doi.org/10.1007/978-3-030-16082-1_34

Introduction

In vivo studies of spinal cord injury (SCI) provide insights into the mechanisms of SCI and assess efficacy of putative therapies. A reliable contusion SCI model should be reproducible, versatile, and clinically relevant. The model must be versatile to produce gradable injuries on different spinal cord segments in multiple species. Reproducibility depends on the accuracy of the injury. Methods to produce a model of contusive SCI include: (1) free drop using different weights and heights (NYU Impactor) [1, 2]; (2) force as the major mechanism (IH System) [3]; (3) controlled tissue displacement (ESCID) [4, 5] and (4) the modified clip [6]. These techniques have been developed specifically for certain species at specified spinal cord segments and are frequently not interchangeable with other animals or at different segmental levels. For example, the NYU device uses a weight drop technique combined with vertebral stabilization to create a rat thoracic SCI. The NYU device produces an excessively severe SCI. Vertebral fixation using the NYU system is accomplished by fixation of the cervical spinous processes (C3-T1) that are extremely short and friable. Thus, the NYU weight drop method designed to produce a SCI in the rat can not be used in mice. A force may be applied to the spinal cord using a highly sensitive force sensor in the IH system that creates a SCI in rats. The IH system becomes less reliable in mice since the force sensor can not accurately record small signals created by the injury force compared to a large noise artifact generated by moving gears.

The development of accurate murine models in SCI research has become paramount as genetically modified mice advance our understanding of molecular mechanisms of SCI. Tissue displacement models used to create experimental SCI avoid the myriad aforementioned pitfalls. Tissue displacement injury methods are reliable and versatile in creating experimental SCI in several animal species and on all spinal cord segments. The ESCID device was the first to create an experimental SCI based on tissue displacement at the injury site that demonstrated an excellent correlation between injury severity, histology, and behavior in the rat [4] and mouse [7]. The Louisville Injury System Apparatus (LISA) contusion model was developed using tissue displacement to create a precise, controllable, and gradable SCI. The tissue displacement method utilizes advanced technology incorporating a laser distance sensor.

Table 1 Characteristics of the Louisville Injury System Apparatus (LISA)

Generates high velocity impacts (up to ≥1.0 m/s) that replicate clinical injury severities
Allows gradable injuries
May be utilized on the thoracic and cervical spinal cord
Produces contusion SCI in mice, rats, and monkeys
Uses different plunger tip sizes to match the spinal cord diameters

The characteristics of the LISA are presented in Table 1. The development of a novel method of vertebral stabilization was necessary to avoid motion of the spine at the moment of impact that may distort the anticipated severity of tissue displacement. In this chapter, we will describe: (1) spine stabilization of targeted vertebrae and (2) a method to create a spinal cord contusion injury based on tissue displacement using the LISA.

Materials

Surgical Preparation

1. *Animals:* Contusion SCI created by the spinal cord tissue displacement method using the LISA may be performed on mice (>17 g), adult Sprague-Dawley (SD) or nude rats (>160 g) and adult rhesus monkeys (5–7 kg). Different sizes of spine stabilizers are available to prevent vertebral motion during distance measurement and performance of the SCI (Fig. 1).

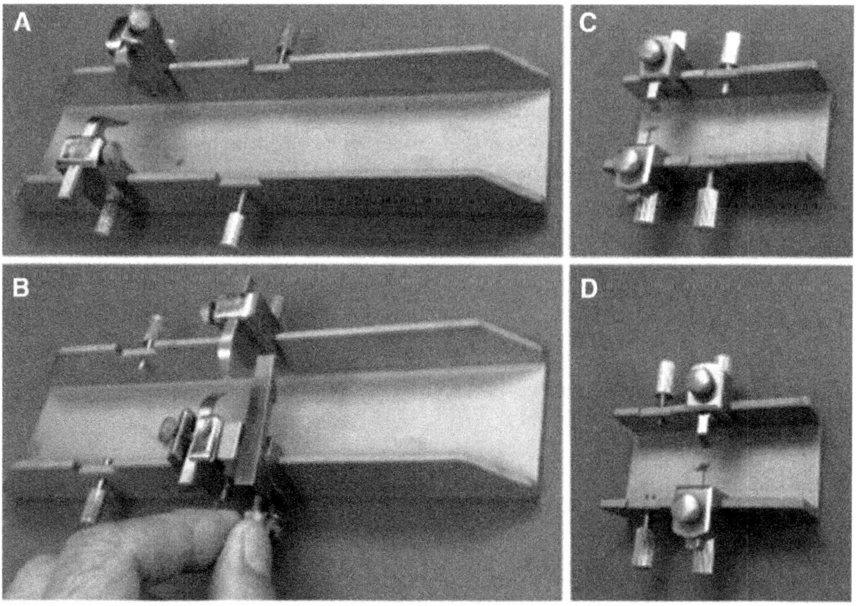

Fig. 1 Stabilizer for (**a**) cervical and (**b**) thoracic vertebral fixation in the rat. Mouse stabilizer for (**c**) cervical and (**d**) thoracic spine fixation. The rodent lies in the trough. In mice an identical pair of steel arms are used for cervical and thoracic spine fixation. Two lengths of arms are utilized for rats: deep ones for the cervical and shallow ones for thoracic vertebrae. (**b**) A metal leveler is attached to the trough to place the spinal cord in a horizontal plane. The leveler is fixed in the desired position by tightening the thumb screws

2. *Anesthesia:* A ketamine and Xylazine mixture is suitable for rats and non-human primates, and Tribromoethanol (Avertin®) is used for mice. Anesthesia complications are more common in nude rats and, therefore, inhalation anesthetics such as isoflurane are utilized. An anesthesia chamber is used for induction of the rat. The rat wears with a gas mask throughout the duration of the procedure.
3. *Equipment and supplies*: Supplies include surgical gloves, #22 and #30 gauge needles, syringes, 4-0 and 5-0 sutures with a swaged needle, #10 and #15 scalpel blades, cotton-tipped applicators, antibiotic ointment, and a device to maintain body temperature at 37 °C. A surgical microscope equipped with a 200–300 mm working distance objective lens is required. Surgical tools include scissors, micro spring-scissors, tissue forceps, micro-retractor and micro-rongeurs for the rat, a Kerrison rongeur for the monkey, and a special forceps-retractor and vertebral forceps for the mouse to facilitate vertebral exposure. Post-operative care requires maintenance of body temperature (37 °C) using a heating pad. Animals are maintained in a clean cage with accessible food and water.

Contusion Device and Technique to Produce a Tissue Displacement Injury

1. *Equipment:* The LISA system is controlled by specially designed software on a personal computer (Fig. 2). The impact velocity can be varied by adjusting the pressure of compressed air that drives the pneumatic piston.
2. *Mechanism to measure the magnitude of tissue* displacement: A detailed method to create a precise contusion SCI in rodents is described in Fig. 3. The same method is used to produce contusion SCI in monkeys with slight modifications. The plunger has direct contact with the spinal cord in rodents, while the injury is

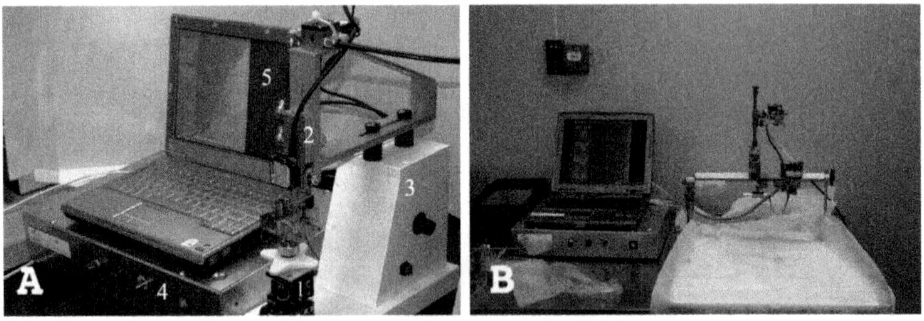

Fig. 2 (**a**) LISA contusion setup for the mouse consisting of a (1) stage, (2) plunger complex that holds a distance sensor that emits a laser beam to measure tissue displacement, (3) base, (4) control box, and (5) LISA software on the computer. A force sensor is placed under the mouse stabilizer to measure force. (**b**) Setup for the monkey is slightly different. The plunger complex is raised to facilitate animal placement under the frame

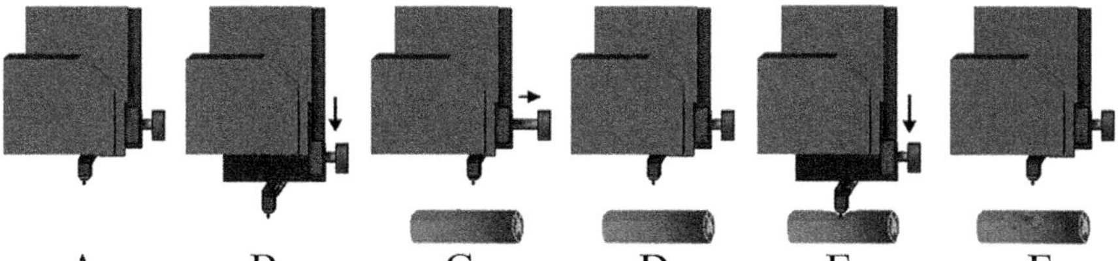

Fig. 3 Cartoons illustrate the mechanism of spinal cord contusion generated by tissue displacement. (**a**) Original position and (**b**) extended position used to calibrate the plunger distance by a laser beam projected to the upper surface of the tip (corresponds to Fig. 4a, c). (**c**) Lateral shift of the tip from the path of the laser beam (corresponds to Fig. 4b) allows measurement of the distance from the laser sensor to the spinal cord and determines the magnitude of tissue displacement. (**d**) By moving the tip back into the laser beam path, the system is prepared to create the contusion. (**e**) The plunger is released to the extended position to contuse the spinal cord. (**f**) The tip is withdrawn from the spinal cord surface after contusion

Table 2 Diameter of plunger tip in various species

Species	Transverse diameter of spinal cord (mm)	Diameter of plunger tip (mm)
Mouse	1.7	1.1
Rat	3.2	2.1
Monkey	5.0	3.2

mediated using an extension rod to reach the deeper location of the spinal cord in monkeys.

The LISA is able to produce a spinal cord contusion in mice, rats, and monkeys. Since the transverse diameter of the spinal cords is different in each species, the diameter of the plunger tip should vary so that it comes in contact with 2/3 of spinal cord diameter (Table 2). It is a simple process to change the diameter of the plunger tip of the LISA device with different species.

Spine Stabilization

The spine stabilizer attaches to the facets bilaterally to fixate the vertebra during creation of a spinal cord contusion. Vertebral stabilization is mandatory for tissue displacement models of SCI since minimal motion of the spine at the moment of impact will affect injury accuracy. Most experimental SCI systems attach to the spinous processes, but our stabilizer clamps to the facets bilaterally prior to the laminectomy. Advantages of facet fixation include: (1) greater stabilization of the cervical spine, (2) avoiding the potential trauma that may be caused by excessive motion of adjacent vertebrae, (3) increased precision of the laminectomy, and (4) elevation of the chest with thoracic spinal facet fixation that prevents respiratory compromise.

Method to Create a Contusion SCI

LISA Device Preparation

1. Turn on the LISA control box, PC, and LISA software (Fig. 2).
2. Aim the laser beam at the center of the top surface of the plunger tip.
3. Set the desired impact velocity (we prefer 1–1.5 m/s) by adjusting the gas pressure. The gas pressure controls the impact velocity. The pressure should be set at approximately 50PSI to prevent possible damage to the pneumatic parts of the device.

Steps

1. Anesthesia is administered intravenously, intraperitoneally, or by gas based on the duration of the procedure. Our preferred anesthetics and doses are: (1) Tribromoethanol (Avertin®) 250 mg/kg via IP injection in mice, (2) Ketamine 90 mg/kg and Xylazine 10 mg/kg mixture via IP injection in rats, and (3) Ketamine 10–15 mg/kg supplemented by xylazine 1.0 mg/kg IM in monkeys. If the procedure takes longer than 25 min, ½ of the initial dose may be added as a supplement. Four percent isoflurane is used for induction and 1% isoflurane for maintenance (30% oxygen in room air). Gas anesthesia is preferred in rats or other animals that require rapid recovery. Under gas anesthesia the animals are allowed to breathe spontaneously through a mask and usually do not require mechanical ventilation.
2. Preparation for surgery requires that hair over the surgical site is shaved and skin is prepped with Betadine solution. Both eyes are covered with ophthalmic ointment (CO) to prevent corneal desiccation. Physiological monitoring (SSEP, MEP) is recommended. An intravenous line is required for surgery in monkeys.
3. Exposure and vertebral stabilization: A midline skin incision is performed using small scissors on mice and a scalpel on rats and monkeys. Paravertebral muscles are dissected from spinous processes and laminar arches bilaterally. Anatomical differences exist between rodents and monkeys and between the thoracic and cervical spine. An incision is made between the trapezius muscles in the neck up to C2 to expose the cervical spine in rodents. Bilateral submuscular fat pads (also called the hibernating gland) are identified caudally, and dissection is performed between them to avoid bleeding. The semispinalis capitis and other paraspinal muscles are dissected down to the T2 spinous process that serves as a reliable landmark. A pair of micro-scissors dissects muscles from the spinous processes and laminae. Muscle dissection begins adjacent to the spinous processes and extends bilaterally with exposure of the lateral facets. Exposure of the cervical vertebrae in monkeys is similar but without a submuscular adipose layer to facilitate exposure. To expose the T9-10 vertebrae in rodents, subcutaneous adipose tissue is pushed rostrally followed by dissection of paraspinal muscles from the spinous processes and laminar arches. Exposure of the thoracic vertebrae in

monkeys involves a similar dissection of the paraspinal muscles bilaterally. After vertebral exposure, the spine is prepared for vertebral stabilization. The accuracy of surgical manipulation is greatly increased after the vertebra is stabilized. The rodent cervical spine is concave, and the thoracic spine is convex. The target spinal cord at each site is placed in a horizontal position using the stabilizing device. In the cervical area the vertebrae must be elevated to eliminate concavity, whereas in the thoracic spine the vertebrae must be depressed to eliminate the convexity. After spinal fixation is complete, a laminectomy is performed to expose the dural tube and spinal cord. The lamina is removed using a micro-rongeur in rodents, and Kerrison rongeurs or a high-speed drill (3 mm diameter fluted ball tip, Medtronic, Memphis, TN) are utilized in monkeys. The underlying ligamentum flavum in rodents is thin and can be easily pushed away from the dura while it is thick in monkeys.

4. The method of creating the contusion SCI using the LISA include: setting the zero level, predicting the injury magnitude, positioning the impactor using the laser sensor, and creating the displacement contusion injury is described in detail in Fig. 4.

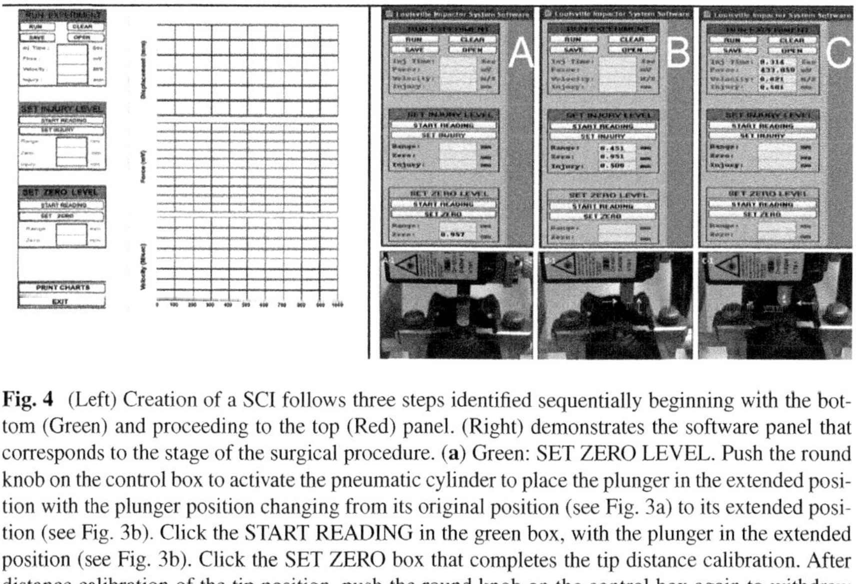

Fig. 4 (Left) Creation of a SCI follows three steps identified sequentially beginning with the bottom (Green) and proceeding to the top (Red) panel. (Right) demonstrates the software panel that corresponds to the stage of the surgical procedure. (**a**) Green: SET ZERO LEVEL. Push the round knob on the control box to activate the pneumatic cylinder to place the plunger in the extended position with the plunger position changing from its original position (see Fig. 3a) to its extended position (see Fig. 3b). Click the START READING in the green box, with the plunger in the extended position (see Fig. 3b). Click the SET ZERO box that completes the tip distance calibration. After distance calibration of the tip position, push the round knob on the control box again to withdraw the plunger back to its original position (**b**). Blue: SET INJURY LEVEL. Loosen the tightening screw to shift the tip laterally (see Fig. 4c) from the laser beam path to measure the distance from the laser to the spinal cord. After the laser beam is focused on the target, the START READING box is clicked and the animal is elevated using the Z-axis micro-driver until the required displacement level is reached. Click the SET INJURY box that sets the injury level at required displacement (see Fig. 3c). (**c**) Red: RUN EXPERIMENT. Shift and lock the plunger tip back into the path of the laser beam (see Fig. 3d). Click the RUN box to activate the plunger to strike the spinal cord to produce the contusion (see Fig. 3e). The dwell time of the plunger against the spinal cord can be controlled by the time relay setting on the control box. The plunger is then automatically removed from the spinal cord based on the predetermined contact duration (see Fig. 3f). The data is saved on the computer software

5. Post-contusion verification and data collection: The spine stabilizer is removed from the injury device, and the lesion is visually verified under the surgical microscope. If there is no active hemorrhage at the site of the contusion, the animal is removed from the stabilizer and the wound is closed in layers. Any active hemorrhage is controlled by placing gelfoam (Pfizer, New York, NY) and/or cotton tips to induce hemostasis. The injury parameters including the magnitude of tissue displacement, plunger velocity, and plunger-cord contact and duration is stored.

Animal Care Following SCI

Postoperative care depends on the type and species of the animal as well as the level and injury severity. Body temperature, respiratory support, and prevention of dehydration are essential. Airway function is assessed by monitoring the rate and depth of respiration to minimize airway obstruction, dyspnea, and hypoxemia. Cyanosis of the paw and nose indicates poor ventilation and hypoxia. Secretions are suctioned from the mouth and trachea that may stimulate respiratory effort. Artificial ventilation is occasionally necessary. Dehydration must be avoided postoperatively. Adequate hydration is obtained by subcutaneous injection of 10 mL in rats and 1 mL in mice. Dehydration is managed by an intravenous injection of 100 mL saline in monkeys. Hypothermia may be caused by anesthesia and poor temperature regulation following SCI. Normal body temperature may be maintained using a heating pad. Postoperative animal maintenance includes soft, nutritious food, bladder expression, and skin inspection to observe for bed sores and autophagy [8].

Calibration of the LISA Device

Calibration must be performed regularly or prior to each study. The proper alignment of the plunger tip and the laser beam must be verified. The reliability of the graded displacements is confirmed by placing a piece of playdough on the spine stabilizer that simulates the spinal cord. To perform a mouse SCI, a series of impacts may be tested on the flat playdough surface by replicating the 0.3, 0.6, and 0.9 mm displacements three times. The plunger tip is moved laterally to check whether the laser beam projects on the middle of the crater created on the playdough. The three cavities on the playdough are confirmed to be 0.3, 0.6, and 0.9 mm deep, respectively.

The reproducibility of the displacement depth must be confirmed. If the displacement depths of the craters are inconsistent, the fastener screw locking the plunger tip to the pneumatic cylinder must be tightened (Fig. 5: steps 1 and 3). A common cause of non-reproducible crater depths is because the plunger tip is not locked

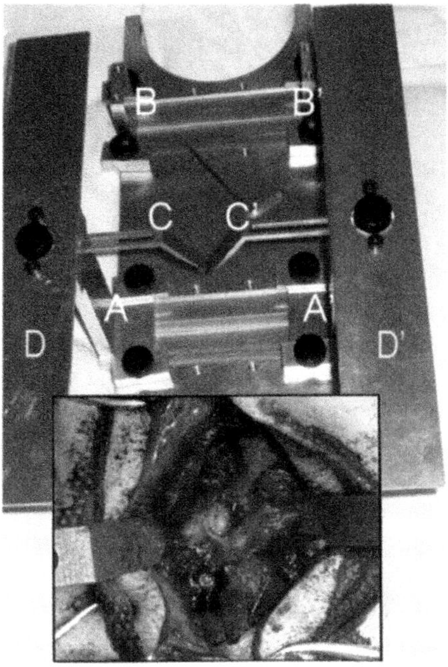

Fig. 5 (A) The monkey spine stabilizer is similar, using a pair of stainless steel arms (C and C′) to attach the facets bilaterally. The monkey stabilizer may be used to create a thoracic SCI with the arms attached at the A-A′ position and cervical fixation with arms attached at the B-B′ position. The steel arms are attached to the metal columns (A and B). The choice of columns depends on the animal size and injury location. Large animals require taller columns in the thoracic area. D and D′ are two bars used by the surgeon to maintain manual stability while operating. The T9 vertebra is stabilized by the attachment of steel arms on both facets. The spinal cord is then exposed

firmly, preventing the laser distance sensor from measuring the distance from the sensor to the crater base accurately. If the laser beam does not focus on the middle of the crater in the playdough, the position of the plunger tip is adjusted or the distance laser sensor (using a 3 mm Allen wrench) is locked to the fastening screws until the laser beam focuses on the middle of the crater.

Results

Using spine stabilization to fixate the facets of the spinal vertebra, an experimental displacement SCI may be created at any level of the spinal cord (cervical, thoracic) in mice, rats, and monkeys with several controllable magnitudes of injury. Using the LISA device, a reliable and reproducible method of generating a contusion SCI has been described. An impact velocity of 1 m/s combined with a displacement of 25% of the AP diameter of the spinal cord will create a mild contusion injury, whereas a displacement causing 60% compression of the spinal cord diameter will produce a severe contusion SCI. This method has been used to create a severe contusion SCI in the non-human primate, as well as graded (mild, moderate, and severe) contusion injuries in rodents.

Discussion

Guidelines for experimental SCI models issued by the International Spinal Research Trust (ISRT) require that descriptions such as the lesion magnitude should be precisely defined [9]. The ideal experimental model should mimic SCI experienced clinically. The method we have described to create an experimental spinal cord contusion closely replicates the injury velocity seen clinically. If a patient falls from a height of 3 m, the velocity at which the body strikes the ground is 7.6 m/s [velocity at the time of impact = square root (2 * g * height); where g (acceleration due to gravity) = 9.8 m/s^2]. In motor vehicle accidents, the impact velocity may attain 11–22 m/s if the vehicle is travelling at 25–50 miles/h (each mile per hour is equivalent to 0.447 m/s). The injury velocity to which the spinal cord is subjected will be considerably less due to the protective role of paraspinal muscles and bony canal surrounding the spinal cord. There is no clinical data available that measures the impact injury velocity or how to convert clinical injuries to metrics that would allow precise replication of an experimental spinal cord contusion. However, the impact velocity must not be so slow that a contusion does not occur. To be clinically relevant an impact velocity of $\geq$1.0 m/s is required to create a meaningful contusion injury. Other methods of creating an experimental SCI subject the spinal cord to an impact velocity of 0.1 m/s (IH system) and approximately 0.148 m/s (ESCID).

The method we have described is flexible and provides a range of tip diameters that correspond to different diameters of the spinal cord in various species. Different severities of SCI can be created in mice, rats, and monkeys with minimal modifications of the setup of the LISA software. We have also used the tip of the mouse plunger to produce a hemi-contusion model of SCI in the rat [10]. We modified this model to create contusion SCI in non-human primates [11] which may be applicable on other experimental animals such as piglets. This method requires stabilization of the target and use of a laser sensor. The various sized spine stabilizers are able to fixate the facets in multiple species. This stabilizing method greatly enhances the accuracy of creating contusion SCI and may be used in conjunction with the NYU and IH systems.

Ordering information

Louisville Impactor System. Contact: cbshields1@gmail.com

Acknowledgments We acknowledge Norton Healthcare for their continued support.

References

1. Constantini S, Young W. The effects of methylprednisolone and the ganglioside GM1 on acute spinal cord injury in rats. J Neurosurg. 1994;80(1):97–111.
2. Noble LJ, Wrathall JR. An inexpensive apparatus for producing graded spinal cord contusive injury in the rat. Exp Neurol. 1987;95(2):530–3.
3. Scheff SW, Rabchevsky AG, Fugaccia I, Main JA, Lumpp JE Jr. Experimental modeling of spinal cord injury: characterization of a force-defined injury device. J Neurotrauma. 2003;20(2):179–93.
4. Stokes BT. Experimental spinal cord injury: a dynamic and verifiable injury device. J Neurotrauma. 1992;9(2):129–31.
5. Zhang YP, Burke DA, Shields LBE, Chekmenev SY, Dincman T, Zhang Y, Zheng Y, Smith RR, Benton RL, Devries WH, Hu X, Magnuson DSK, Whittemore SR, Shields CB. Spinal cord contusion based on precise vertebral stabilization and tissue displacement measured by combined assessment to discriminate small functional differences. J Neurotrauma. 2008;25(10):1227–40.
6. Fehlings MG, Tator CH. The effect of direct current field polarity on recovery after acute experimental spinal cord injury. Brain Res. 1992;579(1):32–42.
7. Jakeman LB, Guan Z, Wei P, Ponnappan R, Dzwonczyk R, Popovich PG, Stokes BT. Traumatic spinal cord injury produced by controlled contusion in mouse. J Neurotrauma. 2000;17(4):299–319.
8. Zhang YP, Onifer SM, Burke DA, Shields CB. A topical mixture for preventing, abolishing, and treating autophagia and self-mutilation in laboratory rats. Contemp Top Lab Anim Sci. 2001;40(2):35–6.
9. Ramer MS, Harper GP, Bradbury EJ. Progress in spinal cord research—a refined strategy for the International Spinal Research Trust. Spinal Cord. 2000;38(8):449–72.
10. Walker CL, Zhang YP, Liu Y, Li Y, Walker MJ, Liu NK, Shields CB, Xu XM. Anatomical and functional effects of lateral cervical hemicontusion in adult rats. Restor Neurol Neurosci. 2016;34(3):389–400.
11. Ma Z, Zhang YP, Liu W, Yan G, Li Y, Shields LBE, Walker M, Chen K, Huang W, Kong M, Lu Y, Brommer B, Chen X, Xu XM, Shields CB. A controlled spinal cord contusion for the rhesus macaque monkey. Exp Neurol. 2016;279:261–73.

Experimental Laceration Spinal Cord Injury Model in Rodents

Yi Ping Zhang, Lisa B. E. Shields, and Christopher B. Shields

Abstract Experimental laceration spinal cord injury (l-SCI) is an important *in vivo* model to investigate mechanisms of axonal regeneration and neurite plasticity following SCI. l-SCI is customarily performed freehand under visual guidance that results in variable lesion depths and shapes. In this chapter, a precise l- SCI model will be described for rodents using the Louisville Injury System Apparatus (LISA). This method incorporates: (1) reliable vertebral stabilization and spinal cord exposure and (2) accurate tissue laceration without contusion.

Keywords Spinal cord injury · Laceration SCI model · Louisville Injury System Apparatus (LISA) · Axonal regeneration

Introduction

The majority of spinal cord injuries are contusions followed by lacerations [1, 2]. Laceration injuries are caused by bone fragments, gunshot wounds, or other sharp objects. In experimental SCI research, the l-SCI is a model to study ascending and descending pathways of the spinal cord and their respective functions [3, 4]. l-SCI models are also used to study methods of axonal regeneration and neurite sprouting. Failure of axon regrowth is caused by intrinsic neuronal mechanisms as well as inhibition by the hostile environment. Known inhibitory molecules in the CNS include chondroitin sulfate proteoglycans [5, 6], myelin derived molecules [7–9], semaphorins [10], and ephrin/Eph [11, 12]. A reproducible and reliable l-SCI model

Y. P. Zhang · L. B. E. Shields (✉)
Norton Neuroscience Institute, Norton Healthcare, Louisville, KY, USA
e-mail: LBES@earthlink.net

C. B. Shields (✉)
Norton Neuroscience Institute, Norton Healthcare, Louisville, KY, USA

Department of Anatomical Sciences and Neurobiology, University of Louisville School of Medicine, Louisville, KY, USA

J. Chen et al. (eds.), *Animal Models of Acute Neurological Injury*, Springer Series in Translational Stroke Research, https://doi.org/10.1007/978-3-030-16082-1_35

is necessary to understand the role of complex intrinsic and extrinsic cues that influence axon degeneration and regeneration.

The International Spinal Research Trust (ISRT) published guidelines for SCI models indicating that the characteristics and magnitude of the lesion must be precisely defined and experimental results would lead to false conclusions if imprecise lesions were used [13]. Precise l-SCI models are important in evaluating mechanisms of injury and screening for novel therapies. Criteria for an optimal l-SCI model include: (1) the laceration should not spare axons to avoid the erroneous interpretation of axon regeneration due to failure to incorporate axons, (2) the laceration lesion should be uniform to prevent other types of injuries such as contusions and should consist of a single cut to avert multiple injuries that will introduce greater lesion variability, and (3) the depth of the laceration must be highly accurate especially when using murine models. Use of genetically modified mice has greatly increased as they allow the study of specific genes or proteins known to regulate axon regrowth [14, 15]. Thus, an ideal rodent l-SCI model creates injuries of precise depths and shapes without contusive or compressive damage.

The creation of a precise l-SCI is difficult to perform. A need for a precise laceration spinal cord model was underscored by contradictory results obtained from three independent research laboratories that studied the Nogo pathway for axon regeneration on gene knockout mice [16–18]. Significant differences in axon regeneration were due to either different strategies used to delete the Nogo gene and the magnitude of SCI. Most l-SCI are performed manually using microscissors, a scalpel, a razor blade, needle tips, or a wire knife [19–21]. This method of creating a freehand laceration is visually controlled, particularly when performed in a bloody surgical field by an experienced surgeon. The lesion shape can be confirmed by postmortem analysis, and imprecise lesions are excluded. Factors that contribute to an inaccurate laceration are presented in Table 1.

Factors that produce unreliable lesions may be avoided by utilizing the LISA. This method uses an oscillating blade combined with a bilateral vertebral stabilizing device [22, 23]. The oscillating blade avoids cord deformation with a similar mechanism to a vibratome. Our system utilizes bilateral facet joint fixation that provides greater stability than cervical spinous process fixation where the spinous processes are fragile [20, 21]. The shape and width of the laceration may be modified by

Table 1 Factors that contribute to an inaccurate spinal cord laceration

Deformity of the spinal cord and vertebral body caused by pressure of the microscissors
Bleeding from the spinal cord during laceration may obscure visual guidance
Repeated attempts may be necessary to attain the required injury [22]
Microscissors are often associated with cord contusion/compression

designing custom-made blades [24]. The depth of the lesion can be predicted to an accuracy of 0.01 mm with addition of microdrivers that avoids the poorly controlled visual methodology.

Steps in Creating a l-SCI Using the LISA Are Described Below

Spine Stabilization

The flexible cervical spine of rodents may decrease the number of steps in surgery such as dural opening, as well as the reliability, reproducibility, and accuracy of the injury. Spine stabilization devices include a small one designed for the mouse and a larger one for the rat. The stabilizing device consists of a pair of stainless steel arms that attach to the cervical or thoracic facets to maintain segmental immobility. Longer arms are designed for cervical spine stabilization and shorter ones for the thoracic spine (Fig. 1).

Fabrication of the Blade

The blade is fabricated by welding a piece of broken razor (10050-00 F.S.T Foster City, CA) to a steel shaft (10035-05 F.S.T. for the mouse spinal cord and 10,035-10 F.S.T. for the rat). The shaft (0.6 mm) may be easily bound with a razor blade (0.1 mm) using a portable 120-volt spot welder (Chicago Electric, Chicago, IL) (Fig. 1). The razor tip is trimmed to the desired width and shape using a mini grinder

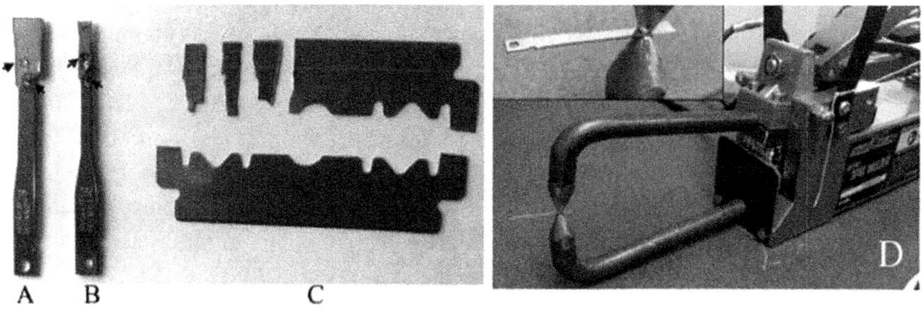

Fig. 1 The blade consists of a metal shaft and cutting tip that are welded together (arrowheads). (**a**) Wide shaft for wide tip and (**b**) narrow shaft for narrow one. (**c**) Illustration of the rough blade pieces from a breakable razor. (**d**). A portable 120-volt spot welder attaches a piece of razor blade with a metal shaft. The razor blade tip is trimmed to the required shape and size by a mini grinder

(Dremel, Racine, WI). The blade width and the magnitude of oscillation influence the lesion width. A greater oscillation amplitude increases the laceration lesion width. The blade is inserted and fixed in the knife holder, and the oscillating amplitude is adjusted by turning the amplitude knob. The recommended vibrating amplitude is ≥0.5 mm.

l-SCI Procedures Using the LISA

Animal Preparation

(A) The following surgical instruments are sterilized: 2–3 pairs of forceps, 2 pairs of microscissors, a 30G needle, suture, needle holder, skin clips, and clip applicator. The spine stabilizer is disinfected.

(B) Rats are anesthetized using an intraperitoneal mixture of Ketamine/Xylazine, and mice are anesthetized using IP Tribromoethanol (Avertin®). Hair is shaved over the target vertebrae.

(C) Skin is prepped with a povidone-iodine solution and 70% alcohol. The animal is moved onto the operating table that is warmed with a heating pad. The eyes are covered with ophthalmic ointment to prevent corneal drying.

(D) When the animal loses the paw pinch reflex, a posterior midline skin incision is made. For a cervical exposure, skin is exposed from the occiput to T2. After the semispinalis capitis muscles are separated, a submuscular fat pad is identified that facilitates dissection in the correct layer. The dissection extends caudally to the T2 spinous process which serves as a reliable landmark. Muscles attached to the T2 vertebra are cut, and the cartilaginous portion the T2 spinous process is removed. Paraspinal muscles are dissected from C2 through the T1 laminae using a pair of microscissors. Muscle dissection begins adjacent to the spinous processes and extends bilaterally to the facet joints.

For the thoracic exposure, the midline skin incision extends from T7 to L1. Paraspinal muscles are dissected from the spinous processes, laminar arches bilaterally, to the facet joints. After the lateral facets are exposed, the animal is placed on the U-shaped frame of the spine stabilizer (Fig. 2). The fixation device consists of a U-shaped metal frame to support the animal and two adjustable stainless steel arms that clamp to both facets. After identifying the target vertebra, the stainless steel arms are placed firmly against the lateral facets and fixed to the frame. This provides excellent immobilization of the target vertebra. After spinal fixation, the spine is slightly elevated to minimize the spine curvature that provides better exposure of the spinal cord.

(E) Once the thumb-screws on the stabilizing frame are tightened and the target spine is immobilized, the spinal cord is exposed. Adjacent laminae overlap in the thoracic spine and, therefore, a laminectomy is required to expose the spinal cord. There is a gap between laminae in the cervical spine, thus, a laminectomy

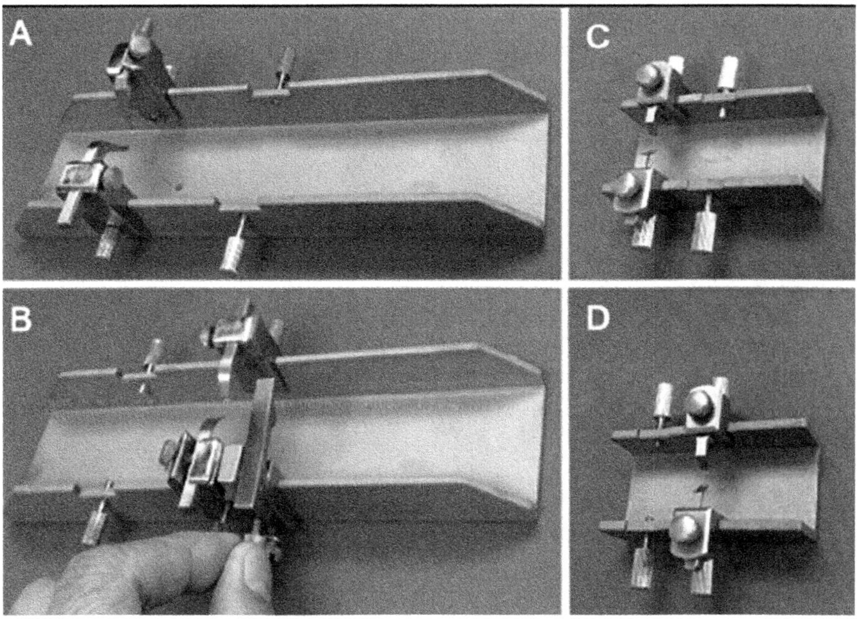

Fig. 2 Rat stabilizing frame with the arms in place for the (**a**) cervical and (**b**) thoracic spine. Mouse stabilizer for (**c**) cervical and (**d**) thoracic spine fixation

is unnecessary. The ligamentum flavum is incised to expose the underlying dura [25]. A durotomy is performed by incising the dura mater between the interlaminar space. A 30G needle is used to create a small durotomy through which microscissors are placed to extend the durotomy. The spinal cord is prepared to undergo the controlled laceration lesion.

Spinal Cord Laceration

After the shape and depth of the lesion is determined, a matching blade is prepared. The cutting blade and blade-holder are attached to the LISA, and the oscillating amplitude is set at ≥0.5 mm to facilitate a sharp laceration.

1. The spine stabilizer and animal are positioned on the LISA stage. The blade-holder is locked in the proper position on the vibrating assembly. The cutting blade is positioned so that it is in contact with the targeted spinal cord using the microdrivers. (Components of the LISA and their functions are described in Fig. 3)
2. The vibrating switch is turned on. Under magnification the oscillating blade is positioned in approximation to the exposed spinal cord so that the oscillating blade covers the required width of the lesion. If necessary, the oscillating amplitude is fine-tuned to accomplish the desired laceration.

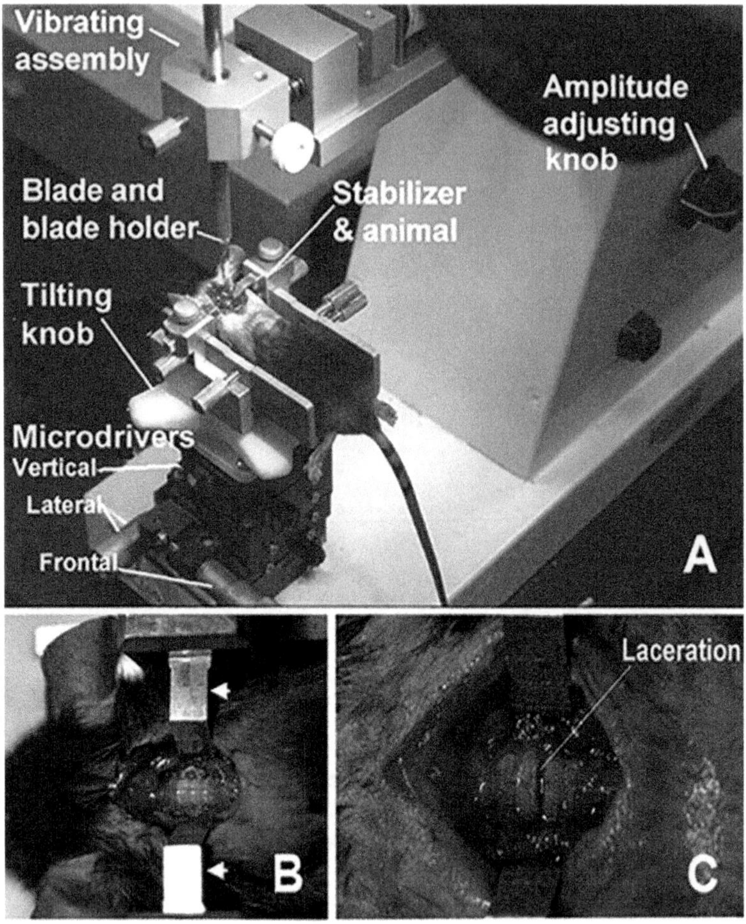

Fig. 3 (**a**) Illustration of the experimental animal on the LISA performing l SCI. Mouse on the frame placed on the stage of the LISA. Posturing of the mouse is controlled by frontal, lateral, and vertical microdrivers and tilt knobs located beneath the animal. The vertical microdriver also controls the lesion depth. The oscillating blade and blade-holder are locked in the vibrating assembly above the animal, with the blade aimed toward the spinal cord. The vibration amplitude is altered by tuning the amplitude adjusting knob. (**b**) Vertebral facet fixation with the stabilizer in place. (**c**) Demonstration of spinal cord laceration

3. The stage supporting the mouse/rat is elevated towards the oscillating blade using the vertical micromanipulator. The "0" position is recorded at the point which the blade barely touches the dorsal spinal cord. The depth of the spinal cord laceration is measured relative to the "0" position. A larger stage is used for the rat compared to the mouse.
4. The vertical microdriver controls the lesion depth to an accuracy of ±0.01 mm. A full turn of the microdriver knob (360°) elevates the stage by 0.25 mm.

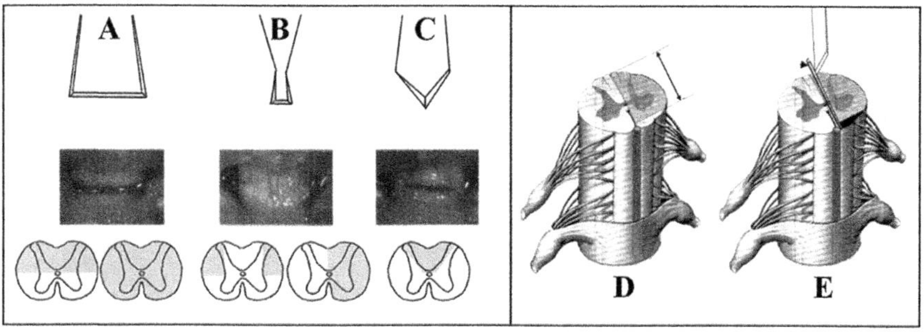

Fig. 4 (**a**) Full blade covering the width of the target spinal cord to create a dorsal hemisection or transection. (**b**) Narrow blade covering the lateral funiculus to create a dorsal-lateral funicular lesion or lateral hemisection. (**c**) Diamond shaped blade to create a dorsal funicular lesion. (**d, e**) Preserving the ventral spinal artery during lateral hemisection is important to produce consistent tissue damage. The first step (**d**) consists of a subtotal unilateral laceration of the spinal cord by keeping the blade a short distance from the anterior spinal artery, followed by (**e**) centrally inserting a guide needle in the gap to protect the anterior spinal artery and the spinal cord on the opposite side. The ventral spinal cord is cut manually using a mini-blade (dark area)

For example, creation of a 1.5 mm dorsal hemisection lesion requires 6 turns of the microdriver knob. While the oscillating blade is cutting the spinal cord, the surgical field is irrigated with saline for lubrication and cooling. During the beginning of the laceration, the knob of the microdriver is turned slowly, cutting through the pia mater. The cutting depth of the spinal cord laceration is controlled by the vertical microdriver and is independent of visual guidance.

5. Once the desired depth is reached, the vibrating switch is turned off. Ideally, the oscillating blade should be positioned in the lesion gap without evidence of tissue deformation or contusion. The stage is then lowered from the cutting blade. Blood and saline are removed from the surgical field using cotton Q-tips®. Hemostasis is easily controlled in <1 min.
6. The animal is released from the spine stabilizer. The dura mater is closed using 10-0 sutures in rats [23] but does not require closure in mice. Paraspinal muscles are approximated using a 6-0 silk suture, and the skin wound is closed using stainless steel Michel clips.

Choice of Injury Shapes and Depths

A unilateral spinal cord laceration may assist in evaluating mechanisms of neuronal plasticity, axon regeneration, or collateral sprouting between the adjacent surfaces of the laceration. Following a unilateral spinal cord laceration, post-injury complications such as quadriplegia and loss of respiratory control are limited which greatly facilitates animal care and improves survival. The anterior spinal artery supplies the ventral 2/3 of the spinal cord. Our method minimizes the risk of anterior spinal

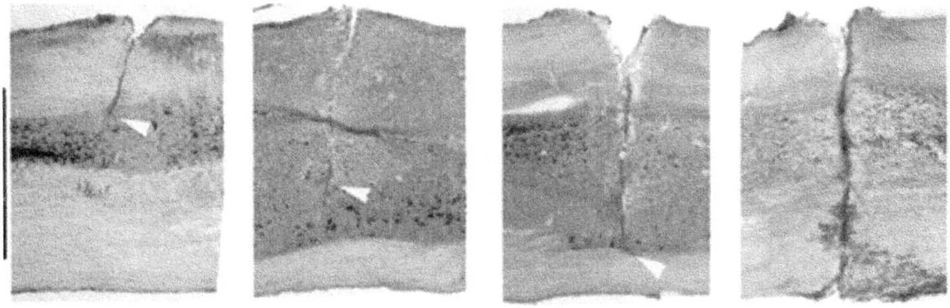

Fig. 5 Four dorsal spinal cord lacerations at depths of 0.5, 0.8, 1.1, and 1.4 mm observed on a sagittal view (cresyl-violet and eosin stain) depicting the high degree of precision

artery damage causing spinal cord stroke. The lateral spinal cord laceration is performed by creating a subtotal lateral hemisection so that a small portion of the spinal cord is preserved ventrally to prevent lacerating the anterior spinal artery. If the experiment requires a complete spinal cord transection, we recommend that the remaining fibers of the ventral cord be cut by a hand held blade to avoid damage to the anterior spinal artery (Fig. 4). The blade laceration made with the LISA is controlled by a microdriver that determines the depth of the lesion. Four laceration depths of SCI ranging from 0.5 to 1.4 mm were highly reproducible (Fig. 5). The precise lesion shape and depth are important to avoid false positive results occurring from lesions that underestimate the true lesion severity. Figure 6 compares a 0.8 mm dorsal hemisection performed manually using microscissors to a lesion made using the LISA device. The freehand laceration SCI was smaller than anticipated both in depth and width (Fig. 6).

The Lesion Gap Following l-SCI

The spinal cord is constantly subjected to longitudinal tension due to traction exerted by nerve roots and the filum terminale. Following a l-SCI, a gap usually develops between the two stumps that is more apparent in the thoracic region than in the cervical spinal cord. The natural curvature of the spinal cord plays a role in creating lesion gaps. A dorsal hemisection in the concave T1-3 segment creates less dorsal tension on the spinal cord than at the convex T9-11. A blood clot develops within the lesion gap creating a greater physical barrier that may include a scar and/or a cyst within the spinal cord. The spinal cord lesion could be made at levels that are less kyphotic to minimize the lesion gap. Alternatively, the free edges of dura may be sutured to approximate the spinal cord stumps [23].

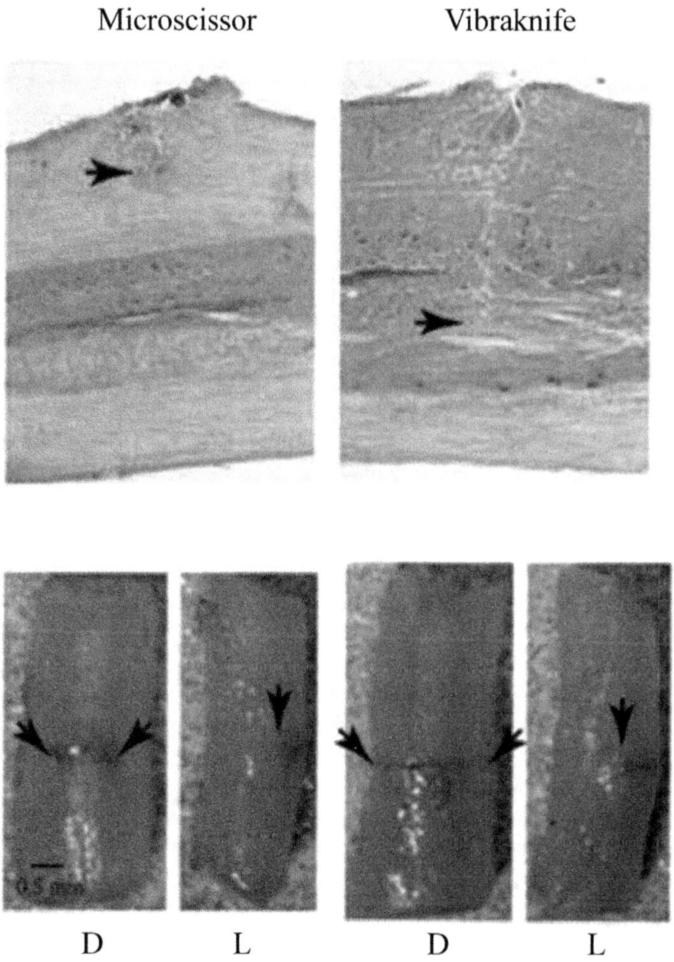

Fig. 6 Depiction of an error made following manually executed laceration using visual guidance. A planned 0.8 mm dorsal laceration was <0.5 mm

Post-injury Animal Care

1. Following the laceration, saline is administered subcutaneously to maintain adequate hydration for the rat (5–10 mL) and mouse (1–2 mL). The animal is placed in a cage on a heating pad maintained at 37 °C until regaining consciousness.
2. Water and soft food are provided *ad lib*, and analgesics (Buprenorphine 3 mg/kg mice, 1 mg/kg rats) are administered for 48 h post-operatively. There is no need for bladder care following dorsal hemisection, lateral hemisection, or a dorsal funicular laceration lesion of the spinal cord. We have not observed evidence of autophagy.

Results

Laceration of the spinal cord using a razor blade or microscissors under manual visual guidance causes quality/quantity lesion variability [3]. This may lead to misinterpretation of outcomes particularly when axonal regeneration of specific pathways is studied. For example, spared dorsal corticospinal axons may be misinterpreted as undergoing regeneration, however, the dorsal corticospinal tract may not have been completely transected at the time of the laceration. Immobilization of the target vertebra is also of great importance in generating precise laceration lesions of the mouse spinal cord. Spine stabilization overcomes the anatomical issues of motion of the spinal cord when short spinous processes are used for stabilization. The depth of the lesion using the LISA is accurate to 0.01 mm [3, 26]. This accuracy is possible because we utilize high precision microdrivers to control the position of the cutting blade without depending on visual control. The high frequency oscillating blade produces a sharp laceration without crushing or contusing the adjacent spinal cord. This method has been used to produce precise spinal cord laceration injuries in rats [23, 24, 27] and mice [3].

Conclusion

A spinal cord laceration method is introduced using an oscillating blade. The spine is fixed by attachment to the facets using a spine stabilizer. The consistency and reproducibility created by this laceration spinal cord lesion avoids conflicting results that are often reported in studies of axon regeneration.

Acknowledgment We acknowledge Norton Healthcare for their continued support.

References

1. Bunge RP, Puckett WR, Becerra JL, Marcillo A, Quencer RM. Observations on the pathology of human spinal cord injury. A review and classification of 22 new cases with details from a case of chronic cord compression with extensive focal demyelination. Adv Neurol. 1993;59:75–89.
2. Quencer RM, Bunge RP. The injured spinal cord: imaging, histopathologic clinical correlates, and basic science approaches to enhancing neural function after spinal cord injury. Spine (Phila Pa 1976). 1996;21:2064–6.
3. Hill RL, Zhang YP, Burke DA, Devries WH, Zhang Y, Magnuson DS, Whittemore SR, Shields CB. Anatomical and functional outcomes following a precise, graded, dorsal laceration spinal cord injury in C57BL/6 mice. J Neurotrauma. 2009;26:1–15.
4. Sherry BA, Alava G, Tracey KJ, Martiney J, Cerami A, Slater AF. Malaria-specific metabolite hemozoin mediates the release of several potent endogenous pyrogens (TNF, MIP-1 alpha, and MIP-1 beta) in vitro, and altered thermoregulation in vivo. J Inflamm. 1995;45:85–96.

5. Bradbury EJ, Moon LD, Popat RJ, King VR, Bennett GS, Patel PN, Fawcett JW, McMahon SB. Chondroitinase ABC promotes functional recovery after spinal cord injury. Nature. 2002;416:636–40.
6. Silver J, Schwab ME, Popovich PG. Central nervous system regenerative failure: role of oligodendrocytes, astrocytes, and microglia. Cold Spring Harb Perspect Biol. 2015;7:a020602.
7. Chen MS, Huber AB, van der Haar ME, Frank M, Schnell L, Spillmann AA, Christ F, Schwab ME. Nogo-A is a myelin-associated neurite outgrowth inhibitor and an antigen for monoclonal antibody IN-1. Nature. 2000;403:434–9.
8. Fournier AE, Gould GC, Liu BP, Strittmatter SM. Truncated soluble Nogo receptor binds Nogo-66 and blocks inhibition of axon growth by myelin. J Neurosci. 2002;22:8876–83.
9. Kim JE, Liu BP, Park JH, Strittmatter SM. Nogo-66 receptor prevents raphespinal and rubrospinal axon regeneration and limits functional recovery from spinal cord injury. Neuron. 2004;44:439–51.
10. Pasterkamp RJ, Verhaagen J. Emerging roles for semaphorins in neural regeneration. Brain Res Brain Res Rev. 2001;35:36–54.
11. Figueroa JD, Benton RL, Velazquez I, Torrado AI, Ortiz CM, Hernandez CM, Diaz JJ, Magnuson DS, Whittemore SR, Miranda JD. Inhibition of EphA7 up-regulation after spinal cord injury reduces apoptosis and promotes locomotor recovery. J Neurosci Res. 2006;84:1438–51.
12. Kempf A, Montani L, Petrinovic MM, Schroeter A, Weinmann O, Patrignani A, Schwab ME. Upregulation of axon guidance molecules in the adult central nervous system of Nogo-A knockout mice restricts neuronal growth and regeneration. Eur J Neurosci. 2013;38:3567–79.
13. Ramer MS, Harper GP, Bradbury EJ. Progress in spinal cord research—a refined strategy for the International Spinal Research Trust. Spinal Cord. 2000;38:449–72.
14. Inman D, Guth L, Steward O. Genetic influences on secondary degeneration and wound healing following spinal cord injury in various strains of mice. J Comp Neurol. 2002;451:225–35.
15. Steward O, Schauwecker PE, Guth L, Zhang Z, Fujiki M, Inman D, Wrathall J, Kempermann G, Gage FH, Saatman KE, Raghupathi R, McIntosh T. Genetic approaches to neurotrauma research: opportunities and potential pitfalls of murine models. Exp Neurol. 1999;157:19–42.
16. Kim JE, Li S, GrandPre T, Qiu D, Strittmatter SM. Axon regeneration in young adult mice lacking Nogo-A/B. Neuron. 2003;38:187–99.
17. Simonen M, Pedersen V, Weinmann O, Schnell L, Buss A, Ledermann B, Christ F, Sansig G, van der Putten H, Schwab ME. Systemic deletion of the myelin-associated outgrowth inhibitor Nogo-A improves regenerative and plastic responses after spinal cord injury. Neuron. 2003;38:201–11.
18. Zheng B, Ho C, Li S, Keirstead H, Steward O, Tessier-Lavigne M. Lack of enhanced spinal regeneration in Nogo-deficient mice. Neuron. 2003;38:213–24.
19. Frisen J, Fried K, Sjogren AM, Risling M. Growth of ascending spinal axons in CNS scar tissue. Int J Dev Neurosci. 1993;11:461–75.
20. Hermanns S, Reiprich P, Muller HW. A reliable method to reduce collagen scar formation in the lesioned rat spinal cord. J Neurosci Methods. 2001;110:141–6.
21. Seitz A, Aglow E, Heber-Katz E. Recovery from spinal cord injury: a new transection model in the C57Bl/6 mouse. J Neurosci Res. 2002;67:337–45.
22. Iannotti C, Zhang YP, Shields LBE, Han Y, Burke DA, Xu XM, Shields CB. Dural repair reduces connective tissue scar invasion and cystic cavity formation after acute spinal cord laceration injury in adult rats. J Neurotrauma. 2006;23:853–65.
23. Zhang YP, Iannotti C, Shields LBE, Han Y, Burke DA, Xu XM, Shields CB. Dural closure, cord approximation, and clot removal: enhancement of tissue sparing in a novel laceration spinal cord injury model. J Neurosurg. 2004;100:343–52.
24. Onifer SM, Zhang YP, Burke DA, Brooks DL, Decker JA, McClure NJ, Floyd AR, Hall J, Proffitt BL, Shields CB, Magnuson DS. Adult rat forelimb dysfunction after dorsal cervical spinal cord injury. Exp Neurol. 2005;192:25–38.
25. Zhang YP, Walker MJ, Shields LBE, Wang X, Walker CL, Xu XM, Shields CB. Controlled cervical laceration injury in mice. J Vis Exp. 2013:50030.

26. Yu P, Zhang YP, Shields LBE, Zheng Y, Hu X, Hill R, Howard R, Gu Z, Burke DA, Whittemore SR, Xu XM, Shields CB. Inhibitor of DNA binding 2 promotes sensory axonal growth after SCI. Exp Neurol. 2011;231:38–44.
27. Sivasankaran R, Pei J, Wang KC, Zhang YP, Shields CB, Xu XM, He Z. PKC mediates inhibitory effects of myelin and chondroitin sulfate proteoglycans on axonal regeneration. Nat Neurosci. 2004;7:261–8.

Rodent Spinal Cord Demyelination Models

Kariena R. Andres, Johnny R. Morehouse, Rachel Cary, Christine D. Yarberry, Nicholas J. Kuypers, and Scott R. Whittemore

Abstract Oligodendrocyte loss and subsequent demyelination is a significant component of the demyelinating diseases such as multiple sclerosis (MS) and traumatic CNS injury such as spinal cord (SCI) or traumatic brain (TBI) injury. Therefore, remyelination, either by enhancing endogenous myelination or engrafting exogenous myelinating cells, is a viable therapeutic target to restore function. To assess specific approaches to facilitate functional remyelination *in vivo*, appropriate injury

Electronic supplementary material: The online version of this chapter (https://doi.org/10.1007/978-3-030-16082-1_36) contains supplementary material, which is available to authorized users.

K. R. Andres · J. Morehouse · R. Cary · C. Yarberry
Kentucky Spinal Cord Injury Research Center, University of Louisville School of Medicine, Louisville, KY, USA

Department of Neurological Surgery, University of Louisville School of Medicine, Louisville, KY, USA

N. J. Kuypers
Kentucky Spinal Cord Injury Research Center, University of Louisville School of Medicine, Louisville, KY, USA

Department of Neurological Surgery, University of Louisville School of Medicine, Louisville, KY, USA

Department of Anatomical Sciences and Neurobiology, University of Louisville School of Medicine, Louisville, KY, USA

S. R. Whittemore (✉)
Kentucky Spinal Cord Injury Research Center, University of Louisville School of Medicine, Louisville, KY, USA

Department of Neurological Surgery, University of Louisville School of Medicine, Louisville, KY, USA

Department of Anatomical Sciences and Neurobiology, University of Louisville School of Medicine, Louisville, KY, USA
e-mail: swhittemore@louisville.edu

J. Chen et al. (eds.), *Animal Models of Acute Neurological Injury*, Springer Series in Translational Stroke Research, https://doi.org/10.1007/978-3-030-16082-1_36

models are needed. This chapter will discuss the strengths and weaknesses of a number of demyelinating lesions of the spinal cord and provide guidelines for choosing which model best suits which experimental condition. Step by step procedures for both creating and assessing the lesion will be provided.

Keywords Spinal cord injury · Demyelination · Ethidium bromide · Lysolecithin · Cuprizone

Introduction

There are a number of methods that can be used to demyelinate axons *in vivo*. These include injection of ethidium bromide (EB), lysolethicin, murine hepatitis virus (MHV), a cocktail of myelin-specific antibodies and complement, or by giving cuprizone in the diet [1, 2]. All of these lesions kill oligodendrocytes in the spinal cord, although cuprizone is much more effective in the corpus callosum than the spinal cord [3] and has the potential for remyelination (from endogenous or exogenously transplanted progenitors). However, depending on a number of factors (i.e. the mode of demyelination, targeted anatomical region, dose, length of administration, and species), other cell types (including axons) can also be lost [1]. Each of these factors must be considered to determine the extent/temporal progression of demyelination, severity of inflammation, and potential for remyelination. All of these lesions ultimately remyelinate by 3–6 months post-demyelination. However, the temporal progression and the extent of remyelination show differences which are discussed below [1, 2]. Importantly, while there is axonal loss in all of these models in rat studies, under optimal injection parameters, that damage is not different from saline injections [1]. Here, we briefly review each demyelinating lesion as well as provide detailed protocol for delivering and assessing focal spinal cord demyelinating lesion via injection of the gliotoxic agents EB and lysolecithin.

When injected into white matter, EB is a gliotoxic compound that intercalates into the small groove of helical DNA and kills both oligodendrocytes and astrocytes, which are highly sensitive to EB toxicity [4, 5] (Fig. 1a–f). While EB intercalates into both chromosomal and mitochondrial DNA, it only affects transcription of mitochondrial DNA [6–8]. At higher doses, neurons and endothelial cells are also killed. Importantly, there are differences as to how the rat and mouse spinal cord respond to EB toxicity. EB lesions in the mouse exhibit a much steeper dose-response curve as there is only a 0.1 mg/mL difference in concentration between little to no demyelination (0.15 mg/mL) and complete axon loss (0.25 mg/mL) [9] whereas the rat exhibits a much wider range of concentrations between little demyelination and axon loss [10–12]. Note that demyelination occurs rapidly and the lesion remains demyelinated at 4 weeks post-EB (Fig. 1a–c).

In contrast to mouse EB lesions where there is axon loss with chronic demyelination [9], there is no evidence in the rat of axon loss following EB lesions at concentrations appropriate for demyelination [13], even with repeated episodes of focal

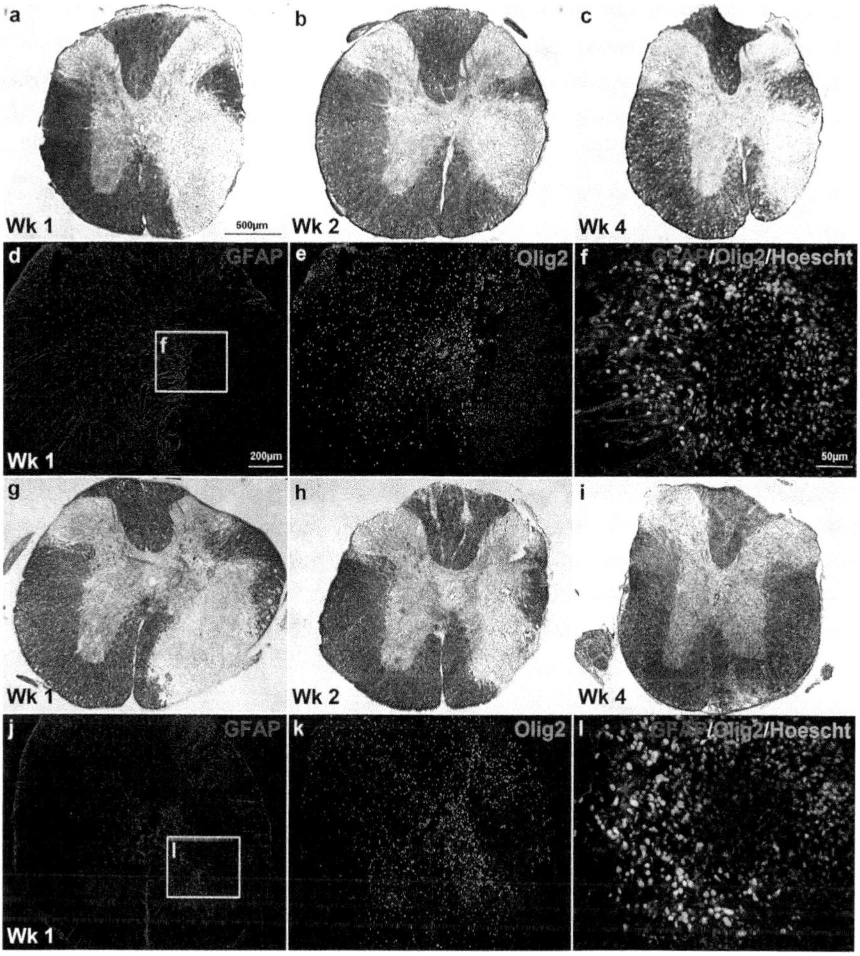

Fig. 1 Representative images of ethidium bromide- and lysolecithin-induced demyelination at the lesion epicenter. Eriochrome cyanine-stained spinal cord section at 1 (**a**, **g**), 2 (**b**, **h**), and 4 (**c**, **i**) weeks post-injection of ethidium bromide (**a–f**) or lysolecithin (**g–l**). Immunofluorescent staining of the astrocyte protein glial fibrillary acidic protein (GFAP; **d**, **j**) or the oligodendrocyte-specific transcription factor Olig2 (**e**, **k**) at 1 week post-injection of (**d–f**) ethidium bromide or (**j–l**) lysolecithin. (**d**, **e**) and (**j**, **k**) are the same field. The boxed regions in (**d**, **e**) are shown at higher magnification in (**f**, **l**), respectively. Magnification bars are indicated

demyelination [11]. However, electrophysiological evidence of spontaneous remyelination in the rat was not observed even up to 8 weeks post-EB [13], indicating the need for physiological evaluation if evidence of functional remyelination is to be concluded. Moreover, a substantial percentage of the chronic remyelination in this model is Schwann cell rather than oligodendrocyte myelination [1]. Importantly, this lesion also kills astrocytes (Fig. 1d) which are essential for oligodendrocyte remyelination [14]. The EB lesion model is best suited for acute remyelination by engrafted exogenous myelinating cells rather than to examine mechanisms of spontaneous, endogenous remyelination. The third major difference between rat and mouse spinal cord EB lesions is the inflammatory response. In the rat, there is a transient inflammatory response that resolves by 1 month post-EB [13], while in the mouse there is a persistent inflammatory response that lasts at least 2 months post-EB [9]. Such chronic inflammatory responses can markedly affect remyelination [15].

At appropriate concentrations and doses, lysolecithin selectively kills oligodendrocytes in the injection site by solubilizing myelin sheaths and facilitating inflammatory responses [1], but spares most axons and astrocytes [16–18]. In contrast to spinal cord EB lesions, lysolecithin-induced demyelinating lesions completely remyelinate spontaneously by 4 weeks after initial treatment (Fig. 1g–l), with only a small fraction of Schwann cell remyelination [1, 19]. Astrocytes are not assignificantly effected by lysolecithin, but there is still loss (Fig. 1j–l). This is an excellent model to study spontaneous remyelination but is more limited for grafting studies due to its robust intrinsic repair.

Antibodies to galactocerebroside C (GalC), an oligodendrocyte glycolipid, specifically target oligodendrocytes for complement-mediated lysis [1, 20]. These lesions also spontaneously remyelinate, with the majority of oligodendrocyte remyelination, but the time course mirrors that of EB, e.g. 3–6 months post-demyelination. The disadvantage to this model is that complement lysis initiates a robust inflammatory response, which as indicated above, can compromise the interpretation of mechanistic data [15].

The cuprizone model is best suited to examine endogenous remyelination *in vivo*. Cuprizone is provided in the diet [21] and it kills oligodendrocytes preferentially by chelating Cu^{2+}, a necessary co-factor for many metalloenzymes [22]. The process of cuprizone remyelination is complex involving an initial OPC propagation phase and a subsequent maturation phase [23]. Demyelination is maximal after 6 weeks on a cuprizone diet and following cuprizone cessation, the peak of oligodendrocyte maturation occurs at 2–3 weeks post-cuprizone with remyelination resolving by 6 weeks of recovery [21, 24, 25]. Note that the corpus callosum is much more effectively demyelinated by cuprizone than is the spinal cord [3].

Experimental autoimmune encephalomyelitis (EAE) and viral-induced demyelination are demyelinating models which closely resemble human MS [26] as each involves an inflammatory component as the cause of demyelination ultimately resulting in neurological deficits [27–29]. EAE is induced via immunization with CNS myelin proteins (i.e. MBP, PLP, MAG, and MOG) along with co-administration of an adjuvant [30, 31]. Viral demyelination is induced by infecting susceptible strains of mice with one of a number of viruses; most commonly Theiler's murine encephalomyelitis virus (TMEV) or murine hepatitis virus (MHV) [32]. Depending on the species, strain, virus, and technical approach, both EAE and viral-induced demyelination can be tailored to model various forms of MS. Moreover, both models display heterogeneous patterns of demyelination which, despite mimicking the clinical pathology observed in MS [30–32], lack the anatomical specificity and reproducibility of the focal demyelination techniques we describe here (i.e. EB and lysolecithin injection).

Materials

Name	Company	Catalog number
1 mL syringes	B-D	309628
10 mL syringes	B-D	309604
60 mL syringes	B-D	309653
15 mL conical tube	Fisher	14-959-70C
#15 blades	F.S.T.	10015-00
4-0 monofilament	Henry Schein	036-626, 4-0
$26G^{5/8}$	B-D	305115
30 guage needle, bent 90°	Henry Schein	036-345
Alpha-dri Bedding	Cincinnati Labs	ADT
Bacitracin	Henry Schein	108-2912
Biosafety hood	Flow Sciences	FS3060m
Castroviejo needle holder	F.S.T.	12565-14
Clippers	Harvard Apparatus	64-0712
Cotton tip applicators	Henry Schein	003-263
Dural scissors	F.S.T.	15002-08
Friedman-Pearson micro	F.S.T.	16221-14
Ronguers, curved tips		
Gauze pads	Henry Schein	006-936
Gel foam	Henry Schein	031-551
Glass micropipette tips	Warner Instruments Corp	G120F-4
Graefe forceps	F.S.T.	11053-10
Heating pad	Henry Schein	263-5521
Jeweler's forceps	F.S.T.	11254-20
Lacrilube	Henry Schein	039-886
Micropipette beveller	Sutter Instrument	BV-10
Micropipette puller	Sutter Instrument	P-97
Mouse decapicone	Braintree Scientific	
Mouse vertebral stabilizer	Louisville Injury System	Handmade (equivalent)
Nitrogen compressed air tank	Scott-Gross	
Picoinjector or nanoinjector		PV800
Scalpel handle	F.S.T.	10003-12
Sponge spears	F.S.T.	18105-03
Spring scissors with fine blades	F.S.T.	15013-12
Staple applicator	F.S.T.	12031-09
Staples	F.S.T.	12032-07
Stereotaxis frame	Benchmark digital	
Straight iris scissors	Henry Schein	012-338
Surgical microscope	Zeiss	
Thoracotomy Scissors	FST	14002-14
Water heating pad	Harvard Apparatus	729715

Reagent	Company	Catalog number
1% lysolecithin	Sigma	L1381, 5 mg
0.9 % sodium chloride	Henry Schein	040-198
2,2,2 tribromoethanol	Sigma	T48402-25G (acute toxicity)
Buprenorphine (0.3 mg/mL)	Reckitt Benckiser	
Derma Chlor	Henry Schein	055-480
Ethidium Bromide (0.2 mg/mL)	Invitrogen	15585-011 (mutagen)
EtOH	Sigma	E7023
Gentamicin (0.6 mg/mL)	Henry Schein	006-913
Ketamine (100 mg/mL)	Hospira	
NaCL	Sigma	S7653
Xylazine (20 mg/mL)	LLOYD laboratories, Akorn Inc.	4811–20 mL

Procedures

All animal procedures were performed in accordance with the Public Health Service Policy on Humane Care and Use of Laboratory Animals, "Guide for the Care and Use of Laboratory Animals, Eighth Edition" (Committee for the update of the Guide for the Care and Use of Laboratory Animals, National Research Council, 2010), and with the approval of the University of Louisville Institutional Animal Care and Use Committee. Adult C57Bl/6 mice 8–10 weeks of age weighing 15–25 g were used for this methods section (Envigo–Harlan; Indianapolis, IN). Rats weighing 180–220 g are also commonly used. Because of the similarities in procedures, details of the mouse surgery are presented below. This section has been divided into five sub-categories ((1) glass micropipette preparation, (2) preparing the micropipette for injection, (3) pre-surgical animal preparation, (4) performing the surgical procedure, (5) post-op animal care) to demonstrate pre-, intra-, and post-surgical preparations.

Glass Micropipette Preparation

1. Prepare micropipette needles by pulling a glass capillary tube with an inner diameter (ID) = 0.53–2.0 mm, using a Sutter Instrument model P-97 micropipette puller (or equivalent). The size of the glass capillary depends on the volume of demyelinating solution. For this procedure, glass capillaries with an ID = 0.69 mm, OD = 1.2 mm, with a length = 10 cm (Warner Instruments, 64-0790) are commonly used. Note: Capillaries contain a filament that helps draw up and expel liquids.
2. Pull glass capillaries to very fine tips using a high temperature setting (600–800 °C) with small force (20 psi) to produce a long tapered tip for deep tissue injection. If using the Sutter Instrument, use program 21-1 (Heat—600, Pull—29, Velocity—37, Time 150).

3. Once pulled, trim the micropipette needles to an ID of 20–30 μm for mice and 30–35 μm for rats using #5 forceps with the aid of a micrometer. Each hash mark on a micrometer represents 5 μm. Note: The ID is based on the viscosity of solution, cell diameter (if cellular therapies are used), and size of the animal being injected. The larger the ID, the less resistance for fluids.
4. Once an ID has been achieved, bevel pipettes using a Sutter Instrument model BV-10 micropipette beveler (or equivalent). Tips should be beveled at 35–45° to facilitate tissue penetration. After beveling, flush pipette needles with 100% ethanol to remove any debris.
5. Always prepare 5–10 pipettes for each operative day to replace broken or clogged needles.
6. Sterilize glass pipettes by laying them on adhesive tape inside a petri dish exposing them to UV light for one hour.

Preparing the Micropipette for Injection

1. Reconstitute lysolecithin to a 1% solution with sodium chloride (NaCl; pH 7.4) or prepare 0.2 mg/mL ethidium bromide solution [9]. Store lysolecithin in small aliquots (75 μL) at −20 °C. Thaw aliquots to room temperature before use. Keep ethidium bromide at room temperature.
2. Handle the pre-pulled glass pipette with extreme caution to avoid damaging the fine tip. Label micropipette needles with hash marks to represent 0.25 μL between each mark.
3. Attach the pipette needle directly to the arm of the stereotactic device.
4. Load the pipette needle by connecting a 60 mL syringe to the inlet of the vacuum compressed air tip with plastic tubing. Make sure seal is tight so that no air escapes.
5. Once connected, lower the pipette needle into your demyelinating solution.
6. Gently pull back on the syringe plunger to slowly fill the pipette needle with glial toxin until the required amount is achieved. Note: Avoid introducing air bubbles into the pipette needle. If air bubbles are present, the pipette preparation must be repeated. Pipettes will be able to inject 4–8 animals before needing to be refilled.
7. Remove the 60 mL syringe from the vacuum inlet and reconnect the pneumatic pico-pump (PV800) to the stereotaxic frame.

Pre-Surgical Animal Preparation

1. Anesthetize mice with an intraperitoneal injection (IP) of ketamine (50 mg/kg) and xylazine (5 mg/kg) or 2.0% avertin (2,2,2-tibromoethenol in 1.25% 2-methyl-2-butanol in sodium chloride solution) at 400 mg/kg using a decapi-cone or by grabbing the scruff of the animal at the shoulder blades. Make sure the animal is safely secured before inserting the needle. Insert the needle with the bevel up and gently pull back on the syringe to test for proper placement. Note: When the

needle is properly placed you will get negative pressure. If blood, feces or urine are pulled into the syringe then remove the needle without injecting, discard the syringe and repeat the procedure again. Avertin is not an FDA approved drug and needs IACUC approval before use.
2. Place the animal in a secured location until they are fully anesthetized. The animal will be under anesthesia for approximately 45 min–1 h.
3. Firmly pinch the toe of the animal to assure that adequate sedation has been achieved. A properly anesthetized animal will not respond to toe pinch. Anesthetic depth should be monitored every 5–10 min throughout the surgical procedure.
4. Shave a 2 in. square along the dorsal spinal column using clippers. Make sure the clipper blade lays flat against the animal's skin to prevent tissue damage.
5. Make sure all clipped hair has been removed from the surgical area.
6. Clean the surgical area with 100% chlorohexidine solution applied to a 2 × 2 gauze pad. Then repeat using a 10% chlorhexidine solution. Carefully apply a small amount of lacrilube to the eyes to prevent drying during the surgical procedure.
7. Administer 1 mL of pre-warmed subcutaneous fluids (SQ) to prevent dehydration.

Performing the Surgical Procedure

To create the desired demyelinating lesion, sterotaxis is used to align the depth of the injection in the ventral-dorsal plane as well as the location of the injection in the medial-lateral plane. To generate demyelinating lesions of the venterolateral funiculus (VLF), the following parameters are used:

- VLF lesions in rats, inject at 1.6 and 1.3 mm depths and 0.7 mm lateral to midline.
- VLF lesions in mice, inject at 1.1 and 0.7 mm depths and 0.45 mm lateral of midline.
- Two unilateral injections spaced 1.75 mm apart were given at a depth of 1.1 and 0.7 mm and 0.45 mm lateral of midline for this demonstration.

Additional Comments: Aseptic technique should be used for all steps during this procedure (gloves, bonnet, mask). Surgical tools should be sterilized before coming in contact with the animal. Each surgical procedure may take 15–20 min. It is recommended to have at least 1–2 people assisting for each surgical procedure. If anesthetic boosting is required during the surgical procedure inject ketamine (50 mg/kg) IP or Avertin (20–40 mg/kg) IP. This surgical procedure may be modified for rats.

1. Move the animal to the surgical table dorsal side up. Surgical tape (optional) may be used to secure the animal to the operating station.
2. Under a surgical microscope with adequate lighting, use small surgical scissors to make a sagittal incision in the skin from T6 to T12.
3. Cut away the connective tissue overlaying the musculature of the dorsal column.

4. Using extra fine spring scissors, separate the muscles from the lamina, exposing the lateral aspects of the facets bilaterally. Adequate exposure is important to allow the vertebral column to be grasped when lifted onto the stabilizer using the facets. Note: The spine stabilizer used here was developed by the Louisville Injury System Apparatus (LISA). However other devices may work well provided the spine is securely immobilized.
5. Using forceps or closed spring scissors, feel for the hard surface of T11 (T11 will protrude slightly higher than the other vertebrae).
6. Once T11 has been located, use spring scissors to make shallow lateral cuts (2–3 mm deep) in the connective tissue between T8–T9 and T9–T10 repeat between T10 and T11. Note: If bleeding occurs hold sponge spears or cotton tipped applicators to area for 15–30 s or until bleeding subsides. Gel-foam may be applied to the area as aid for coagulation.
7. Using Graefe forceps pull the T9 and T10 vertebrae up and off of the column. Removal of the vertebrae will slightly expose the spinal cord so that a laminectomy may be performed.
8. Place the animal in the Louisville Injury System Apparatus or other equivalent device.
9. Use extra fine Graefe forceps to grasp the lateral facets and place the toothed blades of the stabilizer on either side of the facets, thereby securing the vertebral column in the stabilizer by tightening both sides of the stabilizer. Note: The vertebral column must be aligned parallel with the base of the stabilizer.
10. Very gently use #2 laminectomy forceps to remove the lamina from the vertebral column. Make sure the area does not contain any sharp edges which may interfere with pipette placement.
11. Clean the interlaminar soft tissue with a cotton tipped applicator. Open the dura using the tip of a bent (90°) 30 gauge needle. Once the dura has been penetrated expand the opening with laminectomy forceps. Remove any cerebrospinal fluid that may have leaked out when the dura was penetrated. Note: The dural opening should only be large enough to allow the micropipette tip to enter the spinal cord.
12. Transfer the animal in the spine stabilizer to the injection system (see accompanying movie). Place the animal perpendicular to the tip of the micropipette. Align the tip to the midline of the dural opening. Slowly lower the pipette tip until it is just 1–2 mm above the cord.
13. Using the graded measurements of the Y-axis, move the pipette needle 0.4 mm lateral of midline.
14. Slowly move the pipette needle in a downward direction until it just barley touches the spinal cord to obtain a surface reading of x-axis. From this reading, subtract 1.1 mm. Use a quick downward motion to pierce the spinal cord surface and slowly lower the pipette tip until the new measurement is reached.
15. Using the pico-pump, deliver half of the total volume per injection site of glial toxin into the spinal cord. Wait 2 min to allow the demyelinating agent to diffuse.

Note: Lesions depend not only on the target but also on the volume and concentration of the solution, each of which has a specific pattern of diffusion.

16. Using your x-axis reading, add 0.5 mm. Raise the tip to this new measurement. Inject a second bolus of glial toxin into the spinal cord. Wait 2 min. Note: If lesions are required at a different location, refer to a mouse [33] or rat [34] brain atlas for specific injection coordinates.
17. Remove the pipette needle from the spinal cord.
18. Calculate the z-axis to move 1.75 mm caudal to the first injection site. Note: This only applies if a second injection site is needed.
19. Repeat steps 14–17.
20. Loosen the pins of the facet blades and remove the animal from the spinal stabilizer apparatus.
21. Using 4-0 or 5-0 monofilament, tie a single suture through the muscle and adipose tissue overlaying the spinal column.
22. Close the skin incision by applying wound clips using a staple applicator. Apply bacitracin to the incision area. Administer 1 mL of pre-warmed SQ NaCl, and gentamicin (5 mg/kg) SQ. Place the animal in a recovery cage that has been placed halfway onto a warm water circulating heating pad. Note: Only place the cage half way on the heating pad so the animal may escape the warmed area if over-heating occurs.
23. Administer analgesics as directed by the institute's animal care and use program. Animals are fully ambulatory and capable of self-feeding and drinking as soon as they recover from anesthesia.
24. Repeat the procedure on any remaining animals. Note: Between each use the micropipette tip should be submerged in liquid to prevent clogging. The same pipette needle may be used throughout 4–8 surgeries before the needle dulls and has to be replaced. Separate needles should be used when multiple demyelinating agents are being considered. All used micropipette needles should be discarded at the end of surgery. It is not recommended to reuse needles.

Post-op Animal Care

1. Keep animals on a water-heating pad (cage half on half off) overnight in a recovery room. Cages should contain alpha-dri bedding with food pellets laid directly on the alpha-dri for ease of access.
2. Return animals to normal housing condition 24 h post-surgery.
3. Administer pain meds Buprenorphine (0.1 mg/kg) for 48 h post-surgery, or other IACUC approved pain medication.
4. Administer antibiotics Gentamicin (5 mg/kg) SQ once daily for 7–10 days, or other IACUC approved antibiotic. Fluids may also be given as need for dehydration.
5. Keep daily health logs to record each animal's health status for the remainder of the study.

Applications, Complications, and Limitations

(a) When injecting larger volumes into the spinal cord, the volume of fluid itself can compress the surrounding spinal cord or the force of injection can cause fluid to leak out through the needle trajectory. It is recommended two depths be injected so that the volume at each depth is half the total. This prevents undesirable fluid leakage. Each glial-toxin has different diffusion properties as well so the two injection sites also provide a more consistent concentration of toxin throughout the desired target area. The concentration, volume and stereotaxic coordinates for any desired lesion must be optimized empirically.

(b) Micropipette needles may become clogged in between animal injections. To prevent this lower the pipette needle into the glial toxin in between animal injections. If the needle still becomes clogged use the pico-pump to displace the clog. If a needle becomes clogged during injection raise the needle up and out of the spinal cord, gently swab the tip with a cotton tipped applicator that has been moistened with saline and use the pico-pump to displace the clog. Avoid using high pressure settings while the needle is still in the spinal cord. High pressure may cause the needle to flood the cord with too much glial toxin compromising the procedure.

(c) All of these lesions can be used to follow the spatiotemporal progression of remyelination by both endogenous or engrafted myelinating cells, e.g. oligodendrocyte precursor cells (OPCs). If one wants only to look at the contribution of engrafted myelinating cells, the lesion site can be X-irradiated [35, 36] either before or after SCI to prevent endogenous OPCs and Schwann cells from proliferating and migrating to the lesioned area and remyelinating. If endogenous remyelination is to be evaluated, the relative contributions of oligodendrocyte and Schwann remyelination must be quantified as both cells will migrate to the lesion site [11]. Lesions of the dorsal columns remyelinate more efficiently than do lesions of the lateral or ventral white matter.

(d) One must be extremely careful to monitor axonal loss with these models. This can be done by immunostaining for neuron-specific antigens such as the neurofilament proteins. Silver staining can also be used. It is important that longitudinal sections be examined as it is difficult to assess axon continuity on cross sections. The extent and progression of the inflammatory response, which can affect both axonal pathology and remyelination, can be monitored by immunostaining for microglia/macrophage markers, OX42 or ED1 for the rat or Iba1 for the mouse or the leukocyte common antigen CD45 which identifies all rat and mouse leukocytes.

(e) If axons are demyelinated chronically in the mouse, they can degenerate and/or become not patent for remyelination [9]. In all these lesion models, demyelination is complete within 7 days and it is recommended that therapeutic treatments are not delayed more than 1 month post-demyelination.

Illustrative Results

The optimal approach to assess remyelination is to combine multiple outcome measures. We routinely use electrophysiological methods in combination with terminal histological assessment of remyelination. What electrophysiological approach one takes depends on where anatomically in the spinal cord the demyelination lesion was performed. Conduction is measured by somatosensory evoked potentials (SSEPs) for the ascending dorsal columns [37], transcranial magnetic motor evoked potentials (tcMMEPs) for the medial aspects of the VLF [36], and magnetic inter-enlargement reflexes for the mediolateral white matter [38]. A detailed description of these methods is beyond the scope of this chapter, but changes in conduction can be measured by electrophysiological assessment. Histologically, one must quantify both spared white matter and the extent of remyelination and by which cells remyelination occurs. It is imperative that the latter assessment utilize unbiased stereological methods. Figure 1 shows representative results of how EB lesions and lysolecithin lesions may appear after injection. For detailed description of tissue dissections, histological (EC staining) and immunohistochemical methods, see: http://louisville.edu/kscirc/basic-research/laboratories/laboratory-of-molecular-neurobiology/protocols-1.

Conclusions

As outlined above, all of these models have strengths and weaknesses depending on the questions to be asked. None of them faithfully reproduce the clinical spectrum of demyelinating diseases. However, if one's experimental questions are carefully tailored and the inherent caveats are recognized, they provide model systems that enable physiologically relevant questions about CNS remyelination to be addressed.

Acknowledgements Supported by RR15576/GM103507, NS054708, Norton Healthcare, Commonwealth of Kentucky Challenge for Excellence, and the Kentucky Spinal Cord and Head Injury Research Trust.

References

1. Woodruff RH, Franklin RJ. Demyelination and remyelination of the caudal cerebellar peduncle of adult rats following stereotaxic injections of lysolecithin, ethidium bromide, and complement/anti-galactocerebroside: a comparative study. Glia. 1999;25:216–28.
2. Matsushima GK, Morell P. The neurotoxicant, cuprizone, as a model to study demyelination and remyelination in the central nervous system. Brain Pathol. 2001;11:107–16.
3. Herder V, Hansmann F, Stangel M, Skripuletz T, Baumgartner W, Beineke A. Lack of cuprizone-induced demyelination in the murine spinal cord despite oligodendroglial alterations substantiates the concept of site-specific susceptibilities of the central nervous system. Neuropathol Appl Neurobiol. 2011;37:676–84.

4. Blakemore WF. The response of oligodendrocytes to chemical injury. Acta Neurol Scand Suppl. 1984;100:33–8.
5. Blakemore WF. Ethidium bromide induced demyelination in the spinal cord of the cat. Neuropathol Appl Neurobiol. 1982;8:365–75.
6. Desjardins P, Frost E, Morais R. Ethidium bromide-induced loss of mitochondrial DNA from primary chicken embryo fibroblasts. Mol Cell Biol. 1985;5:1163–9.
7. Hayakawa T, Noda M, Yasuda K, Yorifuji H, Taniguchi S, Miwa I, Sakura H, Terauchi Y, Hayashi J, Sharp GW, Kanazawa Y, Akanuma Y, Yazaki Y, Kadowaki T. Ethidium bromide-induced inhibition of mitochondrial gene transcription suppresses glucose-stimulated insulin release in the mouse pancreatic beta-cell line betaHC9. J Biol Chem. 1998;273:20300–7.
8. Hayashi J, Tanaka M, Sato W, Ozawa T, Yonekawa H, Kagawa Y, Ohta S. Effects of ethidium bromide treatment of mouse cells on expression and assembly of nuclear-coded subunits of complexes involved in the oxidative phosphorylation. Biochem Biophys Res Commun. 1990;167:216–21.
9. Kuypers NJ, James KT, Enzmann GU, Magnuson DS, Whittemore SR. Functional consequences of ethidium bromide demyelination of the mouse ventral spinal cord. Exp Neurol. 2013;247:615–22.
10. Crang AJ, Blakemore WF. Remyelination of demyelinated rat axons by transplanted mouse oligodendrocytes. Glia. 1991;4:305–13.
11. Penderis J, Shields SA, Franklin RJ. Impaired remyelination and depletion of oligodendrocyte progenitors does not occur following repeated episodes of focal demyelination in the rat central nervous system. Brain. 2003;126(Pt 6):1382–91.
12. Graca DL, Blakemore WF. Delayed remyelination in rat spinal cord following ethidium bromide injection. Neuropathol Appl Neurobiol. 1986;12(6):593–605.
13. Talbott JF, Loy DN, Liu Y, Qiu MS, Bunge MB, Rao MS, Whittemore SR. Endogenous Nkx2.2+/Olig2+ oligodendrocyte precursor cells fail to remyelinate the demyelinated spinal cord in the absence of astrocytes. Exp Neurol. 2005;192:11–24.
14. Talbott JF, Cao Q, Achim V, Benton RL, Enzmann GU, Mills MD, Rao MS, Whittemore SR. Schwann cell differentiation of adult oligodendrocyte precursor cells engrafted into the demyelinated spinal cord is BMP-dependent. Glia. 2006;54:147–59.
15. Miron VE, Franklin RJ. Macrophages and CNS remyelination. J Neurochem. 2014;130:165 71.
16. Arnett HA, Fancy SP, Alberta JA, Zhao C, Plant SR, Kaing S, Raine CS, Rowitch DH, Franklin RJ, Stiles CD. bHLH transcription factor Olig1 is required to repair demyelinated lesions in the CNS. Science. 2004;306:2111–5.
17. Blakemore WF, Franklin RJ. Remyelination in experimental models of toxin-induced demyelination. Curr Top Microbiol Immunol. 2008;318:193–212.
18. McKay JS, Blakemore WF, Franklin RJ. Trapidil-mediated inhibition of CNS remyelination results from reduced numbers and impaired differentiation of oligodendrocytes. Neuropathol Appl Neurobiol. 1998;24:498–506.
19. Zhu Q, Whittemore SR, Devries WH, Zhao X, Kuypers NJ, Qiu M. Dorsally-derived oligodendrocytes in the spinal cord contribute to axonal myelination during development and remyelination following focal demyelination. Glia. 2011;59:1612–21.
20. Keirstead HS, Blakemore WF. Identification of post-mitotic oligodendrocytes incapable of remyelination within the demyelinated adult spinal cord. J Neuropathol Exp Neurol. 1997;56:1191–201.
21. Kuypers NJ, Bankston AN, Howard RM, Beare JE, Whittemore SR. Remyelinating oligodendrocyte precursor cell miRNAs from the Sfmbt2 cluster promote cell cycle arrest and differentiation. J Neurosci. 2016;36:1698–710.
22. Walshe JM. Copper: not too little, not too much, but just right. J R Coll Physicians Lond. 1995;29:280–8.
23. Kipp M, Clarner T, Dang J, Copray S, Beyer C. The cuprizone animal model: new insights into an old story. Acta Neuropathol. 2009;118:723–36.

24. Morell P, Barrett CV, Mason JL, Toews AD, Hostettler JD, Knapp GW, Matsushima GK. Gene expression in brain during cuprizone-induced demyelination and remyelination. Mol Cell Neurosci. 1998;12:220–7.
25. Jurevics H, Hostettler J, Muse ED, Sammond DW, Matsushima GK, Toews AD, Morell P. Cerebroside synthesis as a measure of the rate of remyelination following cuprizone-induced demyelination in brain. J Neurochem. 2001;77:1067–76.
26. Tsunoda I, Fujinami RS. Two models for multiple sclerosis: experimental allergic encephalomyelitis and Theiler's murine encephalomyelitis virus. J Neuropathol Exp Neurol. 1996;55:673–86.
27. Zamvil SS, Steinman L. The T lymphocyte in experimental allergic encephalomyelitis. Annu Rev Immunol. 1990;8:579–621.
28. Kuchroo VK, Anderson AC, Waldner H, Munder M, Bettelli E, Nicholson LB. T cell response in experimental autoimmune encephalomyelitis (EAE): role of self and cross-reactive antigens in shaping, tuning, and regulating the autopathogenic T cell repertoire. Annu Rev Immunol. 2002;20:101–23.
29. Alley J, Khasabov S, Simone D, Beitz A, Rodriguez M, Njenga MK. More severe neurologic deficits in SJL/J male than female mice following Theiler's virus-induced CNS demyelination. Exp Neurol. 2003;180:14–24.
30. Constantinescu CS, Farooqi N, O'Brien K, Gran B. Experimental autoimmune encephalomyelitis (EAE) as a model for multiple sclerosis (MS). Br J Pharmacol. 2011;164:1079–106.
31. Robinson AP, Harp CT, Noronha A, Miller SD. The experimental autoimmune encephalomyelitis (EAE) model of MS: utility for understanding disease pathophysiology and treatment. Handb Clin Neurol. 2014;122:173–89.
32. Mecha M, Carrillo-Salinas FJ, Mestre L, Feliu A, Guaza C. Viral models of multiple sclerosis: neurodegeneration and demyelination in mice infected with Theiler's virus. Prog Neurobiol. 2013;101–102:46–64.
33. Franklin KBJ, Paxinos G. The mouse brain in stererotaxic coordinates. San Diego: Academic Press; 1997.
34. Paxinos G, Watson C. The rat brain in stereotaxic coordinates. San Diego: Academic; 1998.
35. Loy DN, Talbott JF, Onifer SM, Mills MD, Burke DA, Fajardo LC, Dennison JB, Magnuson DSK, Whittemore SR. Both dorsal and ventral spinal cord pathways contribute to overground locomotion in the adult rat. Exp Neurol. 2002;177:575–80.
36. Loy DN, Magnuson DS, Zhang YP, Onifer SM, Mills MD, Cao QL, Darnall JB, Fajardo LC, Burke DA, Whittemore SR. Functional redundancy of ventral spinal locomotor pathways. J Neurosci. 2002;22:315–23.
37. Hill RL, Zhang YP, Burke DA, Devries WH, Zhang Y, Magnuson DS, Whittemore SR, Shields CB. Anatomical and functional outcomes following a precise, graded, dorsal laceration spinal cord injury in C57BL/6 mice. J Neurotrauma. 2009;26(1):1–15.
38. Beaumont E, Onifer SM, Reed WR, Magnuson DS. Magnetically evoked inter-enlargement response: an assessment of ascending propriospinal fibers following spinal cord injury. Exp Neurol. 2006;201:428–40.

Spinal Cord Lateral Hemisection and Implantation of Guidance Channels

Xiaofei Wang and Xiao-Ming Xu

Abstract Spinal cord laceration injuries such as complete transection, lateral and dorsal hemisection, over hemisection, and dorsal funiculotomy in rodents have been widely used in the studies of neuronal responses to axotomy as well as axonal regeneration through and beyond the site of the injury. The latter approach has utilized transplantation of fetal cord tissues, pieces of peripheral nerves, and biosynthetic guidance channels seeded with different cell types, trophic factors and matrix molecules. In this chapter, we introduce hands-on procedures to perform a spinal cord lateral hemisection, create a lesion cavity on the same side, and transplant a piece of guidance channel that bridges the two ends of the cord stumps. By suturing the dura, cerebrospinal fluid circulation is restored and connective tissue invasion into the lesion site is prevented. With modification, these procedures should also apply to creating above-mentioned laceration injuries when micro-scissors are used.

Keywords Spinal cord injury · Laceration · Hemisection · Transection · Transplantation · Guidance channel

Model Selection

Spinal cord lateral hemisection and guidance channel transplantation is a useful model for studying axonal regeneration following spinal cord injury (SCI). This model was developed based on an earlier model in which the spinal cord was completely transected followed by implantation of either a blind-ended [1] or open-ended [2] semi-permeable guidance channel, seeded with Schwann cells, into a 6–8 mm-long gap created between the two stumps of the transected spinal cord. The lateral

X. Wang · X.-M. Xu (✉)
Spinal Cord and Brain Injury Research Group, Department of Neurological Surgery, Stark Neurosciences Research Institute, Indiana University School of Medicine, Indianapolis, IN, USA
e-mail: wang88@iupui.edu; xmxu0001@louisville.edu; xu26@iupui.edu

J. Chen et al. (eds.), *Animal Models of Acute Neurological Injury*, Springer Series in Translational Stroke Research, https://doi.org/10.1007/978-3-030-16082-1_37

hemisection and guidance channel implantation model was first reported in 1999 [3] and is characterized by a lateral hemisection on one side of the spinal cord, creation of a lesion gap of desired length, implantation of a hollow channel seeded with Schwann cells (or other cell types) into the lesion gap, and restoration of cerebrospinal fluid (CSF) flow by suturing the dura. Major advantages of this model include:

1. Studying regeneration of specific axonal pathways that project unilaterally in the spinal cord, such as the rubrospinal tract, while their counterparts on the intact side serve as internal controls.
2. Allowing the lesion and transplantation to be made at all levels of the spinal cord, including the cervical level as compared to a complete spinal cord transection which is limited by animal survival.
3. Restoring CSF circulation after dura repair ensuring the delivery of nutrients to the graft tissue and prevention of connective tissue invasion to the graft region.
4. Creating more stable cord-graft interfaces than a complete spinal cord transection due to the function of intact hemi-cord.
5. Providing a modifiable channel environment with unlimited potential to test different trophic factors, matrix molecules and cell types in their ability to promote axonal regeneration, modify the graft-host interface, and protect axotomized host neurons from injury induced cell death.
6. Providing a relatively isolated graft environment which is excellent for axonal tract tracing to identify origins of neurons whose axons have regenerated into the graft environment.

Materials

Adult Fisher or Sprague Dawley (SD) rats weighing 160–180 g are the animals of choice. For the surgical procedures, the following materials are recommended (Fig. 1).

1. Rat vertebral stabilizer (Developed by YP Zhang and CB Shields, University of Louisville)
2. 10-0 black monofilament nylon (Ethilon 2888)
3. Retractor (Agricola, F.S.T. 17005-04)
4. Spring scissors with especially fine, small blades (F.S.T. 15002-08)
5. Spring scissors with especially fine, small Blades (F.S.T. 15005-08)
6. Moria forceps (F.S.T. 11154-10)
7. Forceps (Adson, F.S.T. 11027-12)
8. Dumont angled forceps (F.S.T. 11253-20)
9. Dumont forceps (F.S.T. 11251-35)
10. Needle holder (F.S.T. 12500-12)
11. Guidance channel (outer diameter—1.25 mm; length—3 mm)
12. Scalpel handle (F.S.T. 10003-12) and scalpel blades (F.S.T. 10015-00)
13. Spring scissors with especially fine, small blades (F.S.T. 15013-12)

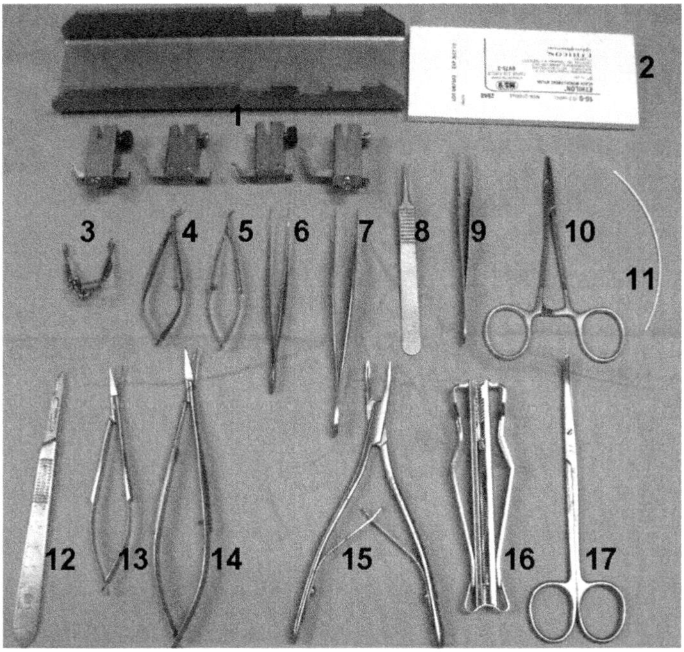

Fig. 1 List of tools (1–17) used for this model

14. Spring scissors with especially fine, small blades (F.S.T. 15370-26)
15. Friedman-Pearson micro rongeur (F.S.T. 16021-14)
16. MikRon 9 mm autoclip
17. Fine iris forceps (F.S.T. 14017-14)
18. All-purpose sponges (Kendall 9022)
19. Cotton tipped applicators (Henry Schein 100-6015)
20. A surgical microscope

Procedures

1. Anaesthetize a rat with sodium pentobarbital (50 mg/kg; i.p.).
2. Inject an antibiotic (such as Ampicillin, 150 mg/kg, i.m.) to prevent infection.
3. Apply Lacrilube ophthalmic ointment (Allergan Pharmaceuticals, Irvine, CA) to eyes to prevent drying.
4. Place the animal on a heating pad to maintain body temperature.
5. Expose the injury site by incising the skin sagittally over the ninth and tenth thoracic vertebrae (T9–10).
6. Perform a laminectomy of the T9–10 vertebrae to expose the underlying cord segments. Remove more of the lamina on both sides of the cord (Fig. 2a) using the micro rongeur (Fig. 1, Item #15).

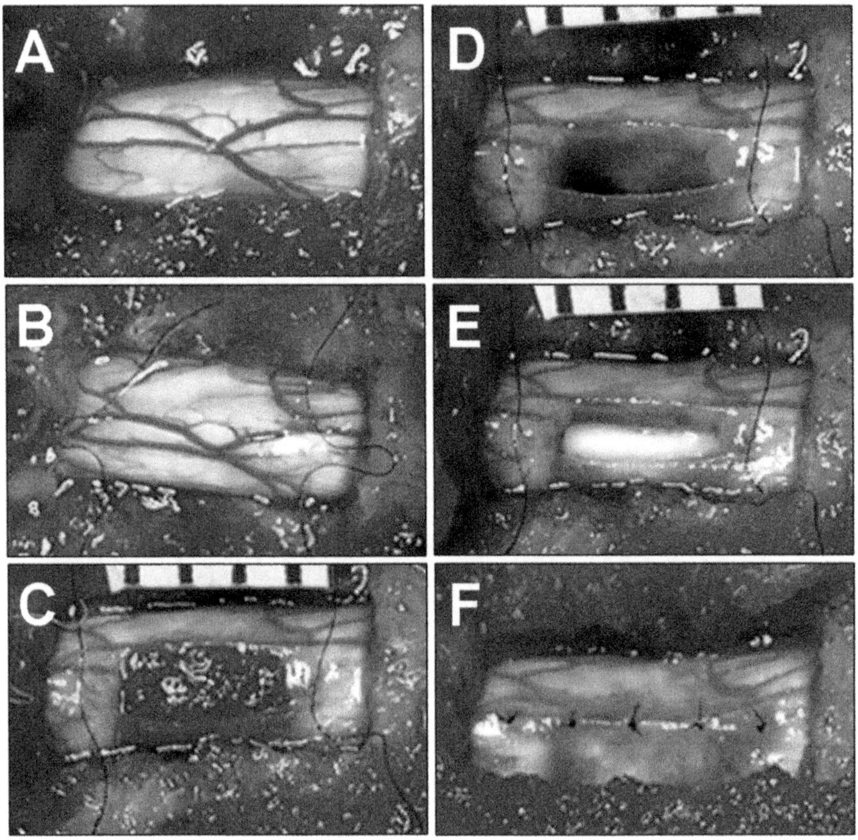

Fig. 2 Step by step procedures: (**a**) Exposure of the T9–10 spinal cord. (**b**) Incision of the dura. (**c**) Hemisection, creation of a lesion cavity, and insertion of a hemostatic sponge to stop bleeding. (**d**) Removal of the sponge to show the lesion cavity. Note that the cutting edge of two cord stumps was sharp and the lateral edge of the incised dura was well-preserved. (**e**) Implantation of a piece of guidance channel. Note the close apposition between the graft and host cord stumps at both ends. (**f**) Suture the dura. The channel and blood vessels underneath the dura can clearly be seen

7. Incise the dura longitudinally along the midline (Fig. 2b) using spring scissors with especially fine blades (Fig. 1, Item #4).
8. Pull the dura laterally, stitch a 10-0 suture thread on each side of the incised dura at both rostral and caudal ends of the incision and leave the ends of the sutures free (Fig. 2b).
9. Make a 3-mm long sagittal cut with a spring scissors (Fig. 1, Item #4) along the midline of the spinal cord. The incision should be made through the entire depth of the spinal cord and the tip of scissors should touch the vertebral canal while cutting.
10. Pull the edge of the cut dura laterally with one angled forceps (Fig. 1, Item #8) and make two transverse cuts, one rostrally and one caudally using a spring scissors (Fig. 1, Item #5), of the hemi-cord to create a gap of 2.5–2.8 mm (Fig. 2c).

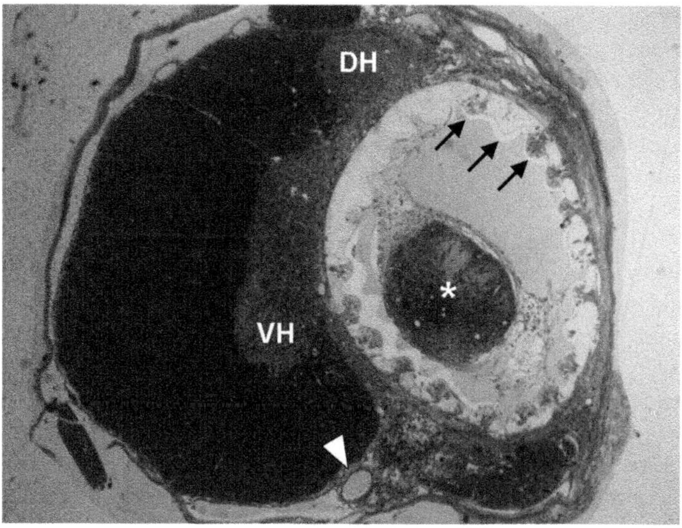

Fig. 3 A Toluidine Blue-stained plastic embedded transverse section through the graft mid-point shows the guidance channel (arrows) and regenerated tissue cable (asterisk) within the channel. Note that the ventral (VH) and dorsal (DH) horns are easily identified in the intact left hemi-cord. An arrowhead indicates the ventral spinal artery

11. Remove the hemisected cord tissue by vacuum aspiration and stop bleeding by inserting a piece of absorbable collagen hemostatic sponge of desired size into the lesion cavity (Fig. 2c).
12. Remove the sponge and make sure that there is no bleeding (Fig. 2d).
13. Trim the guidance channel into a 3 mm-long segment. Insert one end of the channel into the lesion cavity in such a way that the caudal stump of the cord is positioned slightly within the caudal opening of the channel. Push the rostral end of the channel down to the bottom of the cavity, care should be taken not to damage the rostral stump of the cord during this process. Slightly slide the channel rostrally so that the rostral stump of the hemi-cord is also tightly apposed to the rostral opening of the channel (Fig. 2e).
14. Close the dura with 5–6 stitches of 10-0 nylon suture threads (Fig. 2f).
15. Suture the muscle and skin in layers.

Outcome Evaluations

1. Histology: To evaluate whether the spinal cord hemisection is complete, toluidine blue-stained semi-thin transverse sections through the lesion/graft mid-point should be prepared (Fig. 3). Quantitative measures including the number of myelinated axons, blood vessels, and regenerative tissue cable surface areas can be conducted at the light microscopic level. Ultra-thin sections of the same tissue

block can be examined at the electron microscopic level and the ratio between myelinated and unmyelinated axons as well as the total number of regenerated axons can be estimated. Correlations between these parameters and functional assessments after various treatments may also be established.
2. Immunohistochemistry: Immunohistochemistry or immunofluorescence double-labeling can be used to examine the survival of grafted cells, the association between grafted cells and regenerated axons, the astrocytic response to the injury and transplantation, the regeneration of supraspinal axons that express tract-specific transmitters, such as serotonin (5-HT) for the raphe-spinal pathway, and dopamine β-hydroxylase (DBH) for the ceroleo-spinal pathway, and the cavity formation at the graft-host interface [1, 4]. These parameters can also be correlated with functional assessments after various treatments.
3. Retrograde tracing: Retrograde tracers such as Fast Blue (FB) or Fluoro-Gold (FG) may be injected into the graft mid-point or beyond the graft to label neuronal cell bodies whose axons regenerated into or beyond the graft, respectively [1]. If the injection was made distal to the graft, the intact hemicord should be completely transected prior to the tracer injection to prevent the labeling of neurons whose axons course around, rather than thorough, the graft.
4. Anterograde tracing: Axonal regeneration through and beyond the bridge transplant can be examined using anterograde tracers such as Phaseolus Valgaris-Leucoagglutinin (PHA-L) or Biotinylated-Dextran Amine (BDA) by injecting the tracer into specific regions of the cortex, brainstem, spinal cord, or dorsal root ganglia to label corticospinal, brainstem-spinal, propriospinal, or dorsal ascending tracts, respectively. Prior to the injection, the intact hemicord should be completely transected to prevent the labeling of axons course around, rather than thorough, the graft to the distal host spinal cord [5].
5. Behavior assessments: According to our experience, grid walking and footprint analysis are sensitive measures for spinal cord hemisection injuries whereas Basso, Beattie and Bresnahan (BBB) open-filed locomotor rating scale and Louisville Swimming Scale (LSS) are less sensitive (unpublished observations).

Applications, Complications and Limitations

1. The model is ideal for testing axonal regeneration of proprio- and supraspinal pathways. Importantly, it is appropriate for testing the growth-promoting potential of different trophic factors, matrix molecules and cell types since the channel environment can be modified. This model is not appropriate for testing tissue sparing at the site of injury.
2. When the spinal cord is completely transected, a large channel (inner diameter: 2.6 mm) can be placed into the lesion cavity with two cord stumps inserted into the openings of the channel [2]. Evidence for regeneration observed in the complete transection model is more convincing. However, due to the inability

to repair the dura and instability of the host-graft interface in the complete transection model, axonal regeneration is less robust than the hemisection model.

3. The creation of spinal cord hemi-cavity on one side of the cord could also be used for transplantation of other tissues such as segments of peripheral nerves [6] and fetal spinal cord tissues [7].
4. The procedures for creating a spinal cord hemisection without inducing a lesion cavity by using spring scissors or a VibraKnife™ device can also be used for creation of a complete transection [8], lateral over-hemisection [9], dorsal hemisection (transection of the dorsal half of the spinal cord) [10, 11], and dorsal funiculotomy (transection of the dorsal funiculus that contains the corticospinal tract and dorsal ascending tracts in rodents) [12].
5. Keep the midline: To ensure that the hemisection is made exactly through the midline, the spring scissors should be vertically inserted to the vertebral canal and the tips aligned along the longitudinal axis of the spinal cord. This is particularly important if functional assessments are to be conducted which require consistent amounts of spared hemi-cord on the contralateral side. In many cases, dorsal vertebral veins are running irregularly on the surface of the cord. These vessels cannot be used as a landmark for the midline and should be cut if necessary. According to our experience, cutting such vessels does not cause much bleeding and, when bleeding occurs, it could be stopped easily.
6. Prevent cord retraction: After hemisection, the two stumps of the hemisected cord may retract for a short distance. Therefore, the cutting distance between the two stumps should be between 2.5–2.8 mm to ensure a tight apposition between the two stumps and the open ends of a 3 mm-long channel.
7. Avoid damaging the hemisected cord stumps: We found that cutting the cord stumps with spring scissors is better than using scalpel blades because the former induces a sharp laceration. Multiple cuts should be avoided. While introducing guidance channels, care should be taken not to push the channel wall against the cord stumps. If the cord stumps are damaged, they will swell and the regenerative outcomes will be poor.
8. Stop bleeding completely: Bleeding may occur during the surgical procedures. Stop it completely during the procedures, including prior to the transplantation.
9. Suture the dura: We feel that suturing the dura is critical for creating a growth-promoting environment and avoiding connective tissue invasion and scar formation. Extra care should be taken to not damage the dura during the surgical procedures and to suture it carefully after transplantation.
10. The rat vertebral column stabilizer we use was developed by Drs. Christopher B. Shields (cbshields1@aol.com) and Y. Ping Zhang (ypzhan01@louisville.edu) at University of Louisville. If not available, other rat stabilizers may be used.
11. The semi-permeable guidance channels that we use are composed of a 60:40 acrylonitrile/vinylchloride (PAN/PVC) with a nominal molecular weight cut off of 50 kDa. These channels are provided by Dr. Xuejun Wen (xjwen@clemson.edu) at Clemson University. Other biosynthetic channels or membranes can readily be tested using this model.

References

1. Xu XM, Guenard V, Kleitman N, Bunge MB. Axonal regeneration into Schwann cell-seeded guidance channels grafted into transected adult rat spinal cord. J Comp Neurol. 1995;351:145–60.
2. Xu XM, Chen A, Guenard V, Kleitman N, Bunge MB. Bridging Schwann cell transplants promote axonal regeneration from both the rostral and caudal stumps of transected adult rat spinal cord. J Neurocytol. 1997;26:1–16.
3. Xu XM, Zhang SX, Li H, Aebischer P, Bunge MB. Regrowth of axons into the distal spinal cord through a Schwann-cell-seeded mini-channel implanted into hemisected adult rat spinal cord. Eur J Neurosci. 1999;11:1723–40.
4. Iannotti C, Li H, Yan P, Lu X, Wirthlin L, Xu X-M. Glial cell line-derived neurotrophic factor-enriched bridging transplants promote propriospinal axonal regeneration and enhance myelination after spinal cord injury. Exp Neurol. 2003;183:379–93.
5. Bamber NI, Li H, Lu X, Oudega M, Aebischer P, Xu XM. Neurotrophins BDNF and NT-3 promote axonal re-entry into the distal host spinal cord through Schwann cell-seeded mini-channels. Eur J Neurosci. 2001;13:257–68.
6. Houle JD. Demonstration of the potential for chronically injured neurons to regenerate axons into intraspinal peripheral nerve grafts. Exp Neurol. 1991;113:1–9.
7. Bregman BS, Kunkel-Bagden E, Reier PJ, Dai HN, McAtee M, Gao D. Recovery of function after spinal cord injury: mechanisms underlying transplant-mediated recovery of function differ after spinal cord injury in newborn and adult rats. Exp Neurol. 1993;123:3–16.
8. Rabchevsky AG. Segmental organization of spinal reflexes mediating autonomic dysreflexia after spinal cord injury. Prog Brain Res. 2006;152:265–74.
9. Kunkel-Bagden E, Dai HN, Bregman BS. Recovery of function after spinal cord hemisection in newborn and adult rats: differential effects on reflex and locomotor function. Exp Neurol. 1992;116:40–51.
10. Iannotti C, Zhang YP, Shields LB, et al. Dural repair reduces connective tissue scar invasion and cystic cavity formation after acute spinal cord laceration injury in adult rats. J Neurotrauma. 2006;23(6):853–65.
11. Onifer SM, Zhang YP, Burke DA, et al. Adult rat forelimb dysfunction after dorsal cervical spinal cord injury. Exp Neurol. 2005;192(1):25–38.
12. Warden P, Bamber NI, Li H, et al. Delayed glial cell death following Wallerian degeneration in white matter tracts after spinal cord dorsal column cordotomy in adult rats. Exp Neurol. 2001;168:213–24.

Spinal Root Avulsion and Repair

Tak-Ho Chu and Wutian Wu

Abstract Motoneurons undergo retrograde degeneration following axonal injuries. The degree of degeneration varies with types of injuries, proximity of injury site to the cell body, species and age of the animal. Unlike other injuries, spinal root avulsion induces a rapid and massive cell loss of the motoneurons. It not only provides a mean to study the degenerative changes of spinal motoneurons but also resembles the brachial plexus injury in human. Despite they situate in the central nervous system, motoneurons have a strong ability to regenerate under a suitable environment. We herein present step-by-step procedures for performing extravertebral and intravertebral avulsion at seventh cervical segment in rat spinal cord and peripheral nerve implantation to allow regeneration of the motoneurons.

Keywords Spinal cord injury · Root avulsion · Peripheral nerve implantation · Brachial plexus · Motoneurons

Model Selection

Spinal root avulsion is a model to study degenerative changes of motoneurons responding to axonal injuries [1]. A traction force is applied on the ventral root which is then torn off from the surface of the spinal cord. The animal model resembles the most severe type of brachial plexus injuries which could occur during difficult delivery of newborns, sport injuries or car accidents in clinical cases. Unlike peripheral nerve injuries such as axotomy, crush or ligation, root avulsion is a

---

Electronic supplementary material: The online version of this chapter (https://doi.org/10.1007/978-3-030-16082-1_38) contains supplementary material, which is available to authorized users.

T.-H. Chu · W. Wu (✉)
Department of Anatomy, Li Ka Shing Faculty of Medicine, The University of Hong Kong, Hong Kong, China
e-mail: chuth@hkusua.hku.hk; wtwu@hkucc.hku.hk

J. Chen et al. (eds.), *Animal Models of Acute Neurological Injury*, Springer Series in Translational Stroke Research, https://doi.org/10.1007/978-3-030-16082-1_38

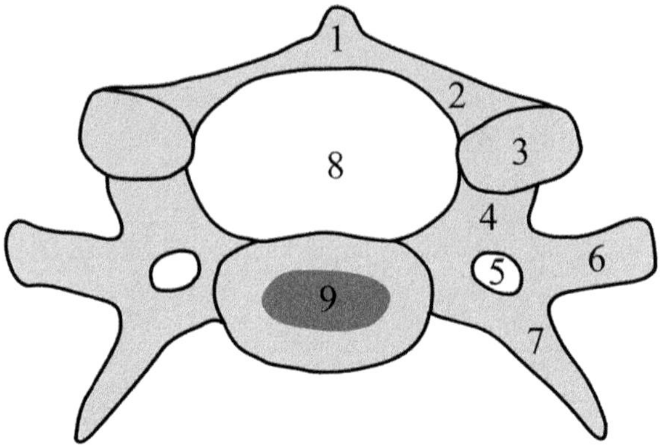

Fig. 1 Caudal surface of the sixth vertebra. 1, spinous process; 2, lamina; 3, articular process; 4, pedicle; 5, transverse foramen; 6, posterior tubercle of transverse process; 7, anterior tubercle of transverse process; 8, vertebral foramen; 9, vertebral body

central nervous system injury. Besides, close proximity of injury site to the cell body, mechanical strain resulted from traction and deprivation of peripheral trophic support contribute to the rapid degeneration of the motoneurons and lead to massive cell loss [1, 2]. Implantation of a peripheral nerve [3, 4] or avulsed ventral root [5–7] can rescue the injured motoneurons from death and allow regeneration to occur in animal studies and human patients. In this chapter, we will describe in details both intravertebral and extravertebral spinal root avulsion in the seventh cervical (C7) segment in adult rat. We will also present a protocol on implanting a peripheral nerve (PN) to allow regeneration of the motoneurons. The advantages of the model are that (1) The C7 root avulsion will not cause any observable functional disturbances in the experimental animals; (2) The surgery can be performed at different levels of spinal segments, most commonly cervical and lumbar segments and (3) It can be done with or without removal of the vertebra [1] (Fig. 1), the latter two considerations give flexibility to the experimental design.

Materials

Surgical instruments common for extra- and intravertebral avulsion are listed below, they could be obtained from various suppliers and some recommendations are given:

Basic Instruments

1. Surgical microscope (6×–40×)
2. Small animal retraction system (FST #18200-50)[1]

---

[1] A simple and economic retraction system can be made by wiring a small metal hook to a pin with a rubber band. Then the animal is placed on a cork board.

Fig. 2 A fire polished hook made of capillary tube. Scale: 1 division = 1 mm

3. 12 cm tissue forceps
4. Scalpel handle #3 with blade
5. Two Dumont #5 forceps
6. Vannas-Tübingen spring scissors (FST #15003-08)
7. Straight fine iris scissors
8. Crile Hemostat
9. Michel suture clips
10. Gel foam (Johnson & Johnson)
11. 3-0 suture (Johnson & Johnson)
12. Autoclaved cotton and 70% ethanol

Additional Tool for Extravertebral Avulsion

Serrated Halsted-mosquito hemostats (FST#13008-12) or equivalent.

Additional Tools for Intravertebral Avulsion

1. Curved Friedman-Pearson Rongeur with 1 mm cup (FST #16021-14 or Roboz #RS-8305)
2. Fire polished avulsion hook (Fig. 2) or a hooked #5 forceps can be used

Additional Tools Needed for Peripheral Nerve Implantation

1. 1.6 mL microfuge tube with normal saline to temporarily store the nerve graft
2. 10-0 and 11-0 sutures (Johnson & Johnson)
3. Micro knife (FST#10056-12) or an angled 27 gauge needle

Procedure

Anesthesia

Adult male Sprague-Dawley rats weighing 280–300 g are used. Anesthetize rats with ketamine (80 mg/kg) and xylazine (8 mg/kg) intra-muscularly or intra-peritoneally.

Extravertebral Avulsion (Fig. 3)

1. Shave the hair on the right chest
2. Lay the animal on surgical platform in supine position
3. Disinfect the skin over the shaved area
4. Make a 2 cm hockey stick incision on the right chest (Fig. 3a)

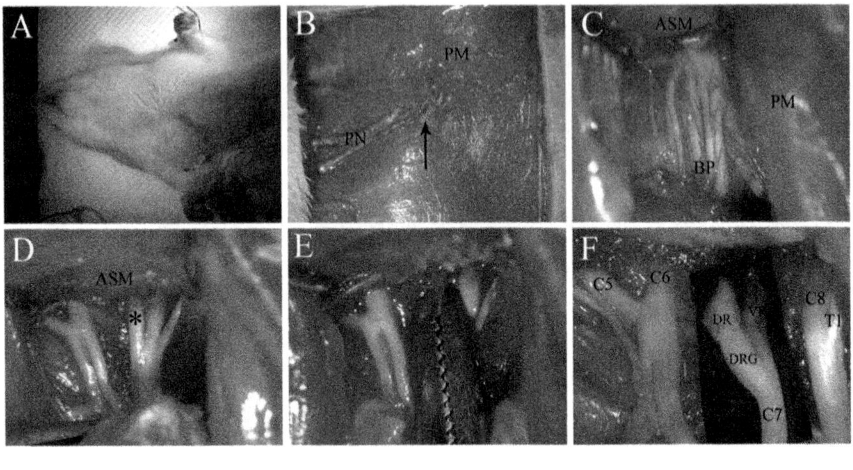

Fig. 3 Extravertebral avulsion: (**a**) An incision line on the right chest. (**b**) Pectoral nerve landmark (arrow). (**c**) Exposure of the brachial plexus. (**d**) Exposure of C5–T1 spinal nerves. Note the C7 spinal nerve (∗). (**e**) Avulsion by a hemostat. (**f**) Examination of the avulsed root. C5–C8 and T1: C5–C8 and T1 spinal nerves; *ASM* anterior scalene muscle, *BP* brachial plexus, *DR* dorsal root, *DRG* dorsal root ganglion, *PM* pectoral muscle, *PN* pectoral nerves, *VR* ventral root

5. The pectoral nerves can be seen (Fig. 3b), blunt separate the greater pectoral muscle where the nerve pierces
6. By tracing the pectoral nerve, further separate the underlying ascending pectoral muscle, retract the muscles laterally to expose the brachial plexus (Fig. 3c)
7. Identify the trunks of the brachial plexus, the C7 spinal nerve is in the middle trunk
8. Dissect between the anterior and middle scalene muscles where the nerve pass through. Retract the anterior scalene muscle to identify the C7 spinal nerve (Fig. 3d)
9. Dissect the nerve free from the surrounding tissue and as close to the vertebra as possible
10. Use a Halsted-mosquito hemostat to grab along a segment of the spinal nerve. While using one forceps to push against the vertebra, slowly pull the spinal nerve out from the intervertebral foramen (Fig. 3e)
11. Stop bleeding by applying a gel foam to the foramen
12. Cut and remove the spinal nerve with the dorsal root ganglion, dorsal and ventral roots to prevent reinnervation
13. Evaluate the success of avulsion by examining the length of the avulsed ventral root (Fig. 3f)
14. Suture close the muscles and skin in layers with 3-0 suture and Michel suture clips

Intravertebral Avulsion (Fig. 4)

1. Locate a prominent protrusion on the center of the back caudal to the neck. It is the spinous process of second thoracic vertebra (T2), shave the hair from the neck to the protrusion
2. Put the rat on the surgical platform and disinfect the shaved area with 70% ethanol
3. Make a 2.5–3 cm midline skin incision (Fig. 4a)
4. Retract the muscles laterally to expose the semispinalis cervicis muscles inserted to the cervical vertebrae
5. Using the T2 protrusion as a bony landmark, identify the sixth cervical (C6) vertebra which encloses the C7 spinal segment (Fig. 4b)
6. While using a tissue forceps to grab the T2 protrusion to stabilize the vertebrae, remove the right sided C6 lamina and articular process by a bone ronguer, in practice, the C5 lamina can also be removed to provide more space for manipulation
7. The dorsal root and its ganglia can clearly be seen beneath the dura, carefully pick up and incise the dura with #5 forceps and spring scissors. Cerebral spinal fluid (CSF) flows out, absorb the fluid before proceeding (Fig. 4c)
8. Incise the dorsal root where it attaches to the spinal cord to expose the ventral root (Fig. 4d)

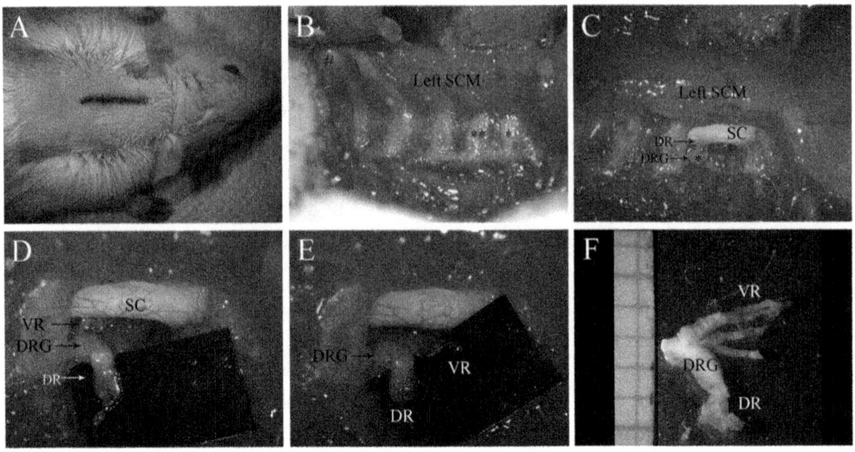

Fig. 4 Intravertebral avulsion: (**a**) An incision line on the neck. (**b**) Exposure of the C5–T2 vertebrae. Note the T2 spinous process (#). C5 (∗) and C6 (∗∗) vertebrae are indicated. (**c**) Exposure of dorsal root ganglion (∗) and dorsal root. (**d**) After dorsal root transection, ventral root (arrow) is exposed. (**e**) Avulsion of ventral root. (**f**) Avulsed spinal nerve. Scale: 1 division = 1 mm. *DR* dorsal root, *DRG* dorsal root ganglion, *Left SCM* left semispinalis cervicis muscle, *SC* spinal cord, *VR* ventral root

9. Use a fine hook to avulse the ventral root, exert a gentle traction almost parallel to the spinal cord (Fig. 4e)
10. Stop bleeding by a gel foam if necessary
11. Together with the ventral root, dorsal root and its ganglion, about 2–3 mm of the spinal nerve is severed and removed (Fig. 4f), a gel foam can be put in the intervertebral foramen if bleeding occurs
12. (Optional) Chemicals such as inhibitors, growth factors or others can be soaked in gel foam and topically applied to the avulsed site
13. (Optional) Suture the dura mater with 10-0 suture
14. Suture the muscles and skin in layers with 3-0 suture and Michel suture clips if no extra procedure is required

PN Graft Implantation (Fig. 5)

1. Shave the medial surface of the right hind leg of the animal and disinfect with 70% ethanol
2. Lay the animal on surgical platform in supine position
3. Saphenous vein should be visible, make a 2 cm skin incision (Fig. 5a)
4. Carefully incise the deep fascia along the nerve and separate the nerve from adjacent saphenous artery (Fig. 5b), remove a 2 cm segment of the nerve, label the nerve at 10 mm from the proximal end with a 10-0 suture on the epineurium (can be done in later step)

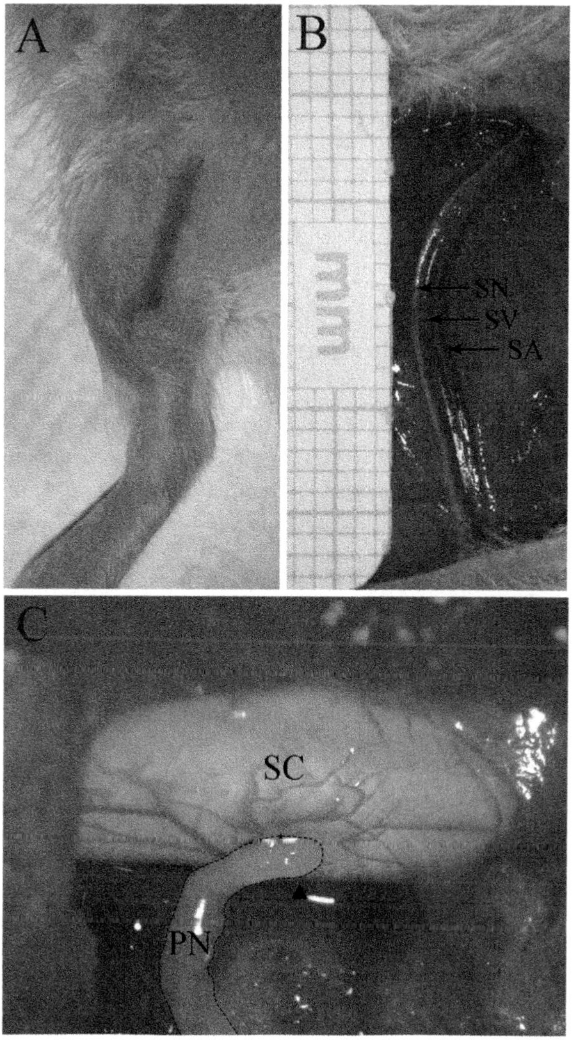

Fig. 5 Peripheral nerve graft implantation: (**a**) An incision line on the right hind leg. (**b**) Exposure of saphenous nerve and its artery and vein. (**c**) Implantation of the nerve graft (dashed line) into the spinal cord. Note the 11-0 suture point (∗) and the nerve insertion (arrow head). *PN* peripheral nerve, *SA* saphenous artery, *SC* spinal cord, *SN* saphenous nerve, *SV* saphenous vein

5. Temporarily store the nerve in chilled normal saline before implantation
6. Suture close the skin with Michel suture clips
7. Following step 11 in intravertebral avulsion, take out the nerve from the saline solution
8. Suture the epineurium (~1 mm to proximal) of the nerve graft on the lateral funiculus of the spinal cord with 11-0 suture
9. Cut the lateral funiculus longitudinally with an angled micro knife or an angled 27 gauge needle of about 1 mm wide and 0.5 mm deep near the 11-0 suture
10. Use a sharp #5 forceps to gently spread apart the cut to create a myelotomy groove, the forceps should not be inserted deeper than 1.5 mm

11. Introduce the peripheral nerve graft almost perpendicularly into the myelotomy groove (Fig. 5c)
12. Suture the distal end of the nerve graft into the juxtaposed muscle with 10-0 suture
13. Suture close muscles and skin in layers
14. To access the regeneration of motoneurons, 2 days before the designated period of survival, 1 μL of 5% Fluorogold or other neuronal tracers can be injected into the peripheral nerve graft at the 10 mm position

Outcome Evaluations

For avulsion injury, the whole avulsed segment can be harvested and serially cut into 40 μm thick cross or longitudinal sections after fixation with 4% formaldehyde. Cresyl violet or neutral red can be used to counterstain the sections. Motoneuron can be easily identified by its location in Rexed's lamina IX in ventral horn, its large soma size and granular appearance of Nissl bodies in the cytoplasm under high magnifications. Number of surviving motoneurons and size of soma can be assessed and compared to ipsilateral side. Typically, 600–900 motoneurons are found at one side of the C7 segment in adult rats. After avulsion, 20% of motoneurons dies at 2 weeks and half of them are lost 4 weeks post-injury.

Regeneration of motoneurons can be assessed by counting the number of fluorescent labeled motoneurons (Fig. 6). Most labeled cells are motoneurons and from the

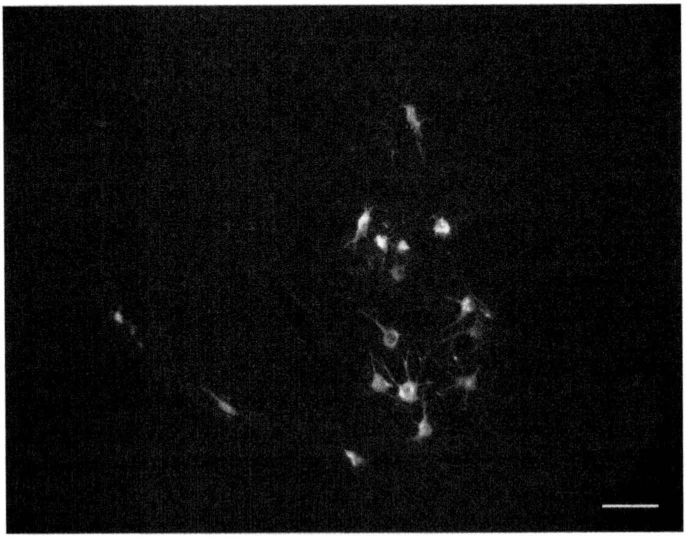

Fig. 6 Regeneration of motoneurons can be examined by application of neuronal tracers. Fluorogold positive motoneurons are found in the ventral horn of the spinal cord. Scale bar = 100 μm

injured segment, however a few from neighbouring segments can be found. Axon number, density and size can be assessed in the cross section of the implanted nerve. Functional recovery can also be assessed after implantation of avulsed roots if a suitable set of ventral roots are avulsed, grooming test and grasping test can be used according to the location of avulsed site and groups of muscle affected.

Applications and Limitations

Applications

Spinal root avulsion is a useful and reproducible model in studying cell death, survival and regeneration of motoneurons in small and large animals.

Cell Death

The model induces a rapid motoneuron death evident at 2 weeks post injury. Structural changes [8] and regulation of genes [9] after injury provide insight into cell death in the CNS.

Cell Survival

Potential therapeutic agents soaked in gel foam can be applied conveniently on the surface of the spinal cord near the ventral root exit zone [10, 11]. Massive cell loss if left untreated can provide a sharp contrast to the survival promoting effect of the drug under testing.

Regeneration

Regeneration of motoneurons under different conditions can be tested in a fairly short observation period in this animal model. Observable functional deficit can be induced by combination of avulsion and ligation thus allowing functional assessment after implantation [7, 12]. More importantly, studying avulsion and regeneration can provide useful cues to develop effective treatment to avulsion injuries in human [13].

Choice of Animals

Experimental avulsion has been performed on larger mammals such as sheeps, monkeys and cats, but a larger pool of data are from small rodents as in rats and mice. Rat is in a reasonable size to perform complicated surgeries such as implantation whereas knockout mice are helpful in studying the underlying molecular pathway. However with the advance of RNA interference, cell type and site specific gene knockdown can be achieved by viral infection [14], thus making rat an ideal animal model to work with.

Limitations

1. If extravertebral avulsion fails, it is impossible to remove the remaining ventral root. If complete avulsion is a concern, intravertebral avulsion should be considered. Implantation can not be done in extravertebral avulsion.
2. Both intravertebral and extravertebral avulsions damage the integrity of blood-spinal cord barrier and result in leakage of CSF.
3. Continuous flow of CSF will dilute the drug applied topically on the avulsed segment, a higher concentration of drug could be used, alternatively, repeated renewal of the drug or implantation of an osmotic pump are justified.
4. Attention has to be paid on laminectomy which could damage the spinal cord if the procedure is not done carefully. Keep the bone ronguer away from the surface of the spinal cord and do not twist the bone inward towards the spinal cord.

References

1. Koliatsos VE, Price WL, Pardo CA, Price DL. Ventral root avulsion: an experimental model of death of adult motor neurons. J Comp Neurol. 1994;342(1):35–44.
2. Wu W. Expression of nitric-oxide synthase (NOS) in injured CNS neurons as shown by NADPH diaphorase histochemistry. Exp Neurol. 1993;120(2):153–9.
3. Wu W, Han K, Li L, Schinco FP. Implantation of PNS graft inhibits the induction of neuronal nitric oxide synthase and enhances the survival of spinal motoneurons following root avulsion. Exp Neurol. 1994;129(2):335–9.
4. Holtzer CA, Marani E, van Dijk GJ, Thomeer RT. Repair of ventral root avulsion using autologous nerve grafts in cats. J Peripher Nerv Syst. 2003;8(1):17–22.
5. Carlstedt T, Grane P, Hallin RG, Noren G. Return of function after spinal cord implantation of avulsed spinal nerve roots. Lancet. 1995;346(8986):1323–5.
6. Hoffmann CF, Marani E, van Dijk JG, vd Kamp W, Thomeer RT. Reinnervation of avulsed and reimplanted ventral rootlets in the cervical spinal cord of the cat. J Neurosurg. 1996;84(2):234–43.
7. Gu HY, Chai H, Zhang JY, et al. Survival, regeneration and functional recovery of motoneurons in adult rats by reimplantation of ventral root following spinal root avulsion. Eur J Neurosci. 2004;19(8):2123–31.
8. Li L, Houenou LJ, Wu W, Lei M, Prevette DM, Oppenheim RW. Characterization of spinal motoneuron degeneration following different types of peripheral nerve injury in neonatal and adult mice. J Comp Neurol. 1998;396(2):158–68.
9. Yang Y, Xie Y, Chai H, et al. Microarray analysis of gene expression patterns in adult spinal motoneurons after different types of axonal injuries. Brain Res. 2006;1075(1):1–12.
10. Wu W, Li L, Yick LW, et al. GDNF and BDNF alter the expression of neuronal NOS, c-Jun, and p75 and prevent motoneuron death following spinal root avulsion in adult rats. J Neurotrauma. 2003;20(6):603–12.

11. Zhou L, Wu W. Antisense oligos to neuronal nitric oxide synthase aggravate motoneuron death induced by spinal root avulsion in adult rat. Exp Neurol. 2006;197(1):84–92.
12. Wu W, Chai H, Zhang J, Gu H, Xie Y, Zhou L. Delayed implantation of a peripheral nerve graft reduces motoneuron survival but does not affect regeneration following spinal root avulsion in adult rats. J Neurotrauma. 2004;21(8):1050–8.
13. Carlstedt T. Approaches permitting and enhancing motoneuron regeneration after spinal cord, ventral root, plexus and peripheral nerve injuries. Curr Opin Neurol. 2000;13(6):683–6.
14. Ralph GS, Radcliffe PA, Day DM, et al. Silencing mutant SOD1 using RNAi protects against neurodegeneration and extends survival in an ALS model. Nat Med. 2005;11(4):429–33.

Index

J. Chen et al. (eds.), *Animal Models of Acute Neurological Injury*, Springer Series in Translational Stroke Research, https://doi.org/10.1007/978-3-030-16082-1

K

L

T

Printed by Printforce, the Netherlands